If structural invention is to allow the efficient solution of the new problems offered daily by the ever-growing activity in the field of building construction, it must become a harmonious combination of our personal intuition and of an impersonal objective, realistic and vigorous structural science.

Future architects, even if they can entrust the final calculation of a structure to a specialist, must themselves first be able to invent it and to give it correct proportions. Only then will a structure be born healthy, vital and, possibly, beautiful

Pier Luigi Nervi

AJ Handbook of Building Structure

EDITED BY Allan Hodgkinson

The Architectural Press, London

Allan Hodgkinson

Consultant editor and authors

The consultant editor for the Handbook is Allan Hodgkinson MEng, FICE, FIStructE, MConsE, Principal of Allan Hodgkinson & Associates, consulting civil and structural engineers. Allan Hodgkinson has been the AJ consultant for structural design since 1951; he is a frequent AJ contributor and is the author of various sections of this handbook.

The authors of each section will be credited at the start of the section of the Handbook in which their material appears. The original *Architects' Journal* articles were edited by Esmond Reid, BArch, and John McKean, BArch, MA, ARIBA, ACIA, ARIAS.

The frontispiece illustration shows one of the most magnificent building structures from the era of the Eiffel Tower, the Forth Bridge and the great railway stations. The Palais des Machines for the Paris Exhibition of 1889 (Contamin, Pierron & Charton, engineers) was a pioneer example of three hinged arches.

Preface to the second edition

There have been considerable changes in some British Standards, Codes of Practice, and Building Regulations since 1974; and unlike the reprints of 1976 and 1977, this is a substantially revised and updated re-issue of the now well-established *AJ Handbook of Building Structure*.

The principal changes are in the sections on Masonry (re-written to take account of the 1976 Building Regulations, and the new BS 5628 'limit state' code of practice); and on Timber (substantially revised to take account of the new timber gradings).

Steel handbooks have been replaced for all types of structural sections; and technical study Steel 3 has therefore been revised accordingly.

In general, the new 'limit state' approach to design is discussed (eg in the section on Masonry); but in view of the rejection of the limit state Codes and draft Codes in their present form, by the majority of practical designers, it has been thought prudent to retain the allowable stress methods of design as the basis of the handbook.

Finally, it should be mentioned that the opportunity has been taken to bring all references in this Handbook up to date; and to correct a number of misprints of the first edition.

ISBN 0 85 139272 5 (clothbound)
ISBN 0 85 139273 3 (paperbound)
First published in book form in 1974 by
The Architectural Press Limited: London

Reprinted 1976, 1977
Second edition 1980
Printed in Great Britain by
W & J Mackay Limited, Chatham

This handbook

Scope

There are two underlying themes in this new handbook on building structure. First, the architect and engineer have complementary roles which cannot be separated. A main object of this handbook is to allow the architect to talk intelligently to his engineer, to appreciate his skills and to understand the reasons for his decisions. Second, the building must always be seen as a whole, where the successful conclusion is the result of optimised decisions. A balance of planning, structure or services decisions may not necessarily provide the cheapest or best solution from any of these separate standpoints, but the whole building should provide the right solution within both the client's brief and his budget.

The handbook provides a review of the whole structural field. It includes sections on movement in buildings, fire protection, and structural legislation, where philosophy of design is discusssed from the firm base of practical experience. Foundations and specific structural materials are also covered, while sufficient guidance on analysis and design is given for the architect to deal with simple structures himself.

Arrangement

The handbook deals with its subject in two broad parts. The first deals with building structure generally, the second with the main structural materials individually.

The history of the structural designer and a general survey of his field today is followed by a section on basic structural analysis. The general part of the handbook concludes with sections on structural safety—including deformation, fire and legislation—and on the sub-structure: foundations and retaining structures.

Having discussed the overall structure, the sections in the second part of the handbook discuss concrete, steelwork, timber and masonry in much greater detail. Finally there are sections on composite structures and on new and innovatory forms of structure.

Presentation

Information is presented in three kinds of format: technical studies, information sheets and a design guide. The technical studies are intended to give background understanding. They summarise general principles and include information that is too general for direct application. Information sheets are intended to give specific data that can be applied directly by the designer.

Keywords are used for identifying and numbering technical studies and information sheets: thus, technical study STRUCTURE 1, information sheet FOUNDATIONS 3, and so on. The design guide is intended to remind designers of the proper sequence in which decisions required in the design process should be taken. It contains concise advice and references to detailed information at each stage. This might seem the normal starting point, but the guide is published at the end of the handbook as it can be employed only when the designer fully understands what has been discussed earlier.

The general pattern of use, then, is first to read the relevant technical studies, to understand the design aims, the problems involved and the range of available solutions. The information sheets then may be used as a design aid, a source of data and design information. The design guide, acting also as a check list, ensures that decisions are taken in the right sequence and that nothing is left out.

5

Contents

AJ Handbook
Building structure
CI/SfB (2-)

Section 1
Building structure: General

Section 1

Building structure: General

Scope
The first section of this handbook consists of two technical
studies which provide an introduction to the subject of
building structure. The first study shows the growth of the
structural designer through history and the role of architect
and engineer up to the present day. The second study
provides a wide review of the subject today, giving the
background on which the architect's knowledge of structure
can build. It provides a frame of reference and guide to the
remainder of the handbook, while also offering knowledge
from practical experience which has not previously been
contained in a structures textbook.

W. Houghton-Evans

Herbert Wilson

Authors
The authors for Section 1 are W. Houghton-Evans and
Herbert Wilson. W. Houghton-Evans AMTPI, RIBA runs a
course in architectural engineering in Leeds University's
Department of Civil Engineering. His buildings include
Leeds Playhouse (AJ 22.12.71 p1428) and his research is in
planning and industrialised building. Herbert Wilson CEng,
FICE MCONSE, is a consulting civil engineer and for many
years was a partner of Norman & Dawbarn, architects and
engineers. Both authors are enthusiastic advocates of
active collaboration between architect and engineer in the
design of building structures.

*Illustration on previous page is a section of Milan
Cathedral from Caesariano's Vitruvius (1521)*

Technical study
Structure 1

Section 1 **Building structure: General**

The structural designer

The major part of this handbook deals in detail with current structural theory and practice. But first it is helpful to understand the role of the structural designer and to

appreciate the development of structural forms. This is the purpose of these first two articles. This first study, by W. HOUGHTON-EVANS, *describes the role of the early architect/engineers and shows how modern structural theory evolved. The second article reviews the various structural forms now available and acts as a guide to the remainder of the handbook*

1 Architects' lack of specialist knowledge

1.01 A building may be regarded simultaneously as a system of spaces for specific uses; a system to control local climate; a system to distribute services and take away wastes; a structural system capable of carrying its own and applied loads to the ground. Each of these is capable of subdivision and elaboration. To be built, a building must be conceived as a constructional system, and during its life may have to facilitate maintenance, alteration, or even removal. In detail it will make its presence felt on those that use it, and as a whole it will affect the town and landscape. **1.02** The unique task of the architect is to propose a solution which simultaneously and in an adequate and appropriate manner satisfies all these roles. The solution is unlikely to 'spring fully-armed' in every detail from his head, and he may need the help of others to develop it in detail. The basic strategy must be capable of tactical elaboration in respect of each role the building will have to play, and therefore the architect must understand every aspect. **1.03** Because of the many sides of any design problem, specialism in designers poses a difficulty. A specialist will tend to see first only that aspect of the task which falls within his specialism, and ignore the others. While special-

ism in architecture is still less well-defined than elsewhere, architects may also be lop-sided in their approach. Their failings will most probably result from an inadequate understanding of specialist matters. It is especially difficult for them to be inventive and to think creatively where their knowledge is superficial and confined to stock solutions. No aspect of design in recent times has created greater difficulties in this regard than structural engineering. This is the more surprising in that throughout history engineering in general—and structural engineering in particular—has been intimately bound up with architecture.

2 Origins of engineering science

Vitruvius and Archimedes

2.01 Vitruvius' *de Architectura* (c 1st century AD) records almost all that is known from antiquity of technical design, and historians of engineering customarily acknowledge Vitruvius as the first writer in their field. In Byzantium, the architects of S Sophia **1** (6th century AD) were Anthemius, a leading mathematician of his age, and Isodorus who wrote a commentary on the works of Archimedes. We are as much indebted to them for the preservation of Hellenic mechanical science as for the most audacious practical demonstration of its validity. **2.02** It was within those very ecclesiastical establishments which embodied the miracle of medieval vaulting, that scholars were appropriately taking up afresh the redis-covered works of Archimedes, and were beginning to make the first significant advances in science since Classical times. Our only architect/author from the Middle Ages, Villard de Honnecourt, shows as lively an interest in machines as in building, and it is a commonplace of architectural history that medieval architecture displays an inventive mastery of structural design **2, 3**.

Leonardo, Alberti and Wren

2.03 The pristine audacity of Renaissance Man, confidently determined to put the entire universe under the sway of human reason, is epitomised in the work of Leonardo da Vinci—painter, sculptor, musician, poet, scientist, inventor: designer of everything from fortifications to birds. He, like Vitruvius, is never absent from histories of architecture and engineering. To his near-contemporary, the architect Alberti, we owe the first great scientific treatises of the age, his mature masterpiece being the 10 books *de re Aedificatoria*. As did Vitruvius, he reviews the entire technology of his time and tries to bring the whole within the scope of scientific principle. Unlike the savants of Antiquity, more-over, this new Universal Man did not affect a patrician disdain of manual craftmanship. Alberti, we are told, 'would

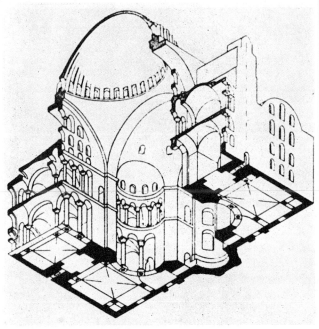

1 *Analytical section of S Sophia*

2 *Rheims cathedral buttresses, as seen by Villard de Honnecourt (c 1230)*
3 *Medieval timber truss structure at Westminster Hall*
4 *Sketch by Domenico Fontana (1543-1607) of his lowering the 327-ton monolith in Rome. Lifting it from plinth strained technical resources of the time to the limit*

learn from all, questioning smiths, shipwrights and shoe-makers lest any might have some uncommon secret knowl-edge'.

So we find recorded in his work some usable rules for dimensioning structural members. But, like rules-of-thumb still in use today, his formulae, eg the thickness of voussoirs and of bridge piers, were probably based upon the accumu-lated experience of centuries, and structural engineering was as yet little more than craft lore. Nonetheless, Alberti did not doubt the responsibility of the architect in technical matters. His definition of the architect as one '. . . who, by sure and wonderful art and method, is able . . . to devise, and, . . . to complete all those works which, by the movement of great weights, and the conjunction and amassment of bodies, can, with the greatest beauty, be adapted to the uses of mankind . . .' compares strikingly with the purpose of civil engineering described by Thomas Tredgold in 1828 as '. . . the art of directing the great sources of powers in nature for the use and convenience of man . . .'.

2.04 For Alberti and Leonardo, and for a century or more after them, the 'movement of great weights and the con-junction and amassment of bodies' continued to be under-taken with little more understanding than in previous centuries **4**. For practical advance, the precision and empiric-ism of modern engineering science was lacking. Brunelles-chi's dome at Florence notwithstanding, the High Renais-

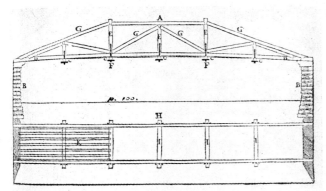

5 *Trussed bridges from Palladio's* Quattro libri dell' architettura

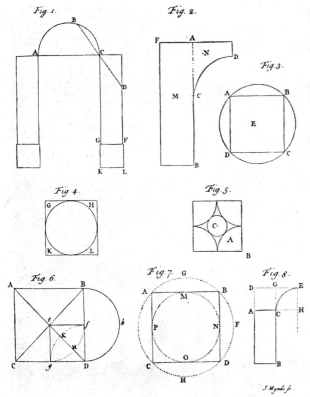

6 *Wren's attempt at scientific analysis of domes and vaults (from his second* Tract on architecture)

sance failed to match in construction its theoretical and spiritual innovations*.

2.05 For some time yet, architecture was able to retain its traditional interest in technology. Palladio, writing in the mid-16th century, although no longer Vitruvian in his range, includes an excellent exposition of trussed bridges among his palaces, temples and piazzas **5**. When, after 1600, Stevinius and Galileo had laid a sound foundation for the whole subsequent development of modern experimental science, architects like Christopher Wren and Robert Hooke (whose celebrated Law is still a corner-stone of structural theory) were among its first devotees. It was not considered remarkable in any way that Wren, a 'natural philosopher' and professor of mathematics, should be asked to design harbour works and fortifications at Tangier as readily as he was commissioned to design St Paul's. Wren writes impatiently of architects who '. . . dwell so much upon this

*It may indeed be that the much-admired art of the period is the most striking expression of precisely its technical limitations. Unable to achieve in reality perfection promised by omniscience and rational outlook, the men of the time sought it most brilliantly in products removed from intractable realities, to the untrammelled realm of the spirit. In his celebrated letter to the Duke of Milan , Leonardo speaks of himself more as inventor than painter, and perhaps without too much exaggeration it may be said that Leonardo the artist is da Vinci the frustrated engineer, creating on paper the mastery over nature which could not as yet be achieved in the real world

ornamental and so lightly pass over the geometrical, which is the most essential part of architecture. For instance, can an arch stand without butment sufficient? If the butment be more than enough, 'tis an idle expence of materials; if too little it will fall; and so for any vaulting . . . the design . . . must be regulated by the art of staticks . . . without which a fine design will fail and prove abortive . . .'. As precise a statement as one could find of what today would be thought the engineer's view of the matter **6**.

Foundations of modern technology

2.06 With Wren and Hooke at meetings of the newly-created Royal Society sat Newton, the scientist and mathematician whose theories and mathematical procedures were to prove an adequate basis for the whole subsequent development of modern technology. Also, in the late 17th century world of commerce and manufacture, there were rapidly growing those tendencies which within a century were to see architecture and civil engineering established as distinct provinces of distinct professions.

2.07 The first formal steps in this direction were taken in France where, in 1716 (following the earlier success of a corps of specially trained military engineers), a civilian corps of engineers for highways and bridges was formed. The word 'engineer' which then begins to enter into common use, derives mainly from the military connotations of the Latin *ingenium*: originally a clever device, later also a war-like instrument. It was not until the latter half of the 18th century that military engineering had a recognisable civilian counterpart in this country*. From these beginnings has grown the modern profession of engineering which, with its many branches and subdivisions, has taken over from architecture much of its classical territory and most of its pursuit of science.

2.08 The schism has never been complete, however, and to this day there remain professionals who are qualified and equally at home in both fields. In some countries, architects and engineers are united in a single professional institution, and in some, elements of joint education persist. Throughout the last 200 years, there have been many who, like Telford, described themselves sometimes as architect, sometimes as engineer. Works on engineering, such as de Belidor's 17th century classics, often used the word 'architecture' to describe their contents, and we still speak of the engineer who designs ships as a 'naval architect'. But in spite of continuing attempts to heal the breach or belittle its significance, differences between engineering and architecture today are only too real, and must be understood if further progress is to be made.

2.09 Early work in modern civil engineering was largely confined to canals and other means of communication essential to developing commerce. Its practitioners were often recruited from the upper level of craftsmen who put their expertise inventively to use. It was not long, however, before new and unfamiliar tasks in the design of hydraulic systems, mills and machinery obliged them to look to contemporary science for aid. In buildings generally, prior to the 19th century, almost all building was accomplished within the use of a few constructional materials: stone (with its mud-based substitutes; bricks and mortar, plaster, concrete) and wood (with reeds, rushes and so on). Even today, most of our building remains within this range.

The new material—iron

2.10 For a time therefore, architecture could be well content to allow others to take over such work-a-day things as

*John Smeaton (1724-92) was first Englishman to call himself 'civil engineer'

roads and machines, while it pursued more elegant goals easily attainable with traditional means. The change was to come only after the engineers began to perfect for their use the great new constructional material of modern times: wrought iron and its derivative, structural steel **7, 8, 9**. Within 50 years of its introduction into fire-proof mill construction, engineers were confidently using the new material to roof great railway station concourses and to span ravines and estuaries. No rule-of-thumb was adequate here. For tasks such as these, full and inventive use had to be made of modern knowledge in the mechanical sciences, and modern structural analysis and design was created as a fundamental component of civil engineering. By 1850, a young man aspiring to be an engineer would have to acquire a scientific and mathematical education if he wished to proceed towards full professional competence.

2.11 What was learnt in terms of iron and steel could with advantage also be applied to other materials. Basing themselves on Galileo's pioneering work, scholars in 18th century France and elsewhere had solved most of the fundamental theoretical problems in structural mechanics. Gregory had shown that the catenary and not the neo-Platonic semi-circular ideal was the 'perfect' curve for the voussoir arch **10**. Mariotte had set the seal upon the classical theory of the bending of beams. Euler had solved the problem of the buckling of columns. Behind the back of architecture there was already developing new experience and knowledge which was to invalidate much of the theoretical basis of its practice. With the coming of the Steam Age, engineers everywhere were quick to use their science to give new shapes to masonry viaducts and timber frameworks. It only needed the invention of reinforced concrete, and the widespread introduction of rolled steel after 1880 to consolidate the claim of structural engineering to re-enter the main stream of architecture after a gap of 200 years during which it made its most fruitful and rapid advances (see cover illustration).

2.12 Today, engineering and architecture confront one another as estranged members of a once-united family. During their long years apart they have acquired strange habits and, if there is to be renewed association, patience, tact and understanding will be called for on both sides. Collaboration is now essential in the many fields they still have in common: structural design, construction technique, environmental science, town planning, the servicing and engineering equipment of buildings. In some of these we may hope for a new breed of architect/engineer which is again capable, across a wide design spectrum. Whatever the outcome of future development, it is necessary for the architect to understand something of modern engineering science, the structural aspects of which are the special concern of this handbook.

3 The engineer's approach

3.01 From the point of view of its behaviour as a structural system, a fully adequate understanding of a building would take account of the way in which *all* elements contributed to strength and stability. Ideally, therefore, it might seem that structural design would tend towards simple, fully

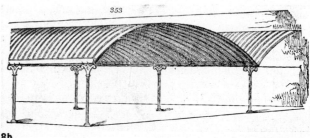

give it the appearance of fig. 349, it is also bent in one general curve in the direction of its length, so as to give it the appearance of fig. 350, we have then an arch of great strength, capable of serving as a roof, without rafters, or any description of support, except at the eaves or abutments. It is evident that, the span of any roof being given, segments of corrugated iron may be riveted together, so as to form such an arch as may be deemed proper for covering it. To every practical man, it will be further evident, that a roof of extraordinary span, say 100 feet, which could not be covered by one arch of corrugated iron without the aid of rafters, might be covered by two or three, all resting on, and tied together by, tie-rods, fig. 351. Further, that in the case of roofs of a still larger span, say 200 feet, a tie-rod might be combined with a trussed iron beam, fig. 352; by which

8a

8b

7 *Interior of Palm House, Kew, by Decimus Burton* (1845)
8a, b *J. C. Loudon's* Encyclopaedia of villa, farm and cottage architecture (1833) *recommends folded plate roofs of corrugated iron*

7

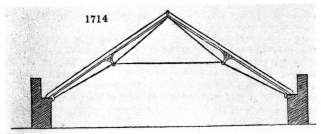

9a

integrated solutions. But structural design is concerned with prediction of probable structural behaviour, and needs to be reasonably certain of its ground. So it is necessary to identify elements likely to play a predictable role throughout the life of the building and regard only *them* as 'structural'. These will then be analysed as components of a 'structural arrangement' which (to make analysis possible) will transmit loads primarily in one of a limited number of definable ways. In practice, much modern building has tended rather to forms in which 'structural' and 'non-structural' elements are separately expressed.

3.02 But the structural engineer is aware that his analytical procedures are necessarily simplifying idealisations, and always keeps his mind open for opportunities to recognise and exploit possible alternatives. He will be aware that inflexible thinking is a disadvantage in structural matters and that over-simple idealisation can lead to wasteful and stodgy design.

3.03 Many new materials are now appearing, and new ways are found of using the old. New techniques of analysis—especially those employing computers—allow prediction to approach much closer to performance and, as a consequence, permit engineering design to move closer to the limit of structural potential.

3.04 But engineering and building are a practical affair. There is no point in designing to impracticable limits or unrealisable tolerances. What is designed must be capable of being built. Construction technique has always been as great a limitation as structural behaviour, and throughout history as much design significance has been attached to the problem of *how* a building was to be built as to any other factor. For instance skeletal frameworks were used from primitive times as permanent scaffolding to support men and materials during as well as after construction, and they have had a profound influence on both architecture and structural engineering. The size of members and assemblies must always be related to transport and lifting capacity. Stability during construction (as most structural failures testify) is as important as upon completion. The sequence of operations and the joining of members have always posed difficult problems in design.

3.05 There will also be important constraints arising from weathering, corrosion and fire resistance. The efficient performance of a structure over a period of time will be dependent upon its protection and maintenance, and for this accessibility will be necessary. Many such considerations are

ribs.
w, or
same.
xtreme
prin-
p, the
ar; *n*,
rough
to the
of the
gutter
arapet
When
kind of
ed. Fig. 1710 shows a side view of the main centre joint:
b b, the
e bottom
711 is a
ne. The
me parts
pletes the
d I shall
of the filling-in rafters.

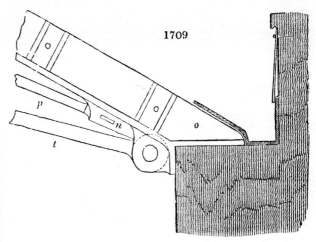

9b

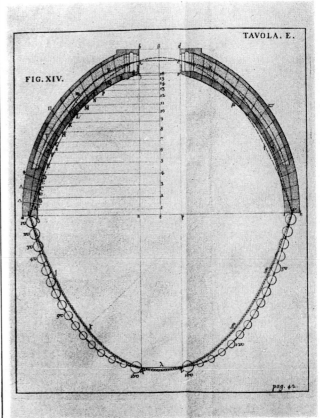

10

9a, b *Iron trusses with tension cables. from Loudon's* Encyclopaedia
10 *Gregory's analysis of voussoir (from Polini's* Memorie istoriche della Gran Cupola del Tempio Vaticano, 1748). *What was learnt in terms of iron and steel could be applied to other materials*

11

12

11 *Palazzetto dello Sport, Rome, by P. L. Nervi*
12 *Mining materials and Metallurgy Building,
Birmingham, by Arup Associates*

reflected in codes and regulations, and much is embodied in legislation.

3.06 While matters of strength and stability pose problems less acute in buildings than in massive works of civil engineering, their involvement in a complex totality can make a satisfactory solution more difficult to achieve. The eloquent clarity of a suspension bridge will rarely be achievable in buildings, and in many it may be appropriate that the structural system should pass unnoticed in the finished product. Structural engineering can nonetheless claim with pride to have descended from the achievements of S Sophia and the Gothic master-masons, and can point to today's achievement in the works of engineer/architects such as Nervi **11**, or architect/engineers' collaboration in the more forward-looking practices **12**.

Technical study
Structures 2

Structural forms, design and materials

Having discussed the development of the structural designer, the handbook turns to a general appraisal of his field today. This article by HERBERT WILSON *discusses structural forms and design principles, and adds an introduction to the various structural materials. This study acts as a framework and guide for the remainder of the handbook*

1 Introduction

1.01 The peculiar division now existing between architect and structural engineer has its roots in the Industrial Revolution; most particularly in this country where 'engineering' became synonymous with 'progress' and where science was omniscient. The engineer, whose impact on society became so intense and diverse during the 19th century, has been reluctant to forgo privilege. However the architect can achieve reciprocal integration and collaboration today through an understanding of the structural engineering objective.

1.02 The words of Thomas Tredgold quoted in the previous study, 'the art of directing the great sources of power in nature for the use and convenience of man', have been embodied in the constitution of the Institution of Civil Engineers to define the role of the engineer. To extend 'use and convenience' into building terms, we have material advantage (maximum worth and pleasure) and personal comfort (efficient environmental properties). The architectural objective cannot be less than this.

2 Maximum worth

2.01 The most recurrent word in all writing or speeches made about building in this century is 'economy'. Overworked and frequently misused it can lead to the demand for minimum first cost rather than maximum worth overall.

2.02 Maximum worth is the supreme test of cohesion both within the professional team, and between the team and its client. The resolution of the individual solutions in which this can be achieved depends upon experience and skill. Design parameters are so complex that a complete scientific or mathematical evaluation is impossible, but attempts are being made to provide more scientific assistance and guidance.

2.03 Reference can be made to recent work at the Imperial College of Science and Technology towards the development of computer-based procedure for economic evaluation and comparison of multi-storied buildings and potential use of findings as a design tool. Although the range of interdependent variables is large, considerable progress has been made and results illustrate the effect design decisions have on running costs, income and capital costs. The project directors consider that 'the main criteria by which any structural system should be judged and towards which any design should be directed are speed of construction and maximum ratio of usable (or lettable) area to gross floor area.'

2.04 Although this investigation is restricted to multi-storied office-type buildings of simple plan shape, it serves to illustrate that the economies of a structural system are not necessarily an expression of maximum building worth, as the choice of an appropriate economic criterion of building has a radical effect on the selection of a structural system **4**.

2.05 This type of analysis, based on a comparison of fluctuating costs of the building elements, is not absolute. The analysis indicates that the most significant cost effectiveness for structure (considered in isolation) is ease and therefore speed of construction. However, the overall building worth depends on decisions made at its inception, rather than on subsequent refinements.

3 Structural forms —solid structures

3.01 The primary building decision is one of structural form. There are (permissibly over-simplified) three basic divisions of structural form: solid construction **1**, skeletal construction **2** and surface construction **3, 5**.

3.02 Solid is the most intuitive form, from cave and rock temple to loadbearing brickwork. During the historical stage of experimentation, the builders of solid structure fully utilised the virtues of stone and its ability to contain compressive loads. Great skill and ingenuity was employed in enclosing space by the transfer of non-vertical reactions through arches, vaults, domes and abutments to vertical forces at foundation level. Solid construction relies on a heavy homogeneous wall mass within which, in the ideal state, compressive forces are uniformly distributed.

3.03 Solid structures perform the function of enclosure, support and protection; but this benefit carries the disadvantage of the highest ratio of mass to unit of enclosure. To some extent this disadvantage is balanced by economies in materials and labour costs, but it could be a significant

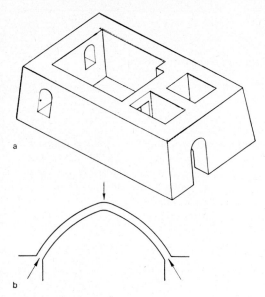

1 *Solid construction;* **a** *loadbearing walls,* **b** *arch as bridging structure*

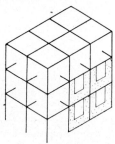

2 *Skeletal construction*

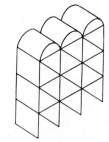

3 *Panel construction*

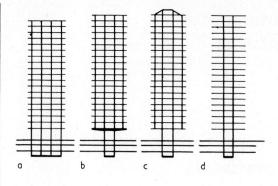

4 *Multi-storey structures. Forms of structure:* **a** *conventional;* **b** *propped;* **c** *suspended;* **d** *cantilevered*

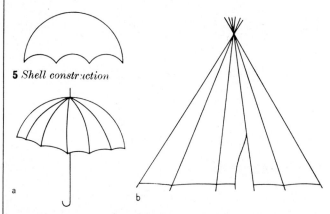

5 *Shell construction*

6 *Skeletal construction, special types with stressed membranes;* **a** *umbrella;* **b** *basic tent;* **c** *Munich Olympic tent by Otto*

6c

factor if choice of foundation is a criterion in the selection of structural form. Modern loadbearing buildings are not true solid structures as they are usually composites of loadbearing walls in the form of perimeter walls (continuous or in sections) with internal walls. These may be cross walls or spine walls, diaphragms and tower structures used in conjunction with slabs, slab beam floors or roofs in various materials. (Structures built up from large flat panels used horizontally and vertically are not 'solid' structures because of the special function of panels as elements of a surface structure, discussed later.)

3.04 'Solid' buildings have structural limitations. Usually they are of modest heights and have short spans (say up to 7·6m). If tall, their forms are confined to those in which each storey has an identical plan. Special consideration is necessary for problems of crack control and differential movements, and these problems will be dealt with in detail in other sections of this handbook. Fire resistance and thermal-insulation properties are good; but their insulation against noise requires specific investigation because of mass transmission effect.

4 Skeletal structures

4.01 Skeletal forms are also traditional, having developed from experience and the availability of materials. The tent is an early, special form of skeletal construction in which the enclosing membrane was stressed in conjunction with an internal framework. Modern tent structures have succeeded in liberating the skeleton from within the skin **6**.

4.02 The trabeated architecture of classical Greece is skeletal in the form of post and beam from which is derived the frame and slab structures (of similar natures but of varying techniques), which are the structural forms of most modern buildings **7**.

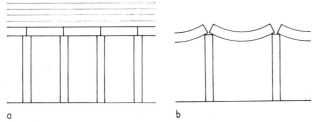

7 *Post and beam structure:* **a** *columns and beams supporting a wall;* **b** *the bending of simple beams on posts;* **c** *the extension of post and beam to a steel framed structure*

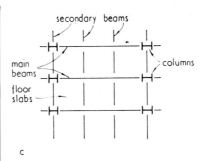

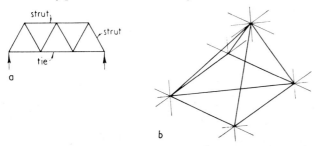

8 *Struts and ties:* **a** *incorporated into a linear frame;* **b** *incorporated into a three-dimensional frame, the element of a space frame*

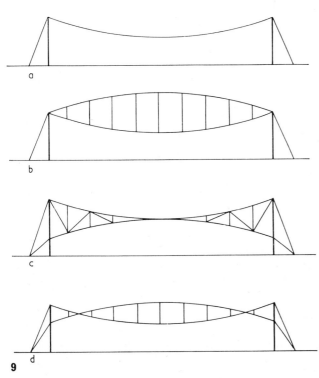

9

9 *Cable structures:* **a** *sections of simple cable;* **b** *cables combined to form girder;* **c** *truss;* **d** *girder-truss*
10 *Cable structures:* **a** *Billingham sports forum;* **b** *and* **c** *hangar for two Boeing 747s*

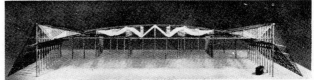

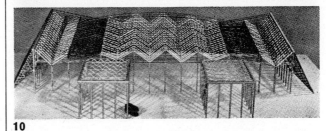

10

4.03 The structural elements of struts, ties and beams were extended to frames by the triangulation of struts and ties, and inevitably to the three-dimensional space frame **8**.
4.04 Skeleton forms avoid the limitations of enclosure imposed on solid forms, and have much more spatial freedom in that the framework can be within the space confined, or between the limits of internal space and external form, or even completely outside the external form. The most outstandingly visual development in skeletal structures has been the exploitation of the qualities of the flexible tie or tendon as used in suspension and tented structures **9**. These range from Nervi's workshops at Mantova (built like a suspension bridge) and cable supported roofs to many aeroplane hangars, **10**; from tendon suspended multi-storey office buildings and cable structures as expressed by Dr Buckholdt, David Jawerth and others, to the controversial

Olympic tent of Professor Behnisch which has now been constructed in Munich **6c** (see AJ 16.2.72, p338).
4.05 The greater use of metals and the growth of appropriate design skills and understanding of the strengths of materials have developed skeletal structures from traditional stone and timber. (The transition from empirical rules to a theoretical statics approach of solid structures was not achieved until the middle of the 18th century, and even modern design methods for such structures retain an empirically-based crudity.)
4.06 The design of skeletal structures has progressed on scientific principles with less restraint from traditional methods. But inevitably there have been pitfalls and one of the most common errors today is the assumption that if the mathematics are in order, then the structure is in order. This is particularly significant when basic assumptions, on

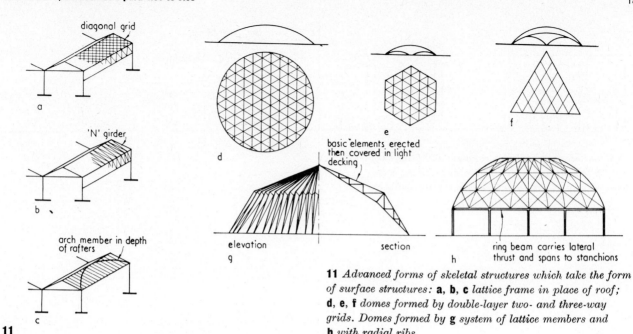

11 *Advanced forms of skeletal structures which take the form of surface structures:* **a, b, c** *lattice frame in place of roof;* **d, e, f** *domes formed by double-layer two- and three-way grids. Domes formed by* **g** *system of lattice members and* **h** *with radial ribs*

which mathematical design is built, are tainted by the arbitrary character imposed on such hypothesis. For these assumptions are the work of earlier compilers of rules and regulations who could not divorce themselves from tradition. Modern design is bold and imaginative, backed by research and aided by modern tools of calculation that enable the designer to consider the building and its foundation as a complex but integral structure.

4.07 The extension of the skeletal frame from horizontal and vertical planes to the three-dimensional frame enabled designers such as Fuller and Zeiss-Dywidag to create structures which almost came within the next category to be described, surface structures **11**.

5 Surface structures

5.01 In these structures the loadbearing surface both defines the space and provides support. In the design of a surface structure an exact understanding of its behaviour and an appropriate scientific analysis is required. Such structures have only recently (in building terms) become practical building realities because of the availability of new materials (reinforced concrete in particular). So there is no background of long experience and therefore no intuitive confidence, but theory and skill of execution are advancing to the stage where our understanding of surface structures is probably more refined than our understanding of any other form of structure.

5.02 The most obvious example is the shell structure in its multiple forms, but such structures are not confined to curved surfaces. The horizontal plate, used as a slab, or the vertical plate used as a wall, panel or beam **12** are forms of surface structures which can be used as 'folded' plate surface structures **13**, or in conjunction with other forms of structure **14**. (See technical study **1 8** for an early example.)

5.03 Surface structures can be constructed in most of the building materials but they are generally limited to the enclosure of space in a single cell or a series of such cells. They are subjected to limitations of construction and architectural conception, engineering theory and practical construction may not be entirely compatible, eg Sydney Opera House.

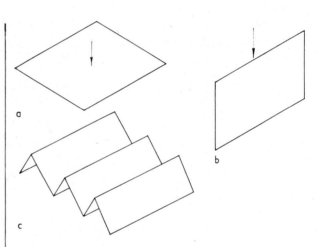

12 *Elements of surface structures derived from flat plates:* **a** *slab;* **b** *panel;* **c** *folded*

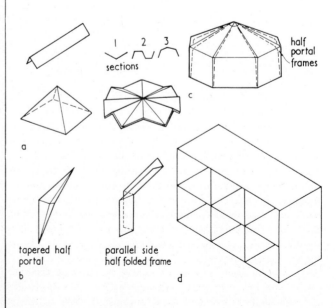

13 *Types of folded structures:* **a** *folded plate;* **b** *folded frame;* **c** *polygonal frame;* **d** *composite from slab and panel*

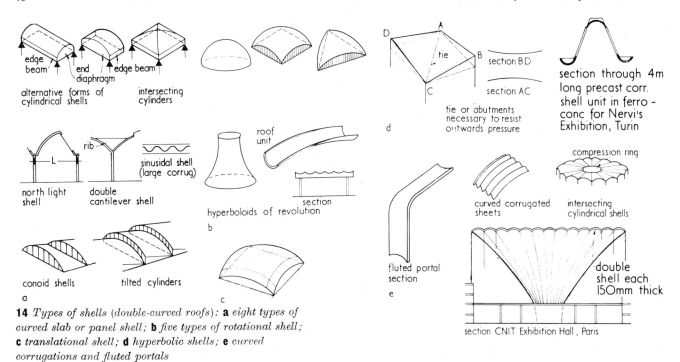

14 *Types of shells (double-curved roofs):* **a** *eight types of curved slab or panel shell;* **b** *five types of rotational shell;* **c** *translational shell;* **d** *hyperbolic shells;* **e** *curved corrugations and fluted portals*

5.04 Most structures are composites of different forms. It is logical to resolve a particular building problem by having a surface roof form supported by a skeleton superstructure with a loadbearing substructure but it is very important to check the structural logic of the points of continuity from one system to another, and where possible to avoid any abrupt changes in the load flow pattern.

6 Structural design—loading

6.01 With established design criteria for structural form based on site limitations, functional requirements and architectural conception, the structural designer moves to the second stage which is concerned with the evaluation of loading and the analysis of loading patterns. The dead loads, or self weights of building and structural elements can be determined with reasonable accuracy at this stage but the values of imposed loads are a matter of judgment. Unless special requirements have to be met and the client is specific about loading in the building, the use to which he will put the building will classify it under Regulations which prescribe floor loadings.

6.02 It should be kept in mind that statutory loadings are attempts to classify a wide variety of buildings load. They govern a minimal condition, and for convenience are expressed as uniformly distributed loads, although this condition is rare in practice. The anticipated disposition of loads in a building should be studied before a rationalised Regulation of uniformly distributed loading is adopted. The capacity of a floor design to spread loads over areas greater than the virtual area of loading application helps towards this rationalisation. A printing machine, a telephone exchange, a magnetometer or a safe are very concentrated loads but require clear working areas around them which can be considered in the spread of load.

6.03 However, in general terms, experience has shown that the classified loadings are not likely to be exceeded unless the building usage changes. For instance, office loading at 2·4 kN/m² plus the acceptable minimum allowance for partitions has proved to be totally inadequate to cover the frequent internal replanning to which such buildings are subjected. The likelihood of these larger loads occurring will have to be assessed, and the probability factor in loading is becoming more prevalent in design practice.

6.04 Not only the size of the loads, but their nature, requires consideration. For instance, the effects of dead load, which is constant, and superimposed load, which is transient, are different in character. The sustained dead load can give rise to creep (ie continuing stress deformation without increase in load) in the materials of construction. Some superimposed loads—such as loads from stores areas—are usually at a constant level and can contribute to this creep effect, which is particularly noticeable and detrimental on long span floor constructions.

6.05 A fluctuating superimposed load can result in vibration, sway, flutter, instability and fatigue. The acceptance of these variable loads and consideration of their effect on overall safety undoubtedly complicates the design processes, but research and experience are creating new guide lines for rationalised design with a greater understanding of what is happening to the building.

6.06 The loads due to natural phenomena which include wind, snow and seismic loads are essentially of uncertain character. Within this category of loading, the user and indeed the legislator are unable to provide any completely satisfactory dictates. The gap between normal loads arising from natural phenomena at fairly frequent intervals, and the exceptional loads of disastrous but infrequent events, is too large to be covered by statutory Regulations and remains a matter of engineering judgment.

6.07 Only actual observations, and tests on complete buildings over a long period of time can provide the information necessary to redraft the Regulations. Any available Regulation notwithstanding, the conditions in the specific locality must always be studied, as, among other things, the effect of wind loading is complex.

Not only does it relate to the building structure as a whole in matters of general stability, but also to the severe local effects on claddings, fastenings and structural elements. In conditions of severe exposure, with tall buildings or complex shape, with buildings having large holes or tunnels through them, and with buildings likely to be effected by the juxtapositioning of other large buildings (as in the case of the cooling towers which collapsed at Ferrybridge), the structural effects can be resolved by intensive investigation only.

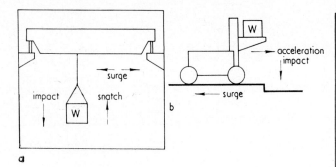

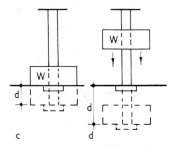

15 a, **b** *Loading owing to impact surge and snatch;* **c** *static loading;* **d** *and dynamic loading causing deformation*

16 *Canning Town, London. Collapse caused by small 'incident'—domestic gas explosion, 16 May 1968*

6.08 Snow loadings, particularly in northern and exposed areas where high winds and low temperatures can prevail, are not adequately covered by Regulations. Efficient thermal insulation can prolong the life of snow on some roofs and this and the retention of frozen snow on concave roof shapes and valleys could extend into a period of exceptional wind load. These combined loadings are probabilities clearly related to the building location.

6.09 Special cases of floor loads may occur not covered by statutory loadings and these may arise from storage, plant, machinery and equipment during construction, installation and the future use of the building. Some of these loadings will give rise to dynamic effects and it should be noted that these can be in vertical and horizontal planes. The acceleration, deceleration and braking of vehicles and cranes will create surge or lateral forces in the structure. The energy release of suddenly applied loads or falling loads create stress effects several times larger than those induced by an equivalent static load **15**.

6.10 Fortunately these effects are, in most cases, of extremely short duration, but they must be considered. The dynamic effects of wind load have been mentioned, and there are classic examples of flutter and resonance in tall structures and long span bridges. The dynamic effects of seismic disturbance are well known, and because of the catastrophic results, much research has been undertaken in this field. The findings of these investigations have been incorporated into design guides for classified earthquake areas, which include the probability factor.

6.11 But there are many more buildings, outside earthquake zones, having complex loadings in which large static loadings are combined with dynamic effects. In industry heavy stamps and presses can produce earthquake-like shocks. Unbalanced reciprocating machines and compressors can induce destructive vibrations and resonance. The release of energy, when a testing machine fractures a specimen, can be transferred through the framework of the machine into the floor and the structure so that isolation or insulation is essential.

6.12 The Building (Fifth Amendment) Regulations 1970 cover the problem of collapse loading in a building over four storeys (including basements) which receives damage due to an incident (the amendment makes no reference to explosion). The purpose of this amendment is to ensure stability if a structural member, ie a section of a beam, a column, a floor slab or a wall, is removed by an 'incident' **16**. Alternatively a pressure of 34 kN/m^2 in any direction must be considered to act in combination with the dead load plus one third the live load and one third the wind load. Floors below the level of the incident must also be capable of carrying the load of debris from above. Under these conditions of loading, a significant reduction in the factor of safety is allowed. For reinforced concrete and steelwork 1·75 times normal stresses may be used and for brickwork 3·5 to 4 times the stresses given in CP 111.

6.13 The fifth amendment of the Building Regulations is brief but is carefully worded to cover all types of structure. However, because of its brevity, it is possible for a number of interpretations to arise. To provide a simple solution early agreement between designer and the responsible authorities is necessary, but the effects of the fifth amendment on fairly substantial buildings have been found less severe than was at first feared.

7 Structural design—foundations

7.01 The ultimate respository of all loads, dead and superimposed, is the ground; and the primary function of structure (of which the foundation is part) is to carry the loads safely and transfer them efficiently to the ground. Within the qualifications of superimposed loads previously referred to, the loads on the structure can be determined. With a properly conducted and efficient subsoil investigation, plus local information on general experience of conditions, the structural qualities of the ground can also be determined.

7.02 However, further decisions must be made in connection with the foundation which could affect the structure. Examples are:

1 structures on shrinkable clays or earths in which loads may have to be collected by foundation beams and/or slabs and then concentrated on piers or piles which will transmit the loads down to levels below the region of climatic change

2 structures over poor or weak loadbearing soils where the total loads have to be widely dispersed by means of rafts or grillages

3 structures requiring piled foundations where column layouts and piling layouts have to be compatible

4 structures of great inherent rigidity which have to be shielded from the effects of normal differential settlement and which require equally rigid foundation units, and possible subdivision into monoliths

5 structures designed to have a large degree of flexibility within isolated units, to accommodate, without structural failure, the effects of subsidence.

7.03 Basements have the dual function of retaining structures and foundation structures which generally are of rigid construction (solid structure). This generates problems in the control of cracks, particularly when located below a water table. Constructional methods, movement joints and water tightness are to be integrated in the design. (Basements to simple buildings will provide efficient foundations on poor soils, but a lightly loaded basement within a water table should be checked for flotation risks, particularly during construction.)

8 Structural design—environment and services

8.01 The environmental requirements of modern buildings have a great influence on structural design. Some requirements are statutory, others either arise from activities within the building, or they may be directives of the client. In general terms the services can be classified in four groups:

1 Environmental services; those directly concerned with control of physical environment, heating, mechanical ventilation, lighting

2 Supply services; those concerned with providing physical materials to meet the needs of building users, hot and cold water, gas, electricity and so on

3 Disposal services; those concerned with removing waste products, refuse, foul and surface water drainage

4 Central plant to provide or generate or motivate the services described above.

8.02 Service layouts have an effect on the structural design, and the following design problems are characteristic:

1 Large ducts and pipes required for ventilation and drainage, are not only space consuming but inflexible, in that tight bends or changes in direction are not always acceptable. Moreover, main distribution lines might have to be planned with consideration for future change of the building's use.

2 Most services, particularly those carrying fluids, are potential noise generators and could be noise transmitters. The transmission of noise and vibration from services and plant will have to be checked against acceptance levels, and

where necessary structural devices incorporated for insulation or isolation of sources of nuisance.

3 Thermal movement of heating and steam pipes can impose heavy loads both on anchorage points and at other points of intentional or accidental restraint. Such movements and loads must be integrated in the structure. Conversely large structural movements, such as deflection of large span beams or floors, must not be transmitted to inflexible services.

4 The choice of structural and cladding materials relates to the standards of thermal insulation necessary in the building envelope, and similar evaluations are required for ventilation, both natural and mechanical, natural lighting and sound insulation.

5 The mechanical devices by which material and people are moved within the building have obvious structural influence in the shape of cranes, hoists, conveyors, lifts and elevators.

9 Structural design—resources

9.01 The question of resources in labour and materials does not frequently arise in the UK in sufficient degree to influence choice of structure. In other countries, however, particularly in 'emergent' countries, inaccessible areas, or areas subject to extreme natural phenomena, the question of resources may be extremely relevant. Local building methods and materials are sometimes worthy of adoption, particularly when 'low-cost' buildings are to be erected. Experience will indicate the most suitable materials and methods to be used, and these will vary from place to place—steel, concrete, bricks and blocks.

10 Structural design—design methods

10.01 During architectural conception of the building, the limits of structural choice have been set by external parameters and the skill and experience of the engineer. Up to this point, mistakes can be expensive and irrevocable if not recognised. But future progress in the structural design is a technical process within the major decisions already taken.

10.02 The structural aims can now be restated and checked against the design concept as:

1 the most efficient structural mechanism, and structural material, with minimum spatial demands within the structural form

2 the best use of structural elements within the chosen mechanism

3 the most efficient use of the chosen material

4 the durability of the structural material

5 the behaviour of the structure in fire

6 the considerations of the site and of construction

7 the economy of the structure.

10.03 To achieve an efficient mechanism, possible patterns of load-flow to the ground must be examined and a system established. In general terms, cost and the spatial demands of structure are related directly to the complexity of the load-flow pattern and the distances covered. The planning decisions already made will help determine the spans, and unobstructed plan areas required. Before an analysis of the relationship of slabs, beams and columns in the load-flow pattern can be made, the matter of structural subdivision conditioned by thermal and shrinkage effects will have to be resolved. Buildings up to say 60 m long may not require complete separation for shrinkage and expansion, but within this length some structural elements such as parapets and brick panels may require special provisions. Buildings of odd shape, or liquid containers and some special purpose buildings (such as cold stores) require further consideration.

Such consideration is also necessary in buildings of composite construction (eg brick walls and concrete slabs have differing characteristics of volumetric change under shrinkage and thermal effects).

10.04 Planning demands may have set limiting or critical dimensions, such as floor thickness and depths of beams vertically, or wall thickness and column sizes horizontally. These considerations have to be balanced against structural deformations under load, which should be checked against statutory limits, but again, acceptance of recommendations should be balanced against experience. Statutory limitations of deflections of slabs and beams are no guarantee against the cracking of walls and partitions carried by those slabs and beams.

10.05 Structural analysis is the subject of this handbook's next section, but an appraisal of design methods is necessary for any understanding of structure. To reconcile safety with economy a correct evaluation of all loads must be followed by a precise evaluation of critical stresses for the structure or its component parts, the relationship between load and critical stress is a measure of safety. Until recently the understanding of the behaviour of materials under load was, in a scientific sense, more the province of the mechanical engineer than of the structural engineer. The traditional knowledge of building was resistant to the new knowledge of materials and because of this a great deal of simplification and rationalisation of structural design, ie practical structural design was inevitable. Some of the assumptions, which were conveniences of design methods, have persisted today and relate directly to timber beams and steel joists.

10.06 Robert Maillart said in an article about the development of flat slab design: 'Previously, rolled steel and timber were available for the construction of long span flat frameworks. Both are materials which cannot be shaped arbitrarily but are only available in beam form where the main dimension is linear and is determined with rolled steel because of the rolling process and with timber because of the growth process. With these materials the single-dimensional basic element struts, beams, piers became so familiar to engineers that any other solution appeared foreign to their minds. Also the calculation process was very simple. Reinforced concrete entered the field and at first nothing changed, one constructed just as with steel and timber with beams spanning from wall to wall or column to column. Set transversely to these main beams were secondary beams and the spans between were spanned by slabs. But instead of being designed and used for their unique structural characteristics, slabs were designed as single strips and were then considered as beams in the old-fashioned manner. Only the steam-ship* designer was in a position to consider the slab as a structural element for which he used the deductions made by Grashof; the structural engineer for the time did not do so.'

10.07 Maillart's point is still valid and structural design follows the convenient assumption that structural elements can be designed in isolation. Furthermore they are completely elastic and this behaviour is ideal. It is convenient to assume that an element will behave as one wishes it to behave—without regard for the behaviour of the complete structure. Such assumptions are not always valid, sometimes because of the structural participation of what are considered non-structural elements; for instance the effect of infill panels of brickwork within a column and beam framework, or the effect of cladding on a lightweight

framework such as a roof truss, or the general stiffening effects of cladding in resisting wind loads.

10.08 The new approach to design is not a rejection of earlier principles which are quite sound; but because there is now a better understanding of behaviour of structure, many of the old restrictions are condemned. The existence of the plastic state beyond the elastic state has been established for a long time but it is only comparatively recently that elasto-plastic design has been generally accepted and has resulted in a more efficient distribution of structural material with considerable progress in the design of highly indeterminate structures. This development has been assisted by research which demonstrated that the previously assumed simple relationships between working stress (in the elastic state) and stress at failure are not valid because of plastic deformation, and in place of the so-called 'factor of safety' a more meaningful relationship (or ratio) is now considered to be that between the working load and the actual load at failure, ie the 'load factor', see **17**, **18**, **19**.

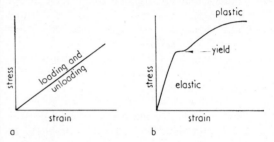

17 *Stress and strain:* **a** *elastic;* **b** *elastic and plastic deformation*

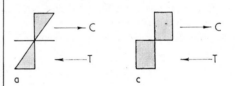

18 *Rectangular steel beam stress diagrams:* **a** *for loading up to yield stress (elastic);* **b** *at yield stress;* **c** *rectangular representation of stress at yield for purposes of calculation*

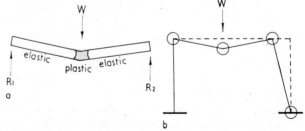

19 a *Failure under simple point load forming a 'plastic' hinge;* **b** *plastic hinge failure in a rigid frame*

*A more recent analogy is the disassociation between designers of aeronautical structures and building structures. The phenomenon and understanding of collapse of box girder bridges is more familiar to aeroplane designers than to bridge designers

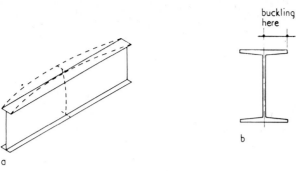

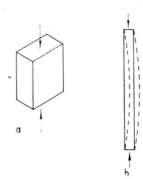

20 a *Buckling of compression flange;* **b** *local buckling of outstanding parts of compression area of a beam*

21 a *Very short strut permissible stress not controlled by length;* **b** *long strut permissible stress controlled by length and stiffness*

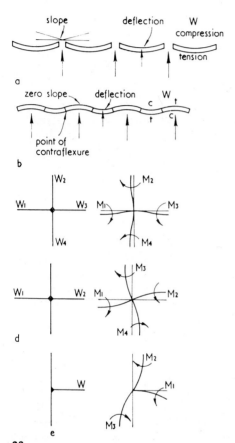

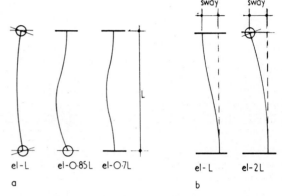

22 *Effective length* (L): **a** *members with ends held in position; and* **b** *with ends free to move*
23 *Continuous joints in frames:* **a** *deflection of a series of simply supported beams; and* **b** *of a fully continuous beam;* **c** *balanced fixed joint members have zero slope at joint;* **d** *beam and interior column junction* M2 = M1; +M3 + M4; **e** *beam and exterior column junction* M1 = M2 + M3

23

10.09 From stress follows strain and it is absolutely essential to have a clear understanding of strain and structural deformation for the proper use of modern design methods. Structural safety is a prime obligation but a structurally adequate building will not be acceptable if structural refinement results in cracked partitions, distorted window and door frames, broken and detached cladding panels, and fractured service mains. Not only strain (the deformation arising as the direct consequence of load) but the dimensional changes in structural materials owing to temperature and chemical changes must be considered. Although unrestrained expansion and contraction because of temperature changes are reversible and cause no damage within the normal range of building temperatures, there will exist a temperature gradient throughout a building mass that will cause differential movements in addition to the differential movements between materials having differing thermal characteristics. The chemical changes in concrete and loss of free moisture results in shrinkage which will persist over the first year of life of the building. The gaps created by such shrinkage will be kept alive thereafter by thermal effects and by changes in the moisture content from atmospheric causes. Other chemical reactions such as the growth of rust on metal, and the expansion of unsound lime can create large volumetric changes as well as unsightly blemishes.

10.10 Stress and strain are mathematical in derivation but 'stability' problems are a combination of theory, practice and engineering skill. The general overall stability of a building is recognised easily and resolved by the principles of equilibrium, but the stability of building elements is less readily understood although it is a familiar phenomenon. Stability controls the design of compression members. Buckling of a strut is a function of its length and cross-sectional stiffness, and failure can occur owing to initial buckling at loads well below the critical compressive stress **20, 21, 22**.

10.11 Once there were many diverse academic theories about the strength of columns, so that by choosing an appropriate one almost any design could be justified. But research has consolidated theory, and rationalised design rules now prevail. The compression zones of beams, cantilevers and plates, acting as vertical girders or diaphragms, are subject to instability. Engineering skill is necessary for the recognition and containment of these physical effects.

10.12 The linear beam (the element of skeleton construction as described by Maillart) in a simply supported condition, and the ideal pin-jointed (or hinged) strut or tie as fabricated into frames are theoretical concepts rarely achieved in practice. A simple support or pin joint implies that the members at the support or joint are free to rotate relative to each other without restraint, but this is obviously an ideal situation and in practice joints have restraint, the members interact with each other (in addition to the purely statical reaction) and moments are developed. The consequences of rigid or partly rigid joints, or in other words the 'continuity' of the structure, have long been recognised and adopted in structural design, but design methods were very mathematical until the original approach of Hardy Cross, see **23**.

10.13 It is significant that Hardy Cross described his method of 'moment distribution' as a physical concept, implying that the deformations of the structure under the various conditions imposed must be visualised. By such methods, and as the words 'moment distribution' imply, the benefits of continuity are a redistribution of the bending effects of load throughout the structure and therefore a more efficient use of the structural material.

10.14 In general terms continuity is a development of linear design extended to two dimensions. The more significant advance (and it is to this that Maillart was referring) was the adoption in structural design of the well known and understood isotropic qualities of homogeneous materials. The ideal structural material should be homogeneous, ie it should have uniform physical characteristics throughout its mass, and it should be isotropic, ie its behaviour under stress should be the same in all directions through the material.

10.15 From this point onwards the understanding of surface structures and the evolution of suitable design methods grew rapidly. The flat slab of Maillart is a classic example. A slab supported along two opposite edges will deform under load to a cylindrical shape, and for design purposes can be considered as a parallel series of linear beams. A slab of material having isotropic qualities supported at the four corners only (no beams) will deform to a shape similar to a spherical figure, and is thus an element of surface structure which cannot with any degree of accuracy be designed on a linear basis, see **24**. Such structures have three possible systems of internal stress: direct loading in tension or compression; shear forces; and bending moments and torsional moments. All these are multi-directional and the shape of the shell governs the relative importance of the three systems **25**. The computer as a design tool can take most of the mathematical load from the design of these structures.

11 Structural materials

Steel

11.01 Steel is an indispensible material both in its own right as a basic structural element and in the supporting role of reinforcement: as a substitute for the homogeneous and isotropic qualities in which concrete and bricks and blocks are deficient. It is provided in varying chemical compositions to fulfil different strength and weathering requirements and, in form, from the thinnest sheet to heavy solid sections or shaped sections.

11.02 Further working can be applied to the steel in rod or bar shape to produce much higher ultimate strengths usually for the reinforcing and pre-stressing of concrete or for the cables of suspension structures. The development of automatic welding and cutting has allowed the fabrication of an even greater variety of shapes and sizes, while the advent of high-strength friction grip bolts has revolutionised methods of jointing, not only for steel work but for all

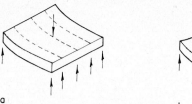
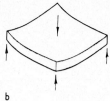

24 a *Normal slab;* **b** *slab acting as an element of surface structure*

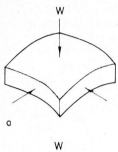
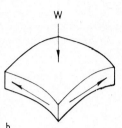
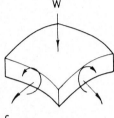

25 *Three possible systems of internal stress in shell structures:* **a** *direct forces;* **b** *shear forces;* **c** *bending moments and torsional moments*

structures. Disadvantages of steel are its relative inflexibility of shape, and the fact that generally protection is required against fire and corrosion attack.

Aluminium

11.03 Aluminium alloys are not at present acceptable substitutes for steel in large structural elements. Greater cost, and the considerable increase in deformation under direct and thermal loadings, outweigh the advantages of lightness in self-weight, and corrosion resistance. However, aluminium may be considered in special circumstances, and particularly when self-weight of structure is a major consideration and the other loads are incidental. It is reasonable to use alloy units of small cross section built up into space frames to roof over large column-free spaces.

Timber

11.04 Timber is one of the oldest materials of construction because it has been readily available, and is easy to work. Although long confined in its use to relatively small linear elements of structure (other than in the construction of ships), it has been developed more recently as a major structural material. This is possible both through a greater understanding of its properties and the development of more comprehensive design methods and is aided by parallel improvements and inventions in the field of adhesives, structural connections and wood-shaping machinery. However, wood is neither a homogeneous nor an isotropic material and the greater expansion of its use, as in surface structures, is possible only because these disadvantages have been minimised by the ability to glue or fasten timber in successive layers with the grain running in different directions. However, it is subject to destruction by insects and fungoidal attack.

Concrete

11.05 Concrete as a mortar was fully understood by the Romans, and it is interesting to speculate how building would have developed if other countries had acquired this skill and developed to the full its quality of unifying small elements into a monolithic structure having considerable tensile qualities. But its structural potential was obliged to wait for the development of acceptable steel reinforcement to provide the special qualities of reinforced concrete, a material which can be designed as though it were homogeneous and isotropic, if imperfect. Being a cast material it has complete freedom of shape but it is a multi-trade material in construction, and under site conditions it demands effective supervision and is a relatively slow building operation.

11.06 These disadvantages have been recognised by the rapid development of the off-site precast industry, but in this, as in all popular acceptance of good ideas, there has been much over-optimism. The period of reappraisal now in being will re-establish the position of precast concrete in the building world. The disadvantages of self-weight and space consuming factors have been overcome by the science of prestress and the development of high-strength concretes, which have practically created a new material different in characteristics and behaviour from 'normal' concrete.

Masonry

11.07 Masonry in the form of natural stone is also one of the oldest of building materials. While stone is still used in minor structures in a loadbearing capacity, the definition now embraces all forms of brickwork and blockwork, unreinforced and reinforced. Statutory Regulations still allow the proportioning of brick structures on an empirical basis but calculated masonry in accordance with the Code of Practice can provide an economic solution even in high buildings where there is a repeated prominent wall system on plan.

Plastics

11.08 The use of plastics as a structural material is still in the experimental stage, and further research and experience is essential to its acceptance in competition with established structural materials. But its development is rapid, and a breakthrough may be imminent.

Conclusion

11.09 The choice of prime material for the structure is largely dependent on structural form. But even here design ingenuity has removed barriers and the prime materials can be used in practically any desired shape. Choice of material is more likely to be conditioned by factors other than structural efficiency.

11.10 Structural form can dictate the material, as can the required foundation, the limitations of resource, site and constructional factors and time available. Freedom of choice of materials where other conditions are equal is restricted to those buildings where the structure is hidden. Although experience, research and design skills may benefit from machine aids to design, such as the computer, they cannot be replaced.

26 *John Hancock Center. Chicago* (SOM). *Example of tapering steel frame building with exposed cross bracing as design feature*

AJ Handbook
Building structure
CI/SfB (2-)

Section 2
Structural analysis

Section 2

Structural analysis

Scope

The later sections on foundations and structural materials are presented in such a way that complex mathematics are not essential to an understanding of structural design. They are intended to give readers a structural awareness, and an appreciation of structural form; the detailed design work being carried out by a consulting engineer.

However, information *is* given on the calculation of simple structural members, and the purpose of this present section is to provide an elementary knowledge of terms and their mathematical basis. Detailed examples on the sizing of members is dealt with under each structural material. This section covers elementary statics, the internal conditions and strengths of materials, beam and strut theory, and analysis of some types of structure.

Author

The author for Section 2 is David Adler BSC DIC MICE, a civil engineer with varied experience in both consulting and contracting.

David Adler

The cover illustration which appears on the previous page compares suspension by a single rope, bearing a total load, with the load borne by two ropes using a pulley. The engraving is adapted from Leoni's English translation (1755) of Alberti's 'Ten books on architecture', and illustrates book VI, chapter VII which concerns simple mechanics.

Technical study
Analysis 1

Section 2 **Structural analysis**

Statics and strengths
of materials

This is the first of two technical studies by DAVID ADLER,
*which provide a simplified mathematical background to
some of the terms used in the later sections on structural
materials, and on foundations and retaining walls*

1 Introduction

1.01 The majority of architects would regard structural
analysis as being entirely pure and applied mathematics.
This is not strictly correct. Mathematics is an important tool
but not the only one. The testing of existing structures,
mock-ups and models, and analysis of the results, as well
as photo-elastic methods, provide an alternative approach.
1.02 In addition, it should be appreciated that the most
abstruse mathematics and the most rigorous analysis can
produce no better answer than the correctness of the basic
assumptions made. Mathematics cannot give all the
answers about a problem any more than an architectural
model will exactly reproduce the appearance of the finished
building. Mathematical analysis, in fact, does construct an
abstract model of the structure. The limitations of the
particular model in each case must always be remembered
by the problem-solver, just as the architect remembers that
the brickwork of his model is actually perspex.

2 Statics

Force
2.01 Force is almost indefinable. It is known mainly by its
effects: acceleration, strain and so on.
2.02 A force has three characteristics: magnitude, direction,
and point of application. In these respects it can be represented by a straight line, its length proportional to the
magnitude of the force **1**.
2.03 Any two or more forces acting at the same point can
be replaced by one combined force called their *resultant*.
The magnitude and direction of the resultant can be
obtained by the method known as the *parallelogram of
forces* **2**.

2.04 Conversely, any force can be 'resolved' into separate,
smaller forces called *components* in given directions. Assuming forces are acting in one plane only, a force P at an
angle ϕ with the horizontal xx axis **3**, can be resolved into
components P cos ϕ for the xx axis and P sin ϕ for the yy
axis in the direction of these axes **4**.
2.05 Resultants and components of forces are *replacements*.
Replacements must not be confused with *equilibriants* which
are described in para 2.12.

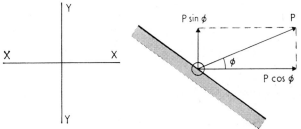

3 *Diagram showing
convention of horizontal* xx
axis and vertical yy *axis*

4 *Resolution of forces. Force
P is resolved into its
horizontal and vertical
components*

Moment
2.06 To understand the concept of 'moment', first consider
a 'plane' system of forces. This is a system which is confined
to two dimensions, and can be represented on a piece of
paper by lines as described in para 2.02. No forces act out
of, or into the plane of the paper.
2.07 The moment of a force about a point is found by
multiplying the magnitude of the force by the distance of
its line of action. Thus in **5**, the moment M of force P about
point A, is the product of P, the magnitude of the force, and
the distance d of its line of action from A measured perpendicular to the direction of the line of action.
Thus Moment = Force × distance, or M = P × d
2.08 But, more usually, forces act in three dimensions, in
which case point A becomes an axis perpendicular to the
plane of the paper, but the same definition applies.

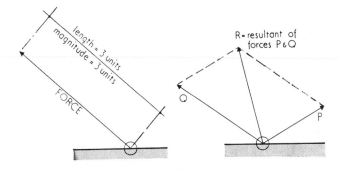

1 *Force represented by
straight line. Length is
proportional to magnitude*

2 *Parallelogram of forces*

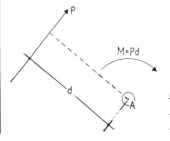

5 *Moment of a force.
Moment is equal to force
P × distance d*

Couples

2.09 A couple is formed by two forces, equal in magnitude P, acting in parallel but opposite directions distance d apart. The moments M of those forces from any point A, distance x from one of the forces, are given in the formula below and shown in **6**.

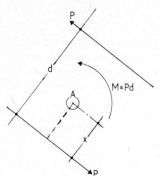

6 *A couple. Two forces equal in magnitude, parallel and opposite in direction*

Moment = force × distance x + force (distance d − distance x)

= force × distance d

or $\quad M = P \times x + P(d − x) = P \times d$

This value is independent of the position of the point about which moments are taken. The two forces form a *couple* or pure moment. The value of a couple in a given plane is constant everywhere in that plane.

Equilibrium

2.10 A force P acting on a free body of mass m in space, will result in that body moving with acceleration $\frac{P}{m}$. If a body does not move (or moves with constant speed in a frictionless environment), then forces acting on that body are said to be in *equilibrium*. That is, the resultant of all the forces is zero.

2.11 First consider a very small body, small enough, in fact, to be a point. This point body is in equilibrium, so that all the forces acting on it must have zero resultant **7a**. Instead of using the parallelogram of forces, described in para 2.03, the polygon of forces is used **7b**. Alternatively, the components of each force can be considered in the x and y directions. These components also have zero resultant, and as they all act in the same direction, the arithmetical sum of all the components is zero, in each direction **8**. Finding the components in direction x and summing to zero is referred to as *resolving in direction* x.

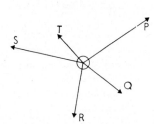

7a *Forces acting at a point in equilibrium, therefore resultant is zero;*

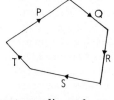

b *corresponding polygon of forces. Magnitude and direction of forces represented by length and direction of lines*

2.12 If the point body is not in equilibrium, then the forces on it have a resultant. A force equal in magnitude to this resultant, but exactly opposite in direction, will cause the system to be in equilibrium, so this force is called the *equilibrant*. For example, a brick weighing 45 N resting on a level table-top has a downwards force of 45 N acting on it **9**. It does not move, so the table is supplying an equilibriant of 45 N upwards. This type of equilibrant is usually called a *reaction*. This proves the first law of statics which is: *Action and reaction are equal and opposite.*

2.13 Now consider a perfectly rigid body of larger dimensions, in space. Forces are acting on this body, but unlike the point body considered earlier, these forces are not all applied at one point, but at a number of positions around, and possibly even inside, the body. These forces can be resolved in the x, y, and z directions*, and the components in these directions will still sum to zero if the forces are in equilibrium. This means that the body will not move up, down, or sideways. There is, however, a further mode of movement not yet investigated: the body could rotate.

2.14 If the body does not rotate, the moments of the forces on that body must also be in equilibrium. In a plane body (of only two dimensions) moments can be taken about any point, and the sum of the moments of all the forces will be zero. In three dimensions, moments can be taken about any axis and the same will be true.

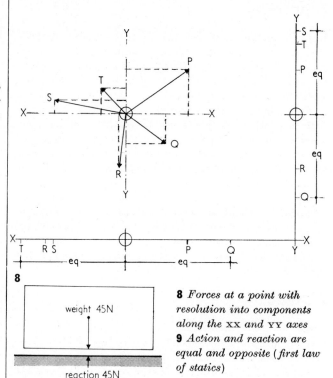

8 *Forces at a point with resolution into components along the* xx *and* yy *axes*
9 *Action and reaction are equal and opposite (first law of statics)*

weight 45N

reaction 45N

9

2.15 Four equations apply to a plane body in equilibrium:
1 *and* 2: Resolution of forces in two different directions
3 *and* 4: Moments taken about two different points.

Only three of these four equations are independent. By the laws of algebra the values of only three unknown quantities can be discovered. If this is not sufficient to find out all the forces on the body, the system is said to be *statically indeterminate*.

2.16 When considering a body in three dimensions it is possible to determine:
1 resolution of forces in three directions
2 moments about three axes.
Of these six quantities, any five will give all the information it is feasible to obtain in this way.

Example

2.17 In the example of the plane body shown in **10**, discover the forces necessary to maintain equilibrium horizontally:
P − 100 = 0 (must have zero resultant to maintain equilibrium)
∴ P = 100

*As previously explained in fig 3 xx and yy are the horizontal axis and vertical axis respectively in one plane only. The zz axis is the third dimension axis, acting perpendicular to the other two

Now take moments about point X, the intersection of forces P and Q:

R × 6 − 100 × 3 = 0
∴ R = 50

Finally, resolve vertically:

R − Q = 0
∴ Q = R
But R = 50
∴ Q = 50

In this example there were three unknown forces, P, Q and R so that three equations were sufficient to find their magnitudes.

Forces and moments: Summary

2.18 Three facts are self-evident and important to remember:
1 Two forces in equilibrium must be equal in magnitude and opposite in direction at the same point of application **11**
2 Three forces in equilibrium must pass through a point (if the lines of action are extended far enough). Their magnitudes must conform to the *triangle of forces*, **12**
3 Any force P acting at a given point can be replaced by a force of equal magnitude acting at any other point in a parallel direction distant d from the original line plus a couple of magnitude Pd, **13**.

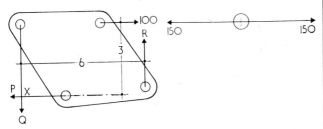

10 *Equilibrium. Example in para 2.17 shows how to calculate forces P, Q and R*

11 *Two forces in equilibrium must be equal in magnitude and opposite in direction*

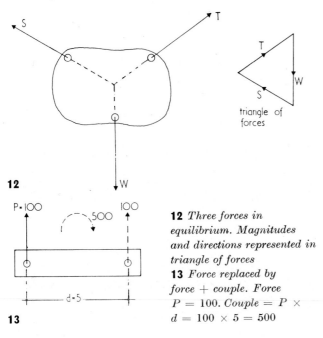

12

12 *Three forces in equilibrium. Magnitudes and directions represented in triangle of forces*
13 *Force replaced by force + couple. Force P = 100. Couple = P × d = 100 × 5 = 500*

13

Rigid body mechanics

2.19 A rigid body in the real world has mass. It is therefore acted on by gravity, which imposes a downward force called its *weight*. If the body is suspended by a string, it will rotate until the tension in the string is in the same line as the line of action of the weight. The lines of action of the weight always pass through one particular point in the body, even when the position of the point where the string is attached is changed **14**. This point is called the *centre of gravity*, and if the mass of the body were concentrated at this point, its behaviour under outside forces would be the same.

Example
2.20 A convenient type of rigid body to use as an example is a retaining wall. Assume that the actual wall is quite long, but that a slice is taken from the middle, of unit length **15**.

2.21 There are only three forces on this slice of wall:
1 its own weight W, acting at the centre of gravity of the slice
2 pressure of the earth behind the wall P: soil mechanics theory indicates that this pressure is likely to act as shown in **15**.
3 the reaction under the foot of the wall R.
2.22 These three forces must pass through one point; the reaction must therefore pass through the intersection of the weight W and the earth pressure P. The magnitude and direction of this reaction is obtained from the triangle of forces (see also **12**). If this reaction as drawn does not pass through the base of the wall, but falls outside it, the wall will fall over.
2.23 It will be shown later that the wall can still fall over even when the reaction passes through the base, if it does not pass through the middle third of that base.

3 Strength of materials

3.01 The previous section on statics considered forces acting on the structure as a whole. The later section on theory of structures will deal with the forces acting within the structure. This present section is concerned, with the forces acting within the materials of which these members are composed.

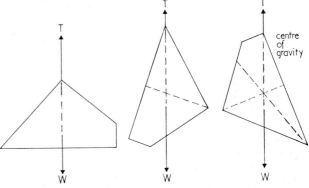

14 *Finding the centre of gravity. The body is suspended from each of its corners (T). Where the* verticals due to its weight (W) converge, there is its centre of gravity or centroid

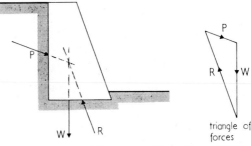

15 *Forces on a retaining wall. Earth pressure P and weight of wall W cause a reaction R under the foot of the wall. Its magnitude and direction is obtained from the triangle of forces (right).*

Procedure: plot P and W to scale and inclination in the triangle of forces. Join ends to obtain R whose magnitude can be scaled and direction determined

Structure

3.02 A structure is any body of material acting in a way which changes the magnitude, position, or direction of natural forces to the advantage of the user **16**.

3.03 A structure is usually composed of structural members. A structural member is nearly always of one, or at the most two homogenous materials, is *prismatic* (see later) and connects two *nodes*, or points where other members converge **17**. The line joining the nodes is called a longitudinal axis, and the *section* of the member is the plane figure produced by cutting the member at right-angles to this axis. A member is said to be prismatic when the section does not vary along its length **18**.

3.04 A structure *can* be composed of an amorphous mass of miscellaneous content, or of non-prismatic members of changing composition, but the vast majority of structures fall into the normal category. The analysis of other types of structure is not within the scope of this section.

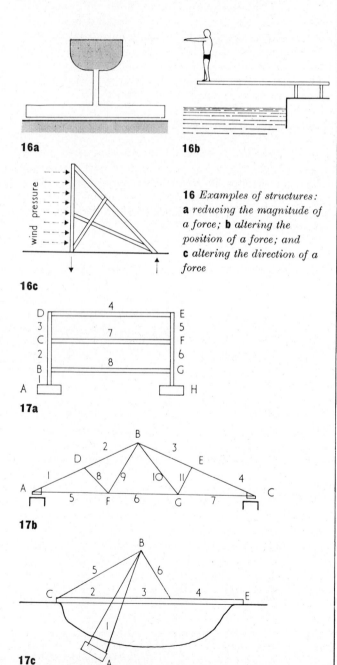

16a

16b

16 *Examples of structures:* **a** *reducing the magnitude of a force;* **b** *altering the position of a force; and* **c** *altering the direction of a force*

16c

17a

17b

17c

17 *Typical structures:* **a** *building frame;* **b** *roof truss;* **c** *cable-stayed bridge. Nodes are lettered. Members are numbered (see para 3.03)*

Stress

3.05 When studying the forces acting on the section of a member it is necessary to understand the concept of *stress*. If a very small area of the section is considered, it can be assumed that the force on it is evenly distributed. The stress on that small area δA is then equal to the force δP on the area, divided by the area (see **19**): stress $= \dfrac{\delta P}{\delta A}$

3.06 Usually the stress will act at some angle to the plane of the section. For convenience, it is resolved into components (see **20**):

1 *direct* stress (f), acts perpendicular, or *normal* to the plane of the section

2 *shear* stress (s) acts parallel to, or *in* the plane of the section.

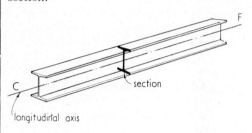

18 *Example of a 'prismatic' member which has a constant cross section. This one is member 7 from* **17a**

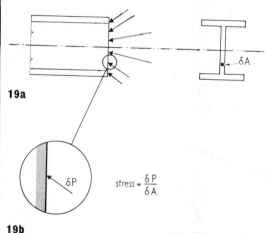

19a

19b

19a *Forces acting on a section;* **b** *forces on a small area* δA
20 *Resolution of stress into direct (f) and shear (s) stresses. (Shear stress is shown on diagrams throughout this section by single headed arrow)*

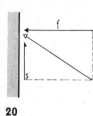

20

3.07 Direct stress can be *compressive* or *tensile* depending on whether it is tending to shorten or lengthen the member. A compressive stress is usually referred to as positive, and a tensile stress as negative **21a**.

3.08 It is important to distinguish between the stress acting on the section, and the stress which is the equal but opposite reaction to it (para 2.12). As defined in para 3.03, the section is produced when the member is cut. This reveals two opposite faces, each face acting on the other with the equal and opposite stress. In this context 'opposite' means 'opposite in direction'. For example, a left-to-right direct stress on the right-hand face implies a right-to-left direct stress on the left-hand face **21b**. However, both of these stresses would tend to shorten the member, and they are therefore both compressive. A downward-acting shear stress on the left-hand face would mean that there was an upward-acting shear stress on the right-hand face **21c**.

3.09 It is usually most appropriate to consider the stress as acting *on* the section, rather than emanating *out* of it. If all the little bits of direct stress acting on the section in this way are added up, the total direct force P on the section is obtained.

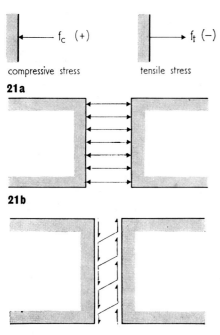

compressive stress   tensile stress

21a

21b

21c

21a *Conventions for showing compressive and tensile stress.* **b** *compressive stress; each force acts on the other with equal and opposite*

stress. **c** *shear stress; downward acting stress on left hand face balanced by similar upward acting stress on the right hand face*

$P = \Sigma f \delta A$

22 *Summation of direct stresses on a section into total direct force P*

3.10 The little bit of force on each area is δP which is equal to f × δA from para 3.05 above. Adding all the little bits together can be written as follows:

$P = \Sigma(f \times \delta A)$.

The symbol Σ means *the sum of all such quantities . . .*, and δA means *a little bit of A*.

3.11 The shear forces can also be summed **23**. As these will vary in direction all over the plane of the section, it is usual to resolve these a second time into the vertical and horizontal directions Y and x thus:

$S_x = \Sigma(s_x \times \delta A)$ and $S_y = \Sigma(s_y \times \delta A)$. (S = total shear)

3.12 All the small forces on the section have now been reduced to three forces at three mutually perpendicular directions: P, S_x and S_y (see **24**). However, the points of application of these forces are not known as these depend on the actual distribution of the stresses across the section. To progress further it is necessary to investigate the geometry of that section.

Section geometry

3.13 Consider the section in **25**. First establish x and Y axes so that any point on the section can be referred to by its co-ordinates. The area of this point is δA.

3.14 If all the little areas are added together the area of the

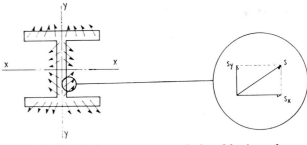

23 *Resolution of shear stresses on a section into*   *vertical and horizontal components*

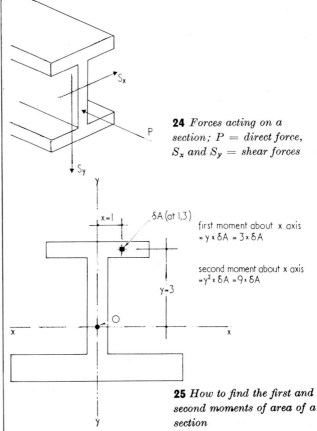

24 *Forces acting on a section; P = direct force, S_x and S_y = shear forces*

25 *How to find the first and second moments of area of a section*

section A is obtained:

$A = \Sigma \, \delta A$

3.15 The value of the moment of the small area about the Y axis is x × δA. All these moments added together give the *first moment of area* of the section, G.

Thus $G_y = \Sigma \, (x \times \delta A)$ and $G_x = \Sigma \, (y \times \delta A)$

Centre of area

3.16 If the section is symmetrical about the x and Y axes, each little area will be balanced by an identical area on the other side of the axis. The values of G_x and G_y will both be zero. Even if the section is not symmetrical, it is always possible to choose x and Y so that the first moments of area about them are zero. The origin of these axes (their intersection) is called the *centre of area* of the section. If the shape of the section were cut out in cardboard, the centre of gravity of the body would be the same point as the centre of area. It follows that if G is zero about two axes through a point, it will always be zero for any other axis at any angle through the same point.

3.17 The longitudinal axis of a symmetrical structural member is usually assumed to run through the centres of areas of the sections of the member.

3.18 A simple horizontal structural member is shown in **26**. A longitudinal axis passes through the centre of area of the section. An x axis, which is horizontal, and a Y axis, which is vertical, pass through the same centre of area.

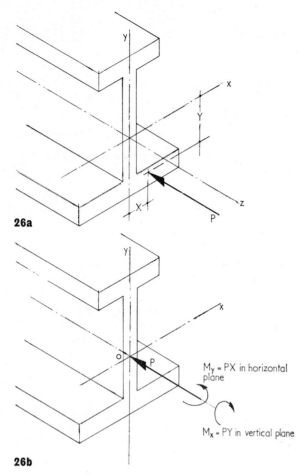

26a

26b

26a *Horizontal structural member subjected to direct force P;* **b** *force P replaced by replacement force acting* *at centre of area creating two couples My and Mx, called bending moments*

Bending moments

3.19 In para 3.12 and **24**, it was seen that the forces on the section could be reduced to three forces in mutually perpendicular directions: P, S_x and S_y. The shear forces S_x and S_y will be considered later; in this section the direct force P will be examined in detail.

3.20 When all the little elements of force on the section were added together, the magnitude of their resultant was found to be P **22**. The position of the point of application of this resultant was not discovered. Assume that this force acts at a point on the section with co-ordinates X and Y **26a**.

3.21 It was shown in para 2.18 that any force at a given point could be replaced by an equal force through another point plus a couple. In this way, the force P at X, Y can be replaced by a force P at O, O (ie at the centre of the section) plus couples PX and PY acting in the horizontal and vertical planes **26b**.

3.22 All the little elements of direct stress on the section have now been replaced by a force P through the centre of area of the section, and two couples. These couples are called *bending moments* and are represented M_y for the moment in the horizontal plane, and M_x for the moment in the vertical plane.

3.23 Each little element of stress makes a contribution to each of these quantities. The contributions can be separated out:

f_a is the part of the stress that adds up to make P acting at the centre of area of the section

f_{bx} is the part that adds up to make M_x

f_{by} is the part that adds up to make M_y.

These separate parts are each stresses acting at the same points, so that
$$f = f_a + f_{bx} + f_{by}$$
3.24 From the definitions in the paragraph above, the following equations can be written (see also **27**):
$$P = \Sigma f \times \delta A$$
$$M_x = \Sigma f \times y \times \delta A$$
$$M_y = \Sigma f \times x \times \delta A$$

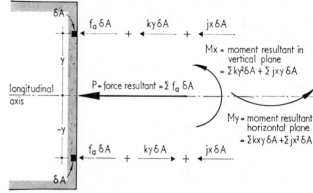

27 *Diagram to illustrate equations described in para 3.26, showing direct force and resultant moments*

3.25 The proofs of the following assumptions are given in para 3.47. It would not be appropriate to give them here, as they rely on the concept of *strain* which has not yet been reached. At this point, these assumptions may simply be accepted:

f_a is constant in value over the whole section.

The value of f_{bx} is proportional to the distance from the x axis of the element in question

ie: $f_{bx} = k \times y$ where k is a constant

similarly the value of f_{by} is proportional to the distance from the Y axis

ie: $f_{by} = j \times x$ where j is a second constant.

3.26 If these values are now put into the equation in para 3.23 it is seen that:
$$f = f_a + k \times y + j \times x$$
and this may be substituted into the three equations in para 3.24 to give:
$$P = f_a \Sigma \delta A + k \Sigma y \times \delta A + j \Sigma x \times \delta A$$
$$M_x = f_a \Sigma (y \times \delta A) + k \Sigma (y^2 \times \delta A) + j \Sigma (x \times y \times \delta A)$$
$$M_y = f_a \Sigma (x \times \delta A) + k \Sigma (x \times y \times \delta A) + j \Sigma (x^2 \times \delta A)$$
These equations are illustrated in **27**.

3.27 In para 3.16 it was demonstrated that if the origin of the x and Y axes was the centre of area of the section, then $\Sigma (x \times \delta A)$ and $\Sigma y \times \delta A$ were both equal to zero.

3.28 The same is not true of $\Sigma (x^2 \times \delta A)$ and $\Sigma (y^2 \times \delta A)$. These are called the *second moments of area* of the section, or sometimes the *moments of inertia;* and are represented I_y and I_x.

3.29 The quantity $\Sigma (x \times y \times \delta A)$ is called the *product of inertia* and is shown as I_{xy}.

3.30 The equations in para 3.26 can therefore be written as follows:
$$P = f_a \times A$$
$$M_x = k \times I_x + j \times I_{xy}$$
$$M_y = k \times I_{xy} + j \times I_y$$
These are the fundamental equations of bending for the general case. In most practical examples of sections, the value of I_{xy} is zero. This is because most sections are symmetrical about at least one axis: and this means that an element of positive $x \times y$ always has a corresponding negative value to balance it. However, a section like the one in **32c** is anti-symmetric about a diagonal axis, and for

*Table 1 Calculation of moment of inertia of section in **28**. See para 3.36 for explanation of calculation*

Area refer- ence	Area = A	First moment of area about reference axis = Ay	Moment of inertia of each part about its own centre of area axis = I_x	Inertia of each part about centre of area axis of whole section = Ar^2	Total moment of inertia of the section
I	$85 \times 25 = 2\ 125$	$2125 \times 12 \cdot 5 = 26\ 562$	$\dfrac{85 \times 25^3}{12} = 110\ 677$	$2125 \times 154 \cdot 5^2 = 50\ 724\ 000$	$50 \cdot 8 \times 10^6$
II	$20 \times 250 = 5\ 000$	$5000 \times 150 = 750\ 000$	$\dfrac{20 \times 250^3}{12} = 26\ 042\ 000$	$5000 \times 17^2 = 1\ 445\ 000$	$27 \cdot 5 \times 10^6$
III	$175 \times 20 = 3\ 500$	$3500 \times 285 = 997\ 500$	$\dfrac{175 \times 20^3}{12} = 117\ 000$	$3500 \times 118^2 = 48\ 734\ 000$	$48 \cdot 8 \times 10^6$
Total	$10\ 625$	$1\ 774\ 062$			$I_x = 127 \cdot 1 \times 10^6$
$y_1 = \dfrac{\Sigma Ay}{\Sigma A}$		$= \dfrac{1\ 774\ 062}{10\ 625} = 167$			

this section there *is* a value of I_{xy}. This will result in a twisting action of the member under bending, which will greatly reduce its calculated strength.

3.31 For any section, axes can be chosen through the centre of area so that I_{xy} for these axes is zero. These are called the *principal axes*. The equations of bending are usually quoted for these axes in the form:

$$P = f_a \times A$$
$$M_x = k \times I_x$$
$$M_y = j \times I_y$$

3.32 In para 3.25 the constants k and j were defined as

$$k = \frac{f_{bx}}{y} \text{ and } j = \frac{f_{ry}}{x}$$

so that the equations above can be written:

$$f_a = \frac{P}{A}$$
$$\frac{f_{bx}}{y} = \frac{M_x}{I_x}$$
$$\frac{f_{by}}{x} = \frac{M_y}{I_y}$$

3.33 If the maximum stress on a section of a member is required, these are the equations that are used to calculate it. Consider the case of bending about the x axis. As the stress is proportional to the distance from this axis, the maximum stress must occur at the point on the section furthest from it.

If y_1 is the distance of that point from the x axis

$$\frac{f_{bx\ max}}{y_1} = \frac{M_x}{I_x}$$
$$\therefore f_{bx\ max} = \frac{M_x}{Z_x} \text{ where } Z_x = \frac{I_x}{y_1}$$

3.34 As pointed out in para 3.31, these equations in this form are only true for sections that are symmetrical about one or other vertical or horizontal axis, such as an I-section, a channel, a solid round, hollow rectangular section and so on. For a completely non-symmetrical section such as an angle, the product of inertia is only zero if the axes of bending are principal axes.

3.35 For steel sections, handbooks issued by the manufacturers will give the values of A, I_x, I_y, Z_x and Z_y. Most other materials are used in simple rectangular or circular shapes for which values of these quantities can be obtained from tables. (A few of these shapes are given in appendix 1 of this handbook). Without going into elaborate mathematics, a few simple rules enable the calculation of the values for some sections not shown in manufacturers' handbooks. I_x (and I_y) are always taken about axes through the centre of area of the section. About another axis parallel to the x axis distant r away, the formula is:

$$I_{rx} = I_x + Ar^2$$

Finding the second moment of area

3.36 To calculate the second moment of area of the section shown in **28**, first find the position of the centre of area. As

the section is symmetrical about a vertical axis, the centre of area must lie on this axis of symmetry; say y_1 from the reference axis at the top of the section. Divide the section into three rectangular areas I, II and III. Table 1 shows the calculations to be done. In the first column, calculate the area of each part, and add to obtain the total area of the section. In the second column, calculate the first moment of each area about the reference axis, ie the area multiplied by the distance of the centre of area of each part. The sum of these quantities will give the first moment of area of the section about the reference axis. If this is divided by the area of the section, it gives the distance of the centre of area from this axis.

3.37 The second part of the calculation starts with calculating the moment of inertia of each part about its own centre of area axis. To this is added the Ar^2 quantity as shown in para 3.35 to obtain the inertia of each part about the centre of area axis of the whole section. In the last column these values are added to obtain the second moment of area, or *moment of inertia* of the section.

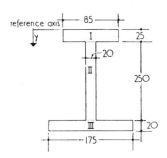

28 *Example to find second moment of area or moment of inertia of a section. See para 3.31 to 3.33 and table 1 for explanation*

3.38 The value of second moment can be obtained by this method for any shape for which values of **A**, **I**, and position of centre of area are available. (A table of shapes is given in appendix 1). Other shapes have to be obtained by calculus methods beyond the scope of this handbook.

Summary

3.39 The following equations for axial stress have been developed:

for bending about the x axis

$$\frac{M_x}{I_x} = \frac{f_{bx}}{y}$$

for the general case

$$f = \frac{P}{A} + \frac{M_x}{I_x} \times y + \frac{M_y}{I_y} \times x$$

3.40 The assumptions on which this equation depends are that:

1 either the x or the y axis or both are axes of symmetry
2 the intersection of these axes is the centre of area of the section (to which must be added the assumptions contained in the equation in para 3.25)
3 the dimensions of the section are small in comparison with the length of the member

4 the section of the member always remains plane
5 the material of the section is homogenous and elastic.

Strain

3.41 A thin metal rod under tension will stretch. This movement under a force is called *strain*, and strain is defined as the extension or contraction of unit length of the member **29**.

$$\text{strain} = \frac{e}{L}$$

29 *Strain is extension or contraction of unit length. L = length of member;*   *e = extension; T = tensile force*

Elastic strain

3.42 One form of strain is called *elastic*. Elastic strain has two properties:

1 Hooke's Law applies. This states that stress is proportional to strain, ie $\dfrac{\text{stress}}{\text{strain}}$ = constant (called Young's modulus)

2 Removal of the stress causes the member to return to its original state.

3.43 Not all materials behave elastically. Some are like modelling clay and deform increasingly under a constant stress, not returning to their original shape after the stress is removed. This behaviour is called *plastic deformation*.

3.44 Most structural materials behave in a manner similar to that shown on the stress strain curve in **30**. For low values of stress the material behaves elastically until the *elastic limit* is reached at point B. The behaviour is then more or less plastic until the breaking point is reached at C. If the stress is reduced during the non-elastic deformation, the material does not return to its original form but retains a permanent deformation as shown by the dotted line.

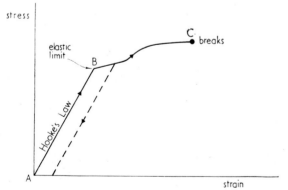

30 *Stress–strain diagram showing elastic limit where Hooke's Law (stress is proportional to strain) no longer operates. Dotted line*   *shows deformation caused when stress is reduced after elastic limit and material does not return to its original form*

Bending stress and strain

3.45 Having introduced the concept of strain, the behaviour of a member under bending can be examined more closely.

3.46 Fig 31 illustrates a member of elastic material under bending. Any section that was plane before bending took place is assumed to remain plane during bending. Two sections distance z apart will then subtend an angle ϕ at the centre of bending such that $z = \phi \times R$. (R is the radius of curvature).

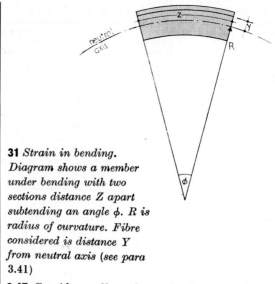

31 *Strain in bending. Diagram shows a member under bending with two sections distance Z apart subtending an angle ϕ. R is radius of curvature. Fibre considered is distance Y from neutral axis (see para 3.41)*

3.47 Consider a fibre above the longitudinal axis of the member through the centre of area of the section (called the *neutral axis*). If this fibre is distance y from the neutral axis, it will be extended during bending to a length of ϕ (R + y); the extension is therefore $\phi \times$ y and the strain:

$$\frac{\phi \times y}{\phi \times R} \text{ or } \frac{y}{R}$$

Now $\dfrac{\text{stress}}{\text{strain}}$ = E (Young's constant) for an elastic material

(see para 3.42). Therefore stress = $\dfrac{E}{R} \times y = k \times y$ as assumed in para 3.25.

Shear

3.48 The detailed theory of shear stress is more complicated than for direct stress. Therefore only an outline of the results is given here rather than a rigorous analysis of the method.

3.49 Shear stresses at each point on the section are resolved into vertical and horizontal components (see para 3.11). The resultants of all these components comprise the vertical and horizontal shear forces on the section S_y and S_x. As for axial stresses, it can be shown that if these forces pass through a certain point some special conditions apply: there will be no torsion or twist of the member. If the resultant of all the shear stresses does not pass through this point, the moment of that resultant about the point is the torsion on the section.

3.50 In the case of shear forces the point is called the *shear centre*. This is not always the same point as the centre of area. In fact, it is the same point only when there are two axes of symmetry or anti-symmetry **32**. A very common case of misconception is the channel shown in **33**. The shear centre is at point S, and a load of 20 N applied at the middle of the flange M will cause a torsion of 1·31 Nm on the section

(ie 28 mm + $\dfrac{75 \text{ mm}}{2} \times$ 20N = 1310 Nmm

or 1·31 Nm)

Distribution of shear stress

3.51 For a symmetrical section with no horizontal shear, a theory can be derived to find the maximum shear stress and its position.

3.52 Consider the section in **34a**. This section is assumed to be subject to bending about the x axis only, with no direct force. An arbitrary shape is chosen to develop a generalised theory before considering special cases like I-sections or rectangular shapes.

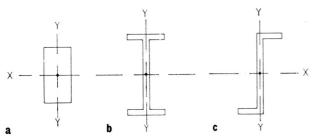

32 *Sections with shear centres at centre of area*

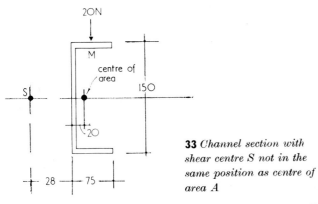

33 *Channel section with shear centre S not in the same position as centre of area A*

3.53 A narrow horizontal strip of section has area δA. The direct stress on this strip is $f = k \times (y)$, where (y) represents the distance of the strip from the neutral axis (the horizontal axis through the centre of area of the section), and k is the constant (from para 3.25).

3.54 Consider the area B_y in **34b**. This area extends from a line distant y from the neutral axis to the extremity of the section. The total force on this area is given by $\Sigma\, k \times (y) \times \delta A$ over this area or $k\, \Sigma\, (y) \times \delta A$. Now $\Sigma\, (y) \times \delta A$ is the first moment of area of the area B_y about the neutral axis, or $B_y \times y_b$ if y_b is the distance of its centre of area from that axis.

Thus: $P = k \times B_y \times y_b$

But $k \times y_b$ is the value of the stress at the centre of area of B_y which can be called f_b.

so $P = f_b \times B_y$

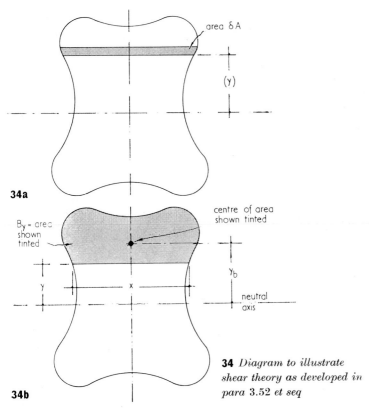

34a

34b

34 *Diagram to illustrate shear theory as developed in para 3.52 et seq*

3.55 Now consider a thin slice of the beam of thickness δz. This is shown in **35** which is not a section of the beam but an elevation. The part of the slice corresponding to the area B_y of the section has been separated from the rest of the slice. It has a force P on each face, and a shear force $s \times x \times \delta z$ on its base from the rest of the slice. These forces must be in equilibrium, so by horizontal resolution:

$$P_1 - P_2 = s \times x \times \delta z$$
$$\therefore\ B_y\,(f_{b1} - f_{b2}) = s \times x \times \delta z$$

Now, $f = \dfrac{M}{I} \times y$ so that $s \times x \times \delta z = \dfrac{y_b \times B_y}{I}\,(M_1 - M_2)$

3.56 By methods outside the scope of these notes, it can be shown that when δz is allowed to get very small, the value of $\dfrac{M_1 - M_2}{\delta z}$ is equal to the vertical shear force S_y on the whole section (see **34b**).

Hence $s = \dfrac{B_y \times y_b}{I \times x} \times S_y$

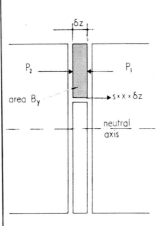

35 *Elevation of beam described in para 3.55 (see also **34**)*

3.57 By the formula above, the value of the longitudinal shear stress in the beam can be determined. That this stress exists can be demonstrated by using two planks to span a wide gap. If the two planks are simply placed one on the other, they sag when the load is applied: each sliding on the other. If the two planks are nailed together this slip cannot occur and the load-carrying capacity of the pair is greatly increased **36**.

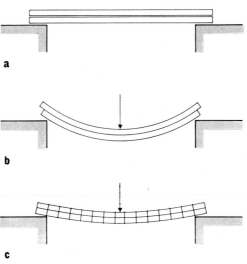

a

b

c

36 *Two planks spanning a wide gap; **a** before application of point load, **b** after application of point load with no fixing between and **c** effect of nailing planks together*

3.58 However, there is also vertical shear stress on the face of the section. If this is shown on a figure similar to that in **35** it becomes as shown in **37**. From the argument in para 3.65, these two shear stresses are equal in value. The formula above, therefore, is also used to find the vertical shear stress at a point on the section, as demonstrated in the following example.

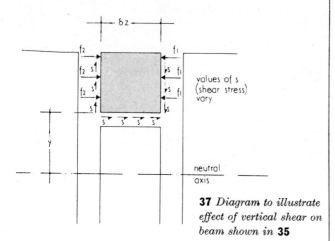

37 *Diagram to illustrate effect of vertical shear on beam shown in* **35**

Example
Using the section shown in **28**, find the shear stress at the join between parts I and II, when the shear force on the section is 20 kN. (All dimensions in mm).

B_y = area of part I = $2 \cdot 125 \times 10^3$
y_b = centre of area of part I from neutral axis = $154 \cdot 5$
I_x = $127 \cdot 1 \times 10^6$ (from table I)
x = breadth of section at point of calculation = 20
S_y = shear on section = 20×10^3
so that:

$$S = \frac{2125 \times 154 \cdot 5}{127 \cdot 1 \times 20} \times 10^{-6} \times 20 \times 10^3$$

$S = 2 \cdot 58 \text{ N/mm}^2$.

Shear on a rectangular section
3.59 Consider a rectangular section depth d and breadth b (see **38**).

$$B_y = \left(\frac{d}{2} - y\right) \times b$$

$$y_b = \frac{y}{2} + \frac{d}{4}$$

$$I = \frac{bd^3}{12}$$

$$x = b$$

thence $s = \dfrac{12\,(d - 2y)\,(d + 2y)\,b}{8\,bd^3 \times b} \times S$

$$= \left[\frac{3}{2b \times d} - \frac{6y^2}{b \times d^3}\right] S$$

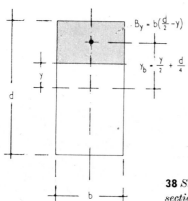

$B_y = b\left(\frac{d}{2} - y\right)$

$y_b = \frac{y}{2} + \frac{d}{4}$

38 *Shear on a rectangular section (see para 3.59)*

3.60 The value of this is obviously greatest when $y = 0$ (that is at the neutral axis) and then:

$$s = \frac{3}{2} \times \frac{S}{b\,d}$$

or the maximum shear stress in a rectangular section is $1\frac{1}{2}$ times the average shear stress, **39**.

3.61 Fig **39** shows the approximate distribution of stress on the rectangular section in graph form. It will be seen that when $y = \dfrac{d}{2}$ that is, on the edge of the section, the stress is zero, as might be expected.

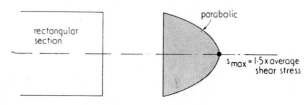

39 *Shear stress in a rectangular section*

Shear on an I-section
3.62 The distribution of shear across an I-section is shown in **40**. It can be seen that the distribution is almost even on the web, and that the influence of the flanges is negligible. The shear stress on the web of the section in the example in para 3.58 can therefore be calculated as:

$$\frac{20\,000}{250 \times 20} = 4 \text{ N/mm}^2$$

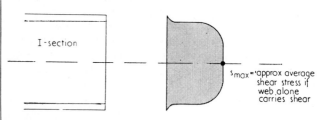

40 *Shear stress on an I-section*

Stress relationships
3.63 Stresses do not exist in isolation. Each form of stress is interdependent on other forms of stress.

Poisson's ratio
3.64 A block of hard rubber if squeezed together will result in a barrel-shape **41**. Similarly, a rubber band when stretched gets thinner. This type of phenomenon happens with all materials: if the strain in the x direction is e_x and the strain in the y direction e_y

$$\frac{e_y}{e_x} = \sigma \text{ (sigma, or Poisson's ratio)}$$

As Hooke's Law also applies it is obvious that, in a similar way

$$\frac{f_x}{f_y} = \sigma$$

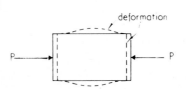

41 *Effect of squeezing a block of rubber*

Shear balance

3.65 In **42a** is shown a small block of material taken from the middle of a structural member, length a, breadth b, and thickness c. On one end there is a shear stress s, or a force s × bc. If resolved vertically, another shear stress s on the other end of the block will prevent vertical movement of the block **42b**, but this will produce a couple of value s × abc. To balance this couple, shear forces s' at the top and base of the block are introduced, producing a couple in the other direction s' × abc, **42c**. Hence s = s'; a vertical shear stress will thus always induce a corresponding horizontal shear stress of equal magnitude.

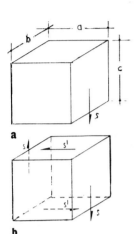

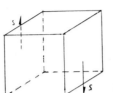

42 Shear balance: **a** shear stress on one end; **b** vertical and opposite force on other end prevents vertical movement but creates a couple; **c** couple balanced by couple in horizontal direction s'

Direct stress induced by shear

3.66 Diagram **43** can be used to discover how shear stress produces direct stress. It is a triangular slice of the block shown in **42**. The length of the hypotenuse is l. Assume an induced direct stress of f and shear stress of s on this hypotenuse. By resolution in the direction of f: f × lb = s' × lb × cos ϕ × sin ϕ + s' × lb × sin ϕ × cos ϕ
∴ f = s' sin 2ϕ
and resolving in the direction of s:
s × lb = s' × lb × sin² ϕ − s' × lb × cos² ϕ
∴ s = s' cos 2ϕ

3.67 The interesting result of this exercise becomes evident when ϕ = 45° because then s = 0 and f = s', that is, a shear stress produces a direct stress of equal value across a plane at 45° to the shear plane. This, of course, is common knowledge **44**.

3.68 The converse is also true: a direct stress induces a shear stress in a plane at 45° to its plane **45**.

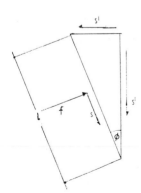

43 Diagram to show how shear stress produces direct stress (see para 3.66)

Principal stresses

3.69 In para 3.67 it was seen that the plane when ϕ = 45° had no shear stress across it. By the theorem in para 3.65 it follows that the plane at right-angles to this plane would also be a plane of zero shear. The direct stresses across these planes are known as *principal stresses*, and one of these will be a maximum direct stress, the other a minimum. The directions of these stresses are called *principal axes*, and if

they are drawn on, say, an elevation of the member, they form stress trajectories.

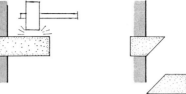

44 *Tension failure due to shear force. Shear stress produces direct stress of*   *equal value across a 45° plane*

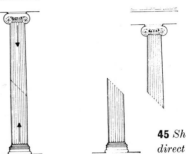

45 *Shear failure due to direct force. Converse of* **44**

Stress trajectories

3.70 Stress trajectories in the side elevation of a cantilever are shown in **46**. The full lines indicate the directions of principal tensile stresses, the dotted lines the directions of principal compressive stresses. These lines cross, as described, at right-angles. The stress magnitudes along these lines do not necessarily stay constant.

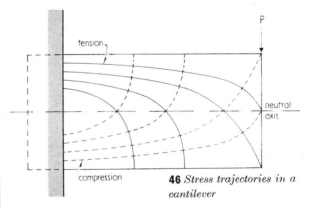

46 *Stress trajectories in a cantilever*

Photo-elasticity

3.71 The directions of principal stresses can also be investigated by means of the curious phenomenon known as *photo-elasticity*. This results from the fact that whenever polarised light passes through certain plastic-based materials, the planes of polarisation are rotated. The amount of rotation has been found to vary with the conditions of stress within the plastic material, so that a model of a structure made of this plastic will show a pattern under polarised light when subjected to forces simulating the loading on the structure **47**. A similar effect is seen when looking at a toughened glass car windscreen through polarised sunglasses.

47 *Photo-elastic patterns in a beam. Note the region of pure bending in the middle, the shear near the ends, and the stress concentrations under the four loads*

Technical study
Analysis 2

Section 2 **Structural analysis**

Structural types

This is the second of two technical studies by DAVID ADLER *which together give a brief mathematical and descriptive background to the later sections of the handbook. This study deals with beams and struts and various types of surface and skeletal structures*

1 Beams

1.01 Beams are either *statically determinate* or *statically indeterminate.* For statically determinate beams bending moments and shears can be determined using statics only. This type of beam is considered first; statically indeterminate beams are dealt with in para 1.24.

1.02 Statically determinate beams are either *simply supported* or *cantilever.*

Simply supported beams

1.03 A simply supported beam is shown in **1a**. It is assumed that the support at each end is a perfect hinge, and that one end is also free to move horizontally. The same beam, conventionally drawn, is shown in **1b**. These perfect conditions are never achieved, but assumptions are necessary to produce a workable theory.

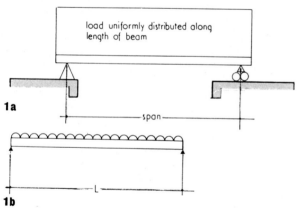

1a

1b

1 *Simply supported beam:* **a** *support at each end assumed to be perfect hinge, right end can move horizontally;* **b** *conventional way of showing beam* **a** *with uniformly distributed* (UD) *load*

1.04 The beam has a span L measured between centres of supports, and has one or more loads on it. In theory, loads are of three types:
1 point loads (P)
2 uniformly distributed loads (W or w)
3 non-uniformly distributed loads

1.05 Point loads are assumed to act at a point, but in practice they always spread a short distance. An example of a point load is a column resting on a beam **2a**. This is represented conventionally in **2b**.

1.06 The forces in **2b** at A and C are *reactions,* and are usually shown R_a and R_c. Taking moments about A for the system:

$20\,000 \times 3 - R_c \times 6 = 0$ (must have zero resultant to maintain equilibrium)

$\therefore R_c = 10\,000$ N or 10kN

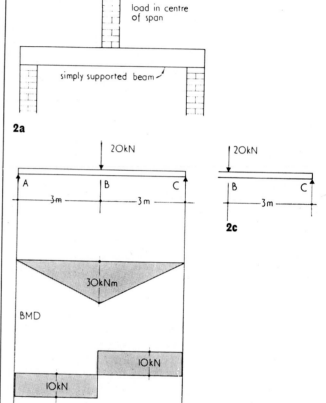

2a

2c

2b

2a *Simply supported beam with point load in centre of span:* **b** *diagrammatic representation of beam with reactions at A and C and load of 20 kN at B, bending moment diagram* (BMD) *and shear force diagram* (SFD) *shown below;* **c** *dimensions used in calculating bending moment under point load*

1.07 Now imagine that the beam is cut immediately to the right of B, where the load is applied. The forces on the half of the beam between this cut and C are:
reaction $R_c = 10$ kN
shear force at B $= S_b$
bending moment at B $= M_b$

1.08 These forces on the part of the beam between the cut and C must be in equilibrium:
$\therefore S_b - R_c = 0$ by vertical resolution (**2b**)
and $M_b - R_c \times 3 = 0$ by moments about B (**2c**).
Thus $S_b - 10 = 0$
$\quad \therefore S_b = 10$ kN
and $M_b - 10 \times 3 = 0$
$\quad \therefore M_b = 30$ kNm

1.09 The bending moment and shear force diagrams (BMD and SFD) for the beam can then be drawn (**2b**).

1.10 The beam shown in **3** has a uniformly distributed load of 1·5 kN/m over a span of 10 m. Using the same analysis as before, or simply by symmetry:

$$R_a = R_c = \frac{1 \cdot 5 \times 10}{2} = 7 \cdot 5 \text{ kN}$$

1.11 Bending moment M_b at the centre of the span is found in the same way:

$M_b - R_c \times 5 + 1 \cdot 5 \times 5 \times 2 \cdot 5 = 0$ (**3b**)

(The effective resultant of the uniform load over half the span acts at the centre of gravity of the half-load, ie the quarter-span point, or 2·5 m from C)

$\therefore M_b = 37 \cdot 5 - 18 \cdot 75 = 18 \cdot 75 \text{ kNm}$

1.12 Shear force at the centre is $S_b = R_c - 1 \cdot 5 \times 5 = 0$ or $S_b = 7 \cdot 5 - 7 \cdot 5 = 0$.

The BM and SF diagrams are shown in **3a**.

1.13 These examples give an indication of methods for solving problems from first principles. Normally formulae obtained from tables are used (see Information sheet ANALYSIS 1) and **2** and **3** would be solved using $\dfrac{WL}{4}$ and $\dfrac{wL^2}{8}$ respectively, where W or w = load and L = span.

Cantilevers

1.14 The cantilever beam (in **4**) and the loads on it are supported at the left-hand end A, where there is a reaction R_a and an *end fixity moment* M_a. End A is often described as *encastre* or built-in.

1.15 It is very simple to calculate R_a and M_a by statics. In the example in **4**:

Vertical resolution

$R_a - 4 \times 3 - 10 = 0$

$\qquad\qquad \therefore R_a = 22 \text{ kN}$

Moments about A

$M_a - 4 \times 3 \times 1 \cdot 5 - 10 \times 3 = 0$

$\qquad\qquad \therefore M_a = 48 \text{ kNm}$

Deflection of beams

1.16 As well as calculating stresses in structural material, it may be necessary to predict how much a structure will move. Then the movement of individual members is assessed; each component can move in three ways:

1 It can lengthen or shorten. This is strain due to direct stress and is readily calculated:

movement = length × strain

$\qquad\qquad$ = length × stress/E (Young's modulus)

2 It can bend. The longitudinal axis can move at right-angles to itself.

3 It can twist. A simple calculation if torsion on the member is known.

1.17 Bending deflection is most difficult to calculate, mainly because the bending stresses are usually not constant along the length of the member and the theory requires the use of calculus. But for simple problems there are two theorems that avoid higher mathematics; their proofs, however, will have to be taken on trust.

THEOREM 1

1.18 The change in the slope of a beam over a given length is equal to the area of the $\dfrac{M}{EI}$ diagram which equals

$$\frac{\text{Bending moment}}{\text{Young's modulus} \times \text{moment of inertia}} \text{ on that length } \mathbf{5a}.$$

THEOREM 2

1.19 In **5b** the vertical distance y of A below the tangent at

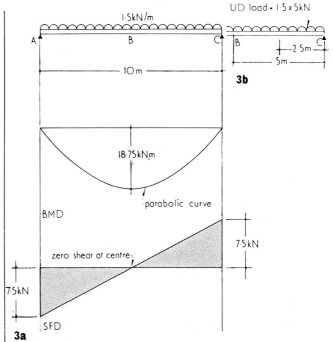

3b

3a

3a *Simply supported beam with uniformly distributed* (UD) *load, showing bending moment and shear force diagrams;*
b *dimensions used in calculating maximum bending moment at centre of span*

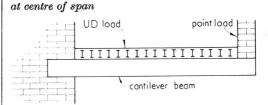

4a

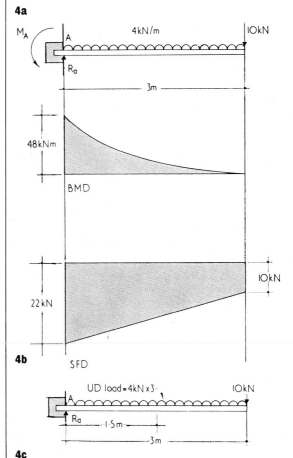

4b

4c

4a *Cantilever beam with* UD *load plus point load at the end;*
b *bending moment and shear force diagrams;* **c** *dimensions used in calculating* BMD *and* SFD

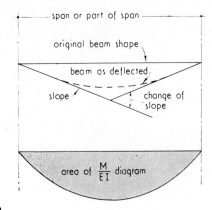

5a

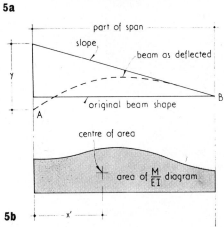

5b

5 *Change of slope and deflection:* **a** *illustrates theorem 1 and shows either complete span or part span where change of slope = area of* $\frac{M}{EI}$ *diagram;* **b** *illustrates theorem 2 and shows only part of a span where* y = *area of* $\frac{M}{EI}$ *diagram* × x'

B is equal to the first moment of area of the $\frac{M}{EI}$ diagram on AB about A.

1.20 In most cases, the material (and therefore E) and the section of the member (and so I) is constant over the length of the member. Therefore it is only necessary to consider the moment variation when using these theorems. The $\frac{M}{EI}$ diagram becomes the M or bending moment diagram, the EI factor being used when the slope or deflection has to be calculated.

1.21 To show how these theorems are used, consider the cantilever in **6**. The M (bending moment) diagram is a triangle of area $\frac{PL^2}{2}$, its centre of area $\frac{2}{3}$ L from the free end A. The beam is horizontal at B, so that the tangent at B is also horizontal. Thus the deflection of A (y_a) is equal to the first moment of area of the $\frac{M}{EI}$ diagram about A:

$$y_a = \frac{PL^2}{2EI} \times \frac{2}{3}L = \frac{PL^3}{3EI}$$

1.22 In practice, cantilevers are often placed at the ends of continuous beam runs. Then the beam is *not* horizontal at B and the deflection of A will be greater by the slope at B (which must be calculated) × length L.

1.23 The deflections of beams are not usually calculated from first principles, because there are tables covering common loadings. The real use of slope and deflection calculations is in the design of structures that are not statically determinate.

Encastré beams*

1.24 The simplest form of statically indeterminate structure is the beam built-in (*encastré*) at both supports **7**. A statically determinate beam can have two simple supports, or one built-in support. With one built-in support and one simple support the beam has one degree of redundancy, because if the bending moment on the *encastre* support were removed, the beam would be statically determinate. The doubly built-in beam has two degrees of redundancy.

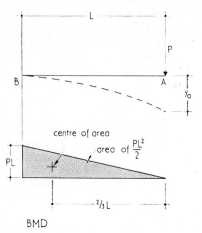

BMD

6 *Deflection in a cantilever,* y_a *is equal to* $\frac{PL^3}{3EI}$ *(ie the first moment of area of the* $\frac{M}{EI}$ *diagram about A)*

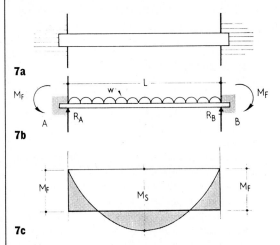

7a

7b

7c

7a *Beam built in (encastre) at both supports;* **b** *moment diagram described in para 1.25.* M_s = *free support moment,* M_F = *end fixity moment* = $\frac{wL^2}{12}$

1.25 The bending moment on a statically indeterminate beam has two components:
1 the free support moment M_s that would occur on the equivalent statically determinate beam
2 the end-fixed moment M_F, which has a straight-line distribution along the beam. This is limited by values of the moments at each end.
From theorem 1, since there is no change of slope along this beam (it is zero at each *encastre* end), the area of the $\frac{M}{EI}$ diagram must be zero.
The area of the free support bending moment diagram is equal and opposite in sign (+ −) to the area of the end-fixity moment diagram.

*Here *encastré* means beams rigidly fixed at both ends and not merely built into brickwork

Because this beam is symmetrical; the moments at each end are equal, and the area of the end-fixity moment diagram is $M_F L$.
Thus:

$M_F L = \dfrac{2}{3} M_S L$ (The area of a parabola, in this case $M_S L$, is two-thirds of its enclosing rectangle)

$\therefore M_F = \dfrac{2}{3} \times \dfrac{wL^2}{8} = \dfrac{wL^2}{12}$

Example: complete analysis of a propped cantilever
1.26 The beam is shown in **8**.

1.27 First find the value of M_F
The tangent at A is horizontal. Because B is propped there is no deflection there and so the first moment of the $\dfrac{M}{EI}$ diagram about B must be zero.

$\therefore M_F \times \dfrac{L}{2} \times \dfrac{2}{3} L = \dfrac{2}{3} \times \dfrac{wL^2}{8} \times L \times \dfrac{L}{2}$

where

$M_F \times \dfrac{L}{2}$ is the area of the triangle formed between R_A and R_B (shown dotted on the BMD in **8**)

$\dfrac{2}{3} L$ is the distance of its centre of gravity from R_B

$\dfrac{2}{3} L \times \dfrac{wL^2}{8} \times L$ is two-thirds of the enclosing rectangle of the parabola, which equals the area of the parabola

$\dfrac{L}{2}$ is the distance of its centre of gravity from R_B.

$\therefore M_F = \dfrac{wL^2}{8} = \dfrac{2 \times 5 \cdot 5^2}{8} = 7 \cdot 56$ kNm

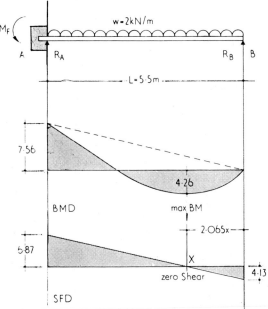

8 *Propped cantilever used in analysis described in paras 1.26 to 1.33. Maximum BM occurs at point of zero shear*

1.28 The vertical reactions R_A and R_B at each end of the beam are calculated by taking moments about each end in turn.
Moments about A

$R_B \times L + M_F - \dfrac{wL^2}{2} = 0$ (Must have zero resultant to maintain equilibrium)

$\therefore R_B = \dfrac{2 \times 5 \cdot 5}{2} - \dfrac{7 \cdot 56}{5 \cdot 5}$

$= 5 \cdot 5 - 1 \cdot 37 = 4 \cdot 13$ kN

Moments about B

$R_A \times L - M_F - \dfrac{wL^2}{2} = 0$

$\therefore R_A = 5 \cdot 5 + 1 \cdot 37 = 6 \cdot 87$ kN

Check by vertical resolution

$4 \cdot 13 + 6 \cdot 87 - 2 \times 5 \cdot 5 = 0$

1.29 Calculation of the maximum positive moment in the span of the beam is easier because the maximum positive moment occurs at the point of zero shear. (Proof of this involves calculus and is outside the scope of these notes.)
1.30 Assume that the point of zero shear is x metres from B towards A. Resolving vertically on XB:

$w \times x - R_B - S_X = 0$ (Where $S_X =$ the shear force at X)

But $S_X = 0$ as X is the point of zero shear, so that $x = \dfrac{R_B}{w}$

$= \dfrac{4 \cdot 13}{2} = 2 \cdot 065$ m

Taking moments at B for XB:

$M_X - \dfrac{wx^2}{2} = 0$ (Where $M_X =$ the bending moment at X)

$\therefore M_X = \dfrac{2 \times 2 \cdot 065^2}{2} = 4 \cdot 26$ kNm (approx)

1.31 Now calculate the maximum stresses in the beam. Assume a beam section of 250×75 mm rectangular.
Area of the section is $250 \times 75 = 18\,750$ mm²
The second moment of area I about the neutral axis is
$\dfrac{bd^3}{12} = \dfrac{75 \times 520^3}{12} = 98 \times 10^6$ mm⁴.

From the general equations of bending developed earlier*

$\dfrac{M}{I} = \dfrac{f}{y}$ (where $f =$ stress, and $y =$ distance of extreme fibres from neutral axis)

$\therefore f = \dfrac{M}{I} y$

The maximum value of M is M_F, and the maximum value of y is at the top and bottom of the section.
In terms of the units by which these values have so far been measured, $f = \dfrac{M \text{ (in kNm)}}{I \text{ (in mm}^4)} \times y$ (in mm)

To make them comparable, M must be multiplied by 1000 (or 10^3), so that all lengths are in mm, and by a further 1000 (a total of 10^6) to reduce kN to N, as the final answer is better expressed in that form. (Conversion to like units is used throughout this section.)

$\therefore f_{max} = \dfrac{7 \cdot 56 \times 10^6}{98 \times 10^6} \times 125 = 9 \cdot 64$ N/mm²

1.32 The maximum shear force is $R_A = 6 \cdot 87$ kN (see **8**). The maximum shear stress on a rectangular section is $1 \cdot 5$ times the average stress†

$\therefore S_{max} = \dfrac{R_A}{\text{area (b} \times \text{d)}} \times 1 \cdot 5$

$\therefore S_{max} = \dfrac{6 \cdot 87 \times 10^3}{18\,750} \times 1 \cdot 5 = 0 \cdot 55$ N/mm²

1.33 The calculation of maximum deflection under load could be very difficult, but the formula $\dfrac{wL^4}{185 \, EI}$ is available from tables and the deflection under load can be computed provided the Young's modulus (E) for the material is known.

*see technical study ANALYSIS 1 para 3.39
†see technical study ANALYSIS 1 para 3.60

2 Struts

2.01 A *strut* is any structural member that is subjected to mainly compressive forces. Often the word is reserved for non-vertical members; vertical struts are described as *column* stanchions or piers, the use depending to some extent on the material (stanchions are mainly iron and steel, and piers brickwork or masonry).

2.02 The theory of struts involves the concept of unstable equilibrium. A system of forces in unstable equilibrium is shown in **9**; this complies with all the requirements for equilibrium, but the smallest change in any of the forces would destroy the system. Systems in unstable equilibrium must be avoided in structural work!

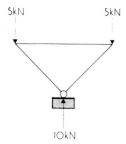

9 *Unstable equilibrium, any change would destroy system*

2.03 Each end of the structural member in **10a** is a pin-joint that can only move in the direction of the length of the member. As long as the member is perfectly straight, the only stress in the member will be pure compression, of magnitude $\dfrac{P}{A}\left(\dfrac{pressure}{area}\right)$. If the member were slightly bowed **10b** so that the deflection of the longitudinal axis at the centre is y, the bending moment at the centre would be Py.

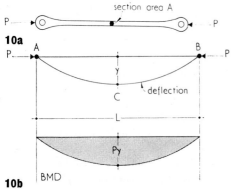

10a

10b

10 *Pressure on a strut:* **a** *pin-jointed member with applied pressures resulting in pure compression;* **b** *strut deflected by pressures. Maximum* BM *at centre is Py*

2.04 A rigorous analysis of the deflection requires calculus but the approximate method used here gives a similar answer. If the deflection of the strut is parabolic the moment diagram follows the same parabola. The moment at any point x is Py. From theorem 2, para 1.19, the deflection of B *above* the tangent at C (the centre of the member) is equal to the first moment of the $\dfrac{M}{EI}$ diagram on CB about B.

Thus $y = \dfrac{2}{3} \times \dfrac{M_C}{EI} \times \dfrac{L}{2} \times \dfrac{5}{16}L = \dfrac{5}{48} M_C \times \dfrac{L^2}{EI}$

(note $\dfrac{5}{16}L$ is distance from B of centre of gravity of that section of the parabola)

Now $M_C = P \times y$

so that $y = \dfrac{5}{48} \times \dfrac{PL^2}{EI} \times y$

This is a rather curious result: either

$y = 0$ and $\dfrac{5}{48} \times \dfrac{PL^2}{EI}$ can have *any* value,

or $\dfrac{5}{48} \times \dfrac{PL^2}{EI} = 1$ and y can have any value.

Thus there will be no appreciable bending of the strut until P, the compression on the strut, approaches $\dfrac{48\,EI}{5L^2}$. Then the deflection of the centre of the strut increases rapidly, and the strut buckles. ($\dfrac{48}{5} = 9 \cdot 6$. A rigorous analysis gives a coefficient of approximately $9 \cdot 87$).

2.05 When designing struts it is therefore important not to exceed the P calculated above. This load is called the *Euler load*. Normally a load factor of at least two is used so that for safety a working load of half the Euler load should not be exceeded. Thus in theorem 2, a figure of half $\dfrac{48}{5}$ or $\dfrac{48}{5 \times 2}$ is used.

2.06 If the compressive stress is f_c then $P = f_c \times A$ so that $f_c \, max = \dfrac{48}{5 \times 2} \times E \times \dfrac{I}{AL^2}$

now $\dfrac{I}{A} = r^2$ (r is the *radius of gyration* of the section)

$\therefore f_c \, max = 4 \cdot 8\,E\left(\dfrac{r}{L}\right)^2$

$\dfrac{L}{r}$ is called the *slenderness ratio* of the member.

2.07 For many materials, values of $f_{C\,max}$ are tabulated against values of the slenderness ratio so that calculation is easier. Of course, the maximum stress must not exceed the safe compressive stress of the material of the member whatever the value of the slenderness ratio.

2.08 The theory of struts is not confined to members in pure compression. The compression flanges of members under bending can also be unstable, and the magnitudes of the compressive stresses on such members are limited by the theory. Here instability depends on a number of factors, and tables are used to determine the safe compressive stresses.

Bending and compression

2.09 Columns in building frames, which are struts, are commonly subjected to both compressive and bending stresses (produced by eccentricities of loading on to the columns). It is therefore important to understand the theory of combined bending and compression.

2.10 The rectangular section (**11a**) 450×250 mm under a load of 300 kN and subject to a moment of 4 kNm can be treated in two ways:

METHOD 1

2.11 First calculate the stress due to pure compressive loading

$f_a = \dfrac{P}{A} = \dfrac{300 \times 10^3}{450 \times 250} = 2 \cdot 67\ \text{N/mm}^2$ (P is multiplied by 10^3 to translate kN to N)

Then work out the tensile and compressive stresses due to bending

$f_c = f_t = \dfrac{M}{I} y$ where $I = \dfrac{bd^3}{12}$ and $y = \dfrac{d}{2}$

$\therefore f_c = f_t = \dfrac{4 \times 10^6}{250 \times \dfrac{450^3}{12}} \times 225 = 0 \cdot 47\ \text{N/mm}^2$

The maximum compressive stress is therefore $2 \cdot 67 + 0 \cdot 47 = 3 \cdot 14\ \text{N/mm}^2$ and the stress on the opposite edge of the section is $2 \cdot 67 - 0 \cdot 47 = 2 \cdot 2\ \text{N/mm}^2$. (See **11b**).

METHOD 2

2.12 Convert the compression force and moment into an eccentric force. If e is the eccentricity:

$$e = \frac{M}{P} = \frac{4 \times 10^3}{300} = 13 \cdot 3 \text{ mm}$$

2.13 The distribution of stresses on the rectangular section of the column is shown in **12a**. The area of this diagram equals the compressive force 300 kN, and the centre of area coincides with the point of action of the force. Thus the stresses on each edge of the section can be found.

This method is important because it covers the behaviour of the structure under changing conditions.

2.14 If the load is increased without changing its eccentricity **12b**, the stresses would all increase in proportion to the increase in load without changing their distribution.

2.15 If the eccentricity of a constant load were to increase **13**, then the stress on the left-hand edge of the section would decrease **13a**, and the stress on the right-hand edge increase. When the stress on the left-hand edge becomes zero **13b**, the stress diagram becomes a triangle, its centre of area one-third of the base length from the right-hand edge, or one-sixth of the base length from the centre-axis. This *middle third* zone of the column is extremely important —if the load does not leave this zone, there is no tension on the section, if the load is outside the middle third, there is tension on part of the section **13c**. This is not always very significant, but some materials have very small tensile strength, and then there is a fundamental change in the behaviour of the section. This will be described in the later sections on structural materials.

2.16 It is possible for the load to have so great an eccentricity as to lie outside the section. Then bending is more significant than compression, and the member should be treated as a beam.

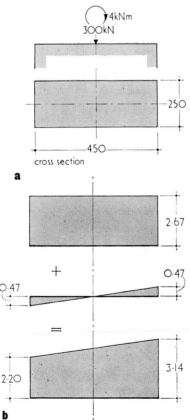

11a *Rectangular section under load with a moment on the section (method 1, see para 2.11);* **b** *(top) stress due to compression loading, (middle) stress due to bending, (below) combined stress*

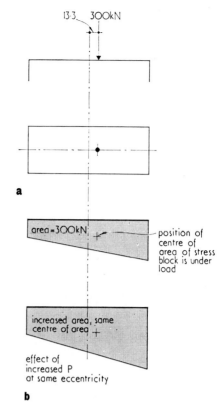

12 *Rectangular section as* **11** *but with load and moment combined to form eccentric load (method 2, see para 2.12):* **a** *distribution of stresses;* **b** *effect of increasing load but not changing eccentricity*

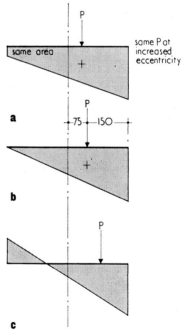

13 *Effect on beam (shown in* **11***) of increasing eccentricity:* **a** *stress on left-hand edge decreases until it reaches* **b***, a triangle where load falls within middle third, and finally* **c** *when tension is produced because load lies outside middle third*

Ties

2.17 Members subject mainly to tension are fairly easy to analyse, but if the member also bends the treatment is similar to that for combined bending and compression. As there is no case of instability, tension members can be very slender; wire or thin strip is often used.

The area of a tension member in the stress calculations is the minimum section. Often this coincides with a hole in the member **14**. For a threaded bolt the minimum section occurs at the root of the thread.

3 Types of structure

3.01 Having looked at stresses in structural materials and the design of structural members methods of analysing structural systems can now be examined.

3.02 These systems are classified:

1 *Skeletal structures:* Pin-jointed (eg roof-truss **15a**) Rigid-jointed (eg Vierendeel frame **15b**)

2 *Surface structures* (eg shell roof **15d**)

3.03 There are also systems which combine elements of each type—thus a continuous beam system is partly rigid-jointed and partly pin-jointed (at the supports) **15e**.

3.04 Mathematical analysis of structures is a kind of model-making; the structural systems analysed are model systems.

3.05 Real structures are, of course, three-dimensional, but for simplicity they are usually split into a number of planar systems. The building frame in **15f** is an example of such treatment.

3.06 Surface structures are often analysed as if they were skeletal structures. The multi-storey flat slab system in **15g** is analysed as two separate but interlocking rigid-jointed skeletal systems.

3.07 Rigid-jointed skeletal structures are often analysed as if they were pin-jointed. A typical example is the common roof truss **15c** where the rafters are large continuous members.

3.08 The successful solution of a structural problem depends on the correct choice of model system. There is a particular method of analysis for each model.

14 *Tension member—a tie; minimum section area is used in calculations*

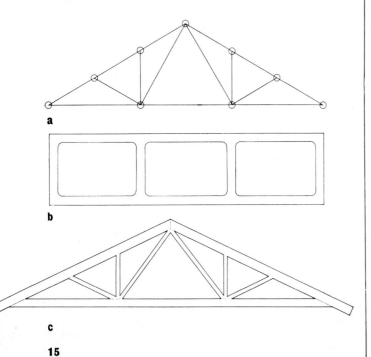

a

b

c

15

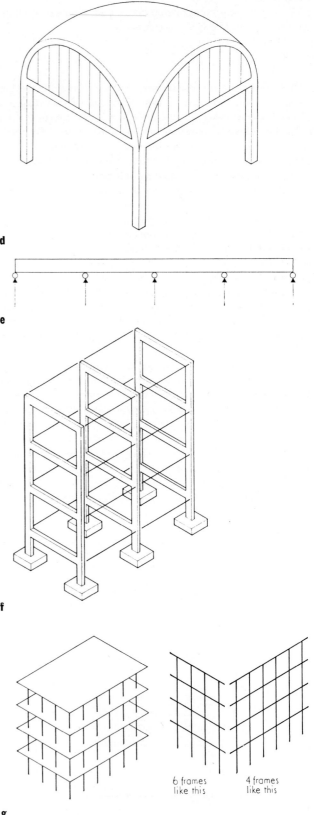

d

e

f

g

15 *Types of structure:* **a** *pin-jointed roof truss (skeleton structure);* **b** *rigid-jointed Vierendeel frame (skeleton structure);* **c** *common roof truss (rigid-jointed skeletal structure often analysed as though pin-jointed);* **d** *Shell roof (surface structure);* **e** *continuous beam (combined skeletal and surface);* **f** *building frame (analysed as series of planar systems);* **g** *multi-storey flat slab system (surface structure analysed as though skeletal)*

4 Isostatic truss

4.01 In an isostatic or statically determinate truss all forces can be determined without consideration of the size or material of the member. It is the archetype of the pin-jointed skeletal system.

4.02 The members of this truss are all straight, joints are all pinned, and all loads and reactions act at the joints. Thus there can be no bending moments in any of the members.

4.03 Forces on the members are purely axial, and can be found by resolution at each joint in turn.

Alternatively the method of section can be used—the truss is assumed to be cut in two and the forces on each part can be resolved, so that the internal forces in the cut members are established.

4.04 For most cases the quickest and easiest way is a graphical method using the polygon of forces*—*Bow's notations.* (See **16**.)

4.05 A letter is given to each of the spaces enclosed by the members and applied forces, **16a**.

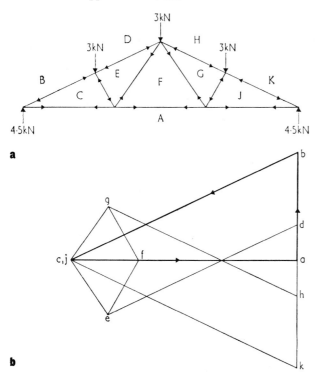

a

b

16 *Graphical method of determining forces in a roof truss:*
a *truss with spaces lettered to identify forces and members;*
b *Bow's notation (a special form of the polygon of forces) which, if drawn to scale, will enable forces in members to be measured directly (see para 4.04 to 4.10). Direction arrows on triangle abc relate to directions of forces around left-hand support in* **a**. *These directions indicate whether the member is in tension (tie) or compression (strut)*

4.06 The Bow's notation diagram **16b** is built up starting from the left-hand support. The reaction of 4·5 kN is an upward force dividing space A from space B, going round the joint in a clockwise direction. Therefore, draw ab vertically upwards of length proportional to 4·5 kN. Continuing round the joint clockwise, the rafter dividing B from C is reached. The force in this member is compressive, acting down the slope towards the joint. Line bc can be drawn parallel to this force in the same direction, although the position of c is still unknown. The next member at this joint is the tie CA—its tensile force acting away from the joint. Drawing ca parallel to this tie establishes the position of c and hence the lengths of bc and ca. The magnitudes of

*see technical study ANALYSIS 1 para 2.11

the forces in BC and CA can therefore be established by measuring lines bc and ca respectively.

4.07 The joint in the middle of the rafter is considered next and the procedure is repeated working clockwise around the joint. cb is already drawn, so this acts as the base for this joint. The magnitude of the load BD is 3 kN, so the position of d can be established. de and ec are drawn parallel to the lines of members DE and EC to establish point e.

4.08 Each joint is treated in this way to complete **16b**. Notice that c and j turn out to be the same point in the diagram. The line bd-dh-hk is the same line in reverse as ka-ab, and represents the external loads on the truss.

4.09 The real truss differs considerably from the model. For example, the compression in rafter BC is found to be 9 kN. The load from the roof is not applied at the joint, but at a number of points along the rafter, and the bending moment due to these loads must be found. The combined bending and axial stresses must also be examined to ensure that they are less than the permitted maxima.

4.10 Thus analysis of the model system is not the end of the calculation, but only one of the stages.

5 Space frames

5.01 To a certain extent the popularity of the space frame is due to the advent of the computer. Nearly all space frames are rigid-jointed, and the analysis of axial, shear, bending and torsion forces in the members and on the joints would be beyond normal hand-methods. In practice the usual assumptions for pin-joints, which are so useful in plane frames, are both dangerous and costly when applied to three-dimensional problems.

5.02 Most space frames are designed by specialist consultants or contractors who have access to the complicated computer programs required.

6 Rigid-jointed frames

6.01 The analysis of a large multi-storey building frame with rigid joints (**17a**) is also a job for the computer. But there is a subdivision process that allows manual methods to be used.

6.02 First, the beams at each floor level are analysed as continuous beams supported on pin joints at each column position. The beams can be designed using results from this analysis **17b**.

6.03 Next, the bending moments imposed on the external columns are estimated using empirical formulae. This information and the axial loads (estimated from the loads of the building fabric and superimposed loading **17c**) are needed when designing the columns.

6.04 Finally, the wind moments in the columns can be calculated by one of a number of semi-empirical methods. This checks that the allowable increase in stress for forces induced by wind is not exceeded **17d**.

7 Continuous beams

7.01 The most common problem in statically indeterminate structures is the continuous beam. This is a combination of a rigid and a pin-jointed skeletal system.

7.02 Various methods exist for solving the problem—the theorem of three moments, virtual work, influence co-efficients, and so on—these can be referred to in standard textbooks. There are also tables that solve the problem for various span ratios and loading conditions.

7.03 The method recommended for analysing continuous beams is the *Hardy-Cross* or *moment distribution* method.

7.04 Assume that all the support joints in a run of con-

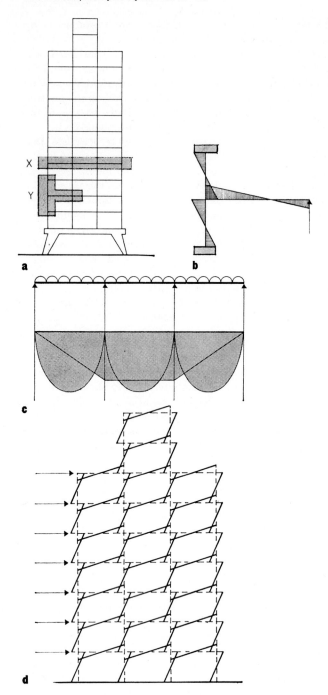

a

b

c

d

17a *Typical multi-storey frame with rigid joints;* **b** *beam-to-column junction at Y;* **c** *beam analysis of typical beam X;* **d** *typical empirical wind analysis of frame in* **17a**

tinuous beams **18a**, are locked so that the ends of each span are horizontal and *encastre*.

7.05 The end moments in each span are calculated.

7.06 Then each joint is unlocked, allowed to take up its free position, and locked again. Each time that a joint is unlocked, the clockwise and anticlockwise moments from the end moments of the adjacent beams are balanced out **18b**.

7.07 There is also a carry-over of moment from the unlocked joint to the locked joints on either side. This must be calculated and allowed for. As a result, each time a joint is unlocked, the previously balanced neighbouring joint has some out-of-balance moment reimposed on it. Thus the distribution method consists of several repetitions or *iterations* of the locking and unlocking, until the residual out-of-balance moment at each joint is very small **18c**.

Theory of moment distribution method

7.08 Apply a moment M_A to the left-hand end of the beam

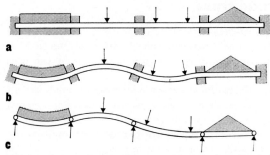

a

b

c

18a *Continuous beam with encastre ends showing loading;* **b** *beam when joints are unlocked;* **c** *final out-of-balance moment at each joint is quite small (see paras 7.04 to 7.07)*

with fixed ends **19a** by rotating it through an angle ϕ. A moment M_B is induced at the right-hand end (see the moment diagram **19b**). From para 1.18 the area of the $\frac{M}{EI}$ diagram is equal to the change in the slope of the beam ie

$$\phi = \frac{\frac{1}{2} \times (M_A + M_B) \times L}{EI}$$

7.09 Also, from para 1.19 the vertical distance of A below the tangent at B is equal to the first moment of area of the $\frac{M}{EI}$ diagram about A

$$\phi = \frac{(M_A + 2M_B) \times L^2}{6EI}$$

hence $M_B = -\frac{1}{2}M_A$

substituting this in the equation above

$$\phi = M_A \times \frac{L}{4EI}$$

or $M_A = \frac{4EI}{L} \times \phi$

7.10 The quantity $\frac{4EI}{L}$ is the *stiffness* of the beam. Stiffness

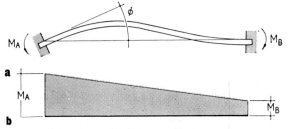

a

b

19a *Beam with fixed ends, with moment applied to left-hand end;* **b** *moment diagram*

Table I *Beam stiffnesses. Stiffness* $= \dfrac{kEI}{L}$

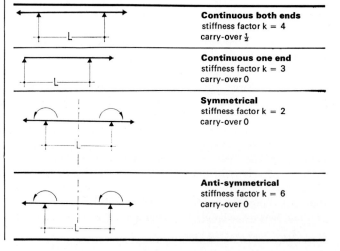

	Continuous both ends stiffness factor k = 4 carry-over $\frac{1}{2}$
	Continuous one end stiffness factor k = 3 carry-over 0
	Symmetrical stiffness factor k = 2 carry-over 0
	Anti-symmetrical stiffness factor k = 6 carry-over 0

is defined as the moment required to achieve unit rotation. Table 1 gives the values of stiffness for a few useful beam conditions.

7.11 Now $M_B = -\frac{1}{2}M_A$, that is, the carry-over moment is half the magnitude of the applied moment. The negative sign indicates that a hogging applied moment produces a sagging carry-over (see **21**). This is not the usual sign convention, and it will be seen later that the carry-over has the same sign as the applied moment.

Continuous beam calculation

7.12 A continuous beam system is shown in **20**.

7.13 *The stiffness ratios* for each pair of spans is calculated using

$$\text{Stiffness} = \frac{kEI}{L}$$

Any quantity (E in this example) that is common to all spans can be ignored because it cancels out in the ratios.

Table II *Stiffness ratios for the beam in* **20**

Span	k	I	L	Stiffness each $\times$ E $\times 10^6$	B	C	D
AB	3	600×10^6	7000	256	0·35		
BC	4	600×10^6	5000	480 $\biggl\{\!\!{}^{736}$	0·65	0·59	
CD	4	500×10^6	6000	333 $\biggl\{\!\!{}^{813}$		0·41	0·57
DE	3	500×10^6	6000	250 $\biggl\}{}^{583}$			0·43

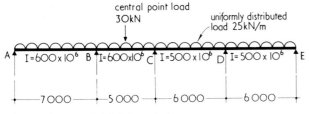

central point load
30kN

uniformly distributed load 25kN/m

A $\quad$ I=600x10⁶ $\quad$ B $\quad$ I=600x10⁶ $\quad$ C $\quad$ I=500x10⁶ $\quad$ D $\quad$ I=500x10⁶ $\quad$ E

|— 7 000 —|— 5 000 —|— 6 000 —|— 6 000 —|

beams of same material, E is constant

a

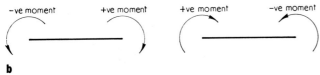

−ve moment $\qquad$ +ve moment $\qquad$ +ve moment $\qquad$ −ve moment

b

20a *Beam loading for example described in paras* 7.12 *to* 7.28; **b** *sign convention for moment distribution*

7.14 The calculation in the second half of table II can be considered in the form:

Stiffness of AB = 256
Stiffness of BC = 480

Therefore, stiffness ratio of AB = $\dfrac{256}{256 + 480} = 0\cdot35$

the stiffness ratio of BC = $\dfrac{480}{736}$
$= 0\cdot65$

7.15 Next calculate the *end-fixity moments* for each span, assuming that it is *encastre* at the supports. The end supports are assumed pinned, both in the calculation of end-fixity moments, and also, as above, in the calculation of span stiffness.

At B, for span AB

$$M_F = \frac{wL^2}{8} \text{ (From table I)} = \frac{25 \times 7^2}{8} = 153 \text{ kNm}$$

at B, in span BC, for uniformly distributed loading

$$M_F = \frac{wL^2}{12} = \frac{25 \times 5^2}{12} = 52 \text{ kNm}$$

and for point load

$$M_F = \frac{PL}{8} = \frac{30 \times 5}{8} = 19 \text{ kNm}$$
$$\text{Total} = 71 \text{ kNm}$$

at C, in span BC, the same as at B by symmetry = 71 kNm
at C and D in span CD

$$M_F = \frac{wL^2}{12} = \frac{25 \times 6^2}{12} = 75 \text{ kNm}$$

at D in span DE

$$M_F = \frac{wL^2}{8} = \frac{25 \times 6^2}{8} = 112 \text{ kNm}$$

Sign convention

7.16 When putting values into the moment distribution table, the correct sign has to be applied to each end-fixity moment. Considering the moment *acting on* the end of the span, clockwise moments are positive, anti-clockwise moments are negative **20b**.

Table III *Moment distribution table*

A	B		C		D		E
	0·35	0·65	0·59	0·41	0·57	0·43	
	+153	− 71	+71	− 75	+ 75	−112	
	− 29	− 53	+ 2	+ 2	+ 21	+ 16	
	+ 1		− 26	+10	+ 1		
	− 1		+ 9	+ 7	− 1		
	+ 4				+ 3		
− 1	− 3				− 2	− 1	
+ 123	− 123		+ 56	− 56	+ 97	− 97	

7.17 For example, consider the first distribution at joint B: out-of-balance moment on the joint $= +153 - 71 = +82$kNm. To balance this, apply -82 kNm, and distribute it between the spans in the ratio of the stiffnesses:
$0\cdot35 \times (-82) = -29$
$0\cdot65 \times (-82) = -53$

7.18 After each distribution at a joint, a line is drawn underneath it. When all joints have been balanced, the carry-over moments are applied, as shown by the arrows in table III. (The carry-over moment is half the applied balancing moment.)

7.19 When the carry-over values have reached a low value everywhere, a long line is drawn, and the moments in each column are summed. The values at each joint should be equal in value but opposite in sign.

Free span bending moments

7.20 Before the bending moment diagram can be drawn, it is necessary to calculate the free span bending moments: in AB

$$M_S = \frac{wL^2}{8} \text{ (from table I)} = 153 \text{ kNm}$$

in BC, from distributed load

$$M_S = \frac{wL^2}{8} = \frac{25 \times 5^2}{8} = 78 \text{ kNm}$$

from point load

$$M_S = \frac{PL}{4} = \frac{30 \times 5}{4} = 38$$

Total = 116 kNm
in CD and DE

$$M_S = \frac{wL^2}{8} = 112 \text{ kNm}$$

Bending moment diagram

7.21 For the bending moment diagram **21**, the normal bending moment sign convention is used—a sagging moment is positive, and a hogging moment negative **22**.
In practice, most midspan moments are positive, and support moments are negative.

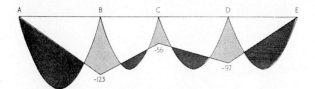

21 *Bending moment diagram for beam shown in* **20**

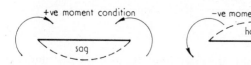

22 *Normal sign convention. Bending moment diagrams are generally drawn to reflect the deflected profile of the beam, ie with positive BM below the line and negative BM above the line*

Calculations of shears and reactions

7.22 Calculation of shears and reactions Consider span AB, and take moments for it about B: (see **23**)

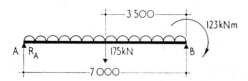

23 *Calculation of shear and reaction of span AB*

$$R_A \times L = w \times L \times \tfrac{1}{2}L - M_B$$

$$\therefore \quad R_A = \tfrac{1}{2}wL - \frac{M_B}{L}$$

$$= 25 \times 7 \times 0 \cdot 5 - \frac{123}{7}$$

$$= 88 - 18 = 70 \text{ kN}$$

7.23 Similarly, taking moments about A

$$S_{BA} = \tfrac{1}{2}wL + \frac{M_B}{L}$$

$$= 25 \times 7 \times 0 \cdot 5 + \frac{123}{7}$$

$$= 88 + 18 \qquad = 106 \text{ kN}$$

And for span BC, moments about C

$$S_{BC} = \tfrac{1}{2}wL + \tfrac{1}{2}P + \frac{M_B - M_C}{L}$$

$$= 25 \times 5 \times 0 \cdot 5 + 30 \times 0 \cdot 5 + \frac{123-56}{5}$$

$$= \qquad 62 \qquad + \quad 15 \quad + \quad 13 \quad = 90 \text{ kN}$$

$$R_B = S_{BA} + S_{BC} = 106 + 90 = 196 \text{ kN}$$

By this method, all the shears and reactions on the beam can be found.

Method used for a symmetrical arrangement of beams

7.24 In many cases the structure is symmetrical, but the loads upon it are not. A typical example is shown in **24**.

7.25 Any system of loads can be built up of a combination of symmetrical and anti-symmetrical systems. This arrangement is split as shown in **25**.

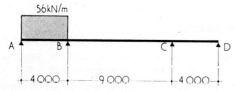

24 *Structure is symmetrical but loads are not*

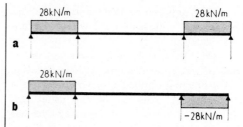

25a *Symmetrical loading system;* **b** *anti-symmetrical loading*

Table IV *Span stiffness—symmetrical case*

Span	k	E	I	L	Stiffness	Ratios
AB & CD	3	—	—	4	0·75 ⎫	0·77
BC	2	—	—	9	0·22 ⎬ 0·97	0·23

7.26 End fixity moment:

At B in span AB $M_F = \dfrac{wL^2}{8} = \dfrac{28 \times 4^2}{8} = 56 \text{ kNm}$

A		B		C
	0·77		0·23	
	+56			
	−43		−13	
	+13		−13	

Table V *Span stiffness— anti-symmetrical case*

Span	k	E	I	L	Stiffness	Ratios
AB & CD	3	—	—	4	0·75 ⎫	0·53
BC	6	—	—	9	0·67 ⎬ 1·42	0·47

7.27 End-fixity moment as before = 56 kNm

A		B		C
	0·53		0·47	
	+56			
	−30		−26	
	+26		−26	

7.28 Reverting to 'normal' sign convention, and combining the load cases to give the loading in figures:

$$M_B = (-13) + (-26) = -39 \text{ kNm}$$
$$M_C = (-13) + (+26) = +13 \text{ kNm}$$

8 Portal frames

8.01 The continuous beam type of structure may be considered as a one-dimensional structure, although the loads and reactions on it are in a second dimension. But as soon as the structure becomes two-dimensional, as in the portal frame, the problem is immediately more complex. There are various methods for dealing with two- and three-dimensional skeletal frames:

1 Moment distribution (as in the previous section)
2 Influence coefficients
3 Slope-deflection
4 Kleinlogels tables
5 Package computer programs

The last two are not strictly methods of analysis but the results of other methods, and can be used to solve a limited number of problems.

8.02 Kleinlogels books of formulae and tables for continuous beams and various portal and multibay frames can be useful, but can involve a lot of work, so engineers tend to use package computer programs. These are available through computer bureaux but the engineer who frequently meets this problem will find a computer terminal installed in his own office, giving access to such programs, will save him much time and effort. If neither of these methods is available one of the first three procedures is used.

8.03 *Moment distribution* is an iterative method. In some cases the number of iterations can be excessive and in extreme cases the method fails to reach a balanced solution. A separate distribution is needed for almost every type of loading but no simultaneous equations are generated and calculations do not have to be taken to many decimal places.

8.04 The *influence coefficients* method is very much more powerful, and can be used on virtually any two- or three-dimensional skeletal structure. Once the flexibility matrix has been found and inverted, the result may be used to calculate any number of loading cases without too much extra work. However, it is not always possible to invert the flexibility matrix by hand. Usually the results depend on small differences in large quantities, and so sufficient decimal places to ensure a high level of accuracy are needed.

8.05 *Slope-deflection* is useful for small structures, but like *influence coefficients* it generally leads to simultaneous equations.

8.06 To demonstrate the analysis of a statically indeterminate frame structure, and give some idea of the work involved in quite simple cases, shortened versions of the analysis of the fixed-foot portal frame **26** by both the *moment distribution* method and the *influence coefficients* method are given below. This example is deliberately non-symmetrical. If either the portal or the load on it are not symmetrical the frame sways **27** and the direction of sway is not always obvious. Particularly in the moment distribution method, it is important to consider sway.

Moment distribution method

8.07 The frame is first translated into a continuous beam **28**. The stiffness of each arm (method para 7.14) and the end-fixity moments in BC (method para 7.16) are determined.

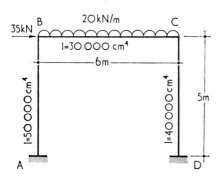

26 *Portal frame example which is analysed by moment distribution and by influence coefficients methods*

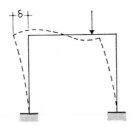

27 *Non-symmetrical portal frame sways*

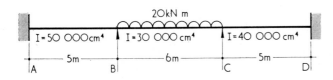

28 *Portal translated into continuous beam*

Table VI *Moment distribution*

A	B		C		D
	0·67	0·33	0·38	0·62	
		−60	+60		
+20	+40	+20	−22·8	−37·2	−18·6
		−11·4	+10		
+3·8	+7·6	+3·8	−3·8	−6·2	−3·1
		−1·9	+1·9		
+0·6	+1·3	+0·6	−0·7	−1·2	−0·6
		−0·4	+0·3		
+0·2	+0·3	+0·1	−0·1	−0·2	−0·1
+24·6	+49·2	−49·2	+44·8	−44·8	−22·4

8.08 These results, table VI, are used to draw the bending moment diagram **29** (B and C are assumed fixed in space; a force H_B is needed to achieve this). Divide the structure into its component members **30**. For BC, taking moments about C:

$$V_A \times 6 = 120 \times 3 + 49 \cdot 2 - 44 \cdot 8$$
$$\therefore V_A = 60 \cdot 73 \text{ kN and } V_D = 59 \cdot 27$$

For CD, moments about D:

$$H_C \times 5 = 44 \cdot 8 + 22 \cdot 4$$
$$\therefore H_C = 13 \cdot 45 = H_D$$

For AB, moments about A:

$$(H_B + H_C) \times 5 = 49 \cdot 2 + 24 \cdot 6$$
$$\therefore H_B = 1 \cdot 31 \text{ and } H_A = 14 \cdot 76$$

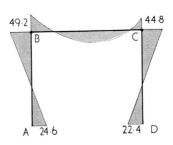

29 *Bending moment diagram for portal frame*

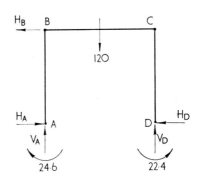

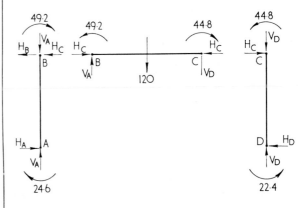

30 *Bending moment diagram divided into component members*

8.09 The force at B to maintain the frame in its original position is $1\cdot31$ kN to the left and the imposed load at this point is 35 kN to the right. Thus the force applied at this *fixed* point is $33\cdot69$ kN. If all the other loads on the frame are now removed, and the restraints at B and C released, the frame will sway to the right a distance d. The rigidity, in the rotational sense, of the joints at B and C is maintained when the restraints on the frame are released.

8.10 Both ends of the member in **31** are *encastre*; B has been deflected through a vertical distance d. From Theorem 1 (para 1.18) the area of the $\dfrac{M}{EI}$ diagram is zero, therefore $M_A = - M_B$.

From Theorem 2 (para 1.19):

$$EI\,d = \tfrac{1}{2}L \times (M_A + M_B) \times \tfrac{2}{3}L - M_B \times L \times \tfrac{1}{2}L$$

$$= \frac{M_A L^2}{6}$$

$$\therefore M_A = M_B = \frac{6EI}{L^2}\,d$$

8.11 These results are used in a second moment distribution to find the end-fixity moments. d is unknown and thus can be given any value at this stage, so put $6EI\,d = 1$. The results of this distribution are shown in **32**. But the moments at the feet of the portal legs are no longer half the moments at the heads because of the sway of the frame.

8.12 As before, the external forces on the frame are found by statics:

$$V_D = - V_A = 1\cdot84 \times 10^{-6}\,\text{kN}$$
$$H_D = 2\cdot7 \times 10^{-6}\,\text{kN}$$
$$H_A = 3\cdot1 \times 10^{-6}\,\text{kN}$$
$$H_B = 5\,8 \times 10^{-6}\,\text{kN}$$

8.13 This establishes that a horizontal force of $5\cdot8 \times 10^{-6}$ kN at B produces a sway of 1/6E m and moments as shown in **32**.

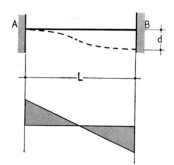

31 *Diagram showing bending moment of encastre member (see **7a**)*

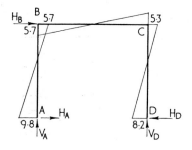

32 *Second moment distribution diagram*

8.14 If these moments are multiplied by $\dfrac{33\cdot69}{5\cdot8 \times 10^{-6}}$, the result will be the moments produced by the applied horizontal force corrected by the sway due to the non-symmetry of the frame. Table IX allows comparison of the values of the unknowns obtained by various methods.

Influence coefficients method

8.15 This method produces a series of simultaneous equations which, for anything other than an extremely simple structure, have to be solved on a computer. The theory behind the method is not really necessary for its understanding, and is beyond the scope of these notes.

8.16 The equation controlling the method is written:

$$X = - G^{-1} \times U$$

X, G and U are not single numbers, but represent *arrays* of numbers called matrices. (A brief outline of the theory of matrices is given in a footnote after 10.)

8.17 X is a column matrix (or vector) representing the restraints; G is the *flexibility matrix* for the structure, and U is the *particular solution* for the given imposed loading.

8.18 The portal frame in **26** is statically indeterminate in the third degree because it has three redundant restraints. The first operation is to make the frame statically determinate by reducing the restraints by three. Methods for doing this are illustrated in **33** but for this example the way shown in **34** is used. The resulting structure is called the *released structure*.

8.19 In **34** a hinge is introduced at A, and a hinge and a roller support at D. The removed restraints are moments at

Table VII *Graphical representation of releases*

Sign	Name	Transmits
⊣├─	cut	nothing
(rollers symbol)	rollers	moment axial force
─○─	hinge	axial force shear
(sleeve symbol)	sleeve	moment shear

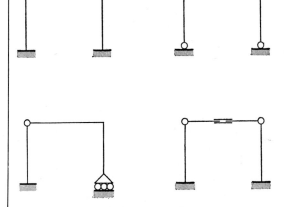

33 *Various forms of released structures. Graphical conventions for these frames are illustrated in table* VIII

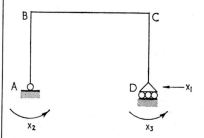

34 *Forms of released structure and restraints chosen for use in the example*

A and D, and a horizontal reaction at D. These are represented by x_1, x_2 and x_3, and since the frame is now statically determinate it is quite simple to draw the bending moment diagrams for $x_1 = 1$, $x_2 = 1$, $x_3 = 1$ and for the imposed loads on the frame **35**. The diagrams are always drawn on the tension side of the member.

8.20 Then integrate together (co-integrate) each pair of diagrams to obtain the influence coefficients that make up the flexibility matrix. For co-integration, consider the member AB in **35a** and **b**. If the volume of the solid produced between **a** horizontally and **b** vertically is calculated, and divided by the value of EI for the member, the result is the co-integral of the member for these two restraints (see **36**). If the moments are on the same side of the member, the co-integral will be positive; on opposite sides, negative. Table VIII lists values for most cases found in practice, and using these considerably simplifies calculations.

Table VIII *Moment diagram co-integration giving values for most cases normally found, see* **36**

A	B	A×B
parabolic M	m	$\frac{2}{3}MmL$
	m	$\frac{1}{3}MmL$
m_1	m_2	$\frac{1}{2}m_1m_2L$
	m_2	$\frac{1}{3}m_1m_2L$
m_2		$\frac{1}{6}m_1m_2L$
parabolic m_1 m_2	m^l	$\frac{1}{3}m^l(2m_1+m_2)L$
	m^l	$\frac{1}{4}m^l(m_1+m_2)L$
m^l		$\frac{1}{2}m^l(5m_1+m_2)L$

8.21 Consider restraints 1 and 2 (**35a** and **b**):

$$g_{12} = \frac{\frac{1}{2} \times 5 \times 1 \times 5}{E \times 5} + \frac{\frac{1}{2} \times 5 \times 1 \times 6}{E \times 3} = +7.5 \text{ (if the}$$

constant divisor E is omitted). To obtain the principal diagonal of the matrix, each diagram is co-integrated with itself. It is only necessary to calculate the values on one side of the diagonal, as obviously $g_{31} = g_{13}$.

8.22 The flexibility matrix that results from all the co-integrations is:

$$G = \begin{vmatrix} g_{11} & g_{12} & g_{13} \\ g_{21} & g_{22} & g_{23} \\ g_{31} & g_{32} & g_{33} \end{vmatrix} = \begin{vmatrix} 68.75 & 7.5 & -8.125 \\ 7.5 & 1.67 & -0.33 \\ -8.125 & -0.33 & 1.92 \end{vmatrix}$$

8.23 This matrix now has to be inverted. A matrix of this size can be done by hand, but a desk calculator makes this easier. Most computers have a cheap matrix inversion program on file; inversion of this example costs 80p on a computer which compares with £4.50 for analysing the frame on a computer program. The inverted matrix is:

$$G^{-1} = \begin{vmatrix} 0.089 & -0.336 & 0.318 \\ -0.336 & 1.889 & -1.095 \\ 0.318 & -1.095 & 1.677 \end{vmatrix}$$

8.24 At this point, and not before, the imposed loading on the frame must be considered. The column matrix of the *particular solution* is calculated by co-integrating the moment diagrams on the released structure produced by the applied loads, together with the restraint diagrams used for finding the flexibility matrix. It is of course essential that the same released structure and restraints are used for the particular solution as for the flexibility matrix.

8.25 For example, co-integrating **35a** together with the sum of **35d** and **e**:

$$u_1 = -\frac{\frac{1}{3} \times 5 \times 175 \times 5}{5} \text{ (member AB)}$$

$$-\frac{\frac{2}{3} \times 5 \times 90 \times 6}{3} - \frac{\frac{1}{2} \times 5 \times 175 \times 6}{3} \text{ (member BC)}$$

$$+ 0 \text{ (member CD)}$$

$$= -1767$$

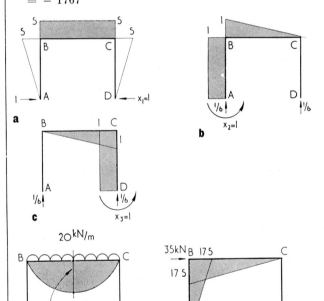

35 *Bending moment diagrams:* **a** *for restraint* 1 ($x_1 = 1$); **b** *restraint* 2 ($x_2 = 1$); **c** *restraint* 3 ($x_3 = 1$); **d** *uniformly distributed load; and* **e** *horizontal load*

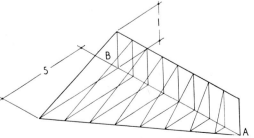

36 *Illustration of co-integration on the member* AB

8.26 Once the particular solution has been calculated the equation can be written as follows:

$$\begin{vmatrix} x_1 \\ x_2 \\ x_3 \end{vmatrix} = \begin{vmatrix} 0\cdot089 & -0\cdot336 & 0\cdot318 \\ -0\cdot336 & 1\cdot889 & -1\cdot095 \\ 0\cdot318 & -1\cdot095 & 1\cdot677 \end{vmatrix} \times \begin{vmatrix} 1767 \\ 264 \\ -118 \end{vmatrix}$$

The signs of the particular solution have been reversed to conform with the negative sign of the controlling equation in para 8.16. It is now a simple matter to calculate the values of the restraints. For example:

$x_3 = 0\cdot318 \times 1767 - 1\cdot095 \times 264 - 1\cdot677 \times 118 = 75$

similarly $x_1 = 31$ and $x_2 = 34$.

8.27 These values may be used to find the values of the moments at any point by using **35**. For example, to find M_B, add together the values for each restraint and the particular solution. The restraints are no longer unity, but have the values computed above: so the diagrams are multiplied accordingly:

$M_B = 31 \times 5$ (restraint 1) $+ 34 \times 1$ (restraint 2)

$\quad\quad + 75 \times 0$ (restraint 3) $+ 0$ (UDL) $- 175$ (horizontal load)

$\quad = 14$ kNm

8.28 Similarly for the vertical reaction at D:

$V_D = 31 \times 0$ (restraint 1) $- 34 \times \dfrac{1}{6}$ (restraint 2)

$\quad\quad - 75 \times \dfrac{1}{6}$ (restraint 3) $+ 60$ (UDL) $+ 29\cdot17$ (horizontal load)

$\quad = 71$ kN

8.29 The quantities found by this method are shown in table IX, where they can be compared with the values found by moment distribution. The final column gives the results obtained using a package computer program.

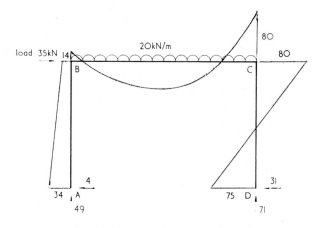

37 *Results of the influence coefficients method of analysis of the portal example*

Table IX *Results of portal frame analysis*

Item	Moment distribution			Influence coefficients	Computer
	Vertical loads	Sway	Total		
M_A	−24·6	+57·0	+32·4	+34	+37·1
M_B	+49·2	−33·2	+16·0	+14	+13·4
M_C	+44·8	+30·9	+75·7	+80	+68·2
M_D	−22·4	−47·7	−70·1	−75	−73·1
V_A	+60·73	−10·7	+50·0	+49	+49·2
V_D	+59·27	+10·7	+70·0	+71	+70·8
H_A	+14·76	−18·0	− 3·2	− 4	− 4·7
H_D	+13·45	+15·7	+29·3	+31	+30·3

9 Surface structures

Cellular structures

9.01 There are two distinct types of surface structures; cellular structures, composed of flat slabs arranged in a box-like fashion, and curved surfaces, shells, domes etc **38**. Cellular structures are essentially stable, provided that the joints between slabs have been designed and constructed to prevent them coming apart, otherwise the structure is like a house of cards and may collapse, as did Ronan Point.

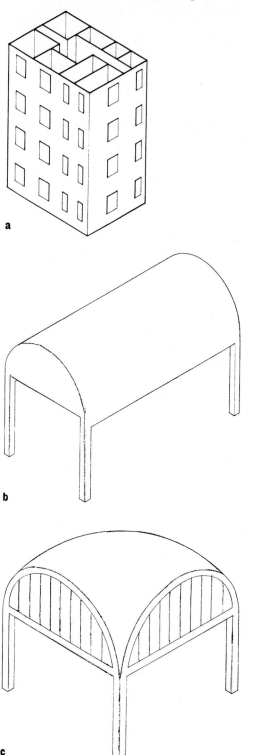

38 *Surface structures:* **a** *cellular structure;* **b, c** *shell structures, cylindrical and two-curved*

9.02 One form of *cellular structure* is load-bearing brick construction which will be discussed in more detail in section 8, MASONRY. Analysis generally involves computing the loading on each vertical panel due to dead and live loads, and assessing the wind loads by semi-empirical methods. The dead and live loads carried by each panel can usually be estimated from the plan of the building. Wind stresses are found by considering the building as a vertically mounted cantilever beam. From the plan of a typical floor, the panels form a complex 'section' of the beam. The neutral axis and section modulus can be found using the method described earlier. These values can then be used to find the maximum stresses in the panels due to wind.

9.03 Analysis of cellular building for wind loading:

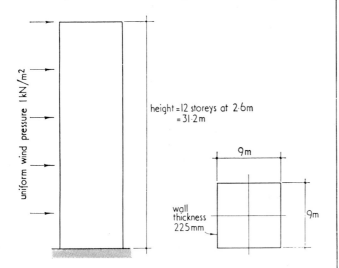

$$A = 4 \times 9 \times 0 \cdot 225 = 8 \cdot 1 \text{ m}^2$$

$$I = \frac{2}{3} \times 9^3 \times 0 \cdot 225 = 109 \text{ m}^4$$

$$Z = \frac{109}{4 \cdot 5} = 24 \cdot 1 \text{ m}^3$$

Loadings:
Dead weight on all floors and roof = $2 \cdot 5$ kN/m²
Live load on all floors and roof = 2 kN/m²
Weight of wall = $4 \cdot 5$ kN/m²
Cantilever moment under wind loading

$$= \frac{wL}{2} = 1 \times 9 \times 31 \cdot 2^2 \times 0 \cdot 5 = 4380 \text{ kN/m}$$

$$\therefore \text{ wind stress} = \pm \frac{M}{Z} = \pm \frac{4380}{24 \cdot 1} = \pm 182 \text{ N/m}^2$$

or $\pm 0 \cdot 18$ N/mm²
Total dead weight of building = $12 \times 2 \cdot 5 \times 9^2$ (floors) + $4 \times 4 \cdot 5 \times 9 \times 31 \cdot 2$ (walls) = 7484 kN.
Live load on building = $12 \times 2 \times 9^2 = 1944$ kN.

$$\therefore \text{ Dead load stress} = \frac{7484}{8 \cdot 1} = 920 \text{ of kN/m}^2 \text{ or } 0 \cdot 92 \text{ N/mm}^2$$

$$\text{Live load stress} = \frac{1944}{8 \cdot 1} = 240 \text{ kN/m}^2 \text{ or } 0 \cdot 24 \text{ N/mm}^2$$

Maximum compressive stress
= dead load stress + live load stress + wind stress
= $0 \cdot 92 + 0 \cdot 24 + 0 \cdot 18 = 1 \cdot 34$ N/mm²
Minimum compressive stress
= dead load stress − wind stress
= $0 \cdot 92 − 0 \cdot 18 = 0 \cdot 74$ N/mm²

9.04 The stresses due to dead load at the same points must be added to the wind stresses. In loadbearing brickwork, a tension cannot be allowed: thus the deadload stress must exceed the tension stress from wind by a reasonable safety margin.

9.05 In precast concrete panel construction, vertical tie rods can be inserted in joints between the panels to carry wind tensions. But for brickwork or concrete, the compressive stresses (the maximum wind compression plus the dead and live load stresses) must be within the carrying capability of the material.

Curved surface structures

9.06 There are many books on *curved surface structures* and it is only possible to outline the subject here. (The reader is referred to the bibliography.) Before considering the strength of curved surfaces, it is necessary to consider their geometric properties.

9.07 Any surface may be considered as a series of lines. A line that is not straight is said to have curvature. For the general non-straight line the curvature varies at each point on the line **39**. For the circular line only the curvature is constant, and is the reciprocal of the radius of the circle:

$$\text{curvature} = \frac{1}{R}$$

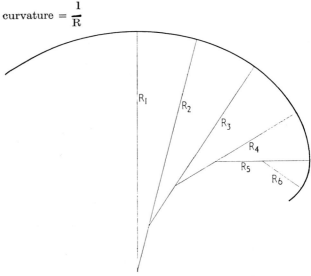

39 *Varying radius of curvature for non-circular line*

9.08 For a small element of curved surface **40** sections can by taken in the X and Y directions and each of these will be lines with curvatures $1/R_x$ and $1/R_y$ respectively. These curvatures define the surface at any point.

9.09 The line ABC in **41** is half a circle with radius R. If this line is moved so that point A travels along the straight line LM, it will generate a surface—a half-cylinder. Many surfaces can be produced by moving one line along another, so that it remains parallel to itself: these are called *surfaces of translation* **42**.

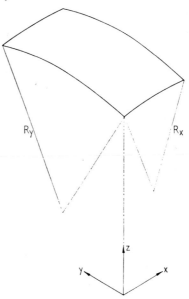

40 *Curvature of a surface*

9.10 However, the moving line does not always have to be kept parallel to itself. The ends may follow two different curves to produce conoidal surfaces **43**. If one end is stationary while the other describes a circle, the resulting surface is one of revolution **44**.

9.11 Having considered the geometry of the surfaces, their use in structures must now be considered. The three basic types are thick shells, thin shells and membranes.

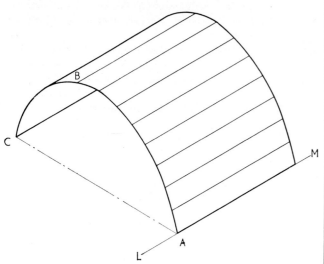

41 *Generation of surface by moving one line (ABC) along another (LM)*

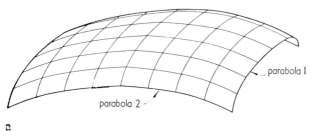

a

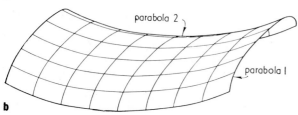

b

42 *Surfaces of translation:* **a** *elliptical paraboloid;* **b** *hyperbolic paraboloid. Lines at 45° to these curves are straight, and so the surfaces can be constructed from straight members*

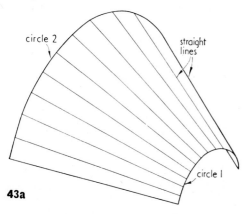

43a

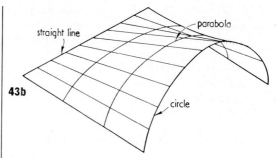

43b

43 *Conoidal surfaces*

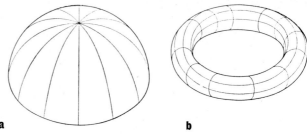

a   **b**

44 *Surfaces of revolution*

Thick shells

9.12 A small element of a thick shell, with the main forces in each face, is shown in **45**. These forces shown are called *stress resultants*; they are the forces per unit length of surface arc. To obtain the stress divide by the thickness t of the shell.

9.13 On each cut face of the shell there are five stress resultants:

T is the direct stress resultant at the neutral axis of the section

S is the tangential component

N is the normal component of the shear stress resultant.

But the direct stress is not constant over the section of the shell, therefore:

M is the bending moment stress resultant of the direct stresses on the section

H is the stress resultant of the variation of shear stress across the section; this is a torsional moment.

These stresses are produced by the applied forces P_x, P_y and P_z. The displacements of the element in the x, y and z directions can be indicated u, v and w respectively. The analysis of these stress resultants requires advanced techniques.

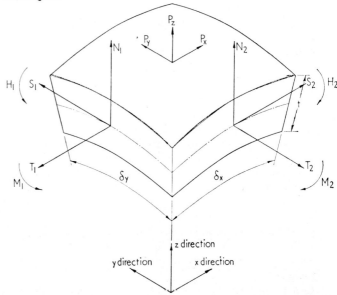

45 *Element of thick shell*

Membranes

9.14 *A membrane* is a shell that is so thin that it has no flexural rigidity. Therefore N, M and H are all zero, leaving only S and T stress resultants on the element **46**. Also because there is no flexural rigidity, the membrane will buckle under compressive stress, so T can only be tensile stress.

9.15 Consider the element of membrane in **46**, the curvatures in the x and y directions are $1/R_x$ and $1/R_y$, and the edges of the element subtend the angles a and β at the centres of the curvature. The imposed pressure on the element is p per unit area normal to the surface. Resolving in this direction:

$p \times aR_x \times \beta R_y = 2T_x \times \beta R_y \times \sin a/2 + 2T_y \times aR_x \times \sin \beta/2$

as a and β are small $\sin a/2 \simeq a/2$,

so it can be shown that:

$$p \cdot R_x \cdot R_y = T_x R_y + T_y R_x$$

$$p = \frac{T_x}{R_x} + \frac{T_y}{R_y}$$

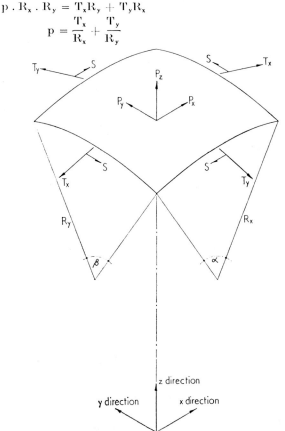

46 *Element of membrane*

9.16 As an example consider a spherical balloon; the material is 0·5 mm thick and the diameter 60 m. If the internal pressure is 1400 N/m², what is the stress in the material?

Since the balloon is spherical $T_x = T_y$ and $R_x = R_y$

thus $p = \dfrac{2T}{R}$ and $T = \frac{1}{2} Rp$

$$= \frac{1}{2} \times 30 \times 1400$$

$$= 21\,000 \text{ N/m}$$

now stress $= \dfrac{T}{\text{thickness}}$

$$= \frac{21\,000}{0\cdot 5 \times 10^3}$$

$$= 42 \text{ N/mm}^2$$

9.17 The shear force resultants do not come into this calculation because they have no component normal to a spherical surface. But with other types of surface these shear forces play a large part in stabilising the membrane against out-of-balance normal loading. Tents, balloons and pneumatic structures are examples of forms susceptible to membrane analysis.

Thin shells

9.18 A *thin shell* has no bending moments or normal shears but is not so thin that it buckles under the slightest compressive stress. The stress resultants are still T and S, but T can be compressive as well as tensile. They are often referred to as *membrane stress resultants*. Most curved structures are thin shells and not thick shells or membranes.

9.19 The analysis of a typical thin shell structure, a concrete cylindrical roof is considered. Near the ends of the shell the non-membrane stresses begin to predominate. If the shell is long, say over four times the radius of curvature, it may be considered as a beam, but the geometry of the form must be maintained. In the case of a cylindrical shell of this type this implies that the gables of the shell are sufficiently rigid. Non-membrane stresses will predominate in this region. For long shells it may be necessary to prevent the straight edges of the surface from spreading transversely due to the arch action. This is done using edge beams.

9.20 The analysis only gives a very approximate idea of the stresses in **47**; the full design should be left to a specialist.

Load per metre run $= \dfrac{\pi \times 6}{2} \times 2\cdot 5 = 23\cdot 6$ kN

$\therefore$ Moment at centre of span $= \dfrac{23\cdot 6 \times 15^2}{8} = 664$ kNm

$A = \pi rt = 3\cdot 142 \times rt$

$I = tr^3 \left(\dfrac{\pi}{2} - \dfrac{4}{\pi} \right) = 0\cdot 298 \ tr^3$

$Z_1 = 0\cdot 83 \ tr^2$

$Z_2 = 0\cdot 47 \ tr^2$

$\therefore f_c = \dfrac{664}{0\cdot 83 \times 0\cdot 075 \times 3^2} = 1185$ kN/m² $= 1\cdot 19$ N/mm²

$\therefore f_t = \dfrac{664}{0\cdot 47 \times 0\cdot 075 \times 3^2} = 2093$ kN/m² $= 2\cdot 09$ N/mm²

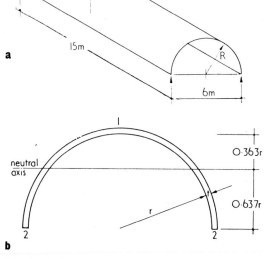

47a, b *Shell examples*

Folded plate

9.21 A similar form is the *folded plate*. This may also be considered as a beam in the longitudinal direction; moreover transverse stresses may be estimated by considering the structure as a run of continuous beams, the folds acting as the supports. So that this can apply to the end plates, the free edges have to be stiffened in some way.

9.22 Example of folded plate shell **48a**. Loading 6 kN/m² on plan.

Longitudinal bending:

$$M = \frac{wL^2}{8} = \frac{6 \times 8 \times 20^2}{8} = 2,400 \text{ kNm}$$

Equivalent section **48b**:

$$Z = \frac{bd^2}{6} = \frac{0 \cdot 46 \times 3 \cdot 46^2}{6} = 0 \cdot 92 \text{ m}^3$$

$$\therefore f_{bt} = f_{bc} = \frac{2400}{0 \cdot 92} = 2 \cdot 60 \text{ N/mm}^2$$

Transverse bending **48c**:

Maximum negative moment (at B and D)
$= 3 \times 4^2 \times 0 \cdot 107 = 5 \cdot 15$ kNm

Maximum positive moment (in spans AB and DE)
$= 3 \times 4^2 \times 0 \cdot 077 = 3 \cdot 70$ kNm

$$Z \text{ per } 1 \text{ m width} = \frac{1 \times 0 \cdot 1^2}{6} = 0 \cdot 00167 \text{ m}^3$$

$$\therefore f_{bc} = f_{bt} = \frac{5 \cdot 15 \times 10^{-3}}{0 \cdot 00167} = 3 \cdot 08 \text{ N/mm}^2$$

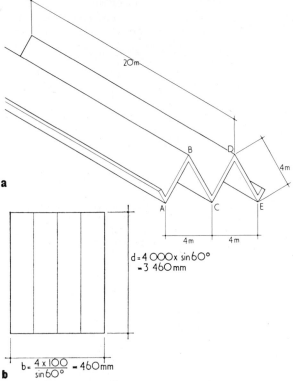

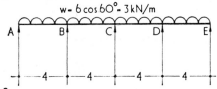

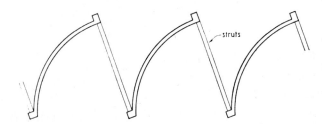

48 *Folded plate example:* **a** *diagram;* **b** *equivalent section;*
c *transverse bending*

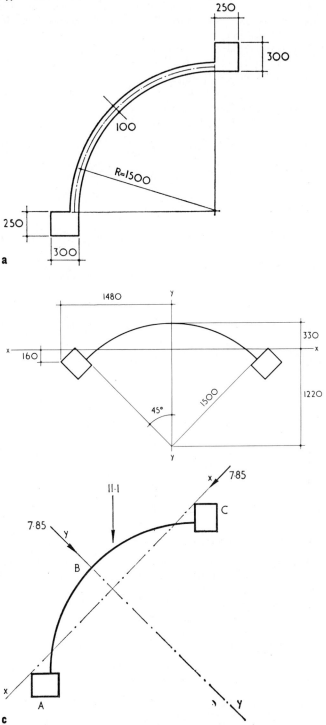

49 *Northlight shells in series*

Northlight shell

9.23 A third example is the *northlight shell*, where the
transverse action is discontinuous. **49** shows a run of such
shells, stabilised with struts at intervals, with gables to
ensure maintenance of the profile. In the analysis of such
a system, the shells can be considered to span as beams.

The section is not symmetrical about the vertical **axis**
through centre of area. The vertical load is resolved into
components parallel to the principal axes. and the stresses
are found.

Span = 12 m free supports.
Snow load $0 \cdot 75$ kN/m^2 on plan.
A $= 385 \times 10^3$ mm^2
$I_{xx} = 15 \cdot 8 \times 10^9$ mm^4
$I_{yy} = 348 \times 10^9$ mm^4

50 *Northlight shell example:* **a** *diagram;* **b** *lengths;* **c** *forces*

Loading:
Total dead weight of shell $= 25 \times 385 \times 10^3 \times 10^{-6}$
$= 9 \cdot 6$ kN/m run

Total snow load $= 0 \cdot 75 \times 2 \cdot 0 = 1 \cdot 5$
Total $= 11 \cdot 1$ kN/m

Resolve this vertical load into components across x and y
axes of shell: 50c

Each component $= 11 \cdot 1 \cos 45° = 7 \cdot 85$ kN/m.

$\therefore$ Bending moment in each direction $= \dfrac{7\cdot 85 \times 12^2}{8}$

$$= 141 \text{ kNm}$$

Maximum compression in shell at B:

$$Z_{xx} = \frac{15\cdot 8 \times 10^9}{330} = 4\cdot 8 \times 10^7 \text{mm}^3$$

$$\therefore f_{bc} = \frac{141 \times 10^6}{4\cdot 8 \times 10^7} = 2\cdot 94 \text{ N/mm}^2$$

Maximum compression at C:

$$Z_{xx} = \frac{15\cdot 8 \times 10^9}{160} = 9\cdot 9 \times 10^7$$

$$Z_{yy} = \frac{348 \times 10^9}{1480} = 23\cdot 5 \times 10^7$$

$$\therefore f_{bc} = \frac{141 \times 10^6}{23\cdot 5 \times 10^7} - \frac{141 \times 10^6}{9\cdot 9 \times 10^7}$$

$$= 0\cdot 60 - 1\cdot 42 = -0\cdot 82 \text{ N/mm}^2 \text{ (tension)}$$

Maximum tension at A:

$$Z_{xx} = \frac{15\cdot 8 \times 10^9}{360} = 44 \times 10^6$$

$$Z_{yy} = \frac{348 \times 19^9}{1200} = 29 \times 10^7$$

$$\therefore fbt = \frac{141 \times 10^6}{44 \times 10^6} + \frac{141 \times 10^6}{29 \times 10^7}$$

$$= 3\cdot 20 + 0\cdot 49 = 3\cdot 69 \text{ N/mm}^2$$

If this stress is assumed constant over whole of beam: tension force $= 3\cdot 69 \times 250 \times 300 \text{ N} = 277 \text{ kN}$. With a steel stress of 200 N/mm² the required area in the beam

$$= \frac{277 \times 10^3}{200}$$

$$= 1385 \text{ mm}^2$$

Hyperbolic paraboloid

9.24 Consider the special case of a *hyperbolic paraboloid* in which parabolas 1 and 2 in **42** are identical but inverted to each other. By symmetry it can be seen that when a vertical load w is applied, the stress in each direction is equal in magnitude but opposite in sign. In **51a** the membrane in the x direction is acting as an arch, and in the y direction as a cable.

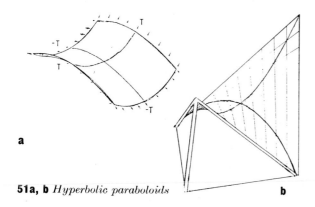

51a, b *Hyperbolic paraboloids*

9.25 From the equation in para 9.15; at the saddle point of the shell:

$$w = \frac{T}{R} + \frac{-T}{-R} = \frac{2T}{R} \therefore T = \tfrac{1}{2} w R.$$

9.26 If the shell is shallow, R does not vary greatly over its surface, and it can be shown that the stress resultants in each direction remain constant. Also there is no membrane shear in the x and y directions. It has been found earlier that planes at 45° to the x and y axes have shear stress resultants of $\tfrac{1}{2}$ w R, but no direct stresses. This means that

instead of being supported by massive beams that can carry the thrusts and tensions produced by direct stresses, the shell edges can be carried by slender beams which have only to take the shears, provided that these beams are at 45° to the axes of the shell **51b**. These beams will also be straight, as described in **42**.

Hemispherical dome

9.27 For the cap of a *hemispherical dome* **52** the compressive membrane stress resultant in meridional direction around the base is T_1. If the unit weight of the material of the dome is w, by resolving in the vertical direction:

$$w \times 2\pi R (1 - \cos \phi) = T_1 \sin \phi \times 2\pi R \sin \phi$$

$$T_1 = \frac{wR}{1 + \cos \phi}$$

9.28 From the equation in paragraph 9.15

$$w \cos \phi = \frac{T_1}{R} + \frac{T_2}{R}$$

thus $T_2 = wR \left(\cos \phi - \dfrac{1}{1 + \cos \phi} \right)$

where T_2 is the hoop stress.

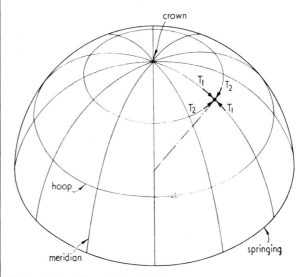

52 *Analysis of hemispherical dome*

9.29 Example: find the stresses in a hemispherical dome spanning 20 m; material thickness 75 mm, loading 2·4 kN/m².

At the crown $\phi = 0$, so that $T_1 = T_2 = \tfrac{1}{2}wR = \tfrac{1}{2} 2\cdot 4 \times 10$
$$= 12 \text{ kN/m}$$

the corresponding compressive stresses are

$$\frac{12\ 000}{1000 \times 75} = 0\cdot 16 \text{ N/mm}^2$$

At the springing, the meridianal stress will be
$$T_1 = wR = 24 \text{ kN/m}$$
and stress 0·32 N/mm² compressive.

The hoop stress will be
$$T_2 = -wR, \text{ so the stress will be } 0\cdot 32 \text{ N/mm}^2 \text{ tensile.}$$

9.30 This hoop stress may cause the edge of the dome to spread, thus altering the geometry of the structure. If this is prevented with a ring beam, non-membrane stresses will be induced in the shell near the beam. Steel reinforcement at a low stress is used to avoid these non-membrane stresses and excessive spreading of the edge.

9.31 If the maximum stress of 200 N/mm² is used, increase in circumference is

$$\frac{\text{stress}}{E} \times \text{original length} = \frac{200}{207} \times 10^3 \times 20\ 000 \times \pi$$

$= 60$ mm, which will cause the diameter to increase by $\dfrac{60}{\pi} = 19$ mm.

Technical study
Analysis 2

Section 2 **structural analysis**

Structural types (part 2)

DAVID ADLER's *analysis of structural types
concludes with a section on tension structures. It is
followed by an appendix to the whole section of the handbook*

10 Tension structures

Cable structures

10.01 The two main types of tension structures are cable
structures and membrane structures. The simplest form of
cable structure is the vertical tie, fixed at the top with a load
on the bottom. The external wall and the outer edges of the
floor slabs are carried on vertical ties suspended from
massive roof-level cantilevers that are carried by the central
service core. The external wall can be very slender in
comparison with more normal compression elements and
the only support at ground level is that for the central core.
The analysis of such a frame is very simple and need not be
considered further.

10.02 Cables can be used to span large horizontal distances,
making the best use of weight. However, the cable must be
reliably anchored and the system often requires considerable
compression elements.

10.03 The cable in **54a** has length L and negligible self-
weight. The horizontal distance between the anchorages
is l, and the sag is h under the single, central load P. From
Pythagoras' theorem:

$$\frac{L^2}{4} = \frac{l^2}{4} + h^2$$

$$\therefore L = l \sqrt{\left(1 + 4\frac{h^2}{l^2}\right)}$$

put $r = \frac{h}{l}$ (r = *sag ratio*)

$$L = l \sqrt{(1 + 4r^2)}$$

if r is small; that is, the sag is small compared to the span
$1 + 4r^2$ is approximately $= (1 + 2r^2)^2$

$$\therefore L = l (1 + 2r^2)$$

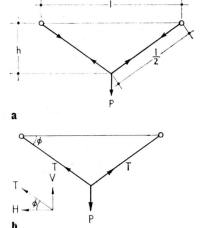

54a *Simple centrally loaded cable;* **b** *forces*

10.04 The results for numbers of loads equally spaced along
the cable **55** can be determined by a similar method:

$$L = l (1 + k r^2)$$

number of loads = 1 2   3   4   5   6 infinite
           k = 2 3 2·5 2·8 2·6 2·7 2·67

10.05 Now, consider the same system **54b** by vertical
resolution at the point of suspension of the load:

$$2T \sin \phi = P$$

the tension in the cable is resolved into horizontal and
vertical components H and V:

$$H = T \cos \phi$$

$$= \frac{P}{2 \tan \phi} \text{ but } \tan \phi = \frac{2h}{l}$$

$$\therefore H = \frac{Pl}{4h}$$

now $\frac{Pl}{4}$ is the midpoint moment (M_s) produced in a simply

supported beam of span l under point load P. It can be
shown that for other loading conditions

$$H = \frac{M_s}{h}$$

Thus, a cable under virtually uniform vertical load for
horizontal increments will take up a parabolic form and

$$H = \frac{w\,l^2}{8\,h}$$

If the cable is heavy, however, it has a uniform vertical load
for increments along the curve of the cable and not for
horizontal increments. The shape is then a catenary.

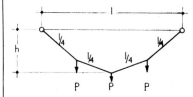

55 *Cable loaded with various equally spaced loads*

10.06 Obviously, a cable with a number of equal loads on it
changes its profile according to the changes in these loads.
Thus, it would not be suitable to support a roof under a
single cable as in **56** because if the wind blew or the snow
drifted in an uneven fashion, the roof would change its shape
violently and suddenly.

10.07 There are three methods of overcoming this :
1 Increase the dead weight of the suspended loads. Despite
obvious disadvantages this makes the effect of incidental
loadings less significant
2 Stay each load with secondary cables. These carry the
incidental loads and so maintain the main system's basic

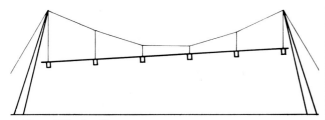

56 *Unstabilised cable structure*

shape. It is not always possible to do this. Also the method is somewhat clumsy

3 Use a multiple cable system in which the supplementary cables have the effect of introducing prestress into main cable. Various system shapes are produced by this method **57**.

10.08 Analysis of such systems introduces the secondary effects of the strain of the cables under load. The natural frequency of the structure as a whole and the natural frequency of each part of the structure is important; if

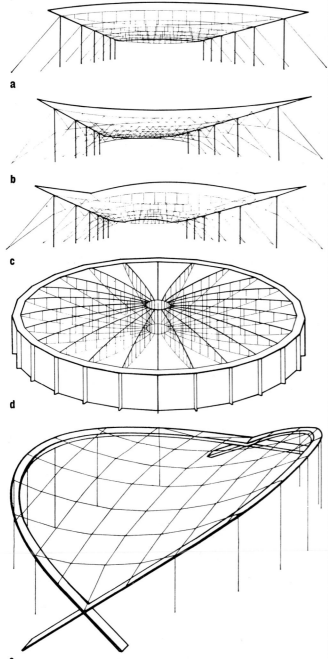

57 *Stabilised cable systems:* **a** *stabilisation cable above suspension cable;* **b** *stabilisation cable under suspension cable;* **c** *stabilisation cable partly below suspension cable;* **d** *bicycle wheel form;* **e** *arch system with cable net*

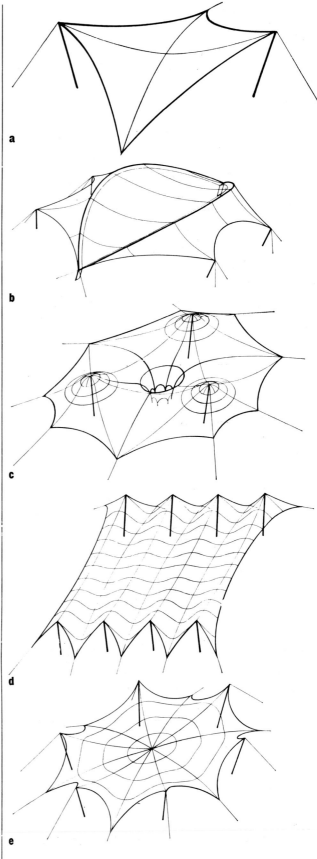

58 *Five types of tension membrane including* **a** *hyperbolic paraboloid*

natural frequencies approach the wind gusting frequency of about three seconds, then the roof, or part of it, may suffer from unpleasant or dangerous vibration.

Tension membranes

10.09 It was said earlier that the *true membrane* can only sustain tensile forces. When dealing with thin shells, these were considered capable of taking a certain amount of compression. Thin shells, however, can be placed wholly in tension by prestressing. **58** shows a hyperbolic paraboloid of this type, as well as other shapes which can also be used in a similar fashion.

11 Bibliography

1 BOMMER, C. M. and SYMONDS, D. A. Skeletal structures: matrix methods of linear structural analysis using influence coefficients. London, 1968, Crosby Lockwood and Son [(2–) (K)] o/p

2 COWAN, H. J. Architectural structures: an introduction to structural mechanics. New York, 2nd edition, 1976, Elsevier [(2–) (K)] £14·00

3 ENGEL, H. Structural systems. London, 1968, Iliffe Books [(2–) (K)] o/p

4 JENKINS, R. S. Theory and design of cylindrical shell structures. London, 1947, published privately by the Ove Arup group of structural engineers, may be in short supply. [(2–) (K)]

5 LISBORG, N. Principles of structural design. London, 1961, Batsford [(2–) (K)] o/p

6 SALVADORI, M. and LEVY, M. Structural design in architecture. Englewood Cliff, NJ, 1967, Prentice-Hall [(2–) (K)] £13·95

7 SHANLEY, F. R. Strength of materials. New York, 1957, McGraw Hill [Yy (K)] o/p

8 Steel Designers' Manual. London, Crosby Lockwood [(2–) Yh2] *Fourth edition (metric)* 1972 £11·00; (paperback £7·00)

Footnote: matrix algebra

1.01 The following is a very short exposition of basic matrix algebra, and should be just enough to understand the working of the portal frame example using the influence coefficients method (para 8.16 of technical study ANALYSIS 2).

1.02 Consider the set of simultaneous equations:

$a_1 x + b_1 y + c_1 z = u_1$
$a_2 x + b_2 y + c_2 z = u_2$
$a_3 x + b_3 y + c_3 z = u_3$

where x, y and z are unknown quantities, and a_1, a_2, b_1, c_1, u_1 etc are coefficients whose values are known. (They are ordinary numbers—6, 1·43, or −17·9).

1.03 The set of equations can be written:

$$\begin{bmatrix} a_1 \, b_1 \, c_1 \\ a_2 \, b_2 \, c_2 \\ a_3 \, b_3 \, c_3 \end{bmatrix} \begin{bmatrix} x \\ y \\ z \end{bmatrix} = \begin{bmatrix} u_1 \\ u_2 \\ u_3 \end{bmatrix}$$

and this can be represented:

$A \times X = U$

1.04 Here A, X and U are not numbers: they are matrices composed of arrays of numbers arranged in a particular way. These matrix quantities obey many of the rules of ordinary arithmetic, and so it is possible to write

$X = A^{-1} \times U$

A^{-1} (the reciprocal of A) can be calculated; but if it is large, say bigger than three elements square, a computer is needed to do the calculation.

1.05 The items within the matrix are known as elements. These elements are arranged in columns and rows and are usually written g_{mn}, where m is the number of the row, and n the number of the column occupied by the element. Thus, the matrix above would be written:

$g_{11} \, g_{12} \, g_{13}$
$g_{21} \, g_{22} \, g_{23}$
$g_{31} \, g_{32} \, g_{33}$

1.06 If g_{mn} is always equal to g_{nm}, then the matrix is symmetrical—this is relatively common. The diagonal with elements g_{mm} is called the principal diagonal.

1.07 To multiply two matrices together, multiply the first element of row one in matrix A by the first element of column one in matrix B, the second element in row one in A by the second element in column one in B, and so on along row one in A and down column one in B. Then add all these quantities together to obtain the first element on row one, column one in the resulting matrix.

1.08 For example if $C = A \times B$

$c_{11} = a_{11} \times b_{11} + a_{12} \times b_{21} + a_{13} \times b_{31} + \ldots$

and $c_{mn} = a_{m1} \times b_{1n} + a_{m2} \times b_{2n} + a_{m3} \times b_{3n} + \ldots$

Note that $A \times B$ does not necessarily equal $B \times A$.

1.09 For a fuller treatment the reader is referred to Bommer and Symonds (see bibliography), or, to Matrix methods of structural analysis by R. K. Livesley (Pergamon Press). The method of influence coefficients was developed at Imperial College by J. C. de C. Henderson and his colleagues.

Appendix 1: Index

This appendix lists and defines the essential symbols and terms used in the text of this section of the handbook. Reference is to the paragraph in which they first appear. *Properties of sections* (table of shapes referred to in technical study ANALYSIS 1 para 3.38) is not now Appendix 1, but forms information sheet ANALYSIS 2.

Symbol or term	Description	Technical study Analysis 1 para	Analysis 2 para
A	area of section through structural member	3.13	
b	breadth of a section through structural member	3.59	
By	area of part of a section of a structural member above a line distance y above the neutral axis	3.54	
beam	structural member under mainly bending stresses		1.01
Bow's notation	a method of finding forces in members of an isostatic truss		4.04
cantilever	a beam supported wholly at one end		1.14
carry-over	the moment produced at the other end of a beam by balancing the moments at the near end in moment distribution method		7.11
centre of area	point on a section where G about any axis through that point is always zero	3.16	
centre of gravity	point in a body through which its weight always acts	2.19	
co-integration	a method of calculating influence coefficients		8.20
component	portion of force acting in another direction	2.04	
continuous beam	beam supported in more than two places		7.01
couple	a pair of equal and opposite forces producing a constant pure moment in a plane	2.09	
d	depth of section of a structural member	3.59	
deflection	displacement of a point on a structural member produced by loading		1.16
dome	a type of membrane structure		9.27
e	eccentricity of load on a column		2.12
e	strain of a structural member $= \dfrac{\text{extension}}{\text{length}}$	3.41	
E	Youngs' modulus $= \dfrac{\text{stress}}{\text{strain}}$	3.42	
elastic limit	the point on a stress-strain curve beyond which strain increases non-linearly with stress	3.44	
encastré	of the end of a beam, built-in, fixed in a horizontal position		1.24
end-fixity moments	in moment distribution method, moments at the ends of spans, assuming them encastré		7.15
equilibrant	force applied to a system of forces not in equilibrium, to cause it to be in equilibrium		2.12
equilibrium	a state in which a body remains at rest, or moving at constant velocity	2.10	
Euler load	maximum load that can be carried on a strut before it fails by buckling		2.05
f	direct stress, compression or tension	3.06	
fa	direct stress due to force at centre of area	3.23	

fbx	direct stress due to bending about the X axis	3.23
fby	direct stress due to bending about the Y axis	3.23
first moment of area G		3.15
force	undefineable phenomenon known mainly by its effects	2.01
free span bending moment	on a beam in a statically indeterminate structure; moment that would occur if the beam were simply supported	7.20
g_{11} etc	influence co-efficients	8.15
G	first moment of area about a given axis	3.15
Gx	first moment of area about the X axis: $= \text{sum of } y \times \delta A$	3.15
G	flexibility matrix in the method of influence coefficients	8.17
Hardy-Cross method	alternative name for moment distribution method	7.03
I	second moment of area of a section about a given axis, sometimes called the moment of inertia	3.28
Ix	second moment about the X axis $= \text{sum of } y^2 \times \delta A$	3.28
Ixy	product of inertia $= \text{sum of } x \times y \times \delta A$ 3.29	
influence coefficients	method of analysis for two-dimensional continuous structure	8.15
isostatic structure	one in which all the forces may be determined solely by using the laws of statics	4.01
j, k	constants used in theory of bending	3.25
L	length of a structural member between nodes	1.03
M	moments of various kinds; a moment is the product of a force magnitude and the distance of its line of action from point in question	2.07
Mx	bending moment about the X axis	3.22
member	a constituent of a structure	3.03
membrane	a shell so thin as to have no flexural rigidity	9.14
middle third	a zone in which a reaction falls if there is no tensile stress on a given area	2.23 2.15
moment distribution	an iterative method of analysis for two-dimensional continuous structure	8.07
moment of inertia	second moment of area of a section	3.28
$N\phi$	hoop force in a dome	9.29
neutral axis	axis through centre of area of section	3.47
node	a point in a structure at which the longitudinal axes of two or more members meet	3.03
P	a force or point load	2.04
photo-elasticity	a phenomenon that renders the stress in a particular material visible	3.71
plastic deformation	constantly increasing deformation under an unchanging stress	3.43
Poisson's ratio (σ sigma)	ratio between perpendicular deformations due to uniaxial stress	3.64
principal stresses	direct stresses in planes with zero shear stresses	3.69

prismatic member	one with cross-section constant along its length	3.03
r	radius of gyration $= \sqrt{\dfrac{I}{A}}$	2.06
R	radius of curvature	3.46
R	reaction force	2.12
released structure	isostatic version of statically indeterminate frame produced by releasing restraints	8.18
resolution	summing the components of forces in a specified direction	2.04, 2.11
restraint	a redundant force, making a structure statically indeterminate	8.18
resultant	a single force with the same effect as a system of forces	2.03
rotation	motion in a circular mode	2.14
s	shear stress	3.06
S	shear force	3.48
second moment of area	moment of inertia of a section	3.28
section	the shape produced in cutting through a member	3.03
shear centre	the point through which a shear force produces no torsion	3.50
slenderness ratio	of a strut $= \dfrac{L}{r}$	2.06
stiffness	of a beam or column $= \dfrac{kEI}{L}$ (k is a constant depending on the type of beam)	7.11
stress	See f	
stress trajectory	a line showing the direction of a principle stress	3.70
strain	$e = \dfrac{\text{extension}}{\text{length}}$	3.41
structure	something that changes the magnitude, direction or position of natural forces	3.02
strut	a structural member under mainly compressive stress	2.01
torsion	a moment in the plane of a section, producing twisting on the member	3.49
U	in the method of influence coefficients, the particular solution for the loading	8.17
w	uniformly distributed load on a beam per unit length	1.13
W	weight of a body, or a point load	
X	in method of influence coefficients, the column matrix of restraints	8.16
X axis	generally a horizontal axis in the plane of section of the member	2.04
Y axis	generally a vertical axis in the plane of section of the member	2.04
Z axis	generally a horizontal axis in the plane of elevation of the member	2.13
Young's modulus	$E = \dfrac{\text{stress}}{\text{strain}}$	3.42
Zx	the section modulus about the X axis $= \dfrac{Ix}{y_{max}}$	3.33

Information sheet
Analysis 1

Section 2 **Structural analysis**

Standard beam conditions

This sheet tabulates formulae and values of moments (M), reactions (R), shear force (S) and deflection (δ) in beams for a number of common loading and support conditions. It covers cantilevers, free support beams, fixed-end beams, and propped cantilevers

1 Cantilevers

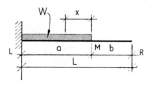

$$M_x = \frac{Wx^2}{2a} \qquad M_{max} = \frac{Wa}{2}$$
$$S_{max} = R_L = W$$
$$\delta_M = \frac{Wa^3}{8EI}$$
$$\delta_{max} = \delta_R = \frac{Wa^3}{8EI} \times \left(1 + \frac{4b}{3a}\right)$$

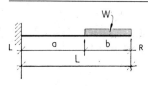

$$M_{max} = W\left(a + \frac{b}{2}\right)$$
$$S_{max} = R_L = W$$
$$\delta_{max} = \delta_R$$
$$= \frac{W}{24EI}(8a^3 + 18a^2 b + 12ab^2 + 3b^3)$$

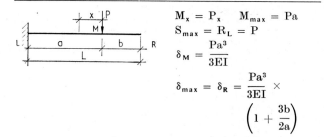

$$M_x = P_x \qquad M_{max} = Pa$$
$$S_{max} = R_L = P$$
$$\delta_M = \frac{Pa^3}{3EI}$$
$$\delta_{max} = \delta_R = \frac{Pa^3}{3EI} \times \left(1 + \frac{3b}{2a}\right)$$

2 Free support beams

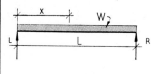

$$M_x = \frac{Wx}{2}\left(1 - \frac{x}{L}\right)$$
$$M_{max} = \frac{WL}{8}$$
$$R_L = R_R = \frac{W}{2}$$
$$\delta_{max} \text{ at centre} = \frac{5}{384} \times \frac{WL^3}{EI}$$

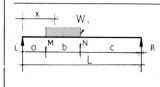

$$M_{max} = \frac{W}{b}\left(\frac{x^2 - a^2}{2}\right)$$
when
$$x = a + R_L \times \frac{b}{W}$$
$$R_L = \frac{W}{L}\left(\frac{b}{2} + c\right)$$
$$R_R = \frac{W}{L}\left(\frac{b}{2} + a\right)$$
if a = c
$$M = \frac{W}{8}(L + 2a)$$
$$\delta_{max} = \frac{W}{384EI} \times (8L^3 - 4Lb^2 + b^3)$$

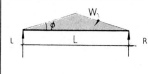

$$M_x = W_x\left(\frac{1}{2} - \frac{2x^2}{3L^2}\right)$$
$$M_{max} = \frac{WL}{6}$$
$$R_L = R_R = \frac{W}{2}$$
$$\delta_{max} = \frac{WL^3}{60EI}$$
If $\phi = 60° \ M = 0\cdot0725 \ wL^3$
$$R = 0\cdot217 \ wL^2$$

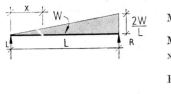

$$M_x = \frac{W_x}{3}\left(1 - \frac{x^2}{L^2}\right)$$

$$M_{max} = 0 \cdot 128WL$$

$$x_1 = 0 \cdot 5774L$$

$$R_L = \frac{W}{3}$$

$$R_R = \frac{2W}{3}$$

$$\delta_{max} = \delta_{x_2} = \frac{0 \cdot 01304WL^3}{EI}$$

$$x_2 = 0 \cdot 5193L$$

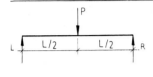

$$M_{max} = \frac{PL}{4}$$

$$R_L = R_R = \frac{P}{2}$$

$$\delta_{max} = \frac{PL^3}{48EI}$$

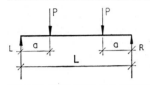

$$M_{max} = Pa$$
$$R_L = R_R = P$$
$$\delta_{max} = \frac{PL^3}{6EI}\left[\frac{3a}{4L} - \left(\frac{a}{L}\right)^3\right]$$

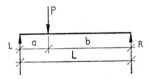

$$M_{max} = P\frac{ab}{L} = M_p$$

$$R_L = \frac{Pb}{L} \qquad R_R = \frac{Pa}{L}$$

δ_{max} always occurs within $0 \cdot 0774L$ of the centre of the beam.

When $b \geqslant a$

$$\delta_{\text{centre}} = \frac{PL^3}{48EI} \times$$

$$\left[3\frac{a}{L} - 4\left(\frac{a}{L}\right)^3\right]$$

This value is always within $2 \cdot 5$ per cent of the maximum value.

$$\delta_p = \frac{PL^3}{3EI}\left(\frac{a}{L}\right)^2\left(1 - \frac{a}{L}\right)^2$$

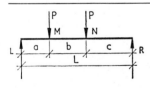

$$M_M = \frac{Pa(b + 2c)}{2L}$$

$$M_N = \frac{Pc(b + 2a)}{2L}$$

$$R_L = \frac{P(b + 2c)}{L}$$

$$R_R = \frac{P(b + 2a)}{L}$$

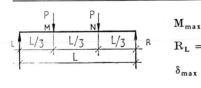

$$M_{max} = \frac{PL}{3}$$

$$R_L = R_R = P$$

$$\delta_{max} = \frac{23PL^3}{648EI}$$

$$M_{max} = M_N = \frac{PL}{2}$$

$$M_M = M_P = \frac{3PL}{8}$$

$$R_L = R_R = \frac{3P}{2}$$

$$\delta_{max} = \frac{19PL^3}{384EI}$$

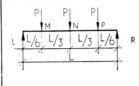

$$M_{max} = \frac{5PL}{12}$$

$$M_M = M_P = \frac{PL}{4}$$

$$R_L = R_R = \frac{3P}{2}$$

$$\delta_{max} = \frac{53PL^3}{1296EI}$$

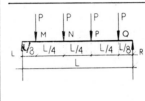

$$M_{max} = P_N = P_P = \frac{PL}{2}$$

$$M_M = M_Q = \frac{PL}{4}$$

$$R_L = R_R = 2P$$

$$\delta_{max} = \frac{41PL^3}{768EI}$$

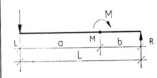

$$M_{ML} = M\frac{a}{L} \qquad M_{MR} = M\frac{b}{L}$$

$$R_A = R_B = \frac{M}{L}$$

when $a > b$

$$\delta_M = -\frac{Mab}{3EI}\left(\frac{a}{L} - \frac{b}{L}\right)$$

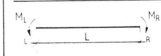

$$R_L = -R_R = \frac{M_L - M_R}{L}$$

when $M_L = M_R$,

$$\delta_{max} = -\frac{ML^2}{8EI}$$

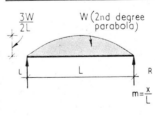

$$M_x = \frac{WL}{2}(m^4 - 2m^3 + m)$$

$$M_{max} = \frac{5WL}{32}$$

$$R_L = R_R = \frac{W}{2}$$

$$\delta_{max} = \frac{6 \cdot 1 WL^3}{384EI}$$

$$m = \frac{x}{L}$$

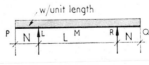

$$M_L = M_R = -\frac{wN^2}{2}$$

$$M_{max} = \frac{WL^2}{8} + M_L$$

$$R_L = R_R = w\left(N + \frac{L}{2}\right)$$

$$\delta_p = \delta_q = \frac{wL^3N}{24EI} \times$$

$$(1 - 6n^2 - 3n^3)$$

$$\delta_{max} = \frac{wL^4}{384EI}(5 - 24n^2).$$

$$n = \frac{N}{L}$$

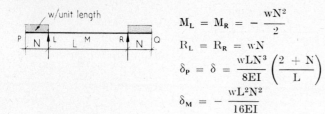

$$M_L = M_R = -\frac{wN^2}{2}$$

$$R_L = R_R = wN$$

$$\delta_P = \delta = \frac{wLN^3}{8EI}\left(2 + \frac{N}{L}\right)$$

$$\delta_M = -\frac{wL^2N^2}{16EI}$$

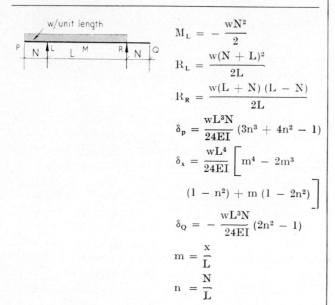

$$M_L = -\frac{wN^2}{2}$$

$$R_L = \frac{w(N + L)^2}{2L}$$

$$R_R = \frac{w(L + N)(L - N)}{2L}$$

$$\delta_P = \frac{wL^3N}{24EI}(3n^3 + 4n^2 - 1)$$

$$\delta_x = \frac{wL^4}{24EI}\left[m^4 - 2m^3 (1 - n^2) + m(1 - 2n^2)\right]$$

$$\delta_Q = -\frac{wL^3N}{24EI}(2n^2 - 1)$$

$$m = \frac{x}{L}$$

$$n = \frac{N}{L}$$

3 Fixed-end beams

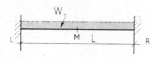

$$M_L = M_R = -\frac{WL}{12}$$

$$M_M = \frac{WL}{24}$$

$$R_L = R_R = \frac{W}{2}$$

points of contraflexure
$0 \cdot 21L$ from each end

$$\delta_{max} = \frac{WL^3}{384EI}$$

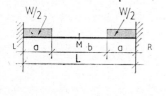

$$M_L = M_R = -\frac{Wa}{12L} \times (3L - 2a)$$

$$M_M = \frac{Wa}{4} + M_L$$

$$R_L = R_R = \frac{W}{2}$$

$$\delta_{max} = \frac{Wa^2}{48EI}(L - a)$$

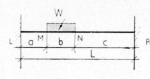

$$M_L = -\frac{W}{12L^2b} \times$$

$$\left[c^3(4 - 3c) - c^3(4L - 3c)\right]$$

$$M_R = -\frac{W}{12L^2b} \times$$

$$\left[d^3(4L - 3d) - a^3(4L - 3a)\right]$$

$$a + b = d$$

$$b + c = e$$

when r = reaction if the beam were simply supported,

$$R_L = r_L + \frac{M_L - M_R}{L}$$

$$R_R = r_R + \frac{M_R - M_L}{L}$$

when $a = c$,

$$\delta_{max} = \frac{W}{384EI} \times$$

$$(L^3 + 2L^2a + 4La^2 - 8a^3)$$

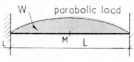

$$M_L = M_R = -\frac{WL}{10}$$

$$M_M = \frac{5WL}{32} - \frac{WL}{10} = \frac{9WL}{160}$$

$$R_L = R_B = \frac{W}{2}$$

$$\delta_{max} = \frac{1 \cdot 3\,WL^3}{384\,EI}$$

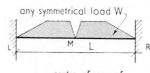

$$M_L = M_R = -\frac{A_S}{L}$$

where A_S is the area of the free bending moment diagram

$$R_L = R_R = \frac{W}{2}$$

$$\delta_{max} = \frac{A_S x - A_1 x_1}{2EI}$$

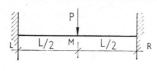

$$M_L = M_R = -\frac{PL}{8}$$

$$M_M = \frac{PL}{8}$$

$$R_L = R_R = \frac{P}{2}$$

$$\delta_{max} = \frac{PL^3}{192EI}$$

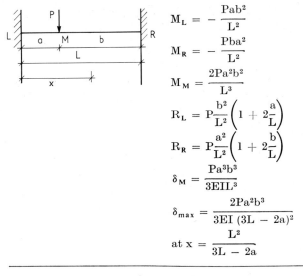

$$M_L = -\frac{Pab^2}{L^2}$$

$$M_R = -\frac{Pba^2}{L^2}$$

$$M_M = \frac{2Pa^2b^2}{L^3}$$

$$R_L = P\frac{b^2}{L^2}\left(1 + 2\frac{a}{L}\right)$$

$$R_R = P\frac{a^2}{L^2}\left(1 + 2\frac{b}{L}\right)$$

$$\delta_M = \frac{Pa^3b^3}{3EIL^3}$$

$$\delta_{max} = \frac{2Pa^2b^3}{3EI(3L - 2a)^2}$$

$$\text{at } x = \frac{L^2}{3L - 2a}$$

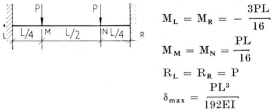

$$M_L = M_R = -\frac{3PL}{16}$$

$$M_M = M_N = \frac{PL}{16}$$

$$R_L = R_R = P$$

$$\delta_{max} = \frac{PL^3}{192EI}$$

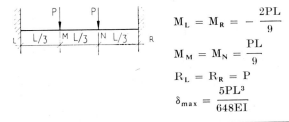

$$M_L = M_R = -\frac{Pa(L - a)}{L}$$

$$M_M = M_N = \frac{Pa^2}{L}$$

$$R_L = R_R = P$$

$$\delta_{max} = \frac{PL^3}{6EI}\left[\frac{3a^2}{4L^2} - \frac{a^3}{L^3}\right]$$

$$M_L = M_R = -\frac{2PL}{9}$$

$$M_M = M_N = \frac{PL}{9}$$

$$R_L = R_R = P$$

$$\delta_{max} = \frac{5PL^3}{648EI}$$

$$M_L = M_R = -\frac{19PL}{72}$$

$$M_N = \frac{11PL}{72}$$

$$R_L = R_R = \frac{3P}{2}$$

$$\delta_{max} = \frac{41PL^3}{5184EI}$$

$$M_L = M_R = -\frac{5PL}{16}$$

$$M_N = \frac{3PL}{16}$$

$$R_L = R_R = \frac{3P}{2}$$

$$\delta_{max} = \frac{PL^3}{96EI}$$

$$M_L = M_R = -\frac{11PL}{32}$$

$$M_N = M_P = \frac{5PL}{32}$$

$$R_L = R_R = 2P$$

$$\delta_{max} = \frac{PL^3}{96EI}$$

4 Propped cantilevers

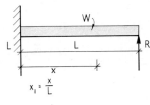

$$M_L = -\frac{WL}{8}$$

$$M_{max} = \frac{9WL}{128} \text{ at } x_1 = \frac{5}{8}$$

$$M = 0 \text{ at } x_1 = \frac{1}{4}$$

$$R_L = \frac{5}{8}W$$

$$R_R = \frac{3}{8}W$$

$$\text{if } m = \frac{1}{3} - x_1$$

$$\delta = \frac{WL^3}{48EI} \times$$

$$(m - 3m^3 + 2m^4)$$

$$\delta_{max} = \frac{WL^3}{185EI}$$

$$\text{at } x_1 = 0\cdot5785$$

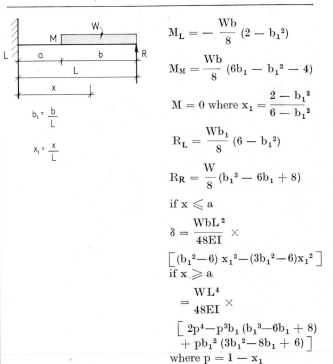

$$b_1 = \frac{b}{L}$$

$$x_1 = \frac{x}{L}$$

$$M_L = -\frac{Wb}{8}(2 - b_1{}^2)$$

$$M_M = \frac{Wb}{8}(6b_1 - b_1{}^3 - 4)$$

$$M = 0 \text{ where } x_1 = \frac{2 - b_1{}^2}{6 - b_1{}^2}$$

$$R_L = \frac{Wb_1}{8}(6 - b_1{}^2)$$

$$R_R = \frac{W}{8}(b_1{}^3 - 6b_1 + 8)$$

$$\text{if } x \leqslant a$$

$$\delta = \frac{WbL^2}{48EI} \times$$

$$\left[(b_1{}^2 - 6)x_1{}^3 - (3b_1{}^2 - 6)x_1{}^2\right]$$

$$\text{if } x \geqslant a$$

$$= \frac{WL^4}{48EI} \times$$

$$\left[2p^4 - p^3b_1(b_1{}^3 - 6b_1 + 8) + pb_1{}^2(3b_1{}^2 - 8b_1 + 6)\right]$$

$$\text{where } p = 1 - x_1$$

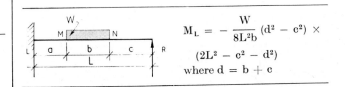

$$M_L = -\frac{W}{8L^2b}(d^2 - c^2) \times$$

$$(2L^2 - c^2 - d^2)$$

$$\text{where } d = b + c$$

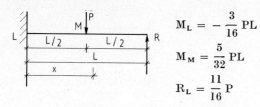

$$M_L = -\frac{3}{16} PL$$

$$M_M = \frac{5}{32} PL$$

$$R_L = \frac{11}{16} P$$

$$R_R = \frac{5}{16} P$$

$$\delta_m = \frac{7PL^3}{768EI}$$

$$\delta_{max} = 0 \cdot 00932 \frac{PL^3}{EI}$$

$$\text{at } x = 0 \cdot 553L$$

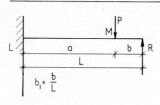

$$M_L = -\frac{Pb}{2}(1 - b_1{}^2)$$

$$(\text{maximum } 0 \cdot 193PL \text{ if } b_1 = 0 \cdot 577)$$

$$M_M = \frac{Pb}{2}(2 - 3b_1 + b_1{}^3)$$

$$(\text{maximum } 0 \cdot 174PL \text{ if } b_1 = 0 \cdot 366)$$

$$R_R = \tfrac{1}{2}\frac{Pa^2}{L^2}(b_1 + 2)$$

$$\delta m = \frac{Pa^3 b^2}{12EI\,L^3} \times (4L - a)$$

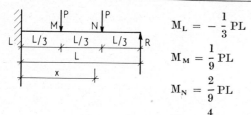

$$M_L = -\frac{1}{3} PL$$

$$M_M = \frac{1}{9} PL$$

$$M_N = \frac{2}{9} PL$$

$$R_L = \frac{4}{3} P$$

$$R_R = \frac{2}{3} P$$

$$\delta_{max} = 0 \cdot 0152 \frac{PL^3}{EI}$$

$$\text{at } x = 0 \cdot 577L$$

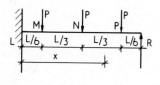

$$M_L = -\frac{19PL}{48}$$

$$M_N = \frac{21PL}{96}$$

$$M_P = \frac{53PL}{288}$$

$$R_L = \frac{91P}{48}$$

$$R_R = \frac{53P}{48}$$

$$\delta_{max} = 0 \cdot 0169 \frac{PL^3}{EI}$$

$$\text{at } x = 0 \cdot 577L$$

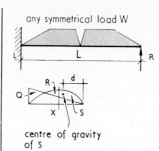

If A_s = area of free bending moment diagram

$$M_L = \frac{3A_s}{2L}$$

$$R_L = \frac{W}{2} + \frac{M_L}{L}$$

$$R_R = \frac{W}{2} - \frac{M_L}{L}$$

δ_{max} at X where area Q = area R

$$\delta_{max} = \frac{\text{area } S \times X \times d}{EI}$$

Information sheet
Analysis 2

Section 2 **Structural analysis**

Properties of sections

This information sheet lists 26 geometrical sections and gives formulae for obtaining their area, distance of extreme fibre from neutral axis, moment of inertia, modulus, and radius of gyration.

Section shape	Area of section A	Distance (y_1) of extremity of section from neutral axis	Moment of inertia about neutral axis $X \times (I_x)$	Modulus $Z_x = \left(\dfrac{I_x}{y_1}\right)$	Radius of gyration $k = \sqrt{\dfrac{I_x}{A}}$
	a^2	$\dfrac{a}{2}$	$\dfrac{a^4}{12}$	$\dfrac{a^3}{6}$	$\dfrac{a}{\sqrt{12}} = 0 \cdot 289\, a$
	bd	$\dfrac{d}{2}$	$\dfrac{1}{12}\, b\, d^3$	$\dfrac{1}{6}\, b\, d^2$	$\dfrac{d}{\sqrt{12}} = 0 \cdot 289\, d$
	$a_1^2 - a_2^2$	$\dfrac{a}{2}$	$\dfrac{a_1^4 - a_2^4}{12}$	$\dfrac{a_1^4 - a_2^4}{6\, a_1}$	$\sqrt{\dfrac{a_1^2 + a_2^2}{12}}$
	$b_1 d_1 - b_2 d_2$	$\dfrac{d_1}{2}$	$\dfrac{b_1 d_1^3 - b_2 d_2^3}{12}$	$\dfrac{b_1 d_1^3 - b_2 d_2^3}{6\, d_1}$	$\sqrt{\dfrac{b_1 d_1^3 - b_2 d_2^3}{12\,(b_1 d_1 - b_2 d_2)}}$
	a^2	$\dfrac{a}{\sqrt{2}} = 0 \cdot 707\, a$	$\dfrac{a^4}{12}$	$\dfrac{\sqrt{2}}{12}\, a^3 = 0 \cdot 118\, a^3$	$\dfrac{a}{\sqrt{12}} = 0 \cdot 289\, a$
	bd	$\dfrac{bd}{\sqrt{b^2 + d^2}}$	$\dfrac{b^3 d^3}{6\,(b^2 + d^2)}$	$\dfrac{b^2 d^2}{6\sqrt{b^2 + d^2}}$	$\dfrac{bd}{\sqrt{6\,(b^2 + d^2)}}$
	bd	$\dfrac{b \sin \phi + d \cos \phi}{2}$	$\dfrac{b\,d}{12}\left(d^2 \cos^2 \phi + b^2 \sin^2 \phi\right)$	$\dfrac{bd}{6}\dfrac{(d^2 \cos^2 \phi + b^2 \sin^2 \phi)}{(d \cos \phi + b \sin \phi)}$	$\sqrt{\dfrac{d^2 \cos^2\phi + b^2 \sin^2\phi}{12}}$
	$\dfrac{3\sqrt{3}}{2}\, a^2$ $= 2 \cdot 598\, a^2$	a	$\dfrac{5\sqrt{3}}{16}\, a^4 = 0 \cdot 541\, a^4$	$\dfrac{5\sqrt{3}}{16}\, a^3$	$a\sqrt{\dfrac{5}{24}}$
	$\dfrac{3\sqrt{3}}{2}\, a^2$ $= 2 \cdot 598\, a^2$	$a\dfrac{\sqrt{3}}{2}$	$\dfrac{5\sqrt{3}}{16}\, a^4 = 0 \cdot 541\, a^4$	$\dfrac{5}{8}\, a^3$	$a\sqrt{\dfrac{5}{24}}$
	$2\, d^2 \tan 22\tfrac{1}{2}^\circ$ $= 0 \cdot 8284\, d^2$	$\dfrac{d}{2}$	$\dfrac{4}{3}\,(4\sqrt{2} - 5)\, d^4$ $= 0 \cdot 8758\, d^4$	$\dfrac{8}{3}\,(4\sqrt{2} - 5)\, d^3$ $= 1 \cdot 7516\, d^3$	$d\sqrt{1 \cdot 057} = 1 \cdot 027\, d$ $\dfrac{bd}{6}\left(\dfrac{d^2 \cos^2 \phi + b^2 \sin^2 \phi}{d \cos \phi + b \sin \phi}\right)$

Section shape	Area of section A	Distance (y_1) of extremity of section from neutral axis	Moment of inertia about neutral axis $X \times (I_x)$	Modulus $Z_x = \left(\dfrac{I_x}{y_1}\right)$	Radius of gyration $k = \sqrt{\dfrac{I_x}{A}}$
Triangle	$\dfrac{bd}{2}$	$\dfrac{d}{3}$	$\dfrac{b \cdot d^3}{36}$	$\dfrac{b\,d^2}{24}$	$\dfrac{d}{\sqrt{18}} = 0.236\,d$
Trapezoid	$\dfrac{a+b}{2}\,d$	$\dfrac{a+2b}{a+b}\dfrac{d}{3}$	$\dfrac{a^2 + 4ab + b^2}{36(a+b)}\,d^3$	$\dfrac{a^2 + 4ab + b^2}{12(a+2b)}\,d^2$	$d\sqrt{\dfrac{a^2 + 4ab + b^2}{18(a+b)^2}}$
Circle	$\dfrac{\pi d^2}{4} = 0.7854\,d^2$	$\dfrac{d}{2}$	$\dfrac{\pi d^4}{64} = 0.0491\,d^4$	$\dfrac{\pi d^3}{32} = 0.0982\,d^3$	$\dfrac{d}{4}$
Hollow circle	$\dfrac{\pi}{4}(d^2 - d_1^2)$	$\dfrac{d}{2}$	$\dfrac{\pi}{64}(d^4 - d_1^4)$	$\dfrac{\pi}{32}\dfrac{d^4 - d_1^4}{d}$	$\dfrac{\sqrt{d^2 + d_1^2}}{4}$
Semicircle	$\dfrac{\pi d^2}{8} = 0.3927\,d^2$	$\dfrac{2\,d}{3\,\pi} = 0.212\,d$	$\dfrac{9\pi^2 - 64}{1152\,\pi}\,d^4 = 0.007\,d^4$	$\dfrac{(9\pi^2 - 64)\,d^3}{192(3\pi - 4)} = 0.024\,d^3$	$\dfrac{\sqrt{9\pi^2 - 64}\,d}{12\,\pi} = 0.132\,d$
Ellipse	$\dfrac{\pi b d}{4} = 0.7854\,bd$	$\dfrac{d}{2}$	$\dfrac{\pi b d^3}{64} = 0.0491\,b\,d^3$	$\dfrac{\pi b d^2}{32} = 0.0982\,b\,d^2$	$\dfrac{d}{4}$
Hollow ellipse	$\dfrac{\pi}{4}(bd - b_1 d_1)$	$\dfrac{d}{2}$	$\dfrac{\pi}{64}(b d^3 - b_1 d_1^3)$	$\dfrac{\pi}{32}\dfrac{b d^3 - b_1 d_1^3}{d}$	$\dfrac{1}{4}\sqrt{\dfrac{bd^3 - b_1 d_1^3}{bd - b_1 d_1}}$
I-section	$(bd - b_1 d_1)$	$\dfrac{d}{2}$	$\dfrac{1}{12}(b d^3 - b_1 d_1^3)$	$\dfrac{b d^3 - b_1 d_1^3}{6\,d}$	$\sqrt{\dfrac{bd^3 - b_1 d_1^3}{12(bd - b_1 d_1)}}$
Channel	$(bd - b_1 d_1)$	$\dfrac{d}{2}$	$\dfrac{1}{12}(b d^3 - b_1 d_1^3)$	$\dfrac{b d^3 - b_1 d_1^3}{6\,d}$	$\sqrt{\dfrac{bd^3 - b_1 d_1^3}{12(bd - b_1 d_1)}}$
Box section	$(bd - b_1 d_1)$	$\dfrac{d}{2}$	$\dfrac{1}{12}(b d^3 - b_1 d_1^3)$	$\dfrac{b d^3 - b_1 d_1^3}{6\,d}$	$\sqrt{\dfrac{bd^3 - b_1 d_1^3}{12(bd - b_1 d_1)}}$
T-section	$(bd - b_1 d_1)$	$\dfrac{bd^2 - 2b_1 d_1 d + b_1 d_1^2}{2(bd - b_1 d_1)}$	$\dfrac{(bd^2 - b_1 d_1^2)^2 - 4bd\,b_1 d_1(d - d_1)^2}{12(bd - b_1 d_1)}$	$\dfrac{(bd^2 - b_1 d_1^2)^2 - 4bd\,b_1 d_1(d - d_1)^2}{6(bd^2 - 2bd\,d_1 + b_1 d_1^2)}$	—
L-section	$(bd - b_1 d_1)$	$\dfrac{bd^2 - 2b_1 d_1 d + b_1 d_1^2}{2(bd - b_1 d_1)}$	$\dfrac{(bd^2 - b_1 d_1^2)^2 - 4bd\,b_1 d_1(d - d_1)^2}{12(bd - b_1 d_1)}$	$\dfrac{(bd^2 - b_1 d_1^2)^2 - 4bd\,b_1 d_1(d - d_1)^2}{6(bd^2 - 2bd\,d_1 + b_1 d_1^2)}$	—
Channel (open)	$(bd - b_1 d_1)$	$\dfrac{bd^2 - 2b_1 d_1 d + b_1 d_1^2}{2(bd - b_1 d_1)}$	$\dfrac{(bd^2 - b_1 d_1^2)^2 - 4bd\,b_1 d_1(d - d_1)^2}{12(bd - b_1 d_1)}$	$\dfrac{(bd^2 - b_1 d_1^2)^2 - 4bd\,b_1 d_1(d - d_1)^2}{6(bd^2 - 2bd\,d_1 + b_1 d_1^2)}$	—
Cross section	$(bd_1 + b_1 d)$	$\dfrac{d}{2}$	$\dfrac{1}{12}(b d_1^3 + b_1 d^3)$	$\dfrac{b_1 d^3 + b d_1^3}{6\,d}$	$\sqrt{\dfrac{bd_1^3 + b_1 d^3}{12(bd_1 + b_1 d)}}$
T (inverted)	$(bd_1 + b_1 d)$	$\dfrac{d}{2}$	$\dfrac{1}{12}(b d_1^3 + b_1 d^3)$	$\dfrac{b_1 d^3 + b d_1^3}{6\,d}$	$\sqrt{\dfrac{bd_1^3 + b_1 d^3}{12(bd_1 + b_1 d)}}$
H-section	$(bd_1 + b_1 d)$	$\dfrac{d}{2}$	$\dfrac{1}{12}(b d_1^3 + b_1 d^3)$	$\dfrac{b_1 d^3 + b d_1^3}{6\,d}$	$\sqrt{\dfrac{bd_1^3 + b_1 d^3}{12(bd_1 + b_1 d)}}$

AJ Handbook
Building Structure
CI/SfB (2)

Section 3
Structural Safety

Scope

Three aspects of structural safety are brought together in this section of the handbook. Building legislation and attitudes towards the analysis of building safety are discussed in the first technical study. The second study considers the movement of a structure either under changing environment or loading. Without discussing details of jointing techniques and sealants, it examines possible sources of movement which must be considered in any structural design. The third topic is fire protection of structures. This problem is discussed generally and by looking at each structural material separately. The study is supplemented by four information sheets which provide tables of 'deemed to satisfy' situations for structural members in concrete, steel, timber and masonry. Much of this information has never been published before and, as far as the AJ can tell, gives the most up to date information available.

The tables on concrete protection have been provided by the Code Servicing Panel of the Institution of Structural Engineers. These tables have since been published in CP 110 (the Unified Code of Practice for concrete). The AJ is grateful to officers of TRADA for general advice and the table on stud partitions, and to H. L. Malhotra of the Fire Research Station (BRE) for general advice and the tables on masonry. It is understood that these masonry tables will also become the basis of a Code of Practice.

Allan Hodgkinson

Author

Allan Hodgkinson, consultant editor to the handbook, has written the studies in this section. The help of the Institute of Structural Engineers, TRADA and the Fire Research Station in the preparation of the fire protection tables is acknowledged above.

The illustration on page 71 is an engraving of the 1666 fire of London from Thornton's 'History of London and Westminster' (1743)

Technical study
Safety 1

Section 3 **Structural safety**

Building legislation and
structural safety

The aim of ALLAN HODGKINSON'S *technical study is to provoke thought on the approach to design safety. He also examines both the traditional and the recent* CIRIA *'limit state' approaches, comparing them in a worked example*

1 Legislation

1.01 Legislation with regard to structure resides directly in Building Regulations and also indirectly in the Codes of Practice in that the latter are named as a 'deemed to satisfy' condition ·of the regulations[1]. Outside London, regulations are administered by local authorities of the UK, and evidence of compliance must be submitted to the local authority's engineer in the form of calculations and working drawings. The engineer may check these in his own office or employ the services of a consulting engineer. Calculations must be clear, concise, indexed, related to drawings and referenced to bibliography where special formulae have been taken from textbooks; otherwise considerable waste of time and manpower will occur all round. The local authority building inspector examines the quality of work during construction.

1.02 The London Building Act is administered by the district surveyors, who are each responsible for a particular area of inner London (old LCC area). They check the calculations and drawings and inspect the work on site. The Constructional Bylaws (1972) were published in a new format early 1973 and rely largely on the Codes of Practice, as do the Building Regulations. While the comments below may refer particularly to the Building Regulations, they are also a reasonable statement of the present state of legislation for London and Scotland.

2 Structural regulations

2.01 The Building Regulations up to April 1970 were comparatively simple, as the fundamental requirements were contained in two clauses D3 and D8.

D3 The foundations of a building shall
a safely sustain and transmit to the ground the combined dead load and imposed load in such a manner as not to cause any settlement or other movement which would impair the stability of, or cause damage to, the whole or any part of the building or of any adjoining building or works; and
b be taken down to such a depth, or to be so constructed, as to safeguard the building against damage by swelling, shrinking or freezing of the subsoil; and
c be capable of adequately resisting any attack by sulphates or any other deleterious matter present in the subsoil.
D8 The structure of a building above the foundations shall safely sustain and transmit to the foundations the combined dead load and imposed load without such deflection or deformation as will impair the stability of, or cause damage to, the whole or any part of the building.

2.02 In April 1970 an amendment (5th Amendment) was issued as a precaution against the collapse of buildings due to accidental damage. This arose directly from the collapse of part of an industrialised building and loss of life as the result of a town gas explosion. The incident received such adverse national publicity that logic was replaced by emotion in the ensuing considerations. The action taken was expounded in a new Regulation, D19 (incorporated within the Building Regulations 1972) which has been bitterly attacked by both individuals and representative engineering opinion.

2.03 This Regulation requires that in a building of five or more storeys, it should be possible to remove any one particular member from the structure with only restricted consequent collapse. Alternatively the member must be capable of supporting a load of 5 psi ($34 \cdot 5$ kN/m²) from any direction (including load transmitted to that member from adjacent elements of structure similarly loaded) in addition to its normal loading, but at enhanced allowable stresses.

2.04 The intention is to avoid progressive collapse of the building, although the damage so sustained may mean rebuilding at a later date. Although the new Regulation arises from an explosion in a special use of structure in industrialised building with large concrete panels, the Regulation must be applied to all forms of structure indiscriminately. Continued efforts by the engineering institutions have secured some relaxation of the requirements in the case of structural steelwork and reinforced concrete frames, and testing work by the British Ceramic Research Association has shown that 178 mm and 229 mm brick internal walls can satisfy the requirements provided a specific vertical load is applied to the walls.

2.05 However, it should be appreciated that there is not the slightest guarantee that the added requirement of D19 produces a structure which would contain a gas explosion, only a structure with a variety of alternative paths of support could hope to have such resistance. It should also be appreciated that a block of dwellings of four storeys could collapse progressively along its length, causing equivalent damage to that of a five-storey structure collapsing progressively vertically.

3 Codes of practice

3.01 The Regulations avoid any direct rules for application to structures of aluminium, steel, reinforced concrete, prestressed concrete, timber and masonry but state that D8 shall be deemed to be satisfied if there is compliance with the Code of Practice or British Standard appropriate to the material. At first sight this appears to be a logical approach. The codes are technical documents containing guidelines and parameters for design. They provide norms on which the structural design can be assessed, and without which the agreement to calculations by an appraising authority might prove impossible.

3.02 However, the procedure does have two unfortunate aspects. First, Codes of Practice are intended to be a general recommendation of good practice and to be implemented by a qualified engineer. The slavish following of rules is not in itself an assurance of a satisfactory whole structure. Secondly, in competitive design there is a tendency just to satisfy the code by using its letter rather than its intent, as there is then the opportunity of sheltering behind its apparent statutory respectability when liability is demanded for failure.

4 Safety

4.01 Generally, safety has been referred to already under the heading of legislation, but how safe is 'safe'? How is safety measured, and is the engineers' definition of safety the same as that of the public who use their structures? The Ronan Point aftermath in 1968 gave the impression that the public considers 'safe' to mean 'impregnable' and it was clearly not understood that total safety from all damage incidence cannot be achieved whatever the cost, though with increasing cost an increasing measure of safety can be provided.

4.02 Until recently, the method of ensuring a measure of safe construction in a particular material was to define safe working stresses in the appropriate Code of Practice or British Standard and then proportion each element of the structure so that the safe working stresses were not exceeded. The relationship of Yield Stress or its equivalent to the safe working stress was the 'factor of safety'.

4.03 In terms of the few failures which occurred in practice, it may be said that this method achieved a considerable record of success, despite constant narrowing of the safety factor as detailed knowledge and testing and control of the materials improved. Nevertheless, the method was not logical, for it treated all structure types as equal, all conditions of loading and quality of construction as equal, and all consequential failure effects as equal.

4.04 A committee of the Institution of Structural Engineers was set up in the mid-1950s to consider safety in structures. Out of the work of this committee has grown the modern statistical approach to all aspects of structural design introduced into a code of practice for the first time in November 1972.

4.05 When considering safety the following aspects require examination:

1 Loading—how accurate is the determination of loading used in analysis, what is risk of overload?

2 Materials—what is the risk that materials of construction do not comply with specification?

3 Design skill—what is the risk of a failure of the design in its basic conception or mistakes in calculation? (This is partly covered by the checking procedure referred to earlier)

4 Inadequate fabrication or construction—what is the risk of a weld in a steel structure being understrength, or the risk of concrete as mixed being understrength and inadequately compacted or reinforcing steel being fixed wrongly? (this is partly covered by the inspection process)

5 Seriousness of failure—what is the risk to life, to continued use of an industrial process, and further consequences of a failure?

6 What is failure? Failure can be collapse, excessive deflection or excessive cracking, all of which in appropriate circumstances can mean extensive repair or rebuilding of the structure.

5 The CIRIA approach

5.01 Obviously the old concept of the safety factor of working stress related to yield stress cannot give either logical or factual discrimination between various combina-

tions of the above-mentioned items. Thus, over the last 20 years there has evolved first the concept of load factors, and more recently the statistical assessment of 'characteristic stresses' and 'characteristic loads'.

5.02 A CIRIA report[2] represents seven years of thinking about the problem, with the object of suggesting an optimum balance between level of safety and overall economy of structural design. Its authors consider that their recommendations can be incorporated in the various Codes of Practice in the next five to 10 years. The new Unified Code of Practice for Concrete, BS CP 110, is the first to appear of the new series.

5.03 The attitude and method recommended by the CIRIA report is nowhere better summarised than in the report's appendix quoted below:

Limit state of collapse
The limit load factor to be adopted in design is to be evaluated in terms of three partial factors, as follows:
γ_1 = to cover unusual and unforeseen deviations of loading from the characteristic loads, and unusual combinations of such loads
γ_2 = to cover unusual and unforeseen deviations of strength from the characteristic strengths of the structure as built
γ_3 = to cover the seriousness of the effects of collapse whether general or partial, sudden or gradual, including danger to personnel and associated economic losses.
Of these partial factors, γ_1, the load variability factor, should for established forms of construction be chosen in the range from $1\cdot2$ to $1\cdot8$; γ_2, the strength variability factor, in the range $1\cdot1$ to $1\cdot6$; and γ_3, the economic factor, in the range $0\cdot9$ to $1\cdot4$.
Having chosen values for these partial factors, the load factor for collapse to be used in design is given by the product $\gamma_1 \times \gamma_2 \times \gamma_3$.

Limit state for local damage
The limit load factor to be adopted in design is to be evaluated in terms of two partial factors, as follows:
γ_4 = to cover the nature of the loads in service, whether static or dynamic and of rare or frequent occurrence
γ_5 = to cover the nature and extent of the damage likely to arise in service.
Of these partial factors, which are to be applied to characteristic loads, γ_4 should, for established forms of construction, be chosen in the range $1\cdot0$ to $1\cdot2$, and γ_5 in the range of $1\cdot0$ to $1\cdot4$.
Having chosen values for these partial factors, the load factor for local damage to be used in design is given by the product $\gamma_4 \times \gamma_5$.

Limit state of excessive deflection
The limit load factor to be adopted in design is to be evaluated in terms of two partial factors, as follows:
γ_6 = to cover the nature of the loads in service, their duration and fluctuation during the life of the structure
γ_7 = to cover the nature and extent of the deflections likely to arise in service.
Of these partial factors, which are to be applied to characteristic loads, γ_6 should, for established forms of construction, be chosen in the range $0\cdot2$ to $1\cdot0$, and γ_7 from $1\cdot0$ to $1\cdot2$.
Having chosen values for these partial factors, the load factor for excessive deflection to be used in design is given by the product $\gamma_6 \times \gamma_7$.

5.04 In limit states, the actual loads in service are to be represented by systems of 'characteristic loads' relating to the particular limit states. The loads will eventually be specified in the codes with the aim of approximating to the *most severe conditions* to be expected in service during the life of the structure concerned. Eventually they will be determined statistically from experience, but initially loads will be based on those currently quoted in BS CP3 chapter V[3] or on greater loads known to the designer.

5.05 Strengths of materials used in the structure are to be represented by 'characteristic strengths' specified in the codes. These will be based on tests and specified with the aim of approximating to the *smallest strength** that is likely to be incorporated in the structure.

5.06 The proposed CIRIA partial factors for steelwork and reinforced concrete building structures are shown in table I.

---

*strength with not more than 5 per cent probability, ie the strength below which not more than 5 per cent of the test results would fall

Table I   Proposed CIRIA *partial factors for reinforced concrete and steelwork compared*

	Limit state	Collapse	Local damage	Excessive deflection
Reinforced concrete	δ_1	1·25 permanent 1·5 imposed 1·25 wind		
	δ_2	1·4 to 1·7 concrete 1·15 steel		
	δ_3	0·9-1·1*		
	δ_4		1·0 all loads	
	δ_5		1·3 concrete 1·0 steel	
	δ_6			1·0 all permanent loads 0·8 short-term imposed and wind together 1·0 short-term imposed and wind separately 0·25 to 1·0 long-term loads†
	δ_7			1·0 concrete 1·0 steel
Steelwork	δ_1·	1·2 permanent 1·5 imposed 1·25 wind		
	δ_2	1·1 continuous frames 1·2 structural elements and non-continuous frames		
	δ_3	0·9* collapse by plastic bending 1·1 to 1·2* collapse by buckling or fracture		
	δ_4		1·0 all loads	
	δ_5		1·0	
	δ_6			1·0 all permanent loads 0·8 short-term imposed and wind together 1·0 short-term imposed and wind separately 0·25 to 1·0 long-term loads†
	δ_7			1·0

*Factor δ_3 depends on probable nature of failure, and frequency or duration of occupation by people and valuable commodities
†Depending on the use of the structure and the likelihood of the characteristic loads acting for long periods

6 The two methods

6.01 Summarising the two approaches to design safety, in the traditional method the unfactored loads are applied to the chosen structure and an analysis determines certain stresses which can be compared with the 'allowable stresses' as set out in the Codes of Practice. The measure of safety is the relationship of the 'allowable stress' to the 'yield stress' or equivalent yield stress of the material in question.

6.02 In the new method, factored characteristic loads are applied in an analysis (which may use either plastic or elastic theories) and the strengths of the chosen structural members are compared with the 'characteristic strengths' attributed to the material in question. The applied factors are then the measure of safety.

6.03 Plastic analysis is perhaps best known to architects at present for calculation of steel frames, and the following examples will illustrate the way in which a simple portal frame can be designed by both methods for safety against collapse.

6.04 A 15 m span warehouse portal is to be designed with a spread load 8·8 kN/m run of portal beam of which 4·4 kN/m is live load **1**. Ignore wind and deflection for this example and assume that a similar steel section will be employed for both beam and stanchion. The calculation of this portal by elastic analysis (BS 449) and plastic analysis (CIRIA recommendation) are compared on the next page.

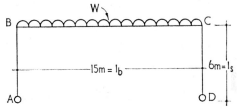

1 *Portal frame example*

6.05 The difference between elastic design (with the allowable working stress approach) and plastic design (in the limit state of collapse) will be carried a stage further in section 6 of the handbook: STEEL. In the problem described, it is shown that the latter method results in a reduced section member representing an economy but giving a greater deflection. Had the portal beam been of reinforced concrete, it would also have been necessary to check for the size of cracks in tension areas if there was the possibility of deterioration of the reinforcing steel. Hence the three checks for collapse, deflection and local damage.

6.06 This leads to a critical assessment of the CIRIA recommendations in practice. Assuming that sufficient evidence is eventually obtained to make the 'characteristic' values meaningful, can the whole process be handled in the design office? The Code CP 110 has certainly brought considerable doubt on this point from practising engineers and time alone will tell whether it will be workable. In any case, it is presumed that Codes CP 111, 112, 114 and 115 will be preserved in their existing form for several years.

Portal frame calculation: comparison of traditional and CIRIA *recommended methods*

Elastic analysis: BS 449

Dead load $4 \cdot 4$ kN/m
Live load $4 \cdot 4$ kN/m

Total $8 \cdot 8$ kN/m

Portal span $l_b = 15$ m, moment of inertia $= I_b$
Stanchion height $l_s = 6$ m, moment of inertia $= I_s$
It can be proved that the bending moment at the knee

$$M_B = M_C = \frac{w l_b{}^2}{4N}$$

where $N = 2K + 3$ and $K = \dfrac{l_s \cdot I_b}{l_b \cdot I_s}$

ie in this case $N = 2\left(\dfrac{6}{15}\right) + 3 = 3 \cdot 8$

$$M_B = \frac{8 \cdot 8 \times 15^2}{4 \times 3 \cdot 8} = 130 \cdot 0 \text{ kN/m}$$

Free moment at centre of portal beam

$$= \frac{8 \cdot 8 \times 15^2}{8} = 247 \cdot 5 \text{ kN/m}$$

Bending moment diagram is therefore:

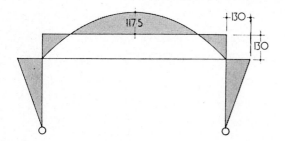

Direct load in stanchion $= \dfrac{8 \cdot 8 \times 15}{2} = 66 \cdot 0$ kN

The allowable working stress $= 165$ N/mm²
from $M = fz$ (see technical study ANALYSIS 1, AJ 25.4.71 for explanation of symbols)

Elastic $z = \dfrac{130\,000 \times 1000}{165} = 789\,000$ mm³ or 789 cm³

Section from Handbook is 406 × 178 by 54 kg/m (z = 922·8)

The stanchion would have to be checked for the combination of the direct load and bending moment

Notional safety factor $= \dfrac{\text{yield stress}}{\text{working stress}}$

$$= \frac{250}{165} = 1 \cdot 52$$

Plastic analysis: CIRIA recommendation

Actual dead load $4 \cdot 4$ kN/m
For dead load γ_1 permanent $= 1 \cdot 2$
            γ_2 continuous $= 1 \cdot 1$
           *γ_3 bending $= 0 \cdot 9$
Factored dead load $= 1 \cdot 2 \times 1 \cdot 1 \times 0 \cdot 9 \times 4 \cdot 4$
$= 5 \cdot 21$ kN/m
For live load γ_1 imposed $= 1 \cdot 5$
            $\gamma_2 = 1 \cdot 1$
           *$\gamma_3 = 0 \cdot 9$
Factored live load $= 1 \cdot 5 \times 1 \cdot 1 \times 0 \cdot 9 \times 4 \cdot 4$
$= 6 \cdot 51$ kN/m
*Frame braced against buckling

Free moment in portal beam $= (5 \cdot 21 + 6 \cdot 51) \times \dfrac{15^2}{8}$

$= 328 \cdot 0$ kNm
Hinges will form at collapse thus:

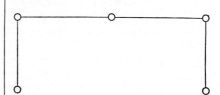

Therefore the bending moment at centre and knee will be
$\dfrac{328}{2} = 164$ kN/m

Bending moment diagram is therefore:

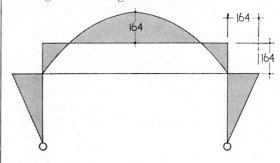

Characteristic stress (yield stress) $= 250$ N/mm². From
$M_p = f z_p$

Plastic $Z_P = \dfrac{164\,000 \times 1000}{250} = 656\,000$ mm³ or 656 cm³

Section from Handbook is 356 × 171 by 45 kg/m (Zp = 771·7)

Reduced plastic modulus for stanchion section (direct load with bending moment) $= 713 \cdot 0$
Both OK
NB the reduced section of the steel members will lead in this case to a larger deflection than the 'elastic design' portal
Notional safety factor (from load factors)
on dead load $1 \cdot 2 \times 1 \cdot 1 \times 0 \cdot 9 = 1 \cdot 19$
on live load $1 \cdot 5 \times 1 \cdot 1 \times 0 \cdot 9 = 1 \cdot 48$

---

7 Design responsibility

7.01 In the event of failure attributable to design error, a client may sue the designer responsible to him for negligence. If the architect is the only person under contract to the client, clearly he will be sued; this liability can only be transferred to the engineer when he too is appointed directly by the client. The architect can, of course, in turn sue for negligence an engineer working to his direction who is not employed by the client. The surest way of minimising failure is for one person to be in charge of the structural design and at all times to appraise the superstructure and foundation as a whole. The practice of having elements of the building designed by various people (who may well be specialists in their own field) leaves the process open both to failures in communication between the parties and failure in the overall achievement of the project conception.

7.02 If an architect considers a project beyond his own structural design capabilities he should appoint a structural engineer or preferably ask the client to make such an appointment. A chartered engineer in the modern contractor's organisation can be as efficient as, and of equal integrity to, his counterpart in the office of the consulting engineer, but his employer, the contractor, may have a commercial axe to grind in the conception of a project. Equally it should be realised that the unlimited liability of a professional consultant is of no greater value than the substance of the consulting engineer: his personal wealth and professional indemnity insurance.

8 Supervision

8.01 As design processes tend to produce structures with minimum mass members and errors of fabrication become of greater potential hazard, supervision at all stages of construction should be an essential contribution to building with an adequate reserve of strength. Neither architect nor engineer can relieve the contractor of his responsibility to supervise the work under construction, although by a system of routine testing and inspection, they can cause a general level of quality to be established. The achievement of this level requires great determination on the part of the supervisor and the client's full support.

8.02 It is of no profit to the client to demand compliance with the programme and a high quality of workmanship if he has already accepted a price for the job which will not yield an adequate return to the contractor. Equally the client should be prepared to pay for the level of supervision required and not expect the architect to supplement the contractors' staff free of charge. It is unfortunate that the various RIBA and ACE (Association of Consulting Engineers) agreements are not as clear as they might be on this matter.

The future

8.03 Relationships between architect, engineer and contractor are changing as both design and fabrication of buildings become more involved. Unless the training of architects can keep pace with the refinements of analysis and the understanding of structural legislation there will be very little structure with which the architect can work without the services of a chartered engineer.

This is a regrettable state of affairs as there are many simple structures where no Building Control Authority should even ask for calculations and many more where the simplest of codes of practice should give an adequate guide to a properly trained architect.

In the search for more design refinement and in the application of more theory and laboratory testing, Code Committees have lost sight of the bread and butter structures of the building industry and have instead created a method of computation in the limit state codes which produces almost the same end result with at least twenty percent more effort. CP 110 has had several years of practical testing in design offices and has been found sadly wanting. It is to be hoped that a lesson has been learned and much simpler limit state guidance will be provided in future codes of practice.

Bibliography

1 The Building Regulations 1976 SI 1676. HMSO £3.30; The Building (First Amendment) Regulations 1978 SI 723. HMSO £0.60p

2 The London Building (Constructional) Bylaws 1972 GLC [(Ajn)]

3 The Building Standards (Scotland) (Consolidation) Regulations 1971 SI 2052 (S218) HMSO £1.30; The Building Standards (Scotland) Amendment Regulations 1975 SI 404 (S51) £0.29p; The Building Standards (Scotland) Amendment Regulations 1973 SI 794 (S65) £0.21p [(A3j)]

4 Construction Industry Research and Information Association Report R30 CIRIA study committee on structural safety, London 1971 [(21) (K) (E2g)] o/p

5 BS CP 110: Parts 1–3 1972 The structural use of concrete

6 BS CP 111: Part 2 1970 Structural recommendations for load bearing walls

7 BS CP 112: Part 2 1971 The structural use of timber

8 BS CP 114: Part 2 1969 Structural use of reinforced concrete in buildings

9 BS CP 115: Part 2 1969 Structural use of prestressed concrete in buildings

Technical study
Safety 2

Section 3 **Structural safety**

Movement in building

Possible sources of movement must be considered in any structural design. Movement of structure under changing environment and applied loading is discussed by ALLAN HODGKINSON *in terms which do not obscure principles with details such as specific jointing techniques*

1 Introduction

1.01 The subject of this study, movement of structure under changing environment and under applied loading, is one which does usually receive appropriate consideration in large engineering structures. Buildings, however, have both the added complications of applied finishes, attached elements and services work, and the contradictions of architectural planning requirements and structural logic. When the problems associated with movement are considered, these conflicting requirements often lead to the replacement of considered common sense by optimism.

1.02 Buildings seldom fail by total collapse, but the adverse effects of movement and excessive deflection can lead to cracking and deformations, which involve repair and maintenance costs and perhaps consequential loss of use. The actual amount of movement is of importance, but differential movement between major portions of the building and building elements is the real problem and a decision has to be made either to prevent such movement or to allow it to take place by incorporating joints.

2 Sources of movement

2.01 Sources of movement can be summarised as shown in table 1. All these sources can have either an overall or a differential resulting movement.

Table 1 Sources of structural movement

Source	Type	Short-term effect	Long-term effect
Active loading	Dead load	●	●
	Steady superimposed load	●	●
	Impact superimposed load	●	
	Wind load	●	
	Seismic load	●	
	Prestress load	●	●
	Chemical expansion of soil or foundation		●
	Freezing expansion of soil	●	
	Mining effects	●	
	Vibration	●	
Passive loading	Soil deflection	●	
	Soil settlement		●
	Mining effects		●
	Water movement		●
Environment	Temperature	●	●
	Moisture	●	●
Materials (almost all building materials)	Creep		●
	Shrinkage		●
	Moisture movement	●	● (to lesser degree)
	Insulation*	●	●
	Conductivity*	●	●
	Elasticity*	●	●
	Coefficient of expansion*	●	●

*controlling properties

3 Major division of the structure

3.01 Major divisions of the structure may have to be made with regard to foundation considerations, overall length of the building and horizontal or vertical variations in building shape. The reasons for soil movement are considered in greater detail in section 4 of this handbook: FOUNDATIONS; here the effect on the building is discussed.

3.02 An isolated infinitely stiff building, loaded reasonably evenly, will settle bodily and if total settlement does not interfere with access or service connections this may not be of consequence. Buildings in Mexico City have settled in amounts ranging from several inches to a whole storey. There is a practical limitation to stiffness however and in the average building, unless founded on rock, there will be differential movements within the building and, in a large development, between different units of the development. Movement joints which were required in a large London development are shown in **1** and **2**.

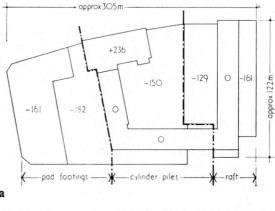

a

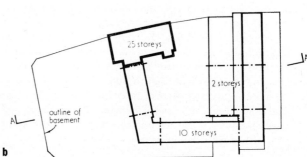

b

1 a *Basement plan, Shell Centre, London; upstream building. Chain lines are joints in both basement and superstructure. Figures are net increases in pressure on soil (ie load of building minus weight of excavated material per unit area) in kN/m²;* **b** *superstructure plan; joints isolate stiff corner areas and stiff lift/stair complexes*

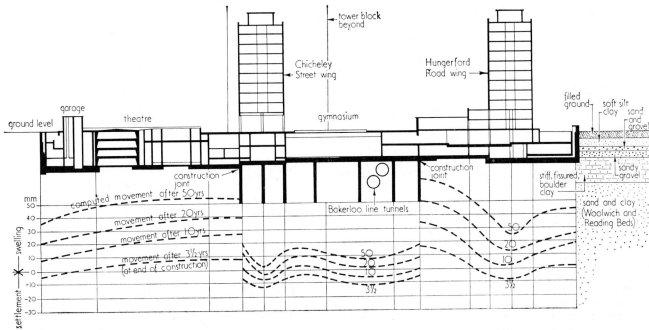

2 *Section through Shell Centre upstream building showing computed settlements*

3.03 Differential movement may arise from uneven soil types; perhaps due to sloping strata, a localised peat layer or a large clay filling of a pocket in chalk. It may also arise from the use of different foundations in adjacent building units, for instance where one building is piled and the next on a pad foundation, or from simple pad foundations sized so as to produce differing ground pressures **3**. A notable example is the Queens Tower of the former Imperial Institute in London, which had a thick raft foundation bearing on London clay at 644 kN/m². The surrounding buildings were equally massive but were carried on thick strip footings with a bearing pressure of 215 kN/m². The buildings achieved a differential settlement of 178 mm producing some very odd effects on the floor levels and in the surrounding buildings when they tried to resist this movement. Tilting can become a problem in high buildings when adjacent ground is loaded unevenly, as **4**, or where the loading within the building is uneven.

3.04 In most cases a vertical separation joint of about 25 mm width between the buildings is desirable. Below ground this may produce problems in keeping out ground water, and above ground in keeping out rain, wind and snow. It may give the architect problems in modelling the elevation, and in the general appearance of the building. Alternatively the structure may be arranged so that during construction there is complete freedom between the high and low areas. On soils where settlement takes place quickly, the connections between the high and low areas can be made after the whole permanent load is in position or, and certainly in the case of long-term settlement, the connections can be designed with provision for articulation.

3.05 The positioning of major movement joints is so fundamental to both the architectural planning and the basic structural concept that decisions must be taken in the earliest stages of the project. The order of settlements can be assessed from the results of ground investigation tests; but both this type of movement and overall longitudinal movement produce different effects in different structural types and the final decision will derive largely from intuition and experience.

3.06 Some advice is given on distortion and settlement in a paper by Skempton and McDonald[1]. They refer to damage

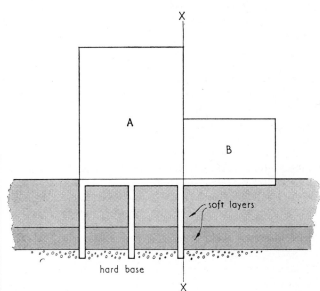

3 *Unless structure is jointed at* xx, *block* B *tries to settle and tilt to right, cracking along* xx

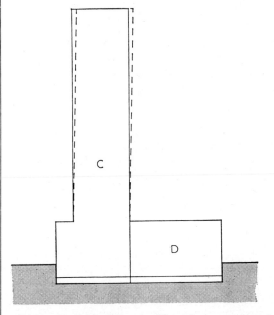

4 *Lighter loaded building* D *still produces asymmetrical settlement contours under* C, *causing a tilt to the right*

limits for loadbearing walls and panels in traditional-type frame buildings (table II):

Table II Damage limits for loadbearing walls and panels[1]

Criterion	Isolated foundations	Rafts
Angular distortion	1/300	1/300
Greatest differential settlement	Clay 44·5 mm	44·5 mm
	Sand 31·8 mm	31·8 mm
Maximum settlement	Clay 76 mm	76 to 127 mm
	Sand 50 mm	50 to 76 mm

3.07 For design purposes they suggest that a factor of safety of 1·25 to 1·50 should be applied. Using the latter figure, an angular distortion limited to 1 in 450 would be considered desirable. The bare frame distortion, ie without stiffening panels, could be double this amount, ie 1 in 225. Other investigators have suggested comparable figures of 1 in 750 and 1 in 150 respectively, indicating the lack of precision possible and therefore the desirability of erring on the safe side.

3.08 In the case of contraction or expansion joints in the length of a building, the object is to avoid damage resulting from the movement of sections which are restrained between stiffer sections. Irrespective of the length, a 150 mm thick concrete slab, if fully restrained, would develop a compressive force between the restraints of about 500 kN/m width of the slab, with a rise in temperature of 17°C. The same slab, if cast in one pour between restraints would produce a tension force on shrinking of the same order. The distance between restraints determines the extent of potential temperature movement. In 30 m this might amount to 6 mm. Obviously the restraints and/or the slab will move or crack under such conditions.

3.09 Summarising the whole question of major movement and joint positioning, building or group of joined buildings will probably crack around certain points. By predeterming these expected cracks in the form of joints, the problems can be faced before the cracks occur. Some examples from experience are shown in **5-9**.

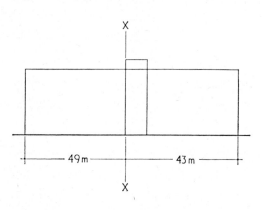

5 *Eight-storey in situ concrete housing structure; with large concrete cladding panels. Structure was built from left to right and joint xx opened 13 mm during construction, as the in situ concrete shrank*

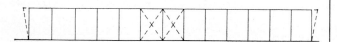

6 *Single-storey steel-framed building 90 m long, braced at centre, will hinge from column bases and expand at roof level. Light cladding allows sufficient flexibility*

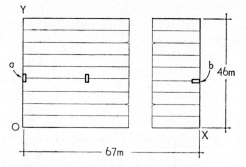

7 *Two-storey car park, approximately 7·6 m square grid with beams running parallel to x axis, slabs spanning in y direction. The columns, internal 610 mm × 229 mm, and external 610 mm × 305 mm parallel to y axis (as a), were successful. But when external columns were turned through 90° for planning reasons (as b), these failed. Stiffness had been increased to four times its original value, and instead of flexing, they attempted to act as buttresses*

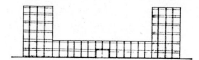

8 *Low flexible buildings should not be trapped between higher rigid buildings*

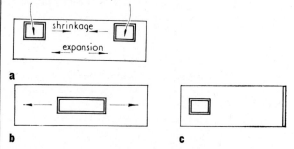

9 *Effect of layout of stiff vertical supports. In typical multi-storey block, positioning of stiff stair and lift complexes as in **a**, although desirable planning, contradicts free movement of structure. Alternative solutions which solve the structural problem either **b** allow movement outwards from the stiff complex, or **c** provide a flexible wall at the end to accept movement*

4 Seismic and mining problems

4.01 Earthquakes are the surface vibration caused either by earth slip along fault lines or by volcanic explosions. Vibrations occur in both horizontal and vertical directions, the former being about five to 10 times the magnitude of the latter. Tremors are occasionally experienced in the UK but are not considered in design. In other parts of the world they present a serious design problem.

4.02 Joints are determined in accordance with the normal procedures outlined previously with the added isolation of building parts dissimilar in mass or rigidity. The joints in such conditions should be much wider than normal to prevent the sections of the building pounding each other during an earthquake.

4.03 Mining problems are found in some areas of the UK, and with increasing scarcity of building land, the use of sites hitherto ignored because of this condition is likely to increase. Two situations arise, one where the mining has already occurred and the other where it will occur after the building has been built. The former becomes a settlement problem of rather large magnitude and follows normal settlement design and detailing procedure. The latter is more complex and requires the help of a mining expert to predict the pattern and order of movement.

4.04 Put in its simplest form, a wave movement occurs so that a part of the building is at one time on the crest, in the trough or riding up and down the slope. As the wave progresses across the site, the foundations will tend to be either torn apart from each other or compressed towards each other. On completion of the mining operation the long-term settlement effect will again occur, depending on the extent to which the mined volume is replaced by fill.

4.05 It is possible to articulate the building by a three-point foundation system, though this is very costly. The entire structure is carried on a triangular frame incorporating levelling jacks at the apexes. More usually the building is divided into small sections, and these are dealt with independently. A shallow raft with a granular underlayer will allow the building to slide over the wave. Where the size is such that normal pad foundations are essential, a high bearing pressure should be employed and the bases tied to each other. The more flexible the superstructure, the less repair will be necessary on completion of the movement. A detailed and useful description by F. W. L. Heathcote[2] has explained both the problems and some methods of overcoming them.

5 Ground floor settlement or heave

5.01 Occasionally, particularly in dock warehouses, the structure is piled and the floor is groundbearing. In such cases, not only must the ground floor be allowed to settle around the pile caps without damaging itself, but the pile caps must not become unstable **10**.

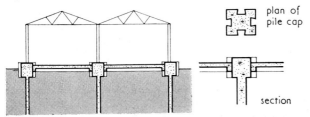

plan of
pile cap

section

10 *Floor is jointed around pile cap. Beam settles with floor in groove in pile cap face, thus stabilising pile caps while avoiding damage to settling floor slab*

5.02 The heated floor of a boiler house can cause shrinkage if there is a cohesive soil underneath, while the cold effect from a cold store can cause expansion of the soil below the floor and therefore heave.

6 Differential movement of building elements

6.01 The obvious movements to be considered are the deflections of beams, floors, walls and columns due to self weight, service load and wind loads. Over the last 30 years a better knowledge of the performance of materials and control in manufacture has led to designed members becoming smaller in section to carry the same load.

6.02 Since the 1930s the allowable bending stress in structural steel has risen from 8 tons psi (123·5 N/mm²) to 10·5 tons psi (162·1 N/mm²) for mild steel, and to 13·5 tons psi (208·4 N/mm²) for high tensile steel. The allowable tensile stress in reinforcing steel has risen from 16 000 lb psi (110·3 N/mm²) to 20 000 lb psi (137·9 N/mm²) for mild steel, 33 000 lb psi (227·5 N/mm²) for medium high tensile steel and even 50 000 lb psi (344·75 N/mm²) for very high tensile steel. In all cases there has been little change in the

modulus of elasticity (the relationship between stress and strain), which results in more deflection when these materials are used with improved structural efficiency.

6.03 The strength of concrete in reinforced concrete structures has risen from a cube strength of 2000 lb psi (13·790 N/mm²) to between 3000 and 6000 lb psi (20·685 and 41·470 N/mm²). The modulus of elasticity does not increase linearly with the strength, therefore the use of high strength concretes tends to produce greater deflection by virtue of the reduced section employed. BS CP 114[3] recognises this by reducing the allowable span to depth ratios of beams and slabs for various combinations of concrete and steel reinforcement stress.

6.04 In pr   essed concrete even higher concrete strengths are demanded, but the deflection becomes a more complex condition. On the one hand the section may be made smaller, but the prestress can keep the whole concrete section uncracked so that the product of second moment of area and elastic modulus may be greater than that of a 'cracked' reinforced concrete section of equal size, therefore having less deflection potential **11**. On the other hand the prestressing force is an active one and will result in an axial shortening of the member and long-term deflections upwards or downwards resulting from further stress variations due to creep. This condition will be examined further on.

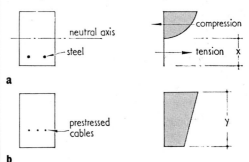

neutral axis

steel

a

prestressed
cables

b

compression

tension   x

y

11 a *Concrete in area* x *is in sufficient tension to be cracked, and therefore is not contributing to second moment of area;* **b** *prestress working condition with load on section entirely in compression, therefore not cracked. Full section, including cable, contributes to second moment of area*

6.05 However, kept within normal limits, the movement of the structure is not harmful and it is only when it is considered relative to the other members which it supports that concern arises. Obviously a deflection large enough to give the impression that the building is about to collapse is undesirable, or a vibration produced by jumping on an over-flexible member can be disquieting. But the major problems arise with other building elements supported on the structure such as partitions, applied finishes, cladding panels and so on. Service conditions also may be unacceptable, such as drainage falls in a roof which become reversed by the deflections.

7 Partitions

7.01 Despite the advice of BS CP 5234[6], partitions still remain a constant source of distress and are a frequent battleground of claims for professional negligence. This need not be so, for partitions can be designed to be compatible with the structure, or vice versa. Unfortunately old traditions die hard despite new materials and design concepts, and it is not unusual to find heavy rigid partitions with hard plaster faces being placed on flexible floors.

7.02 In the 1930s, columns in reinforced concrete invariably had a beam framing into the head on each face as a measure of stability. Relative economics of concrete, formwork and steel indicated a deep beam solution and with this a considerable degree of rigidity. In the early 1950s there arrived the now commonplace concrete slab with 'beam strip' within its depth. Relative economics had changed and a flat slab requiring minimum finishing treatment led to minimum building height and minimum decoration costs. It also led to a more flexible floor both from direct load and wind-loading effects, a property which is still largely unappreciated. Some indication of the cracking which can occur is shown in **12** and **13**, while **14** illustrates the transfer of load from a beam to a non-loadbearing partition.

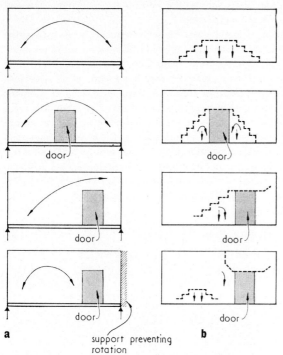

12 *Effect of deflection-prone floors on rigid partitions, with or without doors in them:* **a** *arching action of partitions as supporting floor deflects;* **b** *resultant cracking, downward movement and rotation of cracked areas*

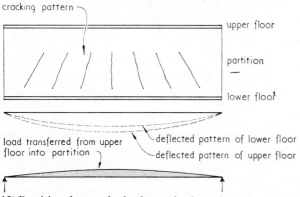

13 *Partition damage by load transfer from floor above. Here partition spans as deep beam, with load transferred from differential deflection between upper and lower floors*

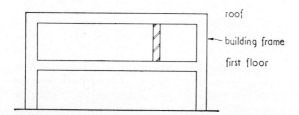

14 *Partition damage by load transfer from beam over. Here partition is constructed too early in building programme, and pinned to roof beam. Roof beam deflects more than first floor beam and transfers load to partition*

8 Other deflection problems arising from applied load

8.01 Cantilevers are often calculated as though they spring from an infinitely stiff structure and no consideration is given for the rotation at the support, which increases or decreases the end deflection **15**. Even when the correct calculation is made, distress can still occur in supported elements **16**, if allowance for the movement is not made. Continuous beams develop tension over the supports, and joints are needed in supported elements **17**.

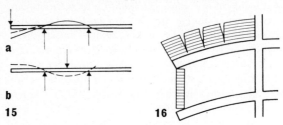

15 *In* **a**, *dotted line is simple cantilever deflection. Rotation at support produces real deflection as shown in solid line. In* **b**, *dotted line shows cantilever lifting as load is applied to next span*

16 *If edge cladding is too stiff or too rigidly connected to the cantilevers it may be damaged, or even fall out. A wall built along cantilever may crack as shown*

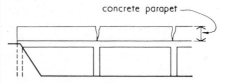

17 *Cracks occur in parapets over column supports to bridge deck. In this case joints had to be sawn in parapet*

8.02 Large panel construction or heavy precast beam erection requires the use of levelling bolts or shims. Unless the shim or bolt is released, or unless a packing capable of carrying the erection load and capable of settling uniformly with the finished structure is employed, 'stress raisers' will occur which will split the supported member **18**. Cladding panels should not be supported in such a manner that they become a prop between floors unless specially designed to do so. It is preferable to have support on one member, and flexible attachment elsewhere **19**. Changes of grid can be another source of cracking **20**. A long span floor can be damaged by the intervention of an extra supporting system and this happens frequently at the ramps in multi-storey car park construction. Problems which arise with the rotation of simply supported beams at support points are shown in **21**, **22** and **23**.

18 *Nylon or plastic washer carries initial erection loads, but it deflects more than the concrete member when further load is applied*

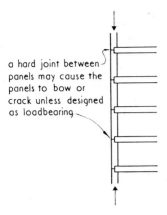

19 *Cracking, caused by hard joints*

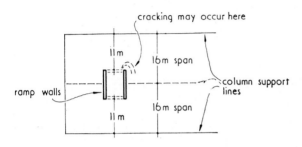

20 *Cracking caused by change of grid*

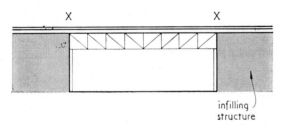

21 *Underground structure, with large span over a theatre or swimming pool. Change in slope in main girder will cause cracking in finishes at x, unless slip surface is introduced*

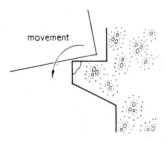

22 *Change in slope at end of heavy girders on brackets or bearings can break the edge. This can be avoided by raising beam on bearing with flexible pad such as neoprene*

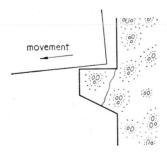

23 *Prestressed beams with high pressure under bearing can pull bracket away as creep occurs*

9 Movement resulting from change in environment

9.01 While all materials react to temperature change, concrete, masonry and timber react also to shrinkage and moisture migration. Despite the use of modern lightweight insulation materials of great efficiency, it is unlikely that all parts of a building will be at the same temperature. Where a structure can be contained entirely within the insulation, the best movement control is effected, but exposed columns and edge beams create transverse deformations which result in the cracking of internal walls or floors **24**. Cladding members, inevitably outside the insulation barrier, should be given an adequate allowance for movement, depending on the size of the member, both in terms of gaps between members and the form of the fixings to the structure.

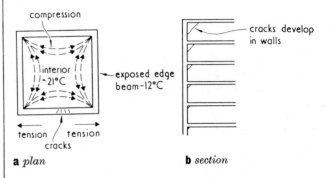

a *plan* **b** *section*

24 *Cracking effects of temperature difference on exposed columns and edge beams:* **a** *floors;* **b** *internal walls*

9.02 Roof beams and slabs will tend to deflect upwards when subjected to heat; this can be a very sudden movement in sunshine where there is poor roof insulation. The roof as a whole can expand and damage the supporting structure, or itself if the supporting structure is stiff. Parapets can be pushed out of place and non-loadbearing partitions can be pulled over laterally or cracked along their length **25**. The effect of direct sun heat on a large concrete panel is shown in **26**.

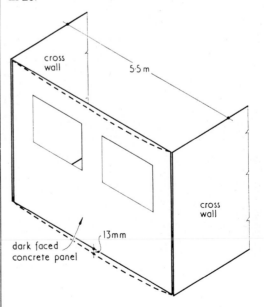

25 *Dark faced aggregate concrete panel bowed 13 mm in sunshine between cross walls 5·5 m apart*

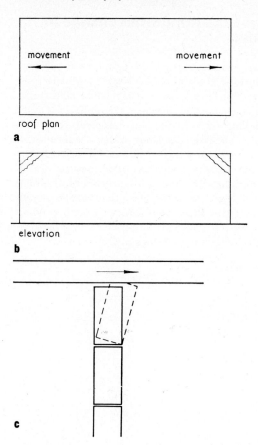

26 *Expansion of roof as a whole:* **a** *plan;* **b** *elevation;*
c *detail of partition pinned to concrete roof which cracks
when roof moves*

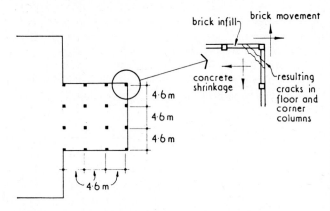

27 *Cracking in concrete floor and corner column where
structure on 4·6 m grid has rigid brick infill cladding*

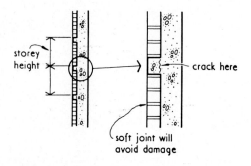

28 *Supporting nib in gable wall sheared off by moisture
movement in brickwork*

9.03 BS CP 111[5] proposes vertical joints in facing brickwork
at intervals not exceeding 15 m and where an external face
of a cavity wall exceeds 9 m in height; the wall should be
supported from the internal structure to avoid disruption
of the wall ties. Combination of brickwork panels in a
concrete structure is complicated by the addition of shrinkage and moisture movement. An actual occurrence where
rigid brick panels expanded within a frame which was
contracting is illustrated in **27**. This resulted in diagonal
cracks in the corners of the floor slabs. Supporting nibs in
gable walls of concrete have been sheared off by the moisture movement of the brickwork when a soft joint has not
been left below the nib, as **28**.

9.04 Shrinkage and creep in concrete will be considered in
more detail later in the handbook (section 5: CONCRETE).
One effect often encountered is where an external concrete
member has tiles, mosaics or precast cladding. Unless
adequate horizontal joints are made, the cladding will be
forced off as the concrete moves downwards under elastic
compression, creep and shrinkage.

10 Conclusion

10.01 The purpose of this study has been to examine the
general aspects of movement but not to solve the detail of
movement joints. In normal circumstances these are costly;
and even more costly if keeping out the weather. Fine
judgment is therefore needed to balance the risk of cracking
against the cost of the joints, and this is an area in which
there should be full understanding between architect and
engineer.

References

1 SKEMPTON, A. W. and MCDONALD, D. H. A comparison of calculated and observed settlements. Stress Conference:
London, 1955, Institution of Civil Engineers [(2–) (L4)]
p318

2 HEATHCOTE, F. W. L. The movement of articulated buildings
on subsidence sites. *ICE Journal*, vol 30, 1965, February
[(2–) (L4)] p347–368

BRITISH STANDARDS INSTITUTION

3 BS CP 114: Part 2 1969 Structural use of reinforced concrete
in buildings
4 BS CP 121: Part 3 1973 Brick and block masonry
5 BS CP 111: Part 2 1970 Structural recommendations for
load bearing walls
6 BS 5234: Code of Practice for internal non-load bearing
partitioning, 1975

Technical study
Safety 3

Section 3 **Structural safety**

Fire protection

ALLAN HODGKINSON'S *final technical study on safety discusses the effects of flame and heat on structural materials. He considers the general principles of protection, and in four information sheets which follow, adds tables of 'deemed to satisfy' conditions for protected structural members in concrete, steel, timber and masonry. This information is the most up to date available on the subject, now largely published for the first time. The technical study is concerned only with general aspects, and protection of structural members*

1 The general problem

1.01 Various statutory regulations classify the use of buildings according to fire risk. They also define fire resistance of structural elements and the resistance of finishes to spread of flame; this is related to the purpose, size and degree of separation of buildings or parts of buildings. The main concern is safety of life but further fire resistance may be necessary for reasons of insurance premium.

1.02 A fire spreads because the radiation and convection from its flames and hot gases heat other combustible materials so that they also ignite. Burning doors and breaking windows may increase ventilation and so help to spread the fire. With smoke and toxic gases a hazard both to escaping occupants and firemen, it is important that this spread should be retarded. The structural material, or finishes applied to it, can help to prevent this spread.

1.03 The temperature may reach 1200°C if a fire continues unchecked. Heat can then be transmitted through walls, floors or roofs, igniting other materials or causing structural members to crack or collapse. If a very hot member is hosed by firemen, the sudden cooling may further impair its strength.

1.04 There are tests that make allowances for these effects and can be used to allot fire resistance gradings to the main structural elements. In addition many commercially sponsored tests on specific products have been carried out by the Joint Fire Research Organisation. The JFRO reports, published from time to time, give guidance on other forms or combinations of materials. From their experience, the officers of JFRO may be able to assess gradings of new structural elements and their advice is usually accepted as the basis of a successful waiver of Building Regulation.

1.05 BS 476[1] defines fire resistance, incombustibility and non-inflammability of building materials and structures, and describes all the requirements for testing. The heating curve given in the standard represents a rapidly growing fire with a smooth temperature profile of maximum duration six hours and maximum temperature 1200°C. Fire resistance is related to the 'standard fire' (as defined in BS 476). Building elements are tested in a furnace where

temperature is controlled so that particular temperatures are reached after periods ranging from $\frac{1}{2}$ hour to 6 hours. Then, for a particular time resistance required, relative to the severity of fire expected, a fire should not be more severe in its effect than the BS 476 test.

1.06 For many years there has been a movement towards basing fire requirements on the actual fire load, ie the calorific value of the buildings' contents. But this would require a special grading procedure and more basic design data. Two JFRO tests, one on multi-storey car parks and the other a two-storey steel-framed building forming part of a multi-storey block of flats, have provided valuable information. The second test established that the position of a stanchion inside the fire compartment has a marked effect on the temperature it reaches. It also showed that the size of the steel section in a casing affects the temperature it reaches and that the degree of protection needed can be calculated. Building Regulations now recognise the effect of the mass of the protected member by relating deemed to satisfy requirements to a minimum weight of steelwork.

2 Choice of structure

2.01 Structure provides support for floors (for habitation) and roofs (for cover). With medium or large spans, the roof is invariably of lightweight construction, but where there are suspended floors, the construction is often repeated at roof level.

2.02 Floors are usually made of reinforced concrete, timber or sheet steel screeded over, roofs have been built in almost every material. Fire hazards exist for all types of structure, and, for those materials lacking the in-built protection of concrete, a careful consideration of protection method is essential.

2.03 Long-span structures—like hangar roofs—are traditionally built in steel or aluminium. But the extra costs of drencher systems, sprinklers and insurance premiums should be assessed before making a final choice. (An aircraft hangar has been built with prestressed precast space frame roof and industrial buildings have employed roof trusses of lightweight concrete.)

3 Reinforced concrete

3.01 Reinforced concrete has the best fire resistance of common structural materials and indeed concrete is used to protect other structures. It does not burn or give off sufficient inflammable vapour to ignite and so may be regarded as incombustible.

3.02 The fire resistance depends largely on the type of aggregate. Siliceous aggregates are the poorest; the class 2 reference of the Building Regulations couples flint-gravel, granite and all crushed natural stones other than limestone. Aggregates which have been subjected to heat during their manufacture give a better performance; the class 1 reference couples foamed slag, pumice, blast furnace slag, pelleted fly ash, crushed brick and burnt clay products, well-burnt clinker and crushed limestone. This group contains the commercial lightweight aggregates.

3.03 Concrete fails in fires because of the differential expansion between exposed hot surface layers and inner cooler layers. The movement of cement, as it shrinks with loss of moisture, compared with the continued expansion of the aggregate as temperature increases, creates another differential and so a further stress.

3.04 Steel reinforcement, exposed by cracking, conducts the heat rapidly and increases the temperature differential. The concrete cracks and spalls and the reinforcement loses strength as its temperature rises. Ultimately the element fails. The insulation value of reinforced concrete, as a structural element or as a casing to steelwork is obviously important; from this point of view lightweight aggregate concretes offer the better value.

3.05 As concrete with sand, ballast, sandstone and limestone aggregates (but not aggregates of igneous rocks) is heated, the colour changes from pink or red at between 300°C and 600°C, to grey between 600°C and 900°C, and to buff at higher temperatures. These colour changes are permanent and so help to identify the extent of the damage. Strength begins to fall rapidly at about 250°C and although the structure may appear sound at 600°C, the strength will have dropped to 40 per cent.

3.06 Repair of a damaged structure is often easy. The temperature is unlikely to have exceeded 800°C if less than one-quarter of the steel surface is exposed by spalling. In the case of mild steel, this represents a permanent loss of strength of 20 per cent of yield strength and 15 per cent of ultimate strength. Work-hardened steels are more seriously affected and it is essential that test pieces are cut from the bars, which is obviously more difficult with prestressing tendons. The steel may revert to its unworked condition, depending on the temperature attained. When the structure is under both superimposed load and fire attack, the critical temperature for ordinary reinforcing steel will be about 550°C and for prestressing tendons about 400°C. At these temperatures the steel retains about half its normal ambient temperature strength.

3.07 A given thickness of concrete protects the reinforcement for a specific time under the worst fire conditions; a total thickness of member controls the temperature rise on the unexposed face and reduces cracking. Different shapes react differently to fire. Variations are needed with different materials, eg less protection thickness for lightweight aggregate, more protection thickness for hard-drawn steel used in prestressed concrete.

3.08 Finishing materials such as plastics, renderings and suspended ceilings add to the protection. Recent investigations showed that end restraint against thermal expansion can substantially increase the fire resistance of a structural element. Until more research results are available, it is proposed that for beams and slabs, built into a structure with restraints to thermal expansion provided at opposite ends, the amount of protective cover to reinforcement may be reduced to that for the next lower period in tables of fire resistance requirements.

3.09 The surrounding structure can be assumed to provide thermal restraint if there are no gaps between it and the ends of floor or beam, no combustible materials used to fill these gaps, and if the surrounding structure can withstand thermal stresses induced by the heated floor or beam.

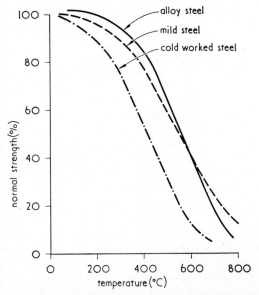

Concrete structures: **1** *rib beams;* **2** *prestressed beams;* **3a** *exposed column;* **3b** *column in wall*

4 Structural steelwork

4.01 Steelwork is incombustible. When heated it expands at a known rate and its strength decreases **4**. Mild steel is not really affected until 300°C but decreases rapidly in yield strength to about 50 per cent at 550°C and to 10 per cent at 800°C. On cooling it will recover about 90 per cent of its initial strength; this is also true of alloy steel. Work-hardened steels, usually cold-worked bars or prestressing wire, deteriorate more rapidly and equivalent yield strength drops to half at about 400°C. On cooling, this steel reverts to the original unworked form. There is considerable permanent loss of strength, and elongation characteristics are affected.

4 *Graph shows changing strength of steel with temperature*

4.02 With a thermal conductivity of 42 W/m °C, there is rapid heat transmission through the element. But there is a high heat capacity, so the temperature of the steel lags behind the environmental temperature. Unprotected members in standard fire tests have reached failure conditions in 10 to 15 minutes, showing that some protection is necessary for even the lowest fire grading.

4.03 The protection system chosen must provide overall value; relative costs of a number of treatments are shown in table I. These are costs of protection per foot of member, assuming conventional I-shaped sections. A protection that follows the profile of the steel should be cheaper with an RHS section.

Table I Costs of various methods of cladding structural steelwork compared with concrete casing

1	Concrete casing	100
2	Composite casing with pvc sheet finish	95
3	Lightweight precast concrete interlocking blocks	85
4	Intumescent paint (½ hr)	75
5	Ex-metal lathing and 'Carlite'	70
6	Preformed vermiculite panels	65
7	Plaster board with 'Carlite' finish	55
8	Lightweight blocks unplastered	40
9	Sprayed asbestos	40

4.04 The Building Regulations suggest a variety of protection methods for stanchions of minimum weight 44·64 kg/m, and beams of minimum weight 29·76 kg/m. These are illustrated in information sheet SAFETY 2.

4.05 The 44·64 kg/m limit excludes the three lowest-weight sections in the universal column range and the eight lowest-weight sections in the universal beam range if the beams are employed as columns. The 29·76 kg/m limit excludes only the lowest-weight section in both the universal beam range and universal column range, and all the six joists in the 5° taper flange range. These excluded sections need special consideration and might be acceptable at a particular fire rating if protection equivalent to the next higher rating is used. The protections illustrated can be supplemented by numerous proprietary assemblies which have either been tested or given a rating by the JFRO.

4.06 Intumescent spray, as opposed to paint, has been used in the US and recently to a limited extent in the UK. With further development, this treatment may provide up to two hours fire protection but the 3 mm to 6 mm thickness is comparatively expensive, unless its decorative value can be exploited.

4.07 Some work has been done in the US and France on hollow columns filled with water. The bases and the tops of the columns are connected by small diameter pipes so that a circuit is formed. This can work as a radiator heating system in reverse when columns are subjected to heat. The heated water moves upwards and is replaced by cooler water, thus keeping the shell of the column at a lower temperature. A header tank can be used, as in a radiator system, to ensure that the system is kept full.

4.08 This might seem to have rather limited application because the bare steelwork is the structural member which is exposed to corrosion. But there are now weathering steels that develop their own protective oxide coating and which, in external situations or on unclad structures, may not reach critical temperature in a fire because the heat is so rapidly dissipated.

5 Aluminium

5.01 Aluminium is not widely used in building structures. Table II compares three properties of steel and aluminium. In a fire, heat would be conducted away from a point more quickly with aluminium but because it melts at 650°C the section would melt before the end of a 30-minute fire test.

Table II Comparison of thermal properties of aluminium and steel

	Thermal expansion °C	Conductivity W/m°C	Melting point °C
Aluminium	29 × 10⁻⁶	230	650
Steel	11 × 10⁻⁶	42	1500

6 Timber

6.01 Unlike the other materials discussed, timber actually burns. But though combustible, the ease of ignition is related to the density and moisture content of the timber and the size of the member.

6.02 In a fire, moisture (which may be from 10 to 20 per cent of the dry weight) is driven from the surface layers of the timber. Little chemical change occurs until the temperature reaches 270 to 290°C when exposed surface layers begin to decompose and the liberated gases can ignite. Flaming continues as long as there is a heat input; without this, heat radiated back towards the wood by the flames is not sufficient to maintain the decomposition process.

6.03 With continued flaming, a layer of charcoal is produced. This shields the inner timber from the effects of the fire. The charcoal is a better insulator than the natural timber, so strength is lost only from the outer layers which are consumed. This charcoal layer is inert up to temperatures of about 500°C, glowing combustion then starts and the charcoal is gradually consumed. A state of steady combustion is achieved and the charring area advances into the unburnt timber at a steady rate **5**.

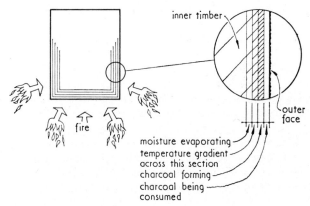

5 *Timber layers under fire attack*

6.04 Laboratory tests have established that for the majority of timber species this rate is 0·64 mm/min, with low-density woods charring quicker than high-density ones. The rate is scarcely influenced by the severity of the fire, and this makes reasonably accurate prediction of the fire resistance of timber members possible.

Sacrificial timber

6.05 A timber member can be designed by the usual methods and then increased in size to allow for the loss of timber expected in a particular period of fire. For example a timber beam **6** bearing its design load, is exposed to fire for 30 minutes on the soffit and two sides. The resulting section for calculation purposes will be $(D - 30 \times 0.64)$ mm deep and $(W - 2 \times 30 \times 0.64)$ mm wide. That is, the section is reduced on each exposed face by the product of the elapsed time (30 min) and the charring rate (0·64 mm/min).

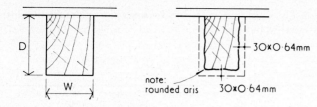

6 *Timber beams showing 'sacrificial' material*

6.06 For timber columns, heat can usually be applied on all sides; an arbitrary charring rate of 0·83 mm/min is used because of the more rapid temperature rise possible under such conditions.

6.07 Beams or columns laminated from smaller timbers using structural adhesives of the resorcinal or phenol-resorcinol type, have charring and strength loss character-istics equal to those of the solid section. But this does not apply to mechanically fastened laminated members, unless the fixings remain within the undamaged timber at the end of the period of required fire resistance. Walls and floors can be calculated in the same way but are generally of such small sections that it would be more economical to consider the protective properties of an applied finish.

7 Masonry

7.01 Masonry, in the form of solid brick, cellular brick, solid concrete blocks, hollow concrete blocks (with either heavy or lightweight aggregate) and aerated concrete blocks, provides considerable resistance to fire.

7.02 Bricks and concrete blocks with hollow cores not exceeding 25 per cent by volume can withstand the four-hour furnace test of about 1100°C face temperature without fusion or spalling from the exposed face. Blocks with larger cavities can have thin internal webs; here the high thermal stresses across the section could lead to fracture. Aerated concrete blocks provide better insulation but as the material loses more strength than other blocks at high temperatures, extra thickness is required.

7.03 As temperature rises, the heated face not only loses strength but, in an axially-loaded wall, creates a condition of eccentricity and thus a further reduction in ultimate load-bearing capacity due to greater instability.

7.04 As the result of a large number of tests which have been carried out on loadbearing walls, the Building Regula-tions give good guidance in 'deemed to satisfy' form. However the latest information will be published in the next revision of CP 121[2].

References

1 BS CP 476 Part 8: 1972   Test methods and criteria for the fire resistance of elements of building construction £5.60

2 BS CP 121: Part 1: 1973 Brick and block masonry (CP 121.201 Masonry walls: CP 121.202 Masonry: rubble walls)

Information sheet
Safety 1

Section 3 **Structural safety**

Fire resistance of concrete structures

This sheet consists of tables of 'deemed to satisfy' conditions for concrete walls, beams and floors

A Reinforced concrete walls

When using lightweight concrete, a reduction of thickness is possible but this must be confirmed by a test. Concrete cover to the reinforcement should not be less than 15 mm for fire resistance up to 1 h and not less than 25 mm for higher periods. Walls containing less than one per cent vertical reinforcement are considered as plain concrete for fire purposes unless a test shows otherwise.

Table I *Fire resistance of reinforced dense concrete walls exposed to fire on one side only*

Description of applied finish	Minimum thickness of concrete in mm to give fire resistance of					
	4	3	2	$1\frac{1}{2}$	1	$\frac{1}{2}$ hours
none	180	150	100	100	75	75
cement or gypsum plaster on one or both sides	180	150	100	100	75	75
vermiculite/gypsum plaster* at least 15 mm thick on both sides	125	100	75	75	65	65

*Vermiculite/gypsum plaster should have a mix ratio in the range from $1\frac{1}{2}$:1 to 2:1 by volume
Walls exposed to fire on more than one face should be regarded as columns.

B Plain concrete walls

From the limited data available, the fire resistance of plain concrete walls can be taken as:
1h for concrete 150 mm thick; $1\frac{1}{2}$h for concrete 175 mm thick

C Reinforced concrete beams

Table II *Fire resistance of reinforced concrete beams*

Description	Dimension of concrete in mm to give a fire resistance of					
	4	3	2	$1\frac{1}{2}$	1	$\frac{1}{2}$ hours
1 dense concrete:						
a concrete cover to main reinforcement	65*	55*	45*	35	25	15
b beam width	280	240	180	140	110	80
2 as 1 with cement or gypsum plaster 15 mm thick on light mesh reinforcement:						
a concrete cover to main reinforcement	50*	40	30	20	15	15
b beam width	250	210	170	110	85	70
3 as 1 with vermiculite/gypsum plaster or sprayed asbestos 15 mm thick:						
a concrete cover to main reinforcement	25	15	15	15	15	15
b beam width	170	145	125	85	60	60
4 lightweight aggregate concrete:						
a concrete cover to main reinforcement	50	45	35	30	20	15
b beam width	250	200	160	130	100	80

*Supplementary reinforcement, consisting of either a wire fabric not lighter than 0·5 kg/m² (2 mm diameter wires at not more than 100 mm centres) or a continuous arrangement of stirrups at not more than 200 mm centres, must be incorporated in the concrete cover at a distance not exceeding 20 mm from the face

Note
● Vermiculite/gypsum plaster should have a mix ratio in the range from $1\frac{1}{2}$:1 to 2:1 by volume. Sprayed asbestos should conform to BS 3590.

● The fire is assumed to attack the soffit and two sides of the beam.
● When the reinforcement is used in more than one layer, the value of the total protective concrete cover is the arithmetic mean of the nominal cover to the tensile reinforcement in each layer. The value of the minimum cover to any bar should not be less than half the value shown in table II for different periods of fire resistance and never less than the value shown under the $\frac{1}{2}$-hour period.
● Alternatively, the average concrete cover may be determined by summing the product of the cross-sectional area of each bar or tendon and the distance from the nearest exposed face, and dividing it by the total area of the steel provided to resist tensile stresses induced by the imposed loads:

$$\text{Average concrete cover} = \frac{As_1c_1 + As_2c_2 \ldots As_nc_n}{As_1 + \ldots As_n}$$

where As_1 = cross-sectional area of the steel bar or tendon
and c_1 = its distance from the nearest exposed face

D Reinforced concrete columns

Table III *Columns with all faces exposed*

Type of construction	Dimension of concrete in mm to give fire resistance of					
	4	3	2	$1\frac{1}{2}$	1	$\frac{1}{2}$ hours
1 dense concrete:						
a without additional protection	450	400	300	250	200	150
b with cement or gypsum plaster 15 mm thick on light mesh reinforcement	300	275	225	150	150	150
c with vermiculite/gypsum plaster* 15 mm thick	275	225	200	150	120	120
d with supplementary reinforcement in concrete cover or limestone aggregate concrete	300	275	225	2C0	200	150
2 lightweight aggregate concrete	300	275	225	200	150	150

*Vermiculite/gypsum plaster should have a mix ratio in the range from $1\frac{1}{2}$:1 to 2:1 by volume. Sprayed asbestos should conform to BS 3590

Note
● The minimum dimension of a column is a determining factor in the fire resistance it can provide. The dimensions given in the table relate to columns which may be exposed to fire on all faces when subjected to characteristic loads. The use of limestone or other calcareous aggregates or the use of supplementary reinforcement in the concrete cover would reduce spalling and allow a reduction in the size of the section. The supplementary reinforcement should consist of steel fabric of not less than 2 mm diameter wire and of mesh not greater than 150 mm, or an equivalent material, and should be placed at mid-cover not more than 20 mm from the face. The concrete cover to the main bars should not exceed 40 mm without the use of supplementary reinforcement. The data in table III are based on a rectangular or a circular cage reinforcement.

Table IV *Columns with only one face exposed*

Type of construction	Dimension of concrete in mm to give a fire resistance of					
	4	3	2	$1\frac{1}{2}$	1	$\frac{1}{2}$ hours
dense concrete:						
a without additional protection	180	150	100	100	75	75
b with 15 mm vermiculite/gypsum plaster* on exposed face	125	100	75	75	65	65

*As footnote to table III

Note
● Columns with their full height built into fire resisting walls may be exposed to fire on one face only. Data in table IV apply when the face of the column is flush with the wall or when that part of the column embedded in the wall is structurally adequate to support the load, provided that any opening in the wall is not nearer to the column than the minimum dimension for the column specified in table IV.

E Reinforced concrete floors

Table V is not exhaustive and the performance of types not shown can either be assessed by analogy or determined by testing

Table v Fire resistance of reinforced concrete floors

Floor type		Dimension of concrete in.mm to give fire resistance of					
		4	3	2	1½	1	½ hours
1 solid slabs	cover to reinforcement	25	25	20	20	15	15
	overall depth*	150	150	125	125	100	100
2 cored slabs, area of cores less than 50 per cent of solid area, cores higher than width	cover to reinforcement	25	25	20	20	15	15
	thickness under cores	50	40	40	30	25	20
	overall depth*	190	175	160	140	110	100
3 hollow box sections with one or more longitudinal cavities	cover to reinforcement	25	25	20	20	15	15
	thickness of bottom flange	50	40	40	30	25	20
	overall depth*	230	205	180	155	130	105
4 inverted T-section beams with hollow infill blocks of concrete or clay having not less than 50 per cent of solid material	cover to reinforcement	25	25	20	20	15	15
	width of T-flange	125	100	90	80	70	50
	overall depth*	190	175	160	140	110	100
5 ribbed floor with hollow infill blocks of clay having less than 50 per cent of solid material and with a 15 mm plaster coating on soffit	cover to reinforcement	25	25	20	20	15	15
	width of T-flange	125	100	90	80	70	50
	overall depth*	190	175	160	140	110	100
6 upright T-sections	bottom cover to reinforcement	65**	55**	45**	35	25	15
	side cover to reinforcement	65	55	45	35	25	15
	width of web	150	140	115	90	75	60
	depth of flange	150	150	125	125	100	90
7 inverted channel sections, radius at intersection of soffits with top of leg not exceeding depth of section	bottom cover to reinforcement	65**	55**	45**	35	25	15
	side cover to reinforcement	40	30	25	20	15	10
	width of web	75	70	60	45	40	30
	depth or thickness at crown*	150	150	125	125	100	90
8 inverted channel sections or U-sections, radius at intersection of soffits, top of leg exceeding depth of section	bottom cover to reinforcement	65**	55**	45**	35	25	15
	side cover to reinforcement	40	30	25	20	15	10
	width of web*	70	60	50	40	35	25
	depth or thickness at crown*	150	150	100	100	75	65

* Non-combustible screeds and finishes may be included in these dimensions
** Additional reinforcement is necessary to hold the concrete cover in position

Note
● In estimating the thickness of concrete, non-combustible screeds or finishes can be taken into account. The effect of the ceiling finish is shown in table VI.

Table vi Effect of ceiling finish on fire resistance of structural suspended floors

Ceiling finish	Thickness of finish in mm to give an increase in fire resistance of				
	3	2	1½	1	½ hours
1 vermiculite/gypsum plaster* or sprayed asbestos applied to the soffit of floor types 1, 2 or 3	25	15	15	10	10
2 vermiculite/gypsum plaster* or sprayed asbestos on expanded metal as a suspended ceiling to floor types 4 or 5	15	10	10	10	10
3 gypsum/sand or cement/sand on expanded metal as a suspended ceiling to any floor type	25	20	15	10	10

*Vermiculite/gypsum plaster should have a mix ratio in the range of 1½:1 to 2:1 by volume. Sprayed asbestos should conform to BS 3590

F Prestressed concrete beams

Table vii Fire resistance of prestressed concrete beams

Description	Dimension of concrete in mm to give a fire resistance of					
	4	3	2	1½	1	½ hours
1 dense concrete: a concrete cover to main reinforcement	100*	85*	65*	50*	40	25
b beam width	280	240	180	140	110	80
2 as 1 with vermiculite concrete slabs, 15 mm thick, used as permanent shuttering: a concrete cover to main reinforcement	75*	60	45	35	25	15
b beam width	210	170	125	100	70	70
3 as 2 with 25 mm thick slabs: a concrete cover to main reinforcement	65	50	35	25	15	15
b beam width	180	140	100	70	60	60
4 as 1 with 15 mm thick gypsum plaster with light mesh reinforcement: a concrete cover to main reinforcement	90*	75	50	40	30	15
b beam width	250	210	170	110	85	70
5 as 1 with vermiculite/gypsum plaster, or sprayed asbestos* 15 mm thick: a concrete cover to main reinforcement	75*	60	45	30	25	15
b beam width	170	145	125	85	60	60

Description	Dimension of concrete in mm to give a fire resistance of					
	4	**3**	**2**	**1½**	**1**	**½ hours**
6 as 5 with 25 mm thick coating:						
a concrete cover to main reinforcement	50	45	30	25	15	15
b beam width	140	125	85	70	60	60
7 lightweight aggregate concrete:						
a concrete cover to main reinforcement	80	65	50	40	30	20
b beam width	250	200	160	130	100	80

*Supplementary reinforcement consisting of either a wire fabric not lighter than 0·5 kg/m² (2 mm diameter wires at not more than 100 mm centres) or a continuous arrangement of stirrups at not more than 200 mm centres must be incorporated in the concrete cover at a distance not exceeding 20 mm from the face

Notes
● Vermiculite/gypsum plaster should have a mix ratio in the range from 1½:1 to 2:1 by volume.
● Sprayed asbestos should conform to BS 3590.
● The protective cover to the tendon for different periods of fire resistance should not be less than in table VII; in no case should the minimum cover to any tendon be less than half the value shown in the table for different periods of fire resistance: it should never be less than the value shown under the ½ hour period.
● When the prestressing tendons to resist tensile stresses due to imposed or working loads are provided in a number of layers, the value of the protective concrete cover is the arithmetic mean of the nominal cover for each layer.
● I-section beams with web thickness of half or less the lower flange breadth require web stirrups amounting to 0·15 per cent of the web area on plan.

G Prestressed concrete floors

Table VIII *Fire resistance of prestressed concrete floors*

Floor types—illustrated as Table V		Dimension of concrete in mm to give fire resistance of					
		4	**3**	**2**	**1½**	**1**	**½ hours**
1 solid slabs	cover to reinforcement	65**	50**	40	30	25	15
	overall depth*	150	150	125	125	100	90
2 cored slabs, area of cores less than 50 per cent	cover to reinforcement	65**	50**	40	30	25	15
of solid area, cores higher than width	thickness under cores	50	40	40	30	25	20
	overall depth*	190	175	160	140	110	100
3 hollow box sections with one or more	cover to reinforcement	65**	50**	40	30	25	15
longitudinal cavities	thickness of bottom flange	65	50	40	30	25	25
	overall depth*	230	205	180	155	130	105
4 inverted T-section beam with hollow infill blocks	cover to reinforcement	65**	50**	40	30	25	15
of concrete or clay having not less than	width of T-flange	125	100	90	80	70	50
50 per cent solid material	overall depth*	190	175	160	140	110	100
5 upright T-sections	bottom cover to reinforcement	100**	85**	65**	50**	40	25
	side cover to reinforcement	100	85	65	50	40	25
	width of web	250	200	150	110	90	60
	depth of T-flange*	150	150	125	125	100	90
6 inverted channel sections, radius at	bottom cover to reinforcement	100**	85**	65**	50**	40	25
intersection of soffits, top of leg not	side cover to reinforcement	50	45	35	25	20	15
exceeding depth of section	width of leg	125	100	75	55	45	30
	depth or thickness at crown*	150	150	125	125	100	90
7 inverted channel or U-sections, radius at	bottom cover to reinforcement	100**	85**	65**	50**	40	25
intersection of soffits, top of leg	side cover to reinforcement	50	45	35	25	20	15
exceeding depth of section	width of leg*	110	90	70	50	45	30
	depth or thickness at crown*	150	150	125	125	100	90

*Non-combustible screeds and finishes may be included in these dimensions
**Supplementary reinforcement is necessary as in table VII or consisting of equivalent expanded metal lath

Notes
● The average cover at a section is assessed as the arithmetic mean of the nominal cover of each equal tendon of prestressing steel in the member below the neutral axis; but the minimum cover to any tendon should not be less than half the value shown under different periods of fire resistance and in no case should it be less than the value shown under the period of ½ hour.
● If the thickness of concrete cover for floor types 4 and 5 exceeds 40 mm, mesh reinforcement must be incorporated in the cover to retain the concrete in position. This is not necessary when ceiling protection of the types shown in section E of this sheet (reinforced concrete floors) is used.
● Similarly the fire resistance of a given form of construction can be improved by using an insulating finish on the soffit or by a suitable suspended ceiling—as described for reinforced concrete floors.

Information sheet
Safety 2

Fire resistance of steel structures

This sheet consists of two tables of 'deemed to satisfy' conditions for structural steel columns and beams

A Fire resistance of protected steel columns

Table 1 Fire resistance of protected steel columns; stanchion weight per metre not less than 44·6 kg

Construction and materials			Minimum thickness (in mm) of protection for a fire resistance of				
			4	2	1½	1	½ hours
Solid protection (unplastered) The casing is bedded close to the steel without intervening cavities and all joints in the casing made full and solid		Concrete not leaner than 1:2:4 mix with natural aggregates: a Concrete assumed not loadbearing reinforced*	50·8	25·4	25·4	25·4	25·4
		b Concrete assumed to be loadbearing reinforced in accordance with BS 449	76·2	50·8	50·8	50·8	50·8
		Solid bricks of clay, composition sand or sand lime	76·2	50·8	50·8	50·8	50·8
		Solid blocks of foamed slag or pumiced concrete reinforced* in every horizontal joint	63·5	50·8	50·8	50·8	50·8
		Sprayed asbestos 144 to 240 kg/m³	44·5	19·1	15·9	9·5	9·5
		Sprayed vermiculite cement	—	38·1	31·8	19·1	12·7
Hollow protection There is a void between the protective material and the steel. A hollow protection must be effectively sealed at each floor level		Solid bricks of clay, composition or sand lime reinforced in every horizontal joint unplastered	114·3	50·8	50·8	50·8	50·8
		Solid blocks of foamed slag or pumice concrete reinforced** in every horizontal joint unplastered	76·2	50·8	50·8	50·8	50·8
		Metal lath with gypsum or cement lime plaster of thickness	—	38·1	25·4	19·1	12·7
		Metal lath with vermiculite or perlite gypsum plaster of thickness	50·8	19·1	15·9	12·7	12·7
		Metal lath spaced 25 mm from flanges with vermiculite gypsum or perlite gypsum plaster of thickness	44·5	19·1	12·7	12·7	12·7
		Gypsum plasterboard with 16 swg binding at 100 mm pitch: a 9·5 mm plasterboard with gypsum plaster of thickness	—	—	—	12·7	12·7
		b 19 mm plasterboard with gypsum plaster of thickness:	—	12·7	9·5	6·4	6·4
		Plasterboard with 16 swg binding at 100 mm pitch a 9·5 mm plasterboard with vermiculite gypsum plaster of thickness	—	15·9	12·7	9·5	6·4
		b 19 mm plasterboard with vermiculite gypsum plaster of thickness	31·8**	9·5	9·5	6·4	6·4
		Metal lath with sprayed asbestos of thickness	44·5	19·1	15·9	9·5	9·5

Table I *continued*

Construction and materials		Minimum thickness (in mm) of protection for a fire resistance of				
		4	2	1½	1	½ hours
	Vermiculite cement slabs of 4:1 mix reinforced with wire mesh finished with plaster skim Slabs of thickness	63·5	25·4	25·4	25·4	25·4
	Asbestos insulating boards of density 513 to 881 kg/m³ (Screwed to 25 mm asbestos battens for ½ hour and 1 hour periods)	—	25·4	19·1	12·7	9·5

*Reinforcement of steel binding wire 13 swg or a steel mesh weighing not less than 0·542 kg/m². Minimum spacing in concrete not less than 150 mm
**Light mesh reinforcement required 12 to 19 mm below surface unless special corner beads are used

B Fire resistance of protected steel beams

Table II *Fire resistance of protected steel beams; joist weight per metre not less than 30 kg*

Construction and materials			Minimum thickness (in mm) of protection for a fire resistance of				
			4	2	1½	1	½ hours
Solid protection (unplastered) The casing is bedded close to the steel without intervening cavities and all joints with casing made full and solid		Concrete not leaner than 1:2:4 mix with natural aggregate: a Concrete not assumed to be loadbearing reinforced* b Concrete assumed to be loadbearing reinforced in accordance with BS 449	63·5 76·2	25·6 50·8	25·4 50·8	25·4 50·8	25·4 50·8
		Sprayed asbestos 144 to 240 kg/m² Sprayed vermiculite cement	44·5 —	19·1 38·1	15·9 31·8	9·5 19·1	9·5 12·7
Hollow protection There is a void between the protective material and the steel. All hollow protection must be effectively sealed at each floor level		Metal lathing: a with cement lime plaster of thickness b with gypsum plaster of thickness c with vermiculite gypsum or perlite gypsum plaster of thickness	— — 31·8	38·1 22·2 12·7	25·4 19·1 12·7	19·1 15·9 12·7	12·7 12·7 12·7
		Gypsum plasterboard with 16 swg wire binding at 100 mm pitch: a 9·5 mm plasterboard with gypsum plaster of thickness b 19 mm plasterboard with gypsum plaster of thickness Plasterboard with 16 swg wire binding at 100 mm pitch: a 9·5 mm plasterboard nailed to wooden saddles finished with gypsum plaster of thickness b 9·5 mm plasterboard with vermiculite gypsum plaster of thickness c 19 mm plasterboard with vermiculite gypsum plaster of thickness d 19 mm plasterboard with gypsum plaster of thickness	— — — — 38·1 —	— 12·7 — 15·9 9·5 12·7	— 9·5 — 12·7 9·5 —	12·7 6·4 — 9·5 6·4 —	12·7 6·4 4·8 6·4 6·4 —
		Metal lathing with sprayed asbestos: 144 to 240 kg/m³ of thickness	44·5	19·1	9·5	9·5	9·5

Construction and materials		**Minimum thickness (in mm) of protection for a fire resistance of**				
		4	2	1½	1	½ hours
floor — asbestos insulating board screw to battens — asbestos battens	Asbestos insulating boards of density 513 to 881 kg/m³ (screwed to 25 mm asbestos battens for ½ hour and 1 hour periods)	—	25·4	19·1	12·7	9·5
floor — slabs reinforced with wire mesh — vermiculite cement slabs — plaster skim coat	Vermiculite cement slabs of 4:1 rise reinforced with wire mesh and finished with plaster skin slabs of thickness	63·5	25·4	25·4	25·4	25·4
floor — woodwool slabs — gypsum sand plaster	Gypsum sand plaster 12 mm thick applied to heavy-duty (type B) woodwool slabs of thickness	—	50·8	38·1	38·1	38·1

*Reinforcement of steel binding wire 13 swg or a steel mesh weighing not less than 0·542 kg/m². Minimum spacing in concrete not less than 150 mm

Information sheet Safety 3

Section 3 **Structural safety**

Fire resistance of timber structures

This sheet consists of two tables of 'deemed to satisfy' conditions for timber floors and structural stud partitions

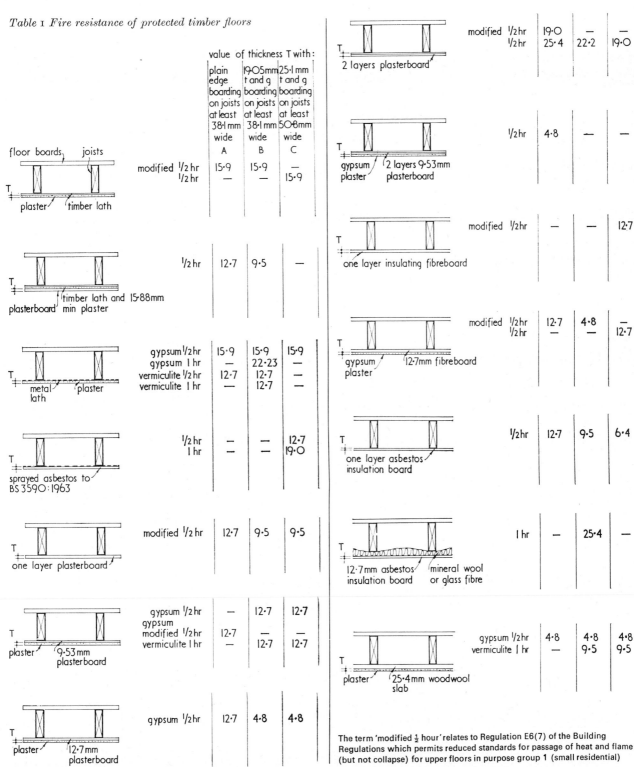
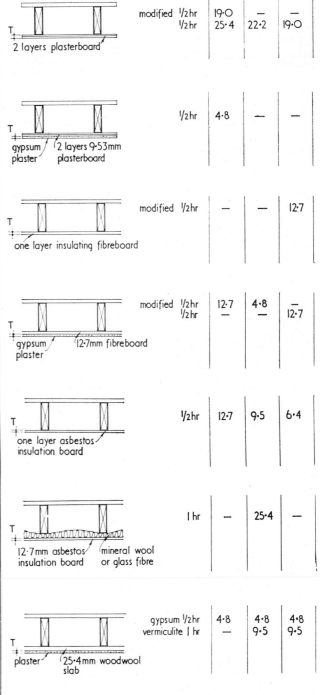

Table 1 Fire resistance of protected timber floors

construction	fire resistance	value of thickness T with:		
		plain edge boarding on joists at least 38·1mm wide **A**	19·05mm t and g boarding on joists at least 38·1mm wide **B**	25·1mm t and g boarding on joists at least 50·8mm wide **C**
floor boards, joists / plaster / timber lath	modified ½ hr	15·9	15·9	—
	½ hr	—	—	15·9
plasterboard / timber lath and 15·88mm min plaster	½ hr	12·7	9·5	—
metal lath / plaster	gypsum ½ hr	15·9	15·9	15·9
	gypsum 1 hr	—	22·23	—
	vermiculite ½ hr	12·7	12·7	—
	vermiculite 1 hr	—	12·7	—
sprayed asbestos to BS 3590:1963	½ hr	—	—	12·7
	1 hr	—	—	19·0
one layer plasterboard	modified ½ hr	12·7	9·5	9·5
plaster / 9·53mm plasterboard	gypsum ½ hr	—	12·7	12·7
	gypsum modified ½ hr	12·7	—	—
	vermiculite 1 hr	—	12·7	12·7
plaster / 12·7mm plasterboard	gypsum ½ hr	12·7	4·8	4·8
2 layers plasterboard	modified ½ hr	19·0	—	—
	½ hr	25·4	22·2	19·0
gypsum plaster / 2 layers 9·53mm plasterboard	½ hr	4·8	—	—
one layer insulating fibreboard	modified ½ hr	—	—	12·7
gypsum plaster / 12·7mm fibreboard	modified ½ hr	12·7	4·8	—
	½ hr	—	—	12·7
one layer asbestos insulation board	½ hr	12·7	9·5	6·4
12·7mm asbestos insulation board / mineral wool or glass fibre	1 hr	—	25·4	—
plaster / 25·4mm woodwool slab	gypsum ½ hr	4·8	4·8	4·8
	vermiculite 1 hr	—	9·5	9·5

The term 'modified ½ hour' relates to Regulation E6(7) of the Building Regulations which permits reduced standards for passage of heat and flame (but not collapse) for upper floors in purpose group **1** (small residential)

Table II Fire resistance of structural timber stud partitions

	Fire resistance (hours)	Notes
57×74 mm 12·7mm plasterboard — joint at post, taped and filled	½	Plasterboard must be either fixed with long dimension horizontal, or stud size raised to 44 mm × 74 mm
12·7mm asbestos insulation board — 9mm asbestos insulation board fillets	½	
two layers, 25mm nominal, mineral wool	½	
74mm 3 layers of 25mm, nominal, mineral wool — 12·7mm plywood, chipboard or medium density hardboard	½	Only tested for medium hardboard. Other linings are assumed satisfactory
almost any kind of external sheathing or cladding, min 44mm — 12·7mm plasterboard 25mm, nominal, mineral wool (not essential)	½	Acceptable for external stud walling
not less than 72×97mm — 2×12·7mm plasterboard, staggered joints joint taped and filled	1	Not suitable for full design load at the given dimensions

Information sheet
Safety 4

Section 3: **Structural safety**

Fire resistance of masonry walls

This sheet consists of two tables of 'deemed to satisfy' conditions for brick and block walls

A Single leaf walls

Table 1 Fire resistance of single leaf masonry walls

Material[1]	Unit	Type	Finish[2]	Minimum thickness, without finish (mm), for notional fire resistance of						
				6	4	3	2	1½	1	½ hours
Fired brickearth, clay or shale	Brick	Solid[3]	None	200	170	170	100	100	90	90**
			vg	170	100	100	90	90	90	90**
		Not less than 75% solid	None and/or sc, sg	—	200	200	170	170	170	100
			vg	200	170	170	170	100	100	90
		Not less than 50% solid	sc, sg	—	—	—	215	215	215	215
			vg	—	215	215	215	215	215	215
		Not less than 40% solid	None and/or sc, sg	—	—	—	—	—	215	215
	Block (outer-web not less than 13 mm thick)	Two cells* not less than 50% solid	sc, sg	—	—	—	100	100	100	100
		Three cells not less than 60% solid	sc, sg	—	150	150	150	150	150	150
Concrete or calcium silicate	Brick	Solid[3]	None	200	190	190	100	100	90	90
			vg	200	100	100	90	90	90	90
Concrete, Class 1 aggregate[1]	Block	Solid	None	150	150	140	100	100	90	90
			vg	150	100	100	90	90	90	90
		Hollow	None	—	—	—	100	100	100	90
Concrete, Class 2 aggregate[1]	Block	Solid	None and/or sc, sg	—	—	—	100	100	90	90
			vg	—	100	100	90	90	90	90
		Hollow, not less than 50% solid	sc, sg	—	—	—	—	—	—	190
			vg	—	—	—	200	200	190	190
Aerated concrete density 480 to 1200 kg/m³	Block	Solid	None	215	180	140	100	100	90	90
			vg	180	150	100	100	90	90	90

* the number of cells in any cross-section through the wall thickness
** suitable for 75 mm brick-on-edge construction with a completely solid unit with plane faces

Note
1 Class I aggregates for concrete blocks can be limestone, aircooled blast furnace slag, foamed or expanded slag, crushed brick, well burnt clinker, expanded clay or shale, sintered pelleted flyash, pumice.
Class II aggregates for concrete blocks include all gravels and crushed natural stone except limestone.
2 The finish shall be not less than 13 mm plaster or rendering on each face of a single skin wall and on the exposed face of a cavity wall.
sc = sand cement plaster with or without lime
sg = sand gypsum plaster with or without lime
sc/sg may be replaced by plaster board of equivalent thickness for fire resistance up to two hours.
vg = vermiculite gypsum plaster in proportions 1½ : 1 or 2 :1 by volume. (Purlite may be substituted in fired clay brickwork or other materials with similar surfaces).
3 Solid brickwork is of bricks without frogs or frogs up to 20 per cent of the brick volume with no through holes or perforations. (This definition differs from BS 3921).

B Cavity walls

Table II *Fire resistance of cavity walls*

Material[1]	Unit	Type	Finish[2]	Minimum thickness, without finish (mm), for notional fire resistance of						
				6	4	3	2	1½	1	½ hours
Fired brickearth, clay or shale concrete or calcium silicate	Brick	Solid[3]	None	100	100	100	100*	100*	90	90
Fired brickearth, clay or shale	Block (outer-web not less than 13 mm thick)	Not less than 50% solid	sc/sg	—	—	—	100	100	100	100
		not less than 70% solid	sc/sg	—	150	150	100	100	100	100
Concrete Class 1 aggregate[1]	Block	Solid	None	100	100	100	100*	100*	90	90
		Hollow	None	—	100	100	100	100	100	90
Concrete Class 2 aggregate[1]	Block	Solid	None	—	—	—	100	100	90	90
Aerated concrete (density 480 to 1200 kg/m³)	Block	Solid	None	150	150	140	100	100	90	90

*may be reduced to 90 mm if load distributed over both leaves 1, 2 and 3 see table I.
numbered footnotes as table 1

AJ Handbook
Building structure
Section 4
Foundations
and retaining
structures

CI/SfB (2-)

Section 4

Foundations and retaining structures

Scope

This section sets out to help an architect understand enough of the principles and methods of foundation engineering to let him both carry out simple designs and appreciate the more complex design problems faced by specialists. For safe and economic design, an engineer is almost always essential, and—as mentioned earlier in the handbook—the same engineer should be responsible for superstructure.

Most of the principles covered apply equally to foundations in general civil engineering, but only building structures are covered in detail. The content falls into two sections: foundations and retaining structures, although there is inevitable overlap between them.

Foundations transmit the loading of a structure to the ground, and their design must depend critically on the underlying soil. This section therefore begins with a technical study on soils engineering (or soil mechanics) which falls into two parts. TS 1a discusses the theoretical behaviour of soils and basic soil types; TS 1b, discusses actual soils and their behaviour in practice. This is followed by a technical study on foundation types and design and information sheets summarising design requirements and information. A technical study on retaining structures discusses the specific soils engineering aspects, and then considers retaining walls, diaphragm walls, cofferdams, etc, individually. Relevant codes, BSS and Regulations with a bibliography, are listed in an appendix.

Author

John Ransford BSc(Eng), FICE, MBCS, a specialist in foundation problems and computer application to civil engineering, has written this section of the handbook. After 10 years as partner in consulting engineering practices, he is now with DOE. The views expressed here are his own and not necessarily those of DOE.

The illustration on the previous page shows the campanile of Pisa Cathedral with a plumb line hung from it; it moved as a result of inadequate soil investigation and foundation design. The engraving is in Josephus Martini's 'Theatrum Basilicae Pisanae' (1728) (from the copy in RIBA library)

Preliminary definitions

Soil in broad terms for foundation engineering is the ground supporting a structure. It may include both rock and material deposited by man, but never the soft 'top soil' (including vegetable matter) found on the surface.

Soils engineering (or soil mechanics) is a science dealing with the behaviour of soil (from the engineering aspect) when subjected to external influences and changes of conditions, eg effects of topography, climate, movement of water, excavation, removal into a new position, treatment in various ways and especially changes in stress. It includes the study of forces exerted by soil on structures with which it is in contact, and also covers ways of obtaining site data by means of a soil investigation.

Rock mechanics is a special branch of soil mechanics dealing with the strongly cemented deposits known as rock.

Foundations are those parts of a structure which directly transmit its loads to the soil. Sometimes the part of a structure which is below the ground surface is called the 'substructure'. This may form an integral part of the foundations (eg a rigid basement) inasmuch as it provides the means of distributing the loads adequately over the parts actually in contact with the soil. From the engineering point of view, a ground bearing slab is a foundation, albeit usually a very lightly loaded one.

Stratum is a layer of soil having a particular set of properties. Geologically it can also refer to a particular layer of deposition by natural causes. (A *fault* is a discontinuity of strata resulting from cracking of the earth's crust due to strain.)

Compaction is compression of soil induced mechanically, eg by rolling or ramming.

Consolidation is compression of soil due to self-weight or applied load.

Settlement is downward movement of soil caused by loading, natural consolidation, decrease in moisture or soil flow.

Swelling (or heave) is movement of fine-grained soils due to decrease of pressure below that pre-existing, or influx of moisture. *Heave* also describes the upward movement due to bodily flow of any soil, and to frost action.

Water table is level at which pore water pressure in soil is zero.

Frost boil is release of large quantities of water when heaved soil thaws, and then softens.

Technical study
Foundations 1

Section 4 **Foundations**

Soils engineering part 1

The first part of this technical study by JOHN RANSFORD *considers basic soil types, their structure and characteristic strengths and deformations. Ground stress and settlement calculations are also covered. Part 2 will consider behaviour of soils in practice*

1 Introduction to soil

1.01 A structure, its foundations and supporting soil, form an interconnected whole, the parts of which tend to interact on one another in a complex way. This whole deforms under stress, while changing conditions may cause other movements. Thus a knowledge of soil behaviour is essential for efficient design of foundations.

1.02 A full treatment of soil behaviour is beyond the scope of this handbook, but the engineering properties of commoner UK soils and their behaviour in practical circumstances will be discussed (in part 2 of this study). Here the effect of particle size and water content on soil properties strength and deformation or movement are considered.

1.03 Soils are aggregates of solid particles, the voids between the particles being filled with air or water. They have been deposited either by nature or man and are comparatively soft, loose, uncemented deposits. Rocks are hard, rigid, strongly cemented substances.

1.04 Soils (excluding solid rock) may conveniently be classified from large to small by particle size—stone (boulders and cobbles), gravel, sand, silt and clay, as **1**. Gravels, sands and silts may be further classified as coarse, medium or fine. Organic soil (peat) has fine particles in a fibrous structure. Fill or 'made ground' (artificial deposits) may include any of the soils classes and man-made material. Natural deposits may be mixtures of two or more classes of soil.

1.05 Unless it is strongly cemented, the engineering properties of a soil depend on particle size, water content, method of deposition and geological history, and to a certain extent on the chemical composition of individual particles and any cementation between them. Similarly the behaviour of a ground mass of various soils which acts as a single support for a foundations system depends not only on the properties of the soil layers but also on their interrelationship, the occurrence of special geological features and any external influences.

2 Normal soil structure and its effects

2.01 The common range of soils, from coarse gravel to clay, have varying sized voids between the solid particles. The nature of the voids (which depends on the size, shape, grading, and degree of packing of the particles) largely determines the amount of water retained and the strength and movement behaviour of the soil.

2.02 Coarser grained soils are more permeable to water and, unless saturated because of site conditions, may have very little water in their voids. They are held together by weight and friction between particles. Their strength is governed by the closeness of packing, or *relative density*. Movement of the soil particles (eg under the action of external forces) can take place quickly.

2.03 Clay soils, with very small voids, are comparatively impermeable and tend to retain water. Clays are usually fully saturated and are held together by *cohesion*, due to the water film between the particles. Their strength depends

1 *Typical particle size distribution curves*

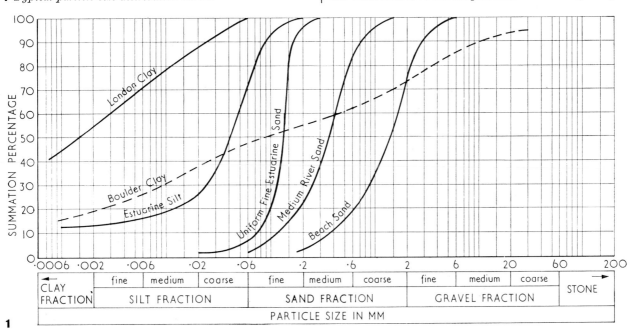

1

largely on the thickness of this water film—the thicker it is the weaker the clay. Any movement of the particles due to change in the volume of voids takes place very slowly, because migration of water through the voids is a slow process. But the behaviour of a clay mass is greatly influenced by the occurrence of fissures or other features which affect its drainage and strength properties. The nature and behaviour of silts is somewhere between these two extremes.

3 Strength and deformation of soils in general

Strength

3.01 The shear strength of a soil governs its ability to support loads without failure. *Failure* is said to occur when indefinitely large movements develop. There may also be limited movements which are too large to be acceptable in practice, but these do not strictly constitute failure. The typical mode of failure of the soil under a loaded foundation block is shown in **2**. When the shear strength is exceeded by the induced stress, there is plastic flow. Theoretical analysis can be used to derive expressions for the ultimate load capacity of different foundations in terms of their size, depth below ground, and the soil properties. Practical experience and experimental testing has led to a development of these formulae. (Some of the more important results are quoted in Information sheets FOUNDATIONS 3 and 7.) The load capacity of a deep foundation may be substantially increased by friction on its sides. This may include adhesion of the foundation to a cohesive soil.

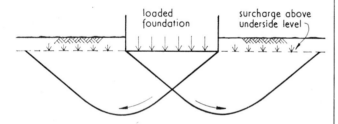

2 *Shear failure of soil under shallow foundation. (For deep foundation, shear resistance of soil higher than foundation base becomes significant)*

3.02 Loading or other in situ tests—especially for piled foundations and in mixed strata—supplement or replace theoretical calculations. The shear strength of normal soils is fundamentally governed by the *angle of internal friction* (usually denoted by ϕ) which depends largely on the degree of packing in coarser grained soils, and on the cohesion (cohesive strength per unit area in shear, denoted by c). Most uncemented coarse grained soils have negligible cohesion, and clays are usually assumed to behave as if $\phi = 0$. Intermediate types like silts are known as 'c − ϕ soils', since they possess both properties. These properties are related by:

$$S = c + q_e \, \text{Tan} \, \phi \qquad \textit{equation 1 (Coulomb's law)}$$

where 'S' is the shear strength and q_e is the effective normal pressure across the shear plane. Water pressure in the soil may reduce the value of q_e (see para 4.02). The implications of this equation in relation to the soil properties measured are considered further in part 2. But even if a soil can support a given loading without failure, its subsequent movement is of interest to the foundation engineer because of the effect on any supported structure.

Deformation

3.03 All soils deform under the action of an applied load, or because of other factors which produce a change in void content. A general flow of soil particles may also occur. The various types of deformation may be classified:

a *Elastic deformation* This is a quick response to changes of stress. It is caused by deformation of the soil structure, and the soil tends to bulge in a direction perpendicular to the applied compressive stress. When the pressure is reduced the soil moves in the opposite direction.

b *Plastic deformation* This can also be quick and is caused by flow of soil particles or by their re-orientation under pressure. It does not reduce appreciably when pressure is removed.

c *Compressive deformation* This describes the behaviour of soils due to 'compressibility', ie the effect of a reduction in the voids between soil particles, which entails expulsion of contained air or water. Air is expelled quickly and any movement resulting from the application of pressure usually follows immediately*. But water movement, as in clay soils, can be much slower and it may take many years before it is effectively completed under the action of a given load. The compression of a soil due to its own weight or an applied load from a structure or earth filling is called *consolidation*; compression induced by mechanical means, *compaction*, is normally only concerned with the expulsion of air. Movement due to a reduction in air voids is irreversible and does not disappear when the load is removed; but water tends to return to the pores of fine grained soils when pressure is reduced. Reducing pressure below the original figure, say by removing overburden by excavation, causes the soil to swell above its original volume.

3.04 *Settlement*, a downward soil movement caused by loading, natural consolidation or decrease in moisture content, may also be caused by a flow of soil away from a point. The upward movement of fine grained soils, due to a decrease of pressure or an influx of moisture for any other reason, is known as *swelling* or *heave*, but the latter term is more properly reserved for upward movement caused by a flow of soil towards a point. Swelling or heave of certain soils may also take place when the included water expands in freezing conditions. The reduction in soil volume caused by removal of water due to climatic conditions or vegetation is known as *shrinkage*.

4 Methods of calculating soil stresses and settlements due to loading

4.01 An important class of limited settlements, including 'negative settlements' or heave, are caused by changes in soil loading. Horizontal movements due to loading are usually unimportant. Other soil movements are hard to calculate; their estimation depends mainly on experience. Before dealing with calculation of settlement, some preliminary ideas which have a wider application in soils engineering theory are introduced.

Effect of pore water pressure

4.02 In general, if water in the voids or pores forms a continuous medium, this water supports both its own weight and any loads transmitted directly to it, and the intergranular pressure or effective pressure between soil

---

*In practice full consolidation of coarse grained soil mass may involve an extended period, because of the effects of slip, degeneration or deformation of the component parts, and external influences such as vibration, especially if voids are large

particles* supports all other loads. Thus if pore water pressure at a given point is u, and effective soil pressure q_e then:

$$q_e = q - u \qquad \textit{equation 2}$$

where q is the total pressure due to all loads above a point (including the weight of water). The pressure q_e used in calculating the true frictional shearing resistance of a soil (see para 3.02) is also effective in producing settlement. This is the average pressure on an area of soil after deducting the hydrostatic pressure between particles.

4.03 In order to calculate u the level of the *water table* must be known. Here ground water conditions are more or less stable, as opposed to temporary percolation of surface water. The water table may become apparent during site investigation as a free standing water level. With clay soils, however, free water will not appear in bore holes unless fissures or more permeable strata are encountered, but pore water pressure is still present and may be measured. When water is struck it may take some time to rise to equilibrium level. If γ_w = the density of water and h_w = the depth of a given point below the water table, then:

$$u = \gamma_w h_w \qquad \textit{equation 3}$$

in the comparatively static conditions under consideration. Negative values of u are possible, above the water table, if moisture is held by capilliary action in the pores of unsaturated fine grained soils. Such water does not contribute to the reduction of effective pressure, but its weight must be included in the total pressure q.

4.04 Application of load to saturated soil usually causes a temporary rise in pore water pressure, which does not revert to its equilibrium value as equation 3 until consolidation is complete. This may take a considerable time with clays.

4.05 If all or part of a structure or foundation is below the level of the water table, its effective weight on the supporting soil will be reduced by *hydrostatic uplift* or 'buoyancy', which eventually equals the weight of water displaced (though the weight of any water supported by the structure must be included in the total load). The entire weight of soil, water and structure is of course finally carried by a completely impermeable stratum at some depth.

4.06 In the case of *submerged soil*, the load from this carried by the underlying soil strata will likewise equal its total weight less the weight of water displaced. Effective soil pressure at depth h due to the overburden is given by:

$$p_0 = \gamma (h - h_w) + \gamma_{sub} h_w \qquad \textit{equation 4}$$
assuming:
$$\gamma_{sub} = \gamma_{sat} - \gamma_w \qquad \textit{equation 5}$$

where γ = bulk density of the soil above the water table (ie dry density plus weight of any moisture in a unit volume), γ_{sat} = fully saturated density of the soil below the water table, γ_{sub} = submerged density of same (ie effective density when calculating load on the underlying strata), γ_w = density of water, h = depth of the point in question below the ground surface, and h_w = depth of the point in question below the water table **3**. If soil conditions vary with depth, average densities are used.

4.07 Because of the smaller load on the underlying soil structure, reduction in effective pressure due to the presence

*Except in certain abnormal conditions eg when equation 3 does not apply due to great variations in ground permeability or movement of water

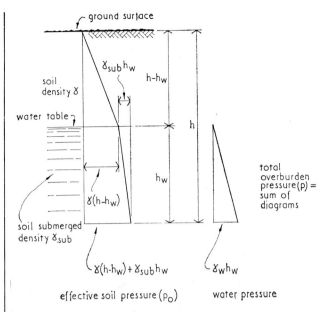

3 *Effect of ground water on overburden pressure*

of pore water can not only reduce the soil strength but also affect settlement. Any lowering of the water table can cause greater settlement because of the increase in effective loading; this is aggravated if there is also a loss in buoyancy of the building structure. At the other extreme it is necessary to check that if the water level rises, buoyancy does not increase to the point where the structure starts to float!

4.08 Overburden pressure can now be defined. *Total overburden pressure*, p, is the total pressure at any point in a horizontal plane due to the weight of the overlying soil, its pore water, and any works which are to be retained before building construction commences. *Effective overburden pressure*, p_0, is the inter-granular pressure corresponding to p. Clearly:

$$p_0 = p - u = p - \gamma_w h_w \qquad \textit{equation 6}$$

In a typical case p_0 is given by equation 4, which also leads to the result (equation 6). The term $\gamma_w h_w$ is omitted if above the water table.

4.09 *Total soil pressure*, q, has already been used when considering the value of q_e in any plane. It will now be taken as the total pressure on a horizontal plane at any point due not only to the overburden p but also to the completed building structure and all associated changes in loading (eg excavation, filling or pore water alterations). Obviously the corresponding *total effective pressure*, q_e, is given by:

$$q_e = q - \gamma_w h'_w \qquad \textit{equation 7}$$

where h'_w is the depth below final water table.

4.10 As a corollary, *increase in total pressure* due to building construction q_n (nett pressure) is:

$$q_n = q - p \qquad \textit{equation 8}$$

The corresponding *increase in effective pressure* (nett effective pressure), q_{ne}, is:

$$q_{ne} = q_e - p_0 \qquad \textit{equation 9}$$
$$= q - p - \gamma_w (h'_w - h_w) \text{ from 6 and 7,}$$

demonstrating the effect of ground water changes. (This = q - p = q_n if the water table does not alter.)

Distribution of ground stress

4.11 Various stresses are produced in a soil mass when a load is applied. These stresses decrease in magnitude the further away from the point of application of the load they are. They are of interest because of their effect on the ability of the soil to safely support a given loading and also because their calculation is a normal preliminary to the determination of settlements produced by loading.

4.12 When calculating the stresses due to applied loading, it is usual to start by evaluating the contact pressures between foundations and their immediate supporting soil. Because undisturbed ground is normally in a state of equilibrium, stresses existing before building begins are, on the whole, of secondary interest. Contact pressures (nett pressures defined by equation 8 above*) and the *changes* in stress which occur during and after construction are more important. Change in effective pressure (equation 9) is only relevant if the water table alters, then the term $\gamma_w\,(h^1_w - h_w)$ is allowed for by taking the *change* in loading as a uniform load. With piled foundations, and in a few other instances, calculations may be based on the total nett loads corresponding to q_n rather than on the contact stresses.

4.13 But further difficulties arise. First of all, the contact pressure under a foundation is rarely uniform. Stresses induced in a uniformly loaded mass of soil tend to be greater at the middle than the edges, and so the characteristic settlement profile is dish shaped. This means that even if the applied loading is distributed uniformly over a foundation, the base will 'dig in' at the edges unless the foundation is completely flexible. Also the edge pressure will be greater than that towards the middle of the base **4** except for foundations near the surface of non-cohesive soils. The final answer depends on the loading and relative stiffness of the foundation and the soil.

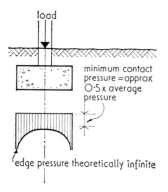

4 *Pressure distribution under a rigid square base on ideal elastic soil*

4.14 If the structural stiffness of a building in relation to the ground is appreciable, similar considerations apply to the distribution of soil reactions under it, even if it is not on a single (raft type) foundation. Unless the analysis of the superstructure allows for settlements on individual foundations, loads may not be those estimated.

4.15 These factors can complicate the design of rafts and certain building structures, and will be considered when discussing methods of computer analysis for complex foundations. Nevertheless, with simple building structures and comparatively small foundations it is sufficiently accurate to assess loads on a simple basis and to assume that pressures are uniformly distributed over individual contact areas.

4.16 But even with a series of pressures applied over various areas and at different levels, determination of realistic stresses at different points in the ground is a

problem. Stresses vary continuously from point to point and are influenced by particular properties and variations of the soil strata, except where pressure is uniform over an area extending far beyond the points in question.

4.17 Stress and strain in the ground are generally matters for the specialist engineer. But with straightforward cases and reasonably uniform ground, experience has shown that a fair idea of the stress distribution may be obtained by treating the soil as a uniform 'elastic' body. The formulae usually employed, those of Boussinesq, assume that the loading is applied at the surface and that the soil extends to an infinite depth. While this is accurate enough for most purposes, others have developed this theory to allow for loadings applied below the surface and other assumptions about the soil. The Boussinesq theory represents the behaviour of clays reasonably well, but sands and gravels tend to produce a lesser degree of lateral distribution of concentrated loads. Formulae and tables for calculating Boussinesq stress distributions due to point loads and uniformly loaded areas of different shapes may be found in standard text books such as Tomlinson[1]. These deal mostly with vertical stresses, but the distribution of horizontal and shear stresses can also be obtained.

4.18 Typical distributions of vertical stress under uniformly loaded flexible circular areas and continuous strips on the ground surface are shown in **5** and **6**. Rigid foundations tend to spread loads farther sideways. The plan continuity of strip footing produces greater stresses at a given depth than an isolated base. But unless individual foundations are sufficiently far apart, they cannot be treated in isolation because stresses produced by them will overlap significantly. The critical distance (in terms of foundation width) is greater for continuous than for isolated footings.

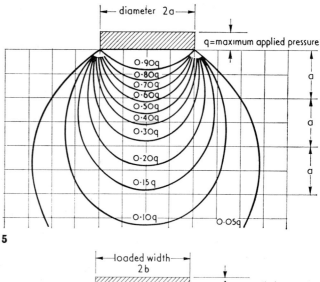

5

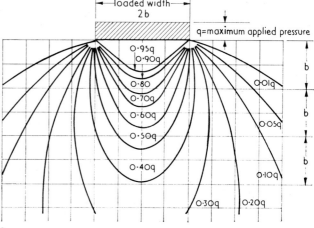

6
Pressure bulbs under **5** *uniform circular and* **6** *strip loads*

*Calculations of long-term consolidation and frictional resistance do also involve total effective pressures

4.19 Soil stress increases that are less than 20 per cent of that applied to the ground immediately under the foundations are not normally considered when calculating the depth of soil involved. Using the Boussinesq theory, it is possible to construct a pressure bulb which encloses all the soil in which the stresses are significant. For an isolated base this bulb usually has a depth 1·5 times the foundation width **7**; for a continuous footing the depth is at least twice as great as the foundation width.

4.20 A much larger bulb may represent the behaviour of a whole group of foundations with several bases close together **8**. This is particularly important when considering the behaviour of rafts and of other foundations with heavily loaded buildings, and when deciding on the depth of the bore holes for a soil investigation. Generally, if foundations are closer than two to three times their average width, the *total* width of the group should be used when determining the depth of soil to be considered. In the case of piled foundations, the widths of pile groups are considered, and the depths of the pressure bulbs are measured from the bottoms of the proposed piles.

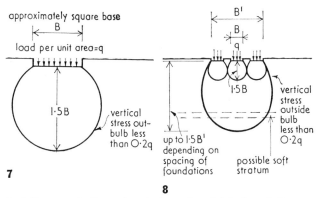

7, 8 *Pressure bulbs under single and multiple foundations*

4.21 A simplified approach to soil stress distribution may be adopted during preliminary calculations. This assumes that loads spread through the soil at an inclination of 1 in 3 to the vertical **9**. For a single square foundation, width of footing B and applied pressure per unit area q, the vertical pressure at depth z below the underside has the average value:

$$q \cdot \frac{B^2}{\left(B + \dfrac{2z}{3}\right)^2} = q \cdot \frac{1}{1 + \dfrac{4}{3}\left(\dfrac{z}{B}\right) + \dfrac{4}{9}\left(\dfrac{z}{B}\right)^2}$$

For a long continuous strip this becomes:

$$q \cdot \frac{B}{B + \dfrac{2z}{3}} = q \cdot \frac{1}{1 + \dfrac{2}{3} \cdot \dfrac{z}{B}}$$

In both cases these results are somewhat below Boussinesq maximum figures (on the centre-line of the foundation) for z/B ratios up to about 0·75. They become steadily greater than Boussinesq as z/B is increased above this value; at a ratio of 2 the overestimate is in the order of 50 per cent. This method is more accurate with non-cohesive soil.

Settlement calculations

4.22 An architect will rarely be required to carry out elaborate settlement calculations and only general principles will be considered here. Formulae and further references are given in Information sheet FOUNDATIONS 3.

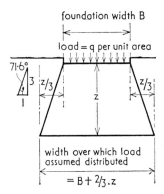

9 *Spread of foundation load for approximate calculations*

4.23 For cohesive soils immediate settlements due to applied loading are usually calculated using Boussinesq theory. A value for the modulus of elasticity E (Young's modulus), obtained either from direct testing or experience, is needed; also a value for Poisson's ratio (ratio of lateral to primary vertical strain) is needed and is usually taken to be 0·5. For reasonably uniform soil conditions, settlements are calculated direct, using foundation sizes and pressures in a standard formula. Where there is a limited compressible stratum, Steinbrenner's method[2] may be used. This method can also be used for the approximate settlements due to a number of strata with different properties, by taking the sums and differences of appropriate Steinbrenner cases for soil of various kinds. For deeper foundations settlement is reduced; the reduction may be estimated using factors of up to 50 per cent suggested by Fox[3]. A reduction may also be made to allow for the rigidity of large foundations.

4.24 If there are a number of separate foundations at different levels, the soil should be divided into suitable layers, and the change in stress at the centre of each layer beneath the point at which a settlement is required computed. This gives the actual settlement produced by each layer. This method is generally used with long-term consolidation, or non-cohesive soils.

4.25 An approximation for the consolidation of a clay layer is the product of the thickness of the layer, the nett stress increase q_n and a coefficient of volume decrease m_v, usually obtained from soil tests. This is then multiplied by a geological factor μ which varies from about 0·2 to 1·2 depending on the type of clay and its geological history. The ultimate consolidation obtained by this method requires an indefinite time for full realisation and so 90 per cent of this value is usually taken as practical completion. The time required for this or any other degree of consolidation to be achieved may be calculated from another measured coefficient, the coefficient of consolidation c_v, provided that the length and nature of the drain path for the clay pore water under load is known. This is often difficult to determine, especially if there are permeable layers and fissures in the clay or if concrete piles are used (concrete being relatively permeable to this sort of water movement). So it is frequently impossible to calculate the rate of consolidation with any accuracy. Both m_v and c_v depend on the average effective pressure in the soil while the load is being applied

ie on the value of $(p_o + \dfrac{q_{ne}}{2})$.

4.26 The settlement of non-cohesive soils under applied loading, though almost immediate, may be complicated because the effective modulus of elasticity normally increases with depth. Where it is necessary to calculate the settlement accurately, the best method is to divide the soil into a number of suitable layers and determine the settlement of each by a formula which involves the layer thickness, its constant of compressibility c, initial effective stress p_o and stress increase q_{ne}[1] .c is determined by penetration tests or by estimation from relative density and experience. (See Technical study FOUNDATIONS 2.)

4.27 Another method for determining non-cohesive settlements is extrapolation from results of in situ plate bearing tests. Some indication of actual settlements can be obtained from Terzaghi's semi-empirical theory[4, 5] of modulus of subgrade reaction, and also from the Terzaghi-Peck[4] basis of using in situ standard penetration test results to determine allowable soil-bearing pressures. These pressures are based on a limitation of total settlement to 25 mm, which ensures that differential settlements between adjacent foundations do not exceed 75 per cent of this limit. In most cases these give very conservative results. Plate bearing tests are expensive, and can only test the soil to a limited depth; they cannot provide information about long-term settlements. But in some cases (eg in very mixed strata or where large stones or boulders prevent the use of any form of penetration test) they may be the only method available. Full-scale loading tests are the most accurate way of determining the immediate settlement of bearing piles **10**, but, as with all loading tests, can give little indication of the settlements produced by the interaction of soil stresses from a complex of adjacent foundations.

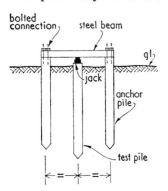

10 *'Anchor pile' method of jacking test pile into ground*

4.28 Effective stress changes from causes other than applied loading eg removal of overburden by excavation, ground filling or variations in the water table, must be included in the calculations. If the nett result is a *decrease* in soil stress, then for cohesive soils care must be taken that the appropriate Young's modulus and m_v for swelling are used (non-cohesive soils do not swell appreciably). Natural consolidation of fills under their own weight depends on many factors, the most important being the degree of initial compaction. Both the amount and rate of consolidation are best estimated on the basis of experience.

4.29 The various methods of calculation assume an ideal soil behaviour; also soil properties vary from point to point in a way that can only be guessed from the limited data provided by a site investigation. Thus settlement calculations can give only approximate results (at best say within 25 per cent of true values). Yet the best possible estimate is often needed if the effect of settlement on the superstructure, substructure or adjoining properties, roads, services etc is to be assessed. So there is an increasing tendency to make settlement calculations more realistic, where possible, by modifying the soil properties obtained from small-scale tests in the light of observed large-scale soil behaviour.

4.30 Foundation settlement may cause cracking of the structure or finishes; it can also influence the actual design and even safety of the building. So an engineer is concerned more with differential settlements between adjacent foundations than with the absolute settlement values. The

characteristics and loading of the structure and its foundations, and variations in soil conditions cause these differentials. Loading effects can be allowed for but without positive site information, effects due to soil conditions can only be estimated. Differentials are sometimes assumed to be half or three-quarters of maximum settlements, but more precise calculations can often reduce these appreciably. They are most usefully quoted as angular distortions eg a difference of 10 mm in a horizontal distance of 10 m represents an angular distortion of 1 in 1000.

4.31 Authorities differ in their views of the amount of angular distortion permissible in a building. Figures quoted range from 1 in 100 for certain open frames to 1 in 1000 or less where brittle finishes, walling, or sensitive machinery are involved. Special structural jointing may be needed to reduce the effects of differential settlement[6]. These may also be minimised by suitable foundations. Due weight must be given to the possible consequences of damage when deciding on what is tolerable in a given case.

4.32 Ground settlement will usually occur beyond the actual building, and particularly with clays, because of the greater depth of soil involved, settlements will be greater for a given ground pressure, the wider the foundations. This can be important if extensive heavy loadings occur on ground bearing floors. These are only another form of foundation, and a soft stratum at a considerable depth affects their behaviour appreciably.

Computer methods

4.33 Suitable computer programs can lessen the drudgery of extensive stress and settlement calculations. But more important, they can be used to integrate the behaviour of the superstructure, substructure, foundations and soil and represent the real interrelation between these elements. This could not be achieved if such techniques were not available. Certainly many complex foundations designed by manual methods have been overdesigned; in other cases the factor of safety has probably been appreciably lower than that imagined by the designer—at any rate there have been defects that could have been avoided if the true overall behaviour had been more exactly appreciated. But sophisticated analyses are not necessary for every building—they are really important for tall or heavy buildings, for rigid, extensive structures, when a large continuous or raft foundation (with or without piles) is used, or when the superstructure is particularly susceptible to differential settlements, as with some forms of precast concrete construction.

4.34 The basic problem is this: the final deflected shape of the complete structure with its foundations must be the same as that of the soil under the foundations. The problem therefore is one of determining the reactions between the foundations and soil which, with the loading on the structure, produce compatible deflections.

4.35 The analysis is in two interrelated parts: the structure and the soil. One approach[7] to the analysis is to divide the foundations into a sufficient number of elements, eg small sections of a raft or individual bases or piles, and to determine their behaviour under load in the particular soil conditions by conventional soils engineering methods; these element settlements must take account of loads on all the elements **11**. Successively closer approximations involving a process of iteration may be necessary. This is a very flexible method, taking account of considerable variations in soil strata, loading and structural conditions, complications such as ground heave and the effects of water, and can be used for quite large problems. In other methods matrices of settlement influence coefficients are

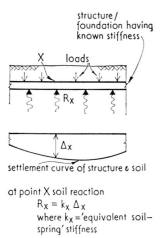

11 *Principle of 'equivalent soil springs' in computer analysis*

used instead of the soil springs. In simple cases spring stiffness can be calculated by a subgrade reaction method but this is essentially a crude method.

4.36 Another approach is the finite element method—the whole mass of soil and the structure is divided up into convenient small elements, to which suitable properties can be assigned. The interaction between all these elements under the given loading system may be calculated. But this approach has the disadvantage of making comparatively large demands on computer time, and so for big problems computer charges may be impracticably high. Theoretical accuracy, however, can be higher than with other methods.

4.37 Theoretical analyses of both structure and soil depend on the accuracy of the assumptions made about the properties and behaviour of the systems concerned. Particularly with soils, these may be quite approximate; only expert judgement can decide the degree of accuracy justified in these analyses.

4.38 In the end, all these methods give a pattern of settlement at a given period in time, the corresponding distribution of ground stresses or loads under the foundations (which may be used to check their safety), and the forces or stresses induced in the structure. A number of programs for such analyses are available on a commercial basis at computer bureaux, but require expert knowledge in application.

5 Soil movements other than those due to loading

Movements primarily caused by water

5.01 These movements, which need not be vertical, are caused by changes in that amount of water in the voids between soil particles not caused by alteration in stress (see para 2.01 *et seq* and some common types of movement mentioned in para 3.04).

5.02 Their effects are generally important with finer grained soils, ie silts and clays. These shrink when dried and swell when wetted, and the resultant movement can be a constant source of trouble with shallow foundations in clay **12**. These movements last as long as the moisture content remains above the shrinkage limit; in the UK shrinkage is worst with stiff fissured heavy clays.

5.03 In shrinkable soils, the most important causes of these movements are climatic conditions and vegetation. The climatic factor—seasonal differences in rainfall and soil temperature—shows as a tendency for the soil near the surface to dry out during the summer and to take up water again in winter. The amount of movement may vary, but in

the UK is unlikely to exceed 25 mm vertically at the surface without vegetation, and probably less than 6 mm at a depth of 1·25 m. In other countries, much bigger movements can occur. Also the building tends to protect the soil beneath it, so that movements are likely to be larger near the perimeter. Paved areas can also afford protection, but this depends on their size and permeability.

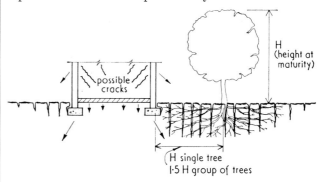

12 *Effect of building with shallow foundations on shrinkable clay subsoil. Arrows indicate direction in which structure tends to move; tension cracks appear in ground surface (shrinkage is less within the building) and soil shrinks away from foundations*

5.04 Plant roots cause more extensive moisture movements, but of a somewhat different type. Large trees and shrubs can cause permanent drying to depths of 5 m during low summer rainfall which is not made good during the winter, except perhaps with grass, although grass can cause drying to depths of about 2 m. Vertical surface movements up to 100 mm can be produced by large trees and other vegetation accentuating the moisture variation caused by climate.

5.05 Soil shrinkage can be both horizontal and vertical; vertical cracks in the surface can cause walls of buildings to move outwards from more protected areas, and allow passage of water. This can seriously soften clay against and even beneath shallow foundations. If vegetation is removed, the soil which had dried out will gradually regain water and swell. This swelling can be considerable and continue over as much as 10 years, if the vegetation was deeply rooted. A similar swelling may occur where ground has been protected by buildings or paved areas which have been removed. This swelling pressure can be strong enough to lift new buildings.

5.06 These movements can cause cracking of a building unless the foundations are taken below the depth of soil likely to be affected. (In the absence of vegetation, 1 m to 1.25 m is usually satisfactory in the UK.) Where vegetation or its removal might cause serious movement, either foundations must be taken deeper or other precautions taken to minimise damage. For example, buildings with shallow foundations in shrinkable soil should not normally be nearer to single trees or shrubs than the height (at maturity) of the tree; this increases to 1·5 times the height of the trees for groups or rows of trees **12**. Similarly trees should not be planted closer than these distances. If vegetation or buildings are removed from within critical distances of a projected building, construction may be deferred until most of the swelling has occurred.

5.07 Cracking of either structure or soil tends to close up during winter months, but not completely—rather it becomes progressively worse each year. Floors supported directly by the ground are also subject to soil movements, and water should not be allowed to accumulate during construction because it could cause long-term swelling, so that under certain conditions floors might have to be suspended. Heating plant can also produce drying shrinkage if the ground is not sufficiently insulated by a well ventilated cavity or other cooling system.

5.08 *Frost heave*, a different kind of movement due to intergranular water, may occur in soils where water can enter voids in the frozen layer near the ground surface and gradually build up an increasing thickness of ice until the soil is disrupted with a considerable lifting of the surface. Silts, sandy clays, chalk and very fine sands are susceptible to frost heave, but not coarse grained soils or clays. The water table must be near the depth to which frost can penetrate—in the UK a maximum of 600 mm during severe winters. Ensuring that there are no susceptible materials under foundations or slabs within the top 500 mm of soil is normally sufficient protection against frost heave. Whatever the soil, there should be no risk of frost heave under heated buildings, but on the other hand cold stores and similar buildings can cause freezing to a considerable depth. The amount of heave in such cases can be very large, and may increase over a period of years, so special insulation or ground heating could be needed; but this is a matter for specialist advice. For example in frost boil the thawing of heaved soil is accompanied by the release of large quantities of water.

5.09 Water can move more rapidly through coarser grained soils than finer grained materials, and this may cause further problems. In these coarser grained soils, a flow of surface water, or seepage of ground water due to excavations, pumping, topographical conditions, flooding or even leakage from sewers and water mains can cause loss of ground. Fine particles may be washed out of an originally well graded and dense sand or gravel, leaving the coarser material in an unstable condition. Alternatively pore water pressure may rise to overcome general intergranular pressure, particularly in fairly loose-packed, fine, uniform sands, possibly because of consolidation due to shock or vibration. This results in a loss of shear strength and is known as a *quicksand* condition. A similar phenomenon, called *piping* or *boiling*, is a lifting of soil caused by heavy seepage of ground water into the bottom of an excavation **13**.

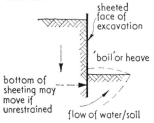

13 *Heave of bottom of excavation due to piping of sand or silt or shear failure of soft ground (see also para 5.26)*

5.10 A related problem occurs when soil of low permeability contains a more permeable layer, such as sand 'partings' in a clay. Then water pressure in the sand can exceed the weight of both the clay and water above and cause flotation or heave of the clay. Water confined at an abnormally high pressure in this way is said to be in an 'artesian' condition **14**, and the sudden change in soil permeability causes a deviation from the normal linear variation of pressure with depth (see section 4 of this study).

Other movements

5.11 Settlement of any soil (and swelling of clays) due to general changes in the water table level have been dealt with as movements caused by loading. Many other large-scale movements are due to some source of major instability, in what could otherwise perhaps be strong and stable soil. They may be naturally or artificially produced, and the

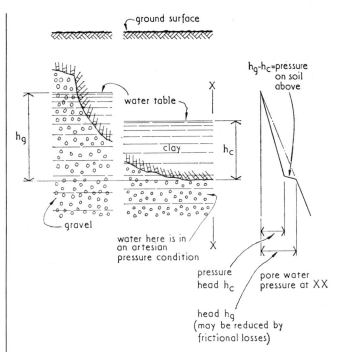

14 *Artesian pressure*

main types, already introduced in Technical study SAFETY 2 (Movement in buildings), are now discussed.

Seismic

5.12 Earthquakes give rise to various kinds of wave motion in the earth's crust and these cause severe vibrations. The long waves which travel along the earth's surface at some 3 to 4 km/s are responsible for most earthquake damage. The magnitude of vibration at a given place is usually expressed in terms of its acceleration, and this varies according to the severity of the shock from perhaps 10 to more than 5000 mm/s/s, ie it can reach more than half the acceleration due to gravity. The amplitude can be 300 mm and the period of a complete vibration as much as one or even two seconds. The design of structures in areas susceptible to earthquakes is a matter for specialists. The necessary precautions concerning general planning and superstructure design are usually covered by local regulations which describe the lateral forces (accelerations) a structure must be designed to resist.

5.13 The general principle is to minimise damage by restricting the amplitude of structural vibrations and limiting differential effects between connected parts. Thus structures and their foundations should be simple and well tied together.

5.14 Piled foundations may be particularly prone to damage. The effects of earthquakes are magnified by local ground weaknesses, such as faults in the underlying rocks, and large-scale movements, such as landslips, can occur. The movements may be in any direction.

Landslides and slips

5.15 The possibility of landslides depends on the local geology; cliffs, steep cuttings and other slopes can give rise to major slides, especially if erosion occurs or there are hidden planes of weakness. Deep-seated rotational slips **15** may occur in an earth embankment (natural or artificial), especially in cohesive soil of low shear strength. The risk of this kind of slip can be analysed theoretically.

5.16 Soft surface soils such as clay can creep slowly down slopes steeper than about 1 in 10. Such movements may be apparent from parallel ridges in the surface or the leaning of trees. The same sort of thing can happen if rock strata dip

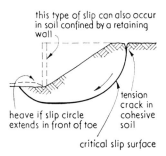

15 *Typical rotational slip of bank. Critical slip surface is approximate circle for which moment about centre of rotation of soil mass exceeds moment of surface's shear resistance by the largest amount*

with the ground surface, and the bedding planes are weak and lubricated by water. Constructional work, particularly if it interferes with the natural drainage, may increase the instability of a slope, and in cases of doubt it is best to avoid building in such areas.

Swallow holes

5.17 Swallow holes or sink holes are cavities that have been eroded in limestone or chalk by the water and which have subsequently been filled by overlying strata. They can be very deep and concealed by superficial soil, but surface depressions due to soil collapse reveal their presence **16**. The infilling material is generally comparatively soft or loose and further subsidence is liable to happen without warning. It may be possible to bridge isolated swallow-holes, but if they are wide or closely spaced it is advisable to resite proposed buildings. Soakaways should be kept well away from buildings in such an area.

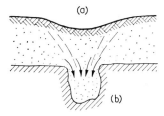

16 *Formation of swallow hole:* **a** *surface depression caused by collapse of soil into cavity;* **b** *water-eroded cavity in chalk or limestone*

Mining subsidence

5.18 Surface depressions similar to swallow holes may be caused by underground mining. In an area with mine workings or minerals that could be extracted in the future, the advice of a consulting mining engineer should be taken on the risks of surface instability.

5.18 In UK, the problem is most commonly associated with coal mining, and information on past, present or likely future workings of seams may be obtained from the Divisional Office of the National Coal Board. Records of long abandoned workings are often non-existent, and local inquiries or special site investigations may have to be made.

5.19 Initial surface subsidence, caused by collapse of roofs of workings as coal is extracted, should almost all have taken place by the time extraction ceases in a particular area. But older methods of working—initially shafts belled out into the coal which were only indifferently filled in, or the later galleries that were driven between pillars of un-worked coal and subsequently either all or partly removed, not only caused somewhat irregular initial surface subsidence, but also gave rise to varying long-term risks of instability.

5.20 The stress changes induced by a new building on the surface can be enough to cause breakdown of gallery roofs not yet fully collapsed or of any remaining pillars. Even where the overburden is enough to prevent this, there is always a risk of subsidence due to deterioration of the old workings. This may be caused by the action of ground water, the later extraction of other seams, pumping out of flooded workings, etc. This subsidence tends to be local, mainly taking the form of surface depressions similar to swallow holes. If the risks of damage from old workings under a proposed building site are appreciable, they might be minimised by high-level foundations that can bridge actual or potential collapsed areas, or by piers or piles taking the building loads down below the suspect workings (with precautions against future horizontal shear and vertical drag-down). It may even be economical to try to fill old workings and shafts by injection of suitable material from the surface.

5.21 Coal is now usually extracted from a continuously advancing long face, the 'longwall' method. The cavity produced is to some degree filled with waste material, and as props are removed or crushed, the roof and the ground surface both slowly subside. The surface movement is a wave that advances with the coal face. The ground in front of the wave is convex and in a state of horizontal tensile strain; behind the wave the ground is concave and in compression. Similar conditions occur near the edges of the face **17**. This illustration also shows the effective angle of draw, commonly 35°. Thus buildings in the area of a subsidence wave are subject to such horizontal ground actions as well as to tilt and vertical movement.

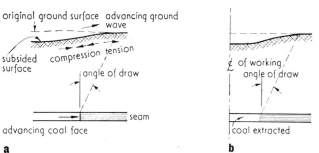

a   **b**
17 *Ground movement due to 'longwall' coal mining:*
a *section parallel to direction of advance;* **b** *parallel to worked face*

5.22 To minimise damage, structural design principles[8] include making buildings as flexible as possible (eg articulated frames), providing shallow rafts to resist horizontal strains, or splitting up large structures into independent units with substantial gaps between them. Jacking against differential settlement may be allowed for and reinforcement and ties should be provided to deal with the effects of this settlement over the structure; brittle finishes should be avoided and services in the ground made flexible. Rigid foundation rafts are generally too costly, except for very important structures. Once the wave has passed, most movement should have ceased, and some of the induced effects will have disappeared.

5.23 Subsidence not only affects buildings directly but also causes other movement by disturbing natural drainage and ground water conditions. Generally an expert can predict the surface movement likely to be caused by underground subsidence. Simultaneous or subsequent working of other coal in the vicinity can complicate the issue, and nearby geological faults will aggravate movement. Vertical subsidence at the surface can amount to many hundreds of millimetres.

5.24 It is generally impracticable to prevent all damage to buildings over underground workings, and the final choice of precautions to minimise this damage is influenced by the likely risks and consequences of any damage and by general economic considerations. Unfortunately the NCB can often only give approximate indications of future workings, and it may be advisable to take legal advice on NCB liabilities[9]. There is also the possibility of arranging, at a cost, for special methods of working or even for sterilisation of areas of coal to reduce or eliminate risk of damage to important structures.

5.25 Underground workings, especially in other countries, can also involve the extraction of minerals other than coal. Pumping of brine, oil or natural gas are among the most common.

Miscellaneous ground movements

5.26 Heave (other than that caused by water) is another movement which may be on a smaller scale than those discussed above. This can be due to pile driving or to movement into the bottom of excavations due to shear failure of the ground; the latter can reach catastrophic proportions. It is accompanied by a corresponding settlement of the ground outside, which also occurs if the sides of the excavation are allowed to move inwards (see also **13**). The effects of vibration can cause appreciable consolidation of sands and sandy gravels, particularly if they are in a loose condition and if the frequency of vibration is in the range 500 to 2500 cycles per min. (approximately 8 to 40 Hz).

The second part of this technical study will discuss actual soils and their behaviour in practice.

References

1 TOMLINSON, M. J. Foundation design and construction. London, 1975, Pitman, third edition [(16)] £15.00; (paperback £9.90)

2 STEINBRENNER, as in Tomlinson above

3 FOX, as in Tomlinson above

4 TERZAGHI, K. and PECK, R. B. Soil mechanics in engineering practice. Second edition, 1968, J. Wiley & Sons, [(L4)] £17.05; (paperback £7.25)

5 TERZAGHI, K. Evaluation of coefficients of subgrade reaction. December 1955, Geotechnique [(L4)]

6 Technical Study SAFETY 2, Movement in buildings [(2–)] AJ 7.6.72, p1293

7 CAPO BIANCO, A. J. Behaviour of complex foundations. Proceedings of symposium on Interaction of foundations and soil. 1971, Birmingham University [(16)]

8 Institution of Civil Engineers. Report on mining subsidence. 1959 [(L4)] o/p

9 The Coal-Mining (Subsidence) Act, 1957. HMSO [(L4)]

A further bibliography appears at the end of this section of the handbook.

Technical study
Foundations 1

Section 4 **Foundations and retaining structures**

Soils engineering, part 2

The first part of JOHN RANSFORD'*s study considered the theoretical behaviour of soils. It is concluded here with a discussion of commoner soils. He considers their properties, how these are determined by site investigation, modified by geotechnical process or preserved by constructional procedure*

6 Behaviour of common soils in practice

6.01 Only an outline of commoner UK soils can be given here, and further details are in BS CP 2001[10] and BS CP 2004[11]. The assessment and identification of the basic soils can be made by simple field tests as in table 1.

Table 1 Soil identification. (Based on BRS *Digest (2nd series)* 64,

Soil type	Field identification	Field assessment of structure and strength
Gravels	Between 2·36 mm and 76·2 mm (retained on 2·36 mm sieve) Some dry strength indicates presence of clay	Loose—easily removed by shovel. 50 mm stakes can be driven well in
Sands	Pass 2·36 mm and retained on 0·075 mm sieve Clean sands break down completely when dry. Individual particles visible to the naked eye and gritty to fingers	Compact—requires pick for excavation. Stakes will not penetrate more than a few inches
Silts	Pass 0·075 mm sieve. Particles not normally distinguishable with naked eye	Soft—easily moulded with the fingers
	Slightly gritty; moist lumps can be moulded with the fingers but not rolled into threads	Firm—can be moulded with strong finger pressure
	Shaking a small moist lump in the hand brings water to the surface	
	Silts dry rapidly; fairly easily powdered	
Clays	Smooth, plastic to the touch. Sticky when moist. Hold together (when dry). Wet lumps immersed in water soften without disintegrating	V. soft—exudes between fingers when squeezed
	Soft clays either uniform or show horizontal laminations	Soft—easily moulded with the fingers Firm—can be moulded with strong finger pressure
	Harder clays frequently fissured, the fissures opening slightly when the overburden is removed or a vertical surface is revealed by a trial pit	Stiff—cannot be moulded with fingers Hard—brittle or tough
Peat	Fibrous, black or brown	Soft—v. compressible and spongy
	Often smelly	Firm—compact
	Very compressible and water retentive	
Chalk	White—readily identified	Plastic—shattered, damp and slightly compressible or crumbly Solid—needing a pick for removal
Fill	Miscellaneous material eg rubble, mineral, waste, decaying wood	

Note: Pocket penetrometer tests are useful in the case of clays

Rock
6.02 Solid rocks are normally excellent foundation material with high loadbearing capacity and negligible settlement under load; the best are *igneous* rocks, eg basalts and granites, formed by solidification of molten material. *Metamorphic* rocks, which are igneous or sedimentary materials altered by heat or pressure, include gneisses (as strong as igneous rock) also slates, schists hard shales which exhibit pronounced cleavage and bedding planes and have a lower loadbearing capacity. *Sedimentary* rocks (solid stratified and consolidated deposits, usually of weathered* material from other rocks) include shales, mudstones, ironstones, coal, chalk, limestones and sandstones. Softer forms behave as normal uncemented soils; massively bedded limestones and hard sandstones have a high bearing capacity, but there are thinly bedded forms whose strength depends on the softer clayey material between the beds. With limestones and chalk there is also the possibility of swallow holes, caves or deep fissures (see para 5.17).
6.03 The loadbearing capacity of all rocks is greatly reduced if they are decomposed, heavily shattered by earth movements or steeply dipping (para 5.16). Indeed the strength of a small homogeneous piece of rock may bear no relation to its strength in situ, which depends on features such as bedding, cleavage, joints, fissures and faults, as well as on any weathering or stratification with other materials. This is particularly so with metamorphic and sedimentary rocks. (Heavy ground water flows can occur in some fissured rocks.) If foundation loads are so heavy or concentrated that the effective strength of 'bed rock' is important, or if there are associated seismic or mining problems, specialist geological and engineering advice will be needed.

Sands and gravels
6.04 Sands and gravels are the principal non-cohesive soils. Shear strength is frictional, and the structural properties depend mainly on closeness of packing (density), which may be reduced by the presence of water. Except for narrow shallow footings on loose sand at or below the water table, the allowable bearing pressure is governed by considerations of settlement (see para 4.22 *et seq*) rather than of strength. Settlement under load is quick and usually small. Loose sand will settle and lose stability under shock or vibration, and a quicksand condition can occur in the presence of water (see section 5).
These soils are very permeable and subject to seepage and piping problems, while fine sand is liable to frost heave through capillary action. Slopes of dry non-cohesive soil will be stable if no steeper than the 'angle of repose', which is approximately the angle of internal friction in the loose state.

*'Weathering' could have occurred very long ago and the material could be deeply buried. Ground water conditions can also affect the degree of weathering

Chalk and limestone

6.05 Chalk, mainly minute lime skeletons and shell fragments, is a form of limestone and subject to the same difficulties with swallow holes etc. According to the degree of weathering, chalk can vary from a soft crumbly or putty-like deposit to a hard, massively bedded rock. The cellular structure holds water but can easily be broken down and the material softened by frost, water or mechanical disturbance. Even at best, the loadbearing capacity is not high compared with most rocks, for settlements can be fairly large and care is needed with foundations on material already weathered or liable to further deterioration.

6.06 For foundation purposes chalk is best treated as a soil, its properties being determined by careful laboratory or in situ tests. Piling operations can cause serious weakening of the material. Ordinary limestones are generally harder and less affected by weathering, so that they do not have many of the difficulties associated with chalk. But the thin-bedded variety may be weaker. Another form is tufa, a spongy mass of calcium carbonate that occurs irregularly and is fairly strong when cemented but is easily broken down.

Coal

6.07 Coal is a soft rock, formed of consolidated plant remains. Foundations are very occasionally placed directly on coal at a comparatively low bearing pressure.

Clays

6.08 The basic properties of clay were discussed in part 1 of this study. Normally they are relatively impermeable, settle slowly under load, and are cohesive, with negligible apparent frictional shear strength.

Because of pore water, they are prone to shrinkage and swelling (para 5.02) and can creep down slopes (para 5.16). Partings or other materials and fissures in certain clays, and the presence of rootlets, can lead to general water seepage, softening and other weaknesses. Because of low shear strengths, rotational slips and excavation heave are possible. Strength and settlement are both criteria in foundation design. Clays frequently contain sulphates which attack Portland cement concrete and aggravate corrosion of buried ferrous metals.

6.09 According to the degree of pressure under which the original fine mud deposit was consolidated, the toughest clays become mudstone or shale. Mudstone may be massively bedded or laminated as a shale, which might contain silt. These materials are prone to considerable deterioration through weathering, and may soften in contact with water. Shales subject to prolonged pressure or heat are metamorphosed into true rock (hard shales or slates). Because of lamination, shales and slates are liable to slip in inclined strata.

6.10 Important clays are the stiff-to-hard boulder clays (which contain random stones and were deposited by glaciers), the 'shrinkable' stiff fissured clays (eg London, Lias, Weald and Oxford clays and those of the Woolwich and Reading beds), the soft alluvial and marine clays and the marls.

6.11 The stiff, fissured and boulder clays which have been subjected to great pressures during their geological history are known as 'over-consolidated' or 'preconsolidated'; other clays are called 'normally consolidated'. The higher the degree of preconsolidation, the less compressible is the material and the lower the 'geological' reduction factor (μ) relating actual to laboratory consolidation (para 4.25). But some soft clays are very *sensitive*, ie they exhibit a large loss of shear strength if disturbed. (Sensitivity is the ratio of undisturbed to remoulded shear strength.)

6.12 Clays containing organic matter or interleaved with it are particularly compressible. Alluvial clays are normally consolidated; this gives them a marked increase in shear strength with depth, compared with preconsolidated clays, but the top metre or so may be partially dried into a stiff surface crust. All clays are of course liable to softening by disturbance and water action and may disintegrate if allowed to dry out completely. Due to their cohesion, firm clays will stand at steeply cut slopes for limited periods. Stability will eventually be impaired by weathering (see also paras 5.15 and 5.26).

6.13 'Marl' normally describes calcareous clays and loams, ie those containing calcium carbonate. Keuper Marl is heavily over-consolidated and very strong when unweathered, but it is often highly fissured and easily softened by water and disturbance. Claystones are hard concretions of clayey material cemented by calcium carbonate, often very large and occurring at certain levels in London and Oxford clays etc.

Silts

6.14 Properties of silts are intermediate between those of sands and clays. Silt is both a cohesive and a frictional material. Windblown deposits are known as 'loess', which includes some English brickearths. Alluvial deposits containing organic matter are called organic silts and are very soft.

6.15 Silts are often difficult foundation materials, with some of the worst characteristics of both sands and clays. They are susceptible to frost heave. Soft silts (glacial or alluvial) retain water but are liable to 'boiling' and heave in excavations (see section 5). Wind-deposited silts are much better foundation material, though prone to structural collapse under the action of water. Loose silts are even worse than soft clays for foundation purposes. A soft mixture of sand, clay and silt, in roughly equal proportions, is called loam.

Peat

6.16 The basic constituent of organic soils, peat may occur with varying quantities of inorganic matter (eg clay or silt). It is a soft, fibrous vegetable material, formed by the decay of plants, and has a characteristic smell. Deposits may be up to several metres thick, or it may appear as layers in alluvial material. It is highly compressible and totally unsuitable for supporting foundation loads. Foundations must, therefore, be taken down through the peat to better soil (eg by using piles), unless the peat can be removed. But peat normally continues to consolidate under its own weight, and this plus surface wastage could eventually expose the foundations. It tends to retain water and therefore shrinks considerably on drying, swelling again when wetted.

Top soil

6.17 Top soil is the soft agricultural surface soil, perhaps supporting vegetation and containing humus (partly decomposed vegetable matter). It is never used for any engineering purpose and is normally removed in constructional areas. Tree roots, of course, can penetrate the ground considerably below the top soil. It may not be practicable to remove old roots completely before building, but foundations should not be placed on soil containing a substantial amount of them.

Fill

6.18 Filling, or made ground, is deposits of natural or manufactured material, both organic and inorganic (eg rubble, soil, colliery or other waste, refuse, old building materials, etc) that have been moved by man. Sometimes it is necessary to consider building on a considerable thickness of such ground. Its properties are determined by those of the constituent materials, the method of placing and the degree of compaction. It may be liable to spontaneous combustion, contain chemicals injurious to foundations, or be otherwise harmful.

6.19 Building on filled ground should be avoided if the ground contains appreciable quantities of soft clay, decomposable or 'collapsible' materials (like tin cans) or anything that could become unstable through the action of water. The main consideration is likely to be settlement rather than strength and because it is normally impossible to compact a fill of any substantial thickness after it has been completed, buildings should never be placed on material unless it is known that it was properly consolidated in thin layers at the time of deposition; unless special treatment can be carried out, or it may be that large settlements are acceptable. It will usually be necessary to recompact the tops of all formations before construction of foundations, possibly after removing clearly unsuitable material.

6.20 Building settlements consist of three components: consolidation of the fill under the structural loading (see para 3.03) and under its own weight, and consolidation of the underlying ground under the combined loading. These will depend to an extent on the height of fill and on the period since its deposition. But there are too many variables to give general rules for fill; here expert advice is essential. Roughly, self-weight consolidation of well compacted good material may take under five years and amount to less than 1 per cent of the depth. In fact the settlement under load could be less than that of many natural soils.

6.21 Because of possible variations in soil compaction and properties, normal soil tests or even plate-bearing tests (see para 4.22 *et seq*) may be of doubtful value. Types of structure and foundations that accommodate or resist considerable differential settlements may have to be considered. These might be continuous foundations which bridge 'soft' spots or pin-jointed superstructures. In bad conditions it may still be possible to build over deep fill by taking foundations (piles) down through it and perhaps also suspending the ground floor.

6.22 Sometimes foundations may be required on new filling. The same general considerations apply, but since the material and method of compaction will be fully controllable (filling should normally be granular and, in accordance with standard practice, of the highest quality)—there should be little doubt about its behaviour.

6.23 Other man-made features which can occur below the ground surface should be mentioned here. These include old foundations, basements and services that may have to be broken out, filled in or bridged. Major works such as tunnels (eg for underground railways) must be dealt with according to the particular circumstances and proposals for construction above agreed with the authorities concerned.

General notes on soil behaviour

6.24 Soil (including filling), which can consolidate appreciably after construction due to either its own weight, surface loading or the constructional procedure, may cause 'negative' skin friction or drag-down on the sides of deep foundations (eg piles) or basement walls. If a soft stratum is overlain by a sufficient thickness of harder soil, it will not increase the risk of shear failure under any foundations in the latter, but may contribute significantly to settlement (see para 4.11 *et seq*). In some cases it may be possible to improve soil properties in situ by geotechnical processes. Soil conditions abroad are often markedly different from those in the UK and specialist/local advice may be needed

7 Site investigations

7.01 A site investigation normally determines the disposition, types and properties of the soils, as well as other ground conditions which might affect design and construction of foundations and substructures. This investigation should be complemented by a report on the feasibility of building on the site, including appropriate general recommendations for design and construction. A comprehensive treatment of site investigations given in *Guide to Site Surveying* edited by Ralph Hewitt[12].

Value and limitations of an investigation

7.02 Every building project should have an early site appraisal. This may be limited to preliminary checks, though in all but the simplest cases, some form of soil investigation is required and becomes essential if there are any complications in the project or in anticipated soil conditions. Thus the extent of investigation depends on the scale and circumstances of the job; it is normally entrusted to a specialist.

7.03 A site investigation should cost less than 1 per cent of the project, and is valuable because it encourages savings in design, and because knowledge of soil conditions leads to confidence about the safety and behaviour of the proposed structure. This knowledge also helps the contractor assess the requirements and risks of his ground work more precisely and so submit a lower tender than might otherwise be possible. Later claims due to unforeseeable conditions may be reduced if not eliminated.

7.04 A soil investigation can lead to erroneous conclusions unless directed and interpreted by an expert. Only a very small proportion of the ground can be sampled and tested, and even test results may be misleading unless procedures and interpretations appropriate to the particular circumstances are applied. When all possible precautions have been taken, there still remains some risk that conditions that could not be foreseen from the results of the investigation may be encountered during construction. When loading on existing foundations is being increased, eg because of upward extension or change of use, a soil investigation may also be needed.

Scope and conduct of a site investigation

7.05 Certain preliminaries should be carried out before any work on site; these include inquiries from the local authority and other relevant sources, checks on any previous or nearby soil investigations and on the construction and condition of adjoining properties, a topographical survey, and consultation of geological and topographical maps and records (including any historical information, aerial photographs or particulars of underground works such as mining). Then, in certain cases—particularly if the proposed buildings are lightweight and simple in form—no more investigation may be required.

7.06 But in many instances further investigation is desirable. This may need several trial holes (or bores) with *in situ* tests and laboratory tests on soil samples. Such work will usually be carried out by specialist firms, but direction of the investigation should be entrusted to the engineer responsible for the report which gives recommendations for design and construction, who may also be responsible for the actual design. He should also advise on obtaining tenders for investigation work. General information on site investigation is given in CP 2001[10].

Trial bores and holes

7.07 The spacing of bores or holes depends on circumstances,

18 *Modern drilling rig*

Contract Name TADLEY, ALDERMASTON					Borehole No. 2	
					Sheet 1 of 3	

Method of boring Shell and Auger		Ground level	
Diameter 200 mm to 3.50 m		Start 24.3.72	
150 mm to 25.0 m		Finish 27.3.72	

Daily progress	Water levels	In-situ tests	Samples	Depth (m)	Reduced level (m O.D.)	Thickness (m)	Description of Strata
	▽		B,U	0.20		0.20	Topsoil
						1.05	Firm brown and grey clay with gravel
		N=79	B	1.25			
			B	1.65			Very dense sand with medium gravel
			U	2.90			
			J				
			U				Soft to firm orange brown fine sand and silt with laminations of green clayey silt
			J	3.50		3.50	
		N=12	B,U J	6.40			
			B				
		N=24	J			2.35	Medium dense grey fine sand and silt
				8.75			
	▽		U				Firm to very stiff dark grey silty clay with laminations of sandy silt Contd........

Notes ▽ Water struck, ▼ Morning water level, —— Casing depth, ---- Borehole depth	

Terresearch Limited	Report No. S.22/511	Appendix 1 Sheet 3

19 *Typical method of recording bore hole information*

but is usually at between 15 m and 50 m centres over the building site. The general depth for ground condition investigations was indicated in para 4.20, but all bore holes need not be taken down to full depth. Generally depths may be reduced if a lower hard stratum of sufficient thickness is known to exist.

7.08 In favourable conditions and to depths of 3 to 4 m, trial holes are suitable since they reveal the strata more clearly than bores. Even when bore holes are employed, some holes may be useful for confirmatory purposes or investigating existing foundations etc. Bottoms of holes may be 'proved' by hand-augering if necessary.

7.09 Two distinct methods of boring are used: 'soft' boring by hand or machine (eg auger or shell methods **18** with steel tube linings where necessary), and rotary machine drilling for strata that are harder than say mudstone or hard marl.

7.10 Whatever method of investigation, soil samples and ground water must be taken for identification and testing and a log **19** kept of conditions encountered, including all strikes and movements of ground water and any in situ tests in borings. Soil samples for testing must be as undisturbed as possible. This requires great care and the use of suitable techniques. Undisturbed sampling of very sensitive clays and of non-cohesive soil is normally not possible, and then in situ testing has to be relied on. (Samples from rotary drilling are known as 'cores.')

In situ and laboratory soil tests

7.11 The range of tests used are summarised in table II. Without describing them in detail, some of their more important aspects are considered. (Where procedures are standardised they are covered by BS 1377[13].)

Table II *Summary of in situ and laboratory soil tests*

Class of test	Purpose	Name of tests
1 To identify types of soils	A necessary preliminary to further tests	Visual inspection Wet or dry sieving Sedimentation test
2 To establish physical, mechanical and chemical properties of soils	To find properties which influence the strength, behaviour under load and chemical reaction when in contact with building materials	Tests to determine: natural moisture content bulk and dry density maximum and minimum densities chemical and index tests
3 To establish strength of soils	General design of foundations and soil retaining structures	Unconfined compression test Triaxial compression test Vane test Menard pressiometer Standard penetration test Dynamic or cone penetration test
4 To establish behaviour of soils under a specific type of loading or stress, or with time	General design of foundations	Consolidation test Triaxial compression test Dynamic or cone penetration test Plate bearing test Full scale pile test
	To discover permeability of ground	Constant head permeameter test Falling head permeameter test Pumping test
5 To supplement boring in establishing continuity of strata	General design of foundations	Geophysical tests

In situ tests

7.12 In situ tests comprise determination of moisture content (occasionally), tests of strength or density (eg in bore holes by vane, pressiometer or various penetration methods), pumping etc and loading tests. Geophysical methods may be included in this category. Vane and cone penetration tests may also be carried out without pre-boring, if the ground is sufficiently soft, and they are then sometimes described as 'deep soundings', as they can be used for rapid checks on soil strength at different positions and various depths on site **20**.

7.13 In vane tests, shear strengths of soft clays are measured by their resistance to the rotation of a vane. However, penetration tests are normally used to determine relative density (and hence shear strength and liability to settlement) of non-cohesive soils and silts; this is done by measuring the force (in 'static' tests) or number of specified blows per unit distance (in 'dynamic' tests) required to cause penetration of a standard cone or sampling tube. The so-called standard penetration test, SPT (see para 4.22 *et seq*) is a dynamic test and perhaps the most important of these tests. It is normally carried out in a borehole, and the blow count (N) per 305 mm penetration is recorded say every 1 to 2 m during boring.

7.14 Pumping tests may be made when movement of ground water is important. Pore water pressure measurements by piezometer can be useful, as, of course, is direct observation of ground water changes by means of perforated tubes inserted in boreholes. Uses and limitations of loading tests have already been mentioned. These are employed to determine both strengths and settlement characteristics of a particular foundation in given ground, and they are relatively expressive.

Geophysical methods
7.15 These methods—electrical, seismic, or magnetic—are highly specialised and only rarely used in building work. They determine change in strata with markedly different properties, eg rock interfaces and discontinuities such as old mine workings or other underground cavities. But for useful information, great expertise is required, and results should normally be confirmed by borings. The most important use of geophysical methods is the carrying out of rapid checks on consistency of strata between bore holes; this establishes if any further investigation is needed. Generally they do not yield direct quantitative data on soil properties.

Laboratory tests
7.16 Laboratory tests may be divided into: identification, moisture content, density, strength and consolidation, permeability and chemical tests.
7.17 Identity and general characteristics of a soil having been established, the most important thing is to measure its strength. In the laboratory this can only be done if the sample is relatively undisturbed, so such tests are mostly

confined to cohesive soils. A simple, quick test that can be applied to clays, which hold together after being cut, is the 'unconfined compression test', and this may even be done on site. It gives a value for the apparent cohesion (see below) equal to approximately half the compressive stress at failure, but this should be used only as a consistency check on more exact determinations. The same applies to the 'pocket penetrometer test', which gives a rough in situ value for unconfined compressive strength by measuring the force required to push a spring loaded plunger a certain distance into the clay.

7.18 The principal laboratory method of measuring strength is the triaxial compression test. By longitudinal crushing of specimens subjected to varying lateral pressures, (and failing in shear) it is possible to determine the values of c and ϕ (para 3.02). In the form of Coulomb's equation, given in para 3.02, c represents the true cohesion and ϕ the true angle of internal friction. These are constants for a given soil under shear, only for the particular density and moisture content that applied at failure. The application of load to a saturated soil usually causes a temporary rise in pore water pressure (para 4.02) until consolidation is complete. This may take a considerable time in the case of clays, and meanwhile the load is partly supported by the excess water pressure and only partly by the inter granular pressure q_e. The true cohesion and angle of internal friction are difficult to measure, and in any case the corresponding (theoretically correct) form of the Coulomb equation would be unsuitable for general use. It is therefore usual to recast the equation in the form

$$s = c + q \tan \phi \qquad\qquad equation~10$$

in which q is the *total* normal pressure on the shear plane and c and ϕ are now called, respectively, the *apparent cohesion* and angle of *shearing resistance* **21**.

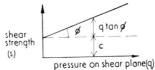

21 *Diagrammatic illustration of Coulomb's law; when* $\phi = o, s = c$

7.19 These quantities are measured under conditions representing as closely as possible those of the practical situation being considered, and are used in any subsequent stability analysis. With normal foundation problems in clay soils, the triaxial test is carried out under 'undrained'

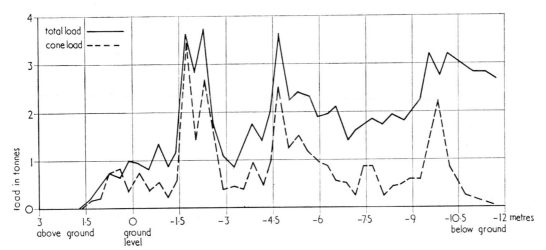

20 *Typical results of cone penetration test*

conditions (ie the specimen is not allowed to lose or gain water as the stress changes). In this test clays usually behave as if the angle of shearing resistance is zero, although their true ϕ may be greater than 20°.

7.20 Triaxial tests are also normally used for 'semi-rocks' such as chalk, marl and clay shale. Both triaxial and unconfined compression tests are sometimes done on hard rocks to determine intact strengths, although these will not usually bear any direct relation to the strength in situ (para 6.03).

7.21 With cohesive soils, consolidation properties may have to be measured to calculate settlement (para 4.22). This is done in an 'oedometer' or one-dimensional consolidation press, over a suitable range of pressure increases (or decreases if swelling properties are to be investigated); m_v and c_v are usually determined for each step in pressure so settlement may be interpolated. Permeability tests may supplement *in situ* drainage tests.

Chemical tests
7.22 These are normally confined to determining soluble sulphate content, pH (acidity) value of soils and ground water, and organic content. The first two tests are used to determine the possibility of injurious attack on foundations and buried services and enough properly distributed samples must be tested. If soil or ground water may be chemically contaminated, a full analysis will be required, especially in filled ground.

Topographical features
7.23 Features observable in a topographical survey—even if only a walk over the site—can give information about existing or future ground conditions. Briefly these are:

a Indications of strata in banks, cuttings or ditches.

b Broken or stepped ground may mean landslips or faults. Terracing of a slope's surface steeper than 1 in 10, or leaning trees on slopes, could be due to clay creep.

c Polygonal surface cracking or crazing in very dry weather probably indicates shrinkable clay. Larger, approximately parallel cracks may be due to deeper sources, such as landslips and mining subsidence. The effects of frost heave may be visible after a cold spell.

d In chalk or limestone, 'swallow holes' are caused by cavities in the underlying rock. Similar depressions or other surface subsidence may be a sign of the extraction of minerals.

e If an area is much lower lying than its surroundings, it might contain soft alluvial deposits (silt, peat etc). This is also true near rivers and some coastal regions, where the possibility of flooding or tidal conditions should also be considered. Other low-lying areas, especially those near dried up or filled in watercourses, are a potential flood risk if there can be a run-off from nearby higher ground.

f Sea, rivers and springs are likely causes of immediate surface erosion.

g Water levels in nearby ditches, ponds and watercourses may give some indication of the ground water table, but this should always be confirmed. Marshy vegetation, poplars and willows, usually mean a high water table.

h Sudden changes in vegetation type may be due to changes in subsoil conditions. Trees and large shrubs will produce surface settlements if subsoil is shrinkable clay, and this may be indicated by cracks or depressions near the trees.

i Type and condition of any buildings (or roads) on or near to the site may give information about likely ground conditions. Severe cracking or movement may simply be due to poor foundation design, but it might be indicative of underground movements such as mining subsidence.

j Any nearby constructional work in progress should be inspected for the possibility of pumping, which could disturb ground water conditions under the site.

k Underground services, or old foundations and other works, may be apparent from surface features.

l The possibility of exceptional climatic conditions should be considered.

The soil report
7.24 A report on the investigations should give both detailed results of the preliminary and site investigations, including in situ and laboratory tests, and conclusions and recommendations for design and construction. This may be supplemented by specialists' reports on particular problems such as underground mining.

7.25 To allow sensible recommendations, the soil report must be prepared with knowledge of all relevant technical and economic requirements for the structure. The report should deal with general feasibility from the soils point of view and alternative solutions should be examined if possible. Design recommendations should give details of the basis to be used, and constructional problems outlined with recommendations for the precautions to be adopted. However, detailed design responsibility remains with the project engineer. Effects of foundations on superstructure design must also be covered and the magnitude of any settlements or other likely movements of the ground predicted.

7.26 Finally, the report must draw attention to any matters which still remain doubtful and which could cause changes in design or construction as more knowledge becomes available. If necessary, further investigation must be recommended; it is not always possible to obtain all required information during the course of a single investigation.

8 Injurious ground conditions

8.01 In certain circumstances, soil or ground water is potentially injurious to buried foundations, structures or services. This possibility should have been revealed by chemical tests in the site investigation (para 7.11 *et seq*).

8.02 The most usual danger is from soluble sulphates, which can occur in many parts of the UK and in solution react with the cement in set concrete. The type and degree of reaction depends mainly on the sulphate, its concentration, and the particular cement used. But the usual effect is the formation of insoluble salts, which on crystallisation cause expansion and disintegration of the concrete surface. The attack can continue with disastrous results if ground water bearing more sulphates comes into contact with the newly exposed concrete.

8.03 The most common sulphates in the ground are calcium, magnesium and sodium. (Calcium sulphate, by far the least soluble, presents the least problem.) Sulphates occur naturally in the UK mainly in the stiff fissured clays (see para 6.08), including Keuper Marl, or in soils contaminated by ground water from nearby clays. They sometimes appear as shiny, partially transparent crystals. Serious sulphate concentrations can also occur in some peats as well as in artificial deposits such as colliery shale, pulverised fuel ash, building rubble or industrial wastes. In some cases sulphuric acid and sulphates in acid solution may be present; this increased risk is indicated by a low pH* value.

8.04 Serious sulphate attack on concrete is only possible if movement of ground water (horizontally or vertically) continually replenishes the soluble salts used up in reacting with the cement. Thus concrete permanently above the water table, or protected against water flow, runs little risk of serious attack. But if the likely sulphate concentration and ground water movements indicate appreciable risk, appropriate precautions must be taken.

8.05 The first essential is a dense, impermeable concrete. Further protection is given by special cements such as sulphate-resisting Portland cement (BS 4027); super-sulphated cement (BS 4248) which may gain strength slowly; and even high alumina cement (BS 915) which should not be used if high temperatures can occur. In extreme cases it may be necessary to provide protective coatings of inert material.

8.06 Table III, based on BRS Digest 90[14], sets out site classifications in terms of sulphate concentrations in soil and ground water, and relevant recommendations for concrete protection. The second column refers to total soil sulphates, the third gives soluble sulphates for classifications 3 and upwards. Supersulphated cement is susceptible to attack by ammonium sulphate, which is very soluble; high alumina cement is unsuitable for very alkaline conditions. It is unwise to use concrete admixtures containing calcium chloride where there are sulphate risks. The BRS digest emphasises that their recommendations are tentative judgements, based on present knowledge.

8.07 Sulphate concentrations in the soil can vary widely and a large number of samples may have to be taken, although testing costs may be reduced by combining samples. There are obvious advantages in classification by ground water analysis only, but samples may be diluted either by surface water or by ground water from less contaminated zones, and the top metre or two of a sulphate-bearing soil may already have been bleached free from any salts. The classification of a site should be made by an experienced engineer on the basis of all available information. In cases of doubt, the use of a particularly good quality concrete, made with sulphate-resisting cement, is a wise insurance at little extra cost.

8.08 Sea water contains considerable sulphates but these are usually not considered particularly aggressive, because of the inhibiting action of sodium chloride.

8.09 Precautions against acid attack on concrete (shown up by pH values) are similar to those against sulphate risks. Acidic ground conditions can arise from organic acids in peaty soils and water, or from carbon dioxide and carbonic acid in soft water, but risks are usually small with a good normal concrete.

8.10 Other problems, beyond the scope of this article, include corrosion of ferrous metals in the ground[11] (sulphates may cause this if certain bacteria are present); deterioration of timber; spontaneous combustion of fills; or general effects of chemical wastes and effluents.

8.11 Finally, disruption of concrete and corrosion of reinforcement can occur due to weathering, eg frost action and attrition, which is perhaps aggravated by alternate wetting and drying and by chemical action. This can usually

*Neutral water has pH value of 7

be prevented with good quality concrete, sound aggregates and adequate reinforcement cover (at least 50 mm).

Table III Sulphates in soils and groundwaters—classification and recommendations. (Based on BRS Digest (2nd series) 90)

Concentration of sulphates expressed as SO_3				Recommendations for types of cement to be used in dense, fully compacted concrete, and special protective measures when necessary (see Note 2). Aggregates should comply with BS 882 or BS 1047
	In soil			
Class	Total SO_3 per cent	SO_3 in 1:1 water extract	In ground-water	
1	Less than 0·2		Less than 30 parts/ 100 000	For structural reinforced concrete work, ordinary Portland cement or Portland-blastfurnace cement. Minimum cement content 280 kg/m³. Maximum free water/cement ratio 0·55 by weight. For plain concrete, less stringent requirements apply
2	0·2-0·5		30-120 parts/ 100 000	(See Note 1) a Ordinary Portland cement or Portland-blastfurnace cement. Minimum cement content 330 kg/m³. Maximum free water/cement ratio 0·50 by weight. b Sulphate-resisting Portland cement. Minimum cement content 280 kg/m³. Maximum free water/cement ratio 0·55 by weight. c Supersulphated cement. Minimum cement content 310 kg/m³. Maximum free water/cement ratio 0·50 by weight
3	0·5-1·0	2·5-5·0 g/litre	120-250 parts/ 100 000	Sulphate-resisting Portland cement, supersulphated cement or high alumina cement. Minimum cement content 330 kg/m³. Maximum free water/cement ratio 0·50 by weight
4	1·0-2·0	5·0-10·0 g/litre	250-500 parts/ 100 000	a Sulphate-resisting Portland cement or supersulphated cement. Minimum cement content 370 kg/m³. Maximum free water/cement ratio 0·45 by weight b High alumina cement. Minimum cement content 340 kg/m³. Maximum free water/cement ratio 0·45 by weight
5	Over 2	Over 10 g/litre	Over 500 parts/ 100 000	Either cements described in 4a plus adequate protective coatings of inert material such as asphalt or bituminous emulsions reinforced with fibreglass membranes, or high alumina cement with a minimum cement content of 370 kg/m³. Maximum free water/cement ratio 0·40 by weight.

This table applies to concrete placed in near-neutral groundwaters of pH 6 to 9, containing naturally occurring sulphates but not contaminants such as ammonium salts. Concrete prepared from ordinary Portland cement would not be recommended in acidic conditions (pH <6). Sulphate-resisting Portland cement is slightly more acid-resistant but no experience of large-scale use in these conditions is currently available. High alumina cement can be used down to pH 4 and supersulphated cement has given an acceptable life provided that the concrete is dense and prepared with a free water/cement ratio of 0·40 or less, in mineral acids down to pH 3·5.

Note

1 The cement contents given in Class 2 are the minima recommended by the manufacturers. For SO_3 contents near the upper limit of Class 2 cement contents above these minima are advised.

2 For severe conditions, eg thin sections, sections under hydrostatic pressure on one side only and sections partly immersed, consideration should be given to a further reduction of water/cement ratio and, if necessary, an increase in cement content to ensure the degree of workability needed for full compaction and thus minimum permeability.

9 Geotechnical processes

9.01 There are various geotechnical processes for improving soil properties in situ, either temporarily or permanently. This is a large and specialised subject, calling for specialised engineers, though commercial firms offer specialist services for many of these processes.

9.02 The desirability, and indeed possibility, of using certain geotechnical methods will normally be indicated by the site investigation. Much will depend on the particle size of the soils concerned, and also on relative economics for which no general rules can be given.

9.03 The methods generally fall into three broad categories:

a Ground water lowering

b Ground water exclusion

c Soil stabilisation, including improvement of loadbearing capacity.

Particular processes may overlap these divisions.

Ground water lowering

9.04 This is a temporary technique for dewatering excavations to allow them and subsequent construction to proceed in the dry—this is not always easy. Concrete can be placed (by bag, skip or tremie pipe, or by grouting prepacked aggregate) and other work carried out underwater if essential, but construction in the dry is almost always preferable. It may be possible to exclude water from an area (see below), but unless a permanent cut-off is required it may be cheaper simply to lower the water table temporarily. This is done by various forms of pumping (from sumps, well points or bored wells) or by electro-osmosis, unless of course ordinary gravity drainage is possible. Ranges of soil particle sizes (mm) for which certain types of drainage are likely to be effective are shown in **22**. (Compressed air is rarely used except in civil engineering works.)

Ground water exclusion (before excavation)

9.05 This may be temporary or permanent, such as a basement in water-logged ground where the exclusion system is incorporated in permanent construction. The latter type will usually take the form of an impermeable

reinforced concrete diaphragm ('cut-off') taken down into a sufficiently impervious stratum. The diaphragm may be an in situ rc wall cast by tremie in a trench temporarily stabilised by a bentonite mud[15], or a continuous bored concrete pile wall. 'Caissons' (watertight structures sunk into ground or water) also come into this category. Temporary methods include the use of sheet piling eg 'cofferdams' constructed on land or in water, grouting methods to form cut-offs by sealing the voids in the soil (see below) and the casings of in situ concrete piles. All water exclusion techniques depend for success on a sufficiently impervious stratum at a reasonable depth, unless it is practicable to seal the ground at the bottom of the excavation after temporarily lowering the water table within the cut-off or by concreting or grouting under water. If all other methods fail, water may be temporarily excluded from an excavation (and the sides stabilised) by freezing. As this is both expensive and slow, it is rarely used. Grouting of soils outside basements is sometimes used to rectify water leaks in the permanent construction.

Soil stabilisation and improvement of loadbearing capacity

9.06 General support of excavations will be considered in TS FOUNDATIONS 3, Retaining structures. Stability of soil adjacent to an excavation, especially under existing foundations, may sometimes be usefully improved by grouting, which can amount to a form of underpinning **23**. Even if the sides of the excavation are closely supported, it is very difficult to prevent some lateral movement, which will then cause settlement of any unstabilised soil. Grouting may also sometimes be used to stabilise the bottom of excavations, while a similar application is the complete or partial filling of underground cavities eg old mine workings. The stability of loose sands etc can be improved by vibration methods (deep or shallow) while freezing is occasionally used in saturated soils, especially silts and clays. Grouting or bolting techniques may succeed in preventing slips in rock, and the stability of excavations generally is usually improved by ground water lowering. Stabilising processes may also be either temporary or permanent.

9.07 Improvement of loadbearing capacity is usually a permanent measure, adopted to permit reductions in the sizes of foundations, to obviate piling or other deep foundations or simply to prevent excessive settlements. There may however be incidental temporary benefits during construction. The processes involved would normally cost more than

22 *Tentative limits within which various groundwater lowering processes can be applied. (Particle sizes from clay to stone are plotted logarithmically)*

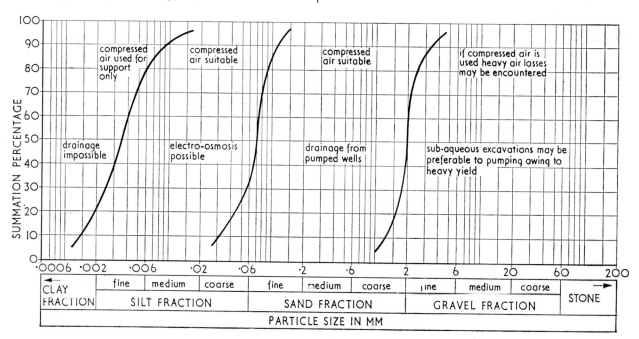

the savings in permanent construction and so ground strengthening of this nature is not undertaken very often. But there are cases in which it is clearly worthwhile, though it should be limited to the particular areas concerned. An obvious method is simply to remove unsatisfactory soil where practicable and replace it by better. Suitable grouting processes are applicable here too, and perhaps one of the most useful techniques is that of deep vibration. This is basically the compaction in depth of loose granular materials—especially clean sands but including gravels and even coarser or finer materials such as miscellaneous fills and silty or clayey sands—by a heavy, high-frequency vibratory unit sunk into the soil. Close pile driving may produce similar but less easily controlled effects. Extra granular material is fed in round the vibratory unit, one variety of which is the water-jetted 'Vibraflot'.

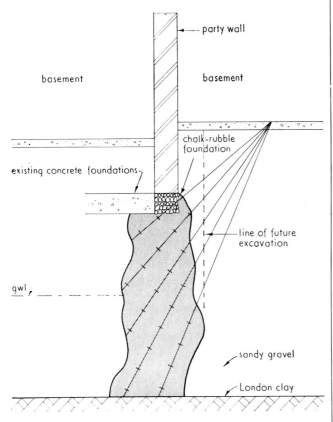

23 *Underpinning with fluid injection grouts*

9.08 Such processes may also be used to form columns of graded stone in soft clays or silts: these can carry direct loads like conventional piles if differential settlement is unimportant, but they also improve the general shear strength and drainage, the latter assisting overall stability and consolidation of the ground (see para 4.22 *et seq*). Similar drainage effects are achieved by vertical (sand) drains installed by boring or driving.

9.09 The surface of soft ground may sometimes be pre-loaded, eg with earth, to increase stability and reduce settlements, perhaps in conjunction with vertical drainage. Shallow compaction (to a depth of roughly 0·5 m) may also be carried out by conventional means such as rolling or vibration, and an alternative surface treatment is to strengthen granular soil directly by mixing it with cement or other suitable material. The permanent lowering of the ground water table, where practicable, will almost always improve soil properties, but before undertaking any such operation, the possible effects on ground settlement and clay shrinkage (see part 1 of this study) should be considered. Adequate surface water drainage can also be important.

9.10 Particular procedures are now discussed. Aspects of those particularly relevant to basement construction—such as ground water lowering, diaphragm walls and sheet piling—are considered in TS FOUNDATIONS 3, Retaining structures while underpinning is dealt with in TS FOUNDATIONS 2, Foundation design.

Electro-osmosis
9.11 In some conditions, soil drainage for temporary de-watering can only be achieved by using an electrical potential to drive the water to negative electrodes placed in wells. The positive electrodes are metal rods driven into the ground **24**. This is a very expensive method, usually confined to silts.

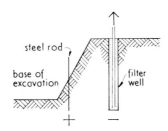

24 *Electro-osmosis causes water to move away from excavation*

Grouting
9.12 It is nowadays possible to improve the properties of most soils by injecting one or other of a large range of grouts. This is essentially a process by which some or all of the water or air in soil voids and fissures is replaced by new material designed for a specific purpose. The latter is usually a reduction in permeability, increase in strength or a combination of both. The practicability of this, and the most suitable method, will depend largely on the particle size distribution of the soil. With an ever-increasing range of grouts available, it is not possible to give detailed information here. However, **25** gives a rough idea of the types of soil in which some of the main processes are applicable. The finer grained soils are most troublesome, and very little can be done about the finest sands and silts except freezing for temporary treatment or trying to force in low-viscosity solutions. This will sometimes compress silts and establish a network of impermeable grout-filled fissures.

9.13 Similar measures may be used for fissured clays, and it is also possible to grout up fissures and cavities in rocks. Electro-chemical hardening, though a process of limited application, may sometimes be used to strengthen clay soil around piles. Coarser materials are easier and cheaper to deal with. Grouts may be suspensions (cement, sand, fly-ash, clay or bituminous emulsion, with fillers such as sand or fly-ash depending on the size of voids, cavities or fissures to be treated), or solutions of varying viscosity (silicates, resins and other chemicals) which gel either after a predetermined time or on completion of a 'two-fluid injection process'.

9.14 All these treatments are relatively permanent, but some grouts tend to be more stable than others; a wide range of properties can generally be provided. Site or laboratory tests may be used to check grouting work. Close site control is required, and it may be necessary to be selective with regard to the strata treated. It is not possible to give any useful general cost information.

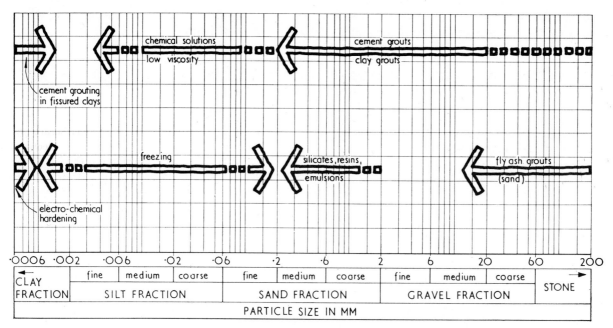

25 *Approximate soil ranges within which various geo-technical processes may be applied*

10 Soil mechanics aspects of site supervision

Responsibilities of site supervisor

10.01 Any person responsible for supervision of construction must, as far as ground conditions are concerned, ensure that everything which is done accords with the intentions of the designer of the permanent works. This includes the type and quality of soil (including ground water conditions) in which construction is carried out, the effect on it of the contractor's operations, or any work which has to be performed on the ground itself. He must also ensure that any temporary works are executed efficiently, all necessary records kept and everything done with a proper regard for safety and avoidance of damage to property. The architect or consultant will normally have only a limited responsibility in this respect.

10.02 Efficient supervision is of particular importance during all work in the ground, because of all the potential problems and unknown quantities. Where the ground works or soil conditions are likely to be at all complicated, the supervisor should be particularly experienced in such matters.

Site supervision check list

10.03 The following items are particularly important :

a Ground conditions should be carefully observed throughout construction, and any significant variation from those anticipated (seen or suspected) should be reported immediately.

b Surface drainage and any lowering and disposal of ground water must be arranged to avoid deterioration in ground conditions. Ground water pumping, whether from sumps, well point tubes or wells, should wherever possible be *away* from excavations. This will prevent surface erosion. Where necessary, filters should be provided so that fine particles are not extracted from the soil. Care must also be taken not to damage adjoining property by pumping, and if the water table is lowered it may be desirable to restore the water in nearby strata by 'recharging wells' to avoid excessive settlement. Watch that rising ground water does not cause flotation of new construction!

c Slopes of excavations are often excessive, and where there are temporary supports to the sides, these are frequently inadequate. Any but the most minor excavations should be carefully watched throughout for signs of instability, eg deterioration of the soil, movement of supports, bottom heave, water seepage or 'boil'. Emergency lowering of the water table may become necessary. The possibility of clay 'flotation' due to artesian pressure must not be overlooked—this can be prevented by vertical 'relief wells'.

d The sequence of work may be important, eg for control of differential settlement or for avoiding undermining of foundations already constructed by further excavations. This is easily overlooked when putting in drain trenches, and it applies equally to work near existing properties, where underpinning of the latter may be necessary, see **26**.

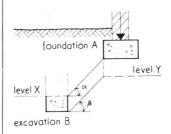

26 *Excavation below level of nearby foundation. Knowing required depth for excavation* B, *existing foundation* A *should be underpinned to level* Y (*based on angle* β), *but if foundation* A *does not yet exist, excavation* B *should be backfilled with concrete at least to level* X (*based on angle* α). *Values for* α *and* β *depend on soil and other circumstances, including local authority requirements. They are both frequently taken as 45°*

e Foundations must be taken down to sufficient depths to suit the conditions actually encountered. 'Soft spots' in the ground must be cut out where necessary and, with any overdeep excavation, filled with suitable material—usually mass concrete.

f All formations for foundations must be inspected immediately prior to construction. The supervisor must look out for any variability in soil conditions and carry out necessary spot checks on quality, eg by pocket penetrometer, as well as ensuring that any officially required tests are made.

g Formations should normally be dry, and must be clean and in good condition. They are easily disturbed (especially clays and chalks) by the action of water and the weather generally and by the contractor's operations, so it is normal practice to leave at least the last 150 mm of excavation to be carried out by hand immediately prior to placing the first portion of the permanent construction—usually 75 mm of mass concrete 'blinding' in the case of reinforced concrete work. It may be necessary first to 'firm-up' the formation by a little light compaction.

h Particular attention should be paid to concrete placed under water. This is most likely to occur with cast-in-place piles.

i It is important that all back-filling to excavations, and also all making up of site levels, should be executed with suitable material, properly compacted in layers. In the case of excavations below adjacent foundations, other than floor slabs (see d) this will normally be mass concrete, and good granular material will usually be employed elsewhere (occasionally a suitable clay). Compaction usually will be by rolling or ramming, but vibratory methods may be used with granular fill[16, 17].

j Expert assistance may be needed in the supervision of specialists' work.

k Tolerances in foundation work inevitably tend to be greater than elsewhere. They do not call for special comment except in the cases of piling and diaphragm walls as discussed in Technical study FOUNDATIONS 3 and Information sheet FOUNDATIONS 7.

References (to part 2 only)

10 BS CP 2001: Site investigations [(11) (A3s)] £3.00

11 BS CP 2004: 1972 Foundations [(16)] £6.90

12 HEWITT, Ralph (editor). Guide to site surveying. Architectural Press [(A3s)] o/p

13 BS 1377: 1967. Methods of testing soils for civil engineering purposes [(L4) (Aq)] £4.40

14 BRE Digest 90 (second series). Concrete sulphate-bearing soils and ground waters. HMSO [Yq (S)]

15 HODGKINSON, Allan. Diaphragm walls AJ Technical study 15.9.71 p593 [(16·2)]

16 BS CP 2003: 1959 Earthworks [(11)] £3.00

17 TRANSPORT AND ROAD RESEARCH LABORATORY Soil mechanics for road engineers. HMSO, 1952 [(L4)] £4.50

A further bibliography appears at the end of this section of the handbook.

Technical study
Foundations 2

Foundation types and design

Technical study FOUNDATIONS 1 *considered the behaviour of the ground under a building and its interaction with the foundations; means of predicting and even controlling this behaviour were discussed.* JOHN RANSFORD *now completes his introduction to the work of a foundation engineer with a survey of foundation types and their uses. He shows how to decide on a suitable type, size and depth for a foundation. Structural design is only outlined, because once ground reactions have been calculated normal structural principles govern the detailed design. This study is supplemented by Information sheets* FOUNDATIONS 1 *to* 7

1 General principles of foundation design

1.01 Good foundation design ensures that structural loads (including weight of foundations) are transferred to the ground: (a) economically, (b) with due safety and (c) without any unacceptable movement — either during or after construction.

1.02 Economy requires the choice of the most suitable foundation (see Information sheet FOUNDATIONS 2), and design assumptions should not be too conservative. There must be an adequate design factor of safety against complete failure of both the structural elements forming the foundations and the ground below them. When loadings have been determined and the foundation type chosen, ground reactions can be calculated to ensure that they are within safe limits (see Information sheet FOUNDATIONS 3).

1.03 Estimated settlements or other movements associated with particular ground conditions (see Information sheet FOUNDATIONS 3) should be compared with those considered acceptable in the circumstances, including movement in the superstructure. Ground movements can affect adjoining structures, roads and services. With some statically indeterminate structures, movements of foundations may cause significant redistribution of the foundation loadings.

1.04 The basic design procedure is summarised:

a Assess site conditions after a suitable investigation (see Technical study FOUNDATIONS 1, section 7)

b Determine structural loadings allowing for self-weights of foundations*; effects of any imposed ground deformations such as in seismic or subsidence design (see Technical study FOUNDATIONS 1, section 5) have to be considered; also determine permissible foundation movements.

c Choose the most suitable type of foundation. The possibility of modifying soil conditions in situ (see Technical study FOUNDATIONS 1, section 9) may be considered here

d Decide permissible ground loadings; these must allow for limiting movements within acceptable values

e Estimate sizes of foundations to suit these loadings, and correct self-weight allowance if necessary

f Check ground stresses against permitted stresses and approximate structural stresses

g Estimate movements, checking these against allowable movement and checking for any effect on foundation loads

h Advise building's designer about necessary movement joints or other structural effects of foundations.

1.05 The last three steps may involve the first stage of structural foundation design or analysis, eg continuous foundations, but in simple cases (g), may be covered by (f). If the results of (f) or (g) are unacceptable or if the design seems over conservative, it may be necessary to return to step (e) (or earlier) with new proposals. Once satisfactory values are obtained, the final structure design can be carried out.

1.06 Ground conditions met during construction should always be checked against those assumed in design. Guidance on matters which may arise after basic design is complete were given in Technical study FOUNDATIONS 1, sections 9 and 10, and this should assist in preparation of the specification. (The above procedure is shown in Information sheet FOUNDATIONS 1.)

2 Choice of foundation type

2.01 Factors governing the choice of foundation include: soil and site conditions (and the degree of certainty about these), type of structure to be supported, its loading, and the amount of movement (total and differential) which can be tolerated. The choice can also be affected by: general economic considerations, resources and time available for construction, the effect of foundation behaviour on the superstructure design, the degree of confidence in design assumptions, constructional problems and the possibility of further development on the site. Sometimes the risks of damage or undesirable movements have to be weighed against the cost of precautions to obviate them. Generally the best solution will be the one that provides the minimum acceptable standard of performance at least cost. However, the relevant factors are influenced by local circumstances, and so this study can only give general guidance. While experience usually provides the answer, alternative schemes may have to be costed before a decision can be made.

*Regulations permit reductions in certain superimposed floor loadings for foundation design. Weight of soil displaced by foundations may be deducted for net loading

Types of foundation

2.02 General selection principles may be illustrated by considering the characteristics of the main types of building foundations—pads, strips, rafts and piles **1**. Usually they are all reinforced concrete, but pads and piles are occasionally mass concrete.

2.03 Pads, strips and rafts are normally *shallow foundations*, ie they transmit loads directly to the soil at depths between 0·5 m and 2 m below the ground surface. Sometimes the foundations are taken to greater depths, either by making them thicker (eg by stooling down their undersides in mass concrete) or occasionally by dropping their levels completely. Then deep pads (piers), supporting high-level beams or slabs, may be a solution, but beyond depths of 4 to 5 m— or sometimes even less—piling is usually cheaper. The depth at which this change of type is economic varies according to circumstances. The basic choice is often between founding at a high level but with a low ground-bearing pressure, or going much deeper to find better soil. Sometimes it is possible to take advantage of a basement requirement and so reduce the cost of the foundations; pads, strips and rafts are often used below general excavation for a sunken structure. A raft may be particularly suitable for heavy loads on compressible soil, and this is sometimes stiffened by a basement substructure.

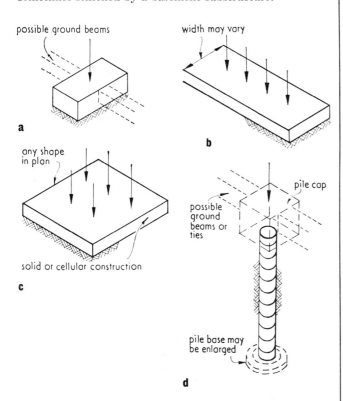

1 *Main types of building foundation. Foundation tops are normally a little below ground level. Loads may be concentrated or distributed and may include horizontal components:* **a** *pad (para 4.02);* **b** *strip (para 4.05);* **c** *raft (para 5.01);* **d** *pile (para 7.01)*

2.04 *Deep foundations*, which may go down to well below 30 m, are usually either bearing piles or basements (which may also bear on piles). Piles are normally concrete, either precast and driven into the ground, or cast in situ in bored holes, driven tubes, or shells. Their loads may be transmitted to the soil partly in end-bearing and partly in *shaft friction*. A high load can be spread over a group of piles joined through a reinforced concrete cap **2**. Piles are often used to support ground beams (foundation beams) carrying walls or other loads.

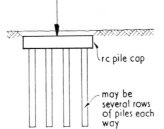

2 *Typical group of piles*

Selection of foundation types

2.05 Before a detailed discussion of these types, consideration of their general characteristics and basic soil and structural principles leads to general conclusions on which type is likely to be most suitable. (These are tabulated in Information sheet FOUNDATIONS 2.) If piling is used, there will be a choice between many available types and systems, and this will also be considered further in this study.

2.06 *Isolated foundations* (pads or piles) are in principle most suitable for comparatively widely spaced loads, such as those transmitted by columns. If, because of loadings or soil conditions, the foundations or pile caps should come close together, it may be more economical to combine these elements into *strips* or *rafts*. Strips are appropriate for loadbearing walls; but if the plan area of separate bases is more than about 70 per cent of the building area, a raft foundation is usually more economical. Strips and rafts are also used when stiffness must be introduced into the foundation system, either to reduce differential settlements or to bridge 'soft spots' or·cavities in the soil. In some cases the supported structure (eg a stiff basement) may wholly or partly provide this stiffness. For heavy buildings, basement rafts are sometimes used as *buoyant foundations*, ie the soil stresses and settlements are reduced considerably because of the weight of soil, and possibly water, displaced.

2.07 Piling is used where, because of soil conditions, economic or constructional considerations, it is desirable to transmit loads to strata beyond the reach of other foundations, for example when there is a high water table which cannot practicably be lowered, even though a good loadbearing stratum occurs not far below it. Piling may also be chosen where a restricted site would make surface foundations more difficult or where there are other unfavourable conditions, such as clay subsoil and wet weather during construction.

2.08 Piles (or piers) and ground beams are sometimes used in place of strip footings to walls, possibly with a suspended ground floor slab; even short piles may be an economical alternative for light loads on a clay site (see Information sheet FOUNDATIONS 7), especially where it is necessary to found below a comparatively deep zone which is subject to shrinkage or swelling (see Technical study FOUNDATIONS 1).

2.09 Suitably reinforced piles act as anchors against hydrostatic buoyancy; they also resist uplift, which may be caused by wind or travelling cranes. Vertical piles resist some horizontal loading, but they may fracture if the strain is great (as in soil slips, see Technical study FOUNDATIONS 1, para 5.15 *et seq*). Inclined or raking piles may be used **3** where there is appreciable horizontal loading.

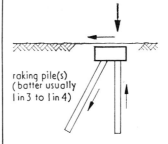

3 *Raking piles to resist large horizontal loads*

2.10 Because of the risk of increased differential settlement, deep and shallow foundations should not be mixed in one structure. For certain special ground conditions the most appropriate foundation types were discussed in Technical study FOUNDATIONS 1: seismic design para 5.12, swallow holes para 5.17, mining subsidence para 5.18, peat para 6.16, and filled ground para 6.18.

3 Items common to most foundation designs

3.01 Once all available information about the structure to be supported, its loading and site conditions (including the results of any soil investigation) have been evaluated, the most suitable type of foundation and its bearing stratum may be chosen. Allowance must be made for inferior lower strata and the ground water level (including possible future fluctuations). Ground water can affect both the method of construction and the loadbearing capacity and settlement characteristics of the soil, especially in permeable ground. At this stage the possibility of ground water lowering—either temporary or permanent—and any other geotechnical processes, plus their likely constructional problems should be considered (see Technical study FOUNDATIONS 1).

3.02 The effect of ground conditions on the durability of the foundations can now be assessed, and the concrete mix and steel cover decided. If there is still doubt about the distribution or concentration of sulphates in the soil, a further and final investigation may be necessary even if the building contract has begun. The next problem is an assessment of allowable ground bearing pressure, and completion of the closely linked settlement calculations.

3.03 Although a site investigation report may have made recommendations on all these points, they remain a fundamental part of the foundation design process. The designer, even if he does not have to make such decisions, does at least have to check recommendations from others.

Allowable ground bearing pressure

3.04 The *ultimate bearing capacity* for a typical foundation base is the average vertical pressure on the ground that leads to failure by shear. For a mass of soil with uniform properties c and ϕ (defined in Technical study FOUNDATIONS 1) this is calculated using formulae developed on a semi-empirical basis; these involve the shape and dimensions of the foundation, its depth below ground surface, the soil density, and the depth of water table. (The most generally used formulae are given in Information sheet FOUNDATIONS 3.) They are usually applied to clays, silts and semi-rocks such as chalk, marl and clay shale. Values of c and ϕ are obtained from undrained triaxial compression tests (Technical study FOUNDATIONS 1, para 7.18). Properties of sands and gravels are normally measured by penetration tests from which ϕ can be determined. But strength formulae are rarely used in such cases because settlement usually governs *allowable bearing pressure*, which is therefore determined directly from penetration results.

3.05 Plate bearing tests are sometimes used for estimating the allowable loadbearing pressure, mainly on the basis of settlement. With rock, experience or specialist advice is required to assess loadbearing capacity. Plate bearing tests and penetration tests are considered later.

3.06 When applying the c–ϕ formulae, samples from a number of points—perhaps at 2 m intervals in bores—in the zone effectively stressed by the foundation are tested (see Technical study FOUNDATIONS 1, para 4.11 *et seq*). The results usually show a 'scatter' **4**, and design values are normally near the lowest measured values unless these are obvious

freaks. This is particularly important with stiff fissured clays (eg London clay), as a large scatter is due mainly to variation of fissures in small samples.

3.07 The ultimate bearing capacity derived by a c–ϕ calculation (or sometimes by plate loading tests, see CP 2001[1]) is reduced to a *net* ultimate bearing capacity by subtracting the total overburden pressure at founding level, before dividing by a factor of safety to give the 'safe bearing capacity' (or 'presumed bearing value'). The factor of safety—to allow for deterioration in the soil strength and uncertainties in soil behaviour—is usually between 2·5 and 3. The higher value applies when conditions are particularly uncertain or maximum design loading is likely. If the soil is uniform and its properties are well known, or if the design loading includes considerable but *unlikely* live load allowances, this safety factor may be reduced to 2 and, for certain temporary works, even less.

3.08 The *allowable bearing pressure*, the maximum allowable net loading intensity on the ground, is then determined. It takes into account safe bearing capacity, expected settlements and restrictions on settlement. It is directly comparable with the actual net pressure q_n. Only with rock or other hard soils is this the same as the safe capacity. Unless settlements are small (or the case very straightforward), they should be checked. Observations of the behaviour of adjoining buildings with the same foundation conditions can be used to confirm the suitability of a proposed bearing pressure.

3.09 If the soil properties are well known from experience then it is possible with small or uncomplicated designs to use allowable bearing pressures proposed by an authority or derived from published tables like those in BS CP 2004[2] and BS CP 101[3]. Building Regulations[4] give acceptable minimum widths of simple strip footings for certain loadings on common soils. The allowable bearing pressure is usually increased by 25 per cent where the corresponding increase in actual pressure is due solely to wind forces on the structure.

3.10 So far the discussion has applied to vertical loading only. Appreciable horizontal loads may be resisted by friction or adhesion of the soil under the foundation and the passive resistance of the ground in contact with its vertical face **5**. This is mentioned in Technical study FOUNDATIONS 3, but is beyond our present scope. Raking piles **3** or other special supports may be necessary.

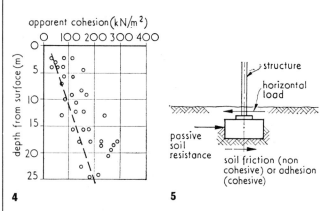

4

5

4 *Typical plot of triaxial test results for stiff fissured (over-consolidated) clay. Shows proposed 'design line' for cohesion at any depth. (Bearing capacity of base or pile is usually calculated from cohesion at its underside)*

5 *Resistance of foundation base to horizontal loading. Forces shown produce overturning moment, resisted by non-uniform vertical pressure on soil under base*

Further notes on choosing a bearing stratum

3.11 Once the allowable pressures on one or more possible strata (considerations mentioned earlier) have been estimated, the level and types of foundations can be decided or at least reduced to two or three choices. The following factors are important during this process:

a *Minimum depth of foundations* They should be below any zone susceptible either to frost heave (0·5 m from the final surface level is generally sufficient in UK) or to shrinkage or swelling. A limit may also be set by structural requirements and the assumed bearing pressure (see Information sheet FOUNDATIONS 3). The relationship to existing or future adjoining foundations or services may also affect the depth (see Technical study FOUNDATIONS 1, **26**).

b *Risk of breaking into underlying softer strata* The pressure on these strata should be roughly checked (using a calculation method discussed below) and compared to the estimated bearing capacity.

c *Minimisation of differential settlement* Normal practice avoids founding on different types of soil or at widely differing depths for one building. Differences in actual bearing pressure should be minimised, and where possible, only one foundation type used. Even with all foundations bearing on the same soil at the same pressure there is usually some differential settlement, but of course this is magnified if soil properties or pressure vary. Different types or sizes of foundations on the same soil will have different safe bearing pressures. For clay, settlements tend to be proportional to the actual pressure for a given size base or to the foundation width for a given pressure; with non-cohesive soils, variation in these factors generally has less effect on settlement.

d *Adjoining buildings* The effect of soil stresses from new foundations overlapping those from foundations of adjacent buildings must be considered. This can affect stability or settlement of both structures **6** (for discussion of underpinning see section 6 of this study).

e *Possible deterioration of bearing capacity during or after construction.* Influences which may have to be considered include water seepage, changes in water table, vibration, climatic effects, constructional disturbance, and risks of underground cavitation or subsidence. (These conditions were discussed in Technical study FOUNDATIONS 1.) Instability of inclined strata may occasionally be a problem. Soakaways should not be constructed near foundations.

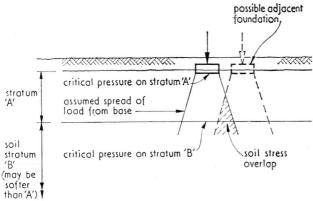

6 *Foundation load dispersal. If soil stress overlap is large, it will require more precise investigation*

Settlements

3.12 The various methods of checking settlement by calculation were outlined in Technical study FOUNDATIONS 1 para 4.22 *et seq*; more detail is beyond the present scope. An architect will probably normally need to do no more than make approximate estimates of the consolidation of clay soils. This is best done by splitting the soil into zones where properties are markedly different; these zones are then divided into layers not thicker than the maximum effective width of loaded area. Then the change in stress q_n due to the loading is calculated for the centre of each layer under the point at which the settlement is required. For this, contact pressure under each foundation element is assumed to be uniformly distributed and spread into the ground below at a fixed side slope. Alternatively a simple Boussinesq distribution may be used where loads can be taken as concentrated on the soil at a number of points. Finally the settlement of each layer is computed using measured or assumed soil properties.

3.13 This method is detailed in Information sheet FOUNDATIONS 3, together with references to other means of calculating stresses and settlements. When the structure is comparatively stiff, a more sophisticated approach that takes account of redistribution of soil stresses due to soil-structure interaction is necessary.

3.14 Settlements in other compressible soils may have to be calculated occasionally. If the relevant soil properties are known, the method described above can be used. For cases such as filled ground this does not hold. However, if plate bearing tests have been carried out, it should be easy to estimate likely short term settlements of the actual foundation bases. These will not include allowances for any interaction between foundations or the influence of different soils below the depth effectively stressed by the plate tests **7**. (Information sheet FOUNDATIONS 3 gives guidance on interpretation of plate tests and on the use of penetration tests for estimating allowable bearing pressures, especially for sands and gravels, on the basis of restricted settlement.)

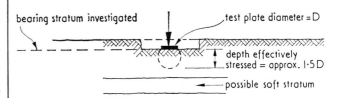

7 *Restricted value of plate bearing pressure in depth. D (test plate diameter) is usually appreciably less than foundation size. A possible soft stratum, though having little effect on test results, would be significant for full size foundation settlements*

3.15 Non-cohesive soils settle quickly; with foundations designed on bearing pressures recommended in published tables, or according to the Terzaghi-Peck standard penetration test relationship (generally considered very conservative), these settlements are usually small enough to be ignored. But in complex conditions settlements should still be checked (probably by an expert), especially if there are more compressible underlying strata. Full scale loading tests for piling may be used.

Clay settlements

3.16 Settlement of very stiff to hard over-consolidated clays (apparent cohesions greater than 150 kN/m²) are so small that they are rarely calculated except for heavy or complex structures. For clays that have half this apparent cohesion value, calculations are usually not required for isolated foundations of up to about 3 m wide, provided the factor of safety against shear failure is of the order of 4. In other cases settlements should normally be checked. Additional movement may occur with soft plastic clays due to gradual lateral displacement.

Silty soil settlement

3.17 As for clay, settlements in plastic silty soil should be checked by consolidation tests. But foundations in very sandy or coarse silts are often designed by penetration test.

3.18 Settlements of rocks, even of the softer varieties, are normally negligible; although plate loading tests may occasionally be desirable on the weathered (softened) type of chalk. Subsidence settlements are a matter for specialist advice.

4 Pad and strip foundations

4.01 Pads and strips are both isolated shallow foundations; a strip is considerably elongated in one direction, either in order to support a line load (as from a wall) or a more or less closely spaced series of concentrated loads.

Pad foundations

4.02 Pads are usually square or rectangular in plan and support one main concentration of loading. Occasionally they may be combined to support two loads to 'tail down' possible overturning on a simple pad because of unavoidable eccentricity of loading **8**. A trapezoidal, or more complicated plan shape may then be economical, but the centroids of total loading and plan area should if possible coincide to avoid increased edge pressure **9**. This increased pressure is produced by any eccentricity of load, eg by an overturning moment on a column. The real ground pressures under a foundation are the result of interaction between the stiffness of the structure or base and of the soil. Their distribution is complex, some typical examples are shown in **10**. Normally it is enough to assume uniform pressure under a pad for centroidal loading or, if there is an eccentricity, that it varies linearly*. (Formulae for pressure calculation are given in Information sheet FOUNDATIONS 4.)

4.03 A pad foundation may then be designed using simple statical principles (for moments and shears). The angle of dispersal of concentrated loads through plain concrete depends both on its tensile strength and on soil characteristics. But if—as is usual—it is taken at 45° to vertical, reinforcement can then be omitted if pad depth is sufficient. Where required, reinforcement is usually a bottom mat or top and bottom mat where there are two loads (see Information sheet FOUNDATIONS 4). Where maximum ground pressure is at the edge, as in **9**, it must be kept within the maximum allowable bearing pressure. But this can be waived if rotational movement is restricted and the total overstress does not exceed about 25 per cent, without extra allowance for wind pressure. Still the *average* ground pressure must not exceed that allowable.

4.04 A reinforced concrete base usually has to be shuttered in its excavation. If depth permits, most of a pad foundation may be mass concrete cast against the excavation faces, with any reinforced concrete confined to a much smaller base at high level, where this proves economical.

Strip foundations

4.05 Strips vary from a simple plain concrete footing for a lightly loaded brick wall, to a heavily reinforced foundation supporting a number of columns. Most of the discussion on pads is also relevant to strip foundation design, but there is the added complication of load distribution to the ground by longitudinal beam action, as in **10b**.

4.06 Where loading is fairly evenly distributed along the length of a strip, contact pressure on the ground is assumed to be uniform or to vary linearly in accordance with the statical principles for pad pressures. Design is then straight-

*Very large or flexible foundations, especially on clay, may require more refined analysis

forward—pressure is always assumed uniform across the width of the foundation, which is, as far as possible, symmetrical about the load line. The foundation should cantilever each side by load spread or the use of bottom transverse reinforcement, as with pads. But where loading is irregular or there is a very stiff foundation on clay, the ground pressure distribution can be much more complicated (see **10b**); design of the foundation as a longitudinal beam then requires considerable skill.

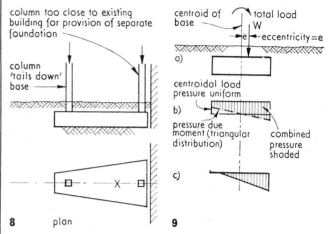

8 *Example of combined pad foundation. Centre of gravity of loads and centroid of foundation (X) should coincide if possible.*

9 *Typical ground pressure under pad due to eccentricity. Eccentricity (e) is equivalent to centroidal load (W) and overturning moment (We), **b** is diagram of ground pressure; **c** is the alternative if moment 'tension' is greater than uniform pressure, ie if W is outside mid-third of base (soil tension not allowed)*

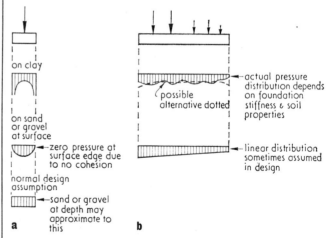

10 *Pressure distribution under spread foundations: **a** pad loaded centrally (assumed rigid); **b** continuous foundation, strip or raft (if very rigid will approximate to pad)*

4.07 Strips may be designed to span or cantilever over possible 'soft spots' in the ground. Then the normal bearing pressure should be particularly conservative. Longitudinal reinforcement (mats or longitudinal rods with transverse links), if required, should be a reasonable minimum at both top and bottom. This covers uncertainty as to where reversals of bending moment might occur. Irrespective of design requirements, the absolute minimum thickness of any strip foundation should be 150 mm, and its width for practical reasons should be at least 50 mm greater than that of the supported structure. On a sloping site strips should be laid horizontal and stepped as necessary. The width should be varied if loading changes considerably.

4.08 It is usually economical to fill strip excavations with mass concrete; this is a possible alternative to a thinner (perhaps reinforced) footing at the trench bottom, with walls or columns taken down to it. A narrow deep strip may be the best solution if it is possible to excavate the trench by machine in stable ground; otherwise inverted TS may be suitable where longitudinal moments are high but soil pressures low **11**.

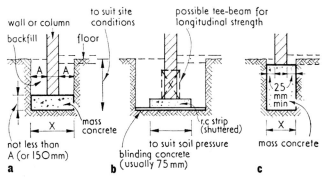

11 *Three types of strip foundation:* **a** *width X may have to be greater than soil pressure requirement if hand dug;* **b** *rc foundation is usually required when wide but shallow, or for longitudinal bending;* **c** *is an alternative to* **a**, *and sometimes also to* **b**. *Width X may be quite narrow if machine dug and the soil pressure permits*

5 Rafts

5.01 Rafts—really two-way strip foundations—are used for one or more of the following reasons:
a To reduce the ground bearing pressure in order to found on poor or unstable ground, or to limit total settlement
b To minimise differential settlements
c To provide a completely tied-together foundation system
d Where separate foundations would occupy a very large part of the total building area.
Rafts are particularly useful in areas subject to mining subsidence, swallow holes or other ground movements. (Stiff basements and ground bearing slabs are particular forms of rafts.)

Design principles

5.02 Rafts, except in very simple cases, are the most troublesome type of foundation from a design point of view. This is because of the difficulty in assessing the distribution of ground pressure (**10b**) and hence of bending moments and shears. Unless great skill and care are exercised, designs will either be uneconomical or potentially unsafe.

5.03 A pressure distribution which corresponds with the loads, as well as with the relative stiffnesses of soil and structure, must be found. (The raft itself may be stiffened by the superstructure, although this is often ignored.) If the raft is flexible the loads will tend to pass almost directly to the soil. As it becomes stiffer, loads are distributed to a greater extent between their points of application, but the overall deformation of the soil begins to impose a further redistribution. In the extreme case of a very rigid raft on clay, there will be a large concentration of pressure near the edges, and hence a large overall bending moment which may even tend to override local spreading moments.

5.04 Almost every engineer has his own answer to this problem. At one time it was very commonly assumed that ground pressure was distributed uniformly or varied linearly according to the eccentricity of total load relative to the raft centroid **9**, **10b**. But this can give too big a spread and yet not allow for overall moment mentioned above; also the moments and shears thus calculated are often quite unrealistic.

An early attempt to allow for the interaction of structure and soil by hand calculation was the 'soil-line method' of Professor A. L. L. Baker[5]. Other manual methods include the assumption of uniform or varying soil 'springs' which have a stiffness sometimes known as the 'modulus of subgrade reaction' (one of the best of these is due to Terzaghi[6]), and various semi-empirical or intuitive ways of distributing ground pressure. Practical hand methods cannot allow sufficiently, however, for interaction of stresses in the soil—and hence its overall deformation—to cover all cases where this is of importance. Fortunately we now have computer facilities for more refined analyses. But since these are essentially based on assumptions of behaviour which may not be fully realised in practice, their results still need skilful interpretation.

5.05 Discussion of the basic principles of raft design cannot be attempted here; the designer inexperienced in this subject should rely on specialist advice. The only exception might be for lightly loaded small rafts; these can safely be designed on the linear pressure distribution of **9** and **10b** (see Information sheet FOUNDATIONS 5). Even so, the quantity of reinforcement should be conservative—a reasonable amount at both top and bottom. The raft's ability to cantilever or span soft spots in the ground should also be checked. The centroids of both raft and loads should, if possible, coincide. These points are also applicable in principle to more complicated designs.

5.06 Settlement rather than safe ground bearing capacity normally governs raft design. Nevertheless preliminary sizing should be based on a conservative bearing pressure (assuming at least a factor of safety of 3) to allow for the possible occurrence of higher pressures. The final pressures at the edges, where shear failure or plastic flow could occur, should be checked against this.

5.07 Although a raft may reduce differential settlements, total settlement can still be high, since the significant pressure bulb is large. However, it may not be as deep as the standard 1·5 times width (see Technical study FOUNDATIONS 1, para 4.11) because if the bearing pressure is lower than normal it will not produce such a significant stress change at depth in relation to existing overburden pressure.

5.08 A raft is normally solid reinforced concrete (possibly forming the lowest floor), of thickness to suit the calculated stresses and assumed stiffness. It should be deep enough to make shear reinforcement unnecessary. Sometimes it may be thickened locally in high stress zones, eg under columns, and also where there are sinkings in it. But every effort should be made to avoid undue disruption by pits or trenches for services. For example, the raft might be dropped a metre or so, providing a separate floor, as in **12**. Highly stressed rafts may be stiffened by beams or designed as cellular structures; often some stiffening is provided by the supported structure.

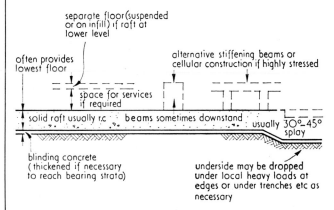

12 *Various raft conditions, showing possible suspended floors, step etc*

But the edges should either be deep enough to be uninfluenced by soil movements due to environmental conditions, or otherwise protected.

5.09 Rafts are occasionally supported by piles if soil conditions necessitate this. The construction, though really an extended pile cap **2** may be designed as a raft, with the loads on the piles taking the place of the pressure on the ground (see also Information sheet FOUNDATIONS 7). Some load from the raft will still be transmitted directly to the soil, but being rather indeterminate this is commonly ignored. Similar considerations apply to piles under strip foundations. If the applied loads are very widely spaced a raft foundation is usually impracticable, and simple piles may then be the most economical foundation.

Floor slabs

5.10 A building's lowest floor is basically part of its foundation system. It is usually a ground bearing concrete slab, but there may be a good case for suspended floors where, for instance, there is a sloping site with very low bearing capacity ground, or where it is otherwise unstable (perhaps because of moisture movements in clay or the possibility of underground cavitation). A compromise solution uses poor ground as a platform on which to cast a concrete slab; but the slab is reinforced so that it can span on to a supporting structure. (These points are considered more fully in the AJ Handbook of Building enclosure[7].)

5.11 Thus a suspended concrete floor can form an integral part of the foundation system or substructure. But the present consideration is the ability of a *ground bearing slab* satisfactorily to transmit its loading to the supporting soil. Here the distinction is that a structural raft is required principally to support building loads at certain fixed positions, whereas the loading on a floor slab is essentially movable and variable.

5.12 Overall floor loads are generally quite light and, if well distributed, within the direct bearing capacity of the underlying ground, even near its surface. But in reality there will be a number of more or less concentrated loads, which the slab must spread to within the allowable ground bearing pressure. The slab can also reduce differential settlements between heavy loads, and may be required to span or cantilever over soft spots or soil cavities. Where concentrated loads are light, the structural design is best derived from experience. (Suggestions for different conditions are given in Information sheet FOUNDATIONS 5.) But in some cases, such as industrial buildings, quite heavy point loads occur. Then design may be based on an assumed load spread, derived from Terzaghi's work on subgrade reaction[6].

5.13 On reasonable ground a single layer of fabric reinforcement in the top of the slab provides crack control, and the thickness of plain concrete is relied on for load spreading. Then Older's formula (see Information sheet FOUNDATIONS 5) or sometimes even the assumption of a simple 45° load spread gives the answer. But for heavy loads on poor ground it may be more economical to work to a wider spread, using the Terzaghi approach with comparatively heavy reinforcement in a thin slab. Top and bottom steel is generally provided throughout in poor soil conditions.

5.14 On most soils a bed of 100 to 150 mm of well compacted hardcore under the slab usually gives a satisfactory formation. This is also an economical way of achieving further load spread on to the ground **13**. If vehicular traffic on a floor is considerable, it may be best designed as a road slab[8].

Basements

5.15 A raft foundation is often used for a structural basement. Where a nominal ground slab joins separate bases,

it may crack if much pressure is transferred because of yielding soil. This problem is overcome by a properly designed raft. A similar situation can arise if there is an appreciable upward water pressure. But a properly designed raft overcomes these problems. A functionally necessary basement (eg under a tall building) can stiffen a heavily loaded raft at comparatively little extra cost if the structural layout is suitable. It can also be an economical way of founding on better, deeper soil. But where a basement is not needed, or stiffening is difficult, piled foundations may be cheaper.

5.16 A structural box with a raft in the bottom may be used effectively as a *buoyancy basement* to reduce net soil stresses and settlement produced by a heavy building **14**. (Such a box used solely for structural purposes is known as a *buoyancy raft*.) Again it may be cheaper to pile than to provide such a structure, or piles may be used under a basement raft where necessary. The design of these foundations is a task for the experienced engineer. Retaining walls enclosing a basement, and related excavation and constructional problems, will be covered in Technical study FOUNDATIONS 3, Retaining structures.

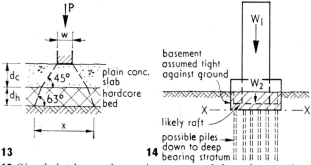

13 *Simple load spread sometimes assumed through ground slab. Where load P is applied over area w × w, the pressure on soil (width X) is approximately given by*

$$\frac{P}{(w + 2d_c + d_h)^2}$$

The 63° in the diagram is in fact a 1 in 0.5 slope

14 *Principle of buoyancy basement. With gross building weight above level XX of W_1, and weight of soil and any water displaced by basement (hatched) of W_2, effective load for stress and settlement at level XX is $(W_1 - W_2)$*

6 Underpinning

6.01 *Underpinning* implies the provision of extra support beneath an existing structure. In the widest sense it is the construction of extra foundation capacity without removing an existing superstructure and is done because of failure or deterioration of old foundations, anticipated undermining of these by a new construction nearby, or simply because it is desired to place extra load on the structure. It can involve highly specialised work[9] in providing supporting beams, stools, bases, piles or jacking arrangements in or adjoining the existing construction. But normally, new material is placed under existing spread foundations to deepen them and perhaps increase their bearing area. This may take the form of a grouting process (see Technical study FOUNDATIONS 1, **23**), or, more usually, the provision of mass concrete stooling. (Such concrete underpinning, carried out by stages to minimise disturbance, is detailed in Information sheet FOUNDATIONS 6.)

6.02 The principle of underpinning in order to avoid undermining by both excavation or construction nearby was mentioned in Technical study FOUNDATIONS 1. In **15**, angle β, commonly taken as 45° in good ground, should be reduced to 30° in wet clay, and zero if the excavation is very close. Tomlinson[9] suggests that in stable ground this angle can be increased to about 63° (2:1 slope), but that underpinning, where necessary, should be taken to the bottom of the new excavation. The 1939 London Building Act[10], following a different principle, broadly required β to be zero within a distance of 3 m but 45° at up to 6 m away.

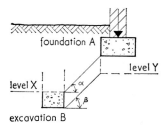

15 *Excavation below level of nearby foundation. Knowing required depth for excavation B, existing foundation A should be underpinned to level Y (based on angle β), but if foundation A does not yet exist, excavation B should be backfilled with concrete at least to level X (based on angle α. Values for α and β are frequently taken at 45°, but see para 6.02*

6.03 The effectiveness of temporary supports to the side of new excavation is crucial; it may be adequate to cut back to a stable slope where a conservative angle of limitation is adopted. There may be occasions when it is worthwhile accepting risk of damage to a building adjoining the excavation rather than incurring the expense of underpinning it.

7 Piled foundations

7.01 Most piling systems are proprietary. They are operated and often designed by specialist firms, and vary quite widely in their characteristics*. Engineers' ideas about behaviour and design of piles, based on research and experience, are continually developing and so final decisions on a job, with regard both to design and to construction, still depend largely on expert skill. So although much could be said about piling, only a brief description of processes and problems, with a few principles concerning choice of pile type, design and construction, are given here.

Types of piles

7.02 The relationship between main types of pile is shown in **16**—*displacement* and *replacement* refer to the fate of the

*Various systems are described in *Specification*[11]

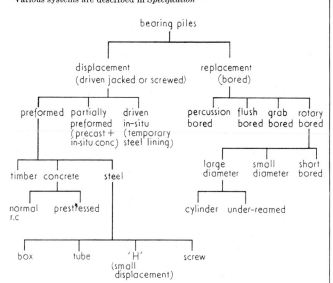

16 *Analysis of basic pile types*

soil which occupied the space finally taken up by the piles. (For uses, characteristics and details of the common types in building work, see Information sheet FOUNDATIONS 7). Sheet piling is normally used for retaining structures and, with various bearing type piles which may also be used for retaining, they are discussed in Technical study FOUNDATIONS 3.

Choice of pile type

7.03 From the constructional point of view, piles differ widely—both in types and in detailed variation; they also behave differently according to the strata in which they are installed. In terms of soil behaviour piles divide into (a) end bearing, (b) frictional, and (c) some combination of these two extreme types. These are illustrated in **17**, which also shows negative skin friction, and a large diameter under-reamed pile. This is much used for heavy buildings founding in firm clay. Not every kind of pile is always suitable or practicable, but there are usually several choices for any particular job, and the decision may only be possible after getting competitive tenders.

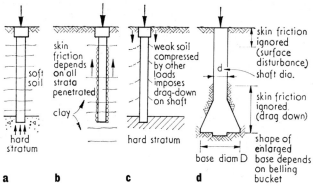

17 *Some piling characteristics:* **a** *end bearing pile;* **b** *frictional pile;* **c** *negative skin friction;* **d** *large diameter under-reamed pile in firm clay. In* **d***, base diameter D can be up to three times shaft diameter d*

Preparation of tender schemes

7.04 Specialist piling contractors can usually undertake piling design but it is generally preferable to use an independent consultant who can advise on the suitability of various types of pile. Contractors naturally prepare schemes that suit their own methods, often making it difficult to assess the relative merits of different proposals. Consultants will try to produce one basic design and elicit truly comparable and competitive tenders from different specialists. Where this is not practicable, alternative schemes may be prepared for which quotations can be obtained.

7.05 Once a contract is placed, the consultant usually agrees final details with the specialist contractor in the light both of the conditions encountered during the work and of the contractor's method of installation; this avoids diminishing the contractor's responsibility. At this stage advantage can be taken of any features of the chosen system which permit economies on basic tender design.

Piling contracts

7.06 Piling for foundations can be executed as part of the main contractor's work, under a separate contract before the main contract begins, or carried out as a nominated subcontract. The main information to be given when inviting tenders is listed in Information sheet FOUNDATIONS 7. This includes a selection of general technical specification clauses, proposed by the Federation of Piling Specialists to produce comparable tenders for proprietary types of cast-in-place driven, or bored piling. These clauses deliberately omit matters which should be in Conditions of Contract, but are very informative on the constructional aspects of this important type of piling.

Pile design

7.07 For preliminary schemes, piles are usually designed in end-bearing by c-ϕ type of formula (para 3.04). With clay, shaft-friction (skin-friction or adhesion) is based on some semi-empirical fraction of the cohesion; for non-cohesive soil it is usually based on an entirely empirical stress. In homogeneous clay, adhesion is developed up to a near-ultimate value at a settlement considerably less than that required to mobilise the full base resistance. However, settlement is often the criterion which determines allowable load per pile.

7.08 A more complicated, if refined, method of designing large bored piles in London clay (with or without under-reams) considers interaction between the shaft and base resistance and the amount of settlement. This was proposed by T. Whitaker and R. W. Cooke in 1966[12]. Closely spaced piles tend to settle more than an isolated pile, and in some circumstances fail at less than average load. This design problem has been extensively investigated by T. Whitaker, and reference may be made to his recent book[13].

7.09 Normally, final design is confirmed by selected load testing; but if the penetration per blow with driven piles is measured, this itself forms a continuous load test. Then the allowable load on a single pile can be estimated by a 'dynamic' formula.

Load tests

7.10 Information on isolated pile behaviour under short term loading may be obtained from load tests. With closely spaced piles, interaction takes place in the soil and settlements may be greatly magnified. (Some suggestions for calculation are made in Information sheet FOUNDATIONS 7). It is usual to load test either test piles or actual working ones. Testing working piles is, of course, essentially confirmatory and often performed where for some reason the quality of a particular pile is suspect. Preliminary testing of special piles to failure could yield information to enable considerable economies in number, size or depth of the working piles.

7.11 On large jobs preliminary tests are usually advisable, except perhaps with large or under-reamed piles because of the cost. On smaller projects, where soil conditions and the use of the particular pile type in similar circumstances are known, testing may often be omitted, especially with driven piles (subject to local authority agreement). But this may involve designing with a greater safety factor, because of the semi-empirical nature of piling design.

References

1 BS CP 2001: 1957 Site investigations. BSI [(11) (AJs)] £3.00

2 BS CP 2004: 1972 Foundations [(16)] £6.90

3 BS CP 101: 1972 Foundations and sub structures for non-industrial buildings of not more than four storeys. British Standards Institution. [(1–)] £1.35

4 THE BUILDING REGULATIONS 1976 SI 1676. HMSO £3.30; The Building (First Amendment) Regulations 1978 SI 723. £0.60p [(A3j)]

5 BAKER, A. L. L. Raft foundations. 1957, Cement and Concrete Association [(16)], third edition o/p

6 TERZAGHI, K. Evaluation of coefficients of subgrade reaction. *Geotechnique*, 1955, December [(L4)]

7 AJ Handbook of Building enclosure. London, 1974, Architectural Press £6.95 paperback; Section 3: EXTERNAL ENVELOPE: Lowest floor and basement [(9–)]

8 TRANSPORT AND ROAD RESEARCH LABORATORY Road note 29. A guide to the structural design of pavements for new roads, third edition, 1971, HMSO [12 (90.22)] £0.70p

9 TOMLINSON, M. J. Foundation design and construction. Third edition, 1975, Pitman [(16) (A3)] £15.00 (paperback £9.90)

10 GREATER LONDON COUNCIL London Building Acts 1930–39; London Building (Constructional) Bylaws 1972, GLC £1.20 (paperback) [(Ajn)]

11 Specification 1978, Architectural Press [Yy (A3)] £16.00

12 WHITAKER, T. and COOKE, R. W. Proceedings of the symposium on large bored piles. ICE 1966 (17.2) o/p

13 WHITAKER, T. The design of piled foundations. Oxford, second edition, 1975. Pergamon Press [(17) (A3)] £7.25 (paperback £4.25)

14 PITT, P. H. and DUFTON, J. Building in inner London. London, 1976, Architectural Press £5.50 paperback

A further bibliography appears at the end of this section of the handbook.

Technical study
Foundations 3

Section 4 **Foundations and retaining structures**

Retaining structures

In this study JOHN RANSFORD *deals with structures designed to retain earth, or more specifically the lateral pressures exerted by soil, including the effects of ground-water and of loads applied to the ground surface. The earlier technical studies and information sheets on foundations provide a helpful background to this discussion*
See also Information sheet FOUNDATIONS 8

1 Earth pressure

Types of earth pressure
1.01 A soil face standing at an angle steeper than its natural angle of repose will exert a pressure on the structure which retains it. There are two extremes of pressure:
a Active pressure: a minimum, developed when the structure moves sufficiently away from the soil to mobilise the full shear strength of the latter against outward movement.
b Passive pressure: a maximum, developed when the structure moves sufficiently into the soil to mobilise its full shear strength against inward movement.
An important intermediate case when the structure moves a negligible amount is known as *pressure at rest*.
1.02 The design pressure on flexible walls, such as free-standing, gravity and cantilever walls, and on all timbering and steel sheet piling, may generally be taken as the active pressure. Pressure at rest applies only to relatively un-yielding walls, eg rc walls, supported by floor slabs. Passive pressure (passive resistance) is used when calculating the resistance of a soil face to an *applied* pressure, eg in preventing the sliding of a wall or base.

Calculation of earth pressure
1.03 Early mathematical theories of earth pressure took soil to be a dry cohesionless material, and assumed that active or passive pressure was realised when the soil moved sufficiently to develop surfaces of failure **1**. In fact the presence of cohesion reduces active pressure but increases passive. Cohesive soils can stand vertically to a limited extent, as shown by tension cracks which form behind a retaining wall under active pressure **2**. These have no connection with the shrinkage cracking in clay. (Technical study FOUNDATIONS 1 section 5.)
1.04 These theories have been developed into practical forms, which allow for inclinations in the surfaces of the

retained soil and in the structure itself, and for friction or adhesion between the structure and the soil. Pressures normal to the wall surface are usually given in the following general forms:
Active $p_a = K_a \cdot p_o - K_{ac} \cdot c_o + p_w$ *equation* 1
At rest $p_h = K_o \cdot p_o + p_w$ *equation* 2
Passive $p_p = K_p \cdot p_o + K_{pc} \cdot c_o + p_w$ *equation* 3
where p_o = effective vertical pressure in soil (see Technical study FOUNDATIONS 1 para 4.06); p_w = pore water pressure $(\gamma_w \cdot h_w)$ (see same para); and c_o = cohesion, ie shear strength at zero normal load (all at the depth considered). Vertical pressure (p_o) is multiplied by K factor to give horizontal pressure. K_a, K_{ac}, K_o, K_p, K_{pc} depend on the soil properties (cohesion c_o and angle of internal friction ϕ), wall friction (or adhesion) and inclinations of the surfaces.
1.05 There may also be components from *soil* pressure acting parallel to the wall surface. Applied loading may be covered by allowing an extra height of soil **3**. Sloping ground surfaces are sometimes treated as a species of surcharge in conjunction with K-values for a horizontal surface. (Surcharge is any applied load on soil surface.)

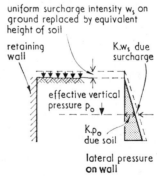

3 *Typical pressure distribution on retaining wall (drawn for non-cohesive soil)*

1.06 With cohesive soil, equation 1 gives a negative soil pressure (excluding p_w) down to a calculable depth. This is normally disregarded because tensile strength is limited and uncertain, but soil pressure is assumed zero for this depth, ie p_w only occurs in tension cracks. The water table in clay soils should never be assumed for design purposes to be lower than the top of the clay, unless the back of the wall is adequately drained.
1.07 With clay near the surface, allowance should be made in equation 2 for swelling (K_o probably at least 1). For a vertical wall and horizontal ground surface, K_o usually lies between $0 \cdot 4$ and $0 \cdot 6$ for non-cohesive soil and for normally consolidated clay at an appreciable depth, K_o is between

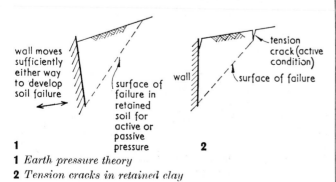

1 *Earth pressure theory*
2 *Tension cracks in retained clay*

0·6 and 0·75. But for undisturbed, over-consolidated clay, K_o may exceed 1·0 (up to about 3·5 for London clay[1]), although 0·5 may be taken for the effect of local loading above the clay considered as a surcharge.

1.08 Design must always account for the possibility of reduction in c_o owing to long term softening of cohesive soil. Indeed, in this context clays may eventually behave as non-cohesive soils ($c_o = 0$; ϕ perhaps of the order of 20°, a value determined from drained triaxial tests) for both active and passive pressures, after the water content has adjusted to the pressure condition.

1.09 General information on earth pressures and design requirements is given in CE CP 2[2]; further useful data is provided by Reynolds[3]. The selection of suitable design values requires considerable skill especially with cohesive soils. But, as a guide, K-values for active and passive pressures on a vertical wall with a horizontal ground surface usually lie in the following ranges (see table I). It is recommended in CE CP2 that p_a for cohesive soil should never be taken as less than 4·7 kN/m² per metre in depth.

Table I Active (K_a) and passive (K_p) pressures on a wall

Non-cohesive soil ($c_o = 0 \therefore K_{ac}.C_o = 0$, etc)	Clay ($\phi = 0$)
$K_a = 0\cdot15$ to $0\cdot50$	$K_a = 1$ $K_{ac} = 2\cdot0$ to $2\cdot8$
$K_p = 2\cdot0$ to $12\cdot0$ (but mostly $< 6\cdot0$)	$K_p = 1$ $K_{pc} = 2\cdot0$ to $2\cdot6$

1.10 Equations 1 to 3 give pressures increasing linearly with depth **3**. With strutted excavations and anchored sheet pile walls, where local yielding of the structure can occur, a redistribution and increase of active pressure is usually made **4**. Some engineers consider this should be extended to rigid walls and 'at rest' pressures.

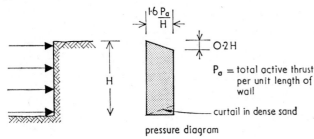

4 *Typical redistribution of pressure on a strutted excavation (sand)*

1.11 Where the retained soil is a fill rather than naturally occurring ground, the design coefficients may have to be assessed semi-empirically. Useful information on this is given by Terzaghi and Peck[4], which also takes into account drainage conditions. In some cases frost action can be a serious problem.

1.12 Vibration or flow of ground water can substantially increase the pressure from certain soils and reduce their passive resistance, as also can the large scale ground movements considered in Technical study FOUNDATIONS 1 section 5.

2 General design requirements

Effect of water in the ground
2.01 Water behind a retaining structure can increase lateral pressure; it may also reduce the bearing capacity of ground under the base and the resistance to sliding and may possibly penetrate the structure, affect durability, disfigure the surface or be otherwise unacceptable. It is therefore common practice to provide drainage in the backfill behind a retaining wall and also at the toe of the base of a freestanding wall if necessary. Detailed guidance is given in CE CP 2.

2.02 Unless the permanent efficiency of the drainage system can be guaranteed it is wise to assume that water will occur behind any retaining structure at some stage in its life— even on dry sites. For example surface or ground water may gradually accumulate due to a damming or ponding effect. It is common practice, assuming soil investigation has not indicated a greater requirement, to design storey height walls for a water table at one-third their height. This may have to be increased for clay.

Basis of structural design
2.03 A retaining structure must be capable of resisting the bending, shears, and compressions or tensions induced by all vertical and lateral forces. When the loads have been evaluated structural design proceeds in accordance with standard principles for the material used, as described elsewhere in the handbook. Concrete mix and reinforcement cover should generally be as for foundations—see also sections below on basement construction and diaphragm walls. Note that tension in brickwork, masonry or plain concrete must be restricted (CE CP 2 permits a limited amount in certain circumstances) and indeed prohibited altogether where water pressure can occur. Tension in concrete reinforcement should be limited to mild steel stresses if the structure is required to be watertight.

2.04 Overall stability of the structure must also be satisfactory, as regards overturning, sliding and pressure on the ground under its foundation. CE CP 2 recommends a factor of safety of 2 against overturning and sliding, although other authorities propose 1·5 especially for temporary conditions. Resistance to sliding may be provided by passive earth pressure, base friction or clay adhesion, as well as by structural support. The possibility of a rotational slip (Technical study FOUNDATIONS 1 para 5.15) must not be overlooked but is normally only a problem with cohesive soil, then the required factor of safety is 1.5.

2.05 The forces acting on typical retaining walls are illustrated **5a, b, c**.

3 Types of retaining structure

3.01 The main types of retaining structure in building construction may be divided into:

a temporary, eg upholding of sides of excavation cofferdams. These are mainly a matter for the contractor, although an architect or his consultant may be required to approve the contractor's proposals.

b permanent, eg basement walls, separate retaining walls. Some structures, such as diaphragm walls, can fulfil both functions.

Upholding of sides of excavations
3.02 Unless excavation can be carried out as open cut with the sides battered to a stable slope, or is in rock, some support must be provided to sides of all but shallow excavations. Even rock, if weakened by fissures, joints or weathering, may require support. Undue risks are sometimes taken in allowing cohesive soil to stand almost vertically for an appreciable depth. It can rapidly become unstable due to the action of weather and water.

3.03 Other problems in excavation work, eg ground water lowering, exclusion of water, dealing with adjoining construction and general precautions which should be observed, have already been discussed in Technical study FOUNDATIONS 1 sections 9 and 10.

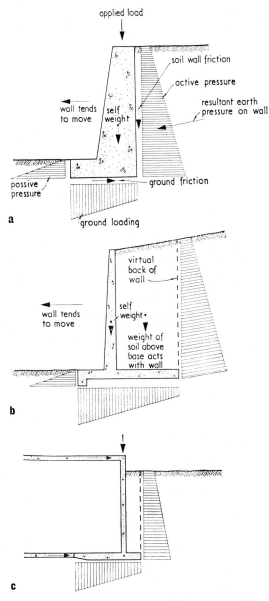

5a *Forces acting on a mass retaining wall*
5b *Forces acting on a cantilever retaining wall*
5c *Forces acting on a basement retaining wall*

3.04 The necessary side support will depend on site conditions. The 'sheeting' to the earth face may be 'open' (discontinuous) for cohesive soil to shallow depths and for some kinds of poor rock; otherwise it must be 'close'. The following are the principal types of support used.

Timber sheeting

3.05 For shallow depths especially, timber sheeting is still sometimes employed. Vertical poling boards can be used when the ground will stand vertically to a sufficient depth for these to be placed **6a**. If it will stand only to a small depth, horizontal boarding is sometimes used instead **6b**. More usually, and especially if the ground is very poor, vertical runners are driven **6c**. In each case supports (usually timber) are added as the excavation becomes deeper, and additional stages of sheeting may be constructed inside the previous. The supports shown assume that the excavation is carried out in trench (leaving an earth 'dumpling' in the centre of a large excavation until a self-supporting retaining structure can be constructed) or that some other means of providing horizontal strutting is available. Otherwise raking shores may be used instead of struts.

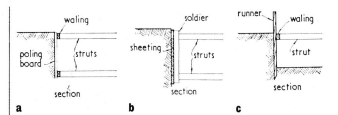

6a *Poling boards held by walings and struts*
6b *Horizontal sheeting supported by soldiers and struts permits excavation by one board depth at a time*
6c *Horizontal members (walings) are put into position first to enable runners to be driven vertically*

Other methods of upholding

3.06 It is now common practice to use steel trench sheeting (light sheet piling) instead of timber runners. In deep excavations heavy sheet piling may be driven (para 3.07) or horizontal sheeting of precast or in situ rc can be used, possibly with steel supports. Deep soldiers may be steel sections either driven into the ground or placed in pre-bored holes. Internal struts may be replaced for convenience with ground anchors placed in the soil behind the sheeting. If near the ground surface these can be simple ties to anchor blocks, but usually holes are bored into the ground at the required levels to receive cables which are grouted in and then generally tensioned both as a test and also to minimise inward movement of the sheeting. Such movement, almost inevitable with internal strutting, can cause serious settlement of surrounding ground[1]. Temporary supports such as sheet piling, bored piles and diaphragm walls may also be incorporated in the permanent work. These can act as water cut-offs, and may be designed to cantilever upwards or be provided with supports as described above.

Sheet piling

3.07 This may be timber or precast concrete, but usually consists of patent interlocking rolled steel sections **7**—particularly for temporary work. These are normally driven to full depth before excavation commences, usually by diesel, steam or compressed air hammer; but vibratory or jacking techniques are sometimes used in order to minimise noise and vibration. Interlocks provide a fair degree of watertightness. Sheet piling may be designed as a cantilever for exposed depths of up to 3 or 4 m, otherwise it is supported by steel walings and ground anchors, internal ties or bracings near the top. Depths of more than 8 to 10 m usually require support at intermediate levels. In addition to any water cut-off requirement the piling must penetrate sufficiently below the exposed depth for stability, and it is normal practice to achieve enough penetration to develop fixity in the ground, even though the sheeting is also to be tied. Typical pressure and bending moment distribution for such a case are shown in **8**. The simplified pressure distribution curve used in design can be determined from the difference of active and passive earth pressures, once the point of contraflexure (zero bending moment) has been assessed. However, sheet piling design requires specialised skills and is therefore the province of the engineer. Guidance for preliminary designs may be obtained from handbooks published by specialist firms and these also give useful information on practical details such as corrosion problems. Piling may be used in up to about 18 m lengths but penetration into hard ground can be limited by driving resistance. For this reason, the cross-section may have to be greater than that calculated from the earth pressure; also it has to provide a margin against corrosion.

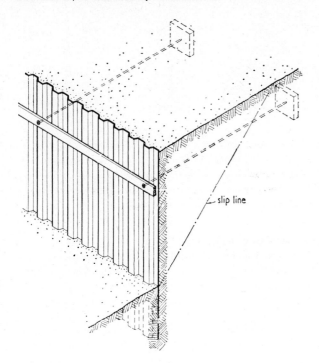

7 *Sheet piling*

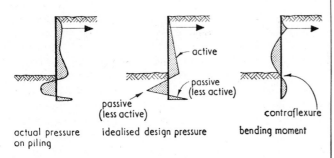

actual pressure on piling   idealised design pressure   bending moment

8 *Typical pressure and moment diagrams for a sheet pile wall fixed at bottom and anchored at top*

Cofferdams

3.08 These are usually temporary structures, built in waterways or waterlogged ground, to exclude water from constructional work. They may be up to 15 m in depth and are used either to enclose fairly large areas or else as relatively small box-type structures within which the work is carried out. The most common material is steel sheet piling, but solid embankments, cribs (described below), bored piling and sheet piling in other materials can be used. The typical, cofferdam shown in **9** is piled and comprises double walls in steel with an infill for stability.

Bored piles

3.09 Bored in situ concrete piles are used to form retaining structures in both temporary and permanent work. They may abut or interlink and have similar characteristics to the more commonly used diaphragm walls.

Diaphragm walling

3.10 This is now generally taken to mean a rc wall cast in a deep trench **10**. The work, carried out by a specialist firm, can involve one of several systems. The basic procedure is to construct the wall in panels up to about 6 m in length. This is done by machine excavating a trench to the required depth and immediately replacing the soil removed by a bentonite mud—a suspension of a special clay having thixotropic properties; in other words it behaves like a liquid when agitated but when at rest forms a gel capable of supporting the sides of the trench. The required reinforce-

ment is then placed in the form of a cage, and concreting carried out by tremie pipe from the bottom upwards so as to displace the bentonite, which may then be reused to a limited extent. In parts of the wall the concrete can be blocked off, and reinforcement provided for future structural connections. Some progress has recently been made in the application of precast panels to this system. Design for the temporary condition after excavation is generally similar to that for sheet piling. If incorporated in the permanent work a diaphragm wall may then usually be treated like a normal basement or retaining wall. 'At rest' earth pressures are assumed if the wall has unyielding lateral support above its 'toe-in'. The concrete quality should not be less than $25 \cdot 0$ N/mm² 28 day cube strength, and the minimum reinforcement cover 75 mm. Nominal thicknesses can vary from 500 mm to about 1000 mm and the depth up to about 35 m. A vertical tolerance of 1 in 80 is usual; bulges in the wall may occur in certain soil conditions. The surface exposed by excavation will probably require trimming and facing-up unless covered by a separate skin or wall in the permanent construction.

3.11 A diaphragm wall can act as a water cut-off but joints between panels may have to be specially treated to prevent seepage. In cases where there is a lining wall, seepage water can be drained from the cavity. For purely temporary work steel sheet piling, probably withdrawn after use, is cheaper. But diaphragm walls can be economical if they also form part of the permanent work, and have the advantage that they can be installed without undue noise or vibration; also they can be positioned very close to existing buildings, which usually requires construction in short lengths, rather like underpinning[5].

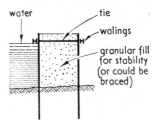

9 *Double wall sheet pile cofferdam*

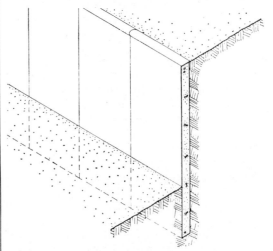

10 *Diaphragm wall*

Retaining walls

3.12 These are usually freestanding unless acting as part of a larger structure. They may be classified as:

a 'Mass' (or gravity) walls **11**. They resist overturning and bending stresses largely by virtue of their mass and are generally of brick, stone or plain concrete up to 2 m high. One form of mass wall seldom used in this country is the crib. It comprises a built-up series of open box-like sections of timber or precast concrete, into which earth is placed so that the whole acts as a retaining wall (see CE CP 2).

b Flexible cantilever walls **12** are normally of in situ rc and up to about 6 m high. Lateral earth pressure is transmitted to the ground beneath the base by the flexural strength of the rc. Where possible part of the base should be extended under the backfill so that its weight can act to increase stability.

c Counterfort walls **13** are normally used for retaining walls of greater depth, and act like a flexible cantilever except that the wall stem is stiffened by counterforts built off the base. Further stiffening can be achieved by the addition of horizontal ribs, and further stability by providing lateral anchorages or ties. Occasionally flexible walls are prestressed vertically, mainly to prevent movement away from the earth and resulting settlement of the ground behind. Concrete may be permanently shuttered or otherwise faced.

Basement walls

3.13 These are the most common retaining wall in buildings and may be diaphragm or bored pile walls, or may be similar to separate retaining walls, except that the base is usually part of a floor or raft construction **5c**. If shallow, then suitably waterproofed brickwork can be used for economy, otherwise they are normally constructed of in situ rc. Building details may require a permanently cantilevered construction, but where possible advantage is taken of the occurrence of upper floors to provide lateral support. The lowest floor or raft may have to act as a restraint against upward water pressure as well as a foundation.

3.14 An interesting method of constructing a basement from the top downwards, using the floors to provide both temporary and permanent support to diaphragm (or bored pile) walls, is shown in **16**.

Watertight concrete

3.15 Detailed information on functional, as distinct from structural, requirements of basement walls, may be found in the AJ Handbook of Building enclosure[6]. But some extra notes on the problem of making concrete walls and basements watertight are included here. Given good workmanship and provided certain precautions are taken reinforced concrete should resist the penetration of water without the provision of a waterproof membrane. It will not be entirely vapour-proof, but in practice can be made reasonably 'damp-tight'. It is unusual to achieve a 100 per cent watertight basement at the first attempt, but the small number of leaks which develop can be permanently sealed by methods such as pressure grouting. The presence of a membrane tends to obscure the location of leaks in the structural concrete and to make permanent repairs more difficult.

3.16 Design of watertight concrete is work for an experienced engineer and the contractor should be responsible for its practical performance. The following are some of the more important design requirements for the avoidance of porosity, cracking and leaking joints:

1 Concrete should be vibrated and designed for a minimum 28 day cube strength of 27 N/mm^2 and a minimum cement content of 280 kg/m^3. Additives may be used to improve workability—but none containing calcium chloride.

2 The usual below ground reinforcement cover should be used; stresses not to exceed those allowed for mild steel in normal work above ground.

3 Profiles should be as simple as possible and without sudden changes in thickness, except possibly at movement joints; 230 mm is taken as the minimum allowable concrete

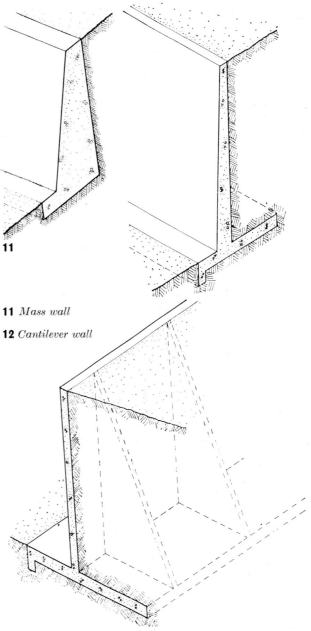

11 *Mass wall*

12 *Cantilever wall*

13 *Counterfort wall*

thickness.

4 Shuttering should be rigid, grout-tight and without through-bolts.

5 Keep construction joints square, with the reinforcement running through and the concrete surface slightly roughened before the next stage is placed against it. Opinions differ as to the desirability of providing water bars in construction joints. The internal dumb-bell type tends to displace and interfere with placing and compaction. However, water bars can be used with the external type shown in **15** particularly to vertical-face joints.

6 In the unlikely event of movement joints occurring in basement construction, they will need special treatment.

7 The spacing of construction joints and the sequence of construction should be arranged to minimise the effects of shrinkage. It is difficult in practice entirely to eliminate cracking or opening of joints, and it may be wise to concentrate cracking at pre-determined positions; eg in long walls by forming vertical holes at about 6 m centres which may subsequently be grouted. The number of horizontal construction joints in walls should be limited by concreting in as large complete lifts as possible. The joint between wall and floor should be at least 225 mm above the floor, eg by using an upstand 'kicker' cast integrally with the floor itself.

References

3.17 For references to this study, see appendix, section 4, below.

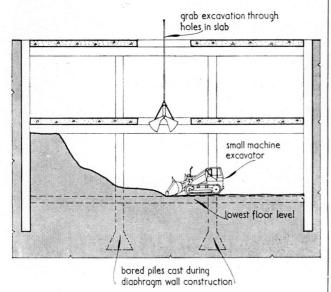

16 *Basement construction from top downwards using diaphragm wall*

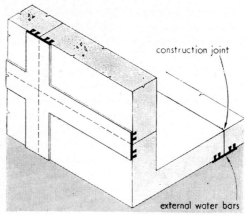

15 *External water bars in basement construction*

is now out-dated and under revision.

1.03 The codes and standards relating specifically to materials used in the actual structure of the foundations are covered as appropriate, elsewhere in the handbook.

cp 2001 is a mine of information not only on site investigations itself (including preliminary inquiries) but also on engineering soils and geology.

Appendix

Published at the conclusion of the final technical study in this section of the handbook, this appendix refers to the whole of section 4 Foundations and retaining structures.

1 Codes and standards

Foundations

1.01 The principal Codes of Practice* relating to foundation work are:

(a) bs cp 2004[35] is a revised and metricated version of ce cp no 4[7], issued after this present text (the contents of which it does not substantially alter), and gives general recommendations covering all types of foundations, associated ground works, and work under water, in connection with the normal range of buildings and engineering structures. It does not deal specifically with the design of earth retaining structures.

(b) bs cp 101[22] This deals with the simplified requirements for the foundations and substructure of non-industrial building under five storeys in height, such that the foundation loads are dispersed evenly on the ground.

(c) bs cp 2001[15] Site investigation normally required prior to foundation work is covered in detail, and is supplemented by bs 1377[20] for all standardised in situ and laboratory soil tests. For safety precautions in large diameter piling see cp 2011[27].

Retaining structures

1.02 ce cp 2[2] is the principal code of practice relating to earth-retaining structures. This is out-dated in some respects and under present revision. bs cp 2004 covers sheet piling and cofferdams and bs cp 2003[8] deals with upholding and other excavation problems. bs cp 102[9] is concerned with the general protection of buildings against ground water, bs cp 2007[10] covers the design of the reinforced concrete but is primarily concerned with a rather different type of construction from ordinary basements. It

2 Regulations

2.01 In Great Britain foundation work is normally subject to the requirements of the local authority and in particular to the following acts and regulations.

England

2.02 The Building Regulations[28] apply to all of England and Wales excepting inner London. Foundations requirements are governed by regulations d3 to d7. d3, the general functional specification is divided into three requirements:

(a) ability to transmit loads to the ground safely and without any movement which would damage or impair the stability of the building or any adjoining work,

(b) construction to be such as to safeguard the building against damage by swelling, shrinking or freezing of the subsoil and

(c) capability of resisting attack by any deterious matter in the subsoil.

2.03 Regulations d4 to d6 designate 'deemed-to-satisfy' codes of practice eg ce cp 4 for regulation d3 generally, cp 114[29] for d3 (a) as regards the use of reinforced concrete and cp 101 for d3 (a) in respect of buildings (other than factories and storage buildings) having not more than four storeys.

2.04 d7 given 'deemed-to-satisfy' provisions for simple strip foundations on common soils in respect of regulation d3 (a). Part c of the Regulations deals with both 'Preparation of site and resistance to moisture' and the construction of floors on the ground.

Scotland

2.05 The Building Standards (Scotland) (Consolidation) Regulations[30] apply generally throughout Scotland. Functional requirements for foundations are given in regulation c2, and part g covers 'Resistance of sites and resistance to passage of moisture'. Again ce cp 4 and cp 101 are designated as 'deemed-to-satisfy' codes of practice.

London

2.06 The London Building Acts[31] together with their constructional by-laws, govern building in the City of London and the 12 inner-London Boroughs. Requirements for Foundations and other ground works are covered specifically in Part v of the by-laws, but in most cases the work comes under the jurisdiction of the GLC District Surveyors, who have wide powers under the Acts. Part 6 of the 1939 Act contains a special statutory code governing party structures and rights of adjoining owners including under-pinning. A somewhat similar code applies in Bristol—elsewhere common law rules usually apply.

2.07 No attempt has been made to encompass other national or local legislation which might have effect on foundation works; for example building over existing drains or sewers and on certain filled ground is controlled by Public Health Acts but dealt with when application is made for Building Regulation approval. [The Coal-Mining (Subsidence) Act[18] was mentioned in Technical study FOUNDATIONS 1.]

3 General references

3.01 The most useful general references are:

(a) BS CP 2001, BS CP 2004 and BS CP 101 (*see above*)

(b) *Guide to site surveying* for site investigation[19]

(c) M. J. Tomlinson's book[11] is probably the best general reference for the whole subject, and particularly valuable from the constructional view point.

(d) Terzaghi and Peck's work[4] is a standard text, but rather specialised and occasionally out-dated.

(e) *Soil mechanics for road engineers*[32] gives useful general information on soil mechanics, and is a valuable reference on methods of soil compaction. The latter is also covered in CP 2003[8], together with other aspects of engineering earth-works.

(f) Donovan H. Lee[12] has written a good text book on deep foundations, sheet piling and cofferdams. Reynolds[3] gives perhaps the most useful direct data for the design of rc retaining structures, and Terzaghi and Peck may be consulted for details on soil behaviour.

3.02 BRS Digests 63, 64 and 67 outline the principles of soil behaviour, the ways in which it can affect the foundations of buildings and the choice of building sites, the investigations necessary to determine soil conditions and the types of foundation appropriate. BRS Digest 95 considers different types of piling and tells which to use when.

3.03 These digests present a readable summary of the basic problems confronting the foundation designer, advising on their solution in simple cases. The Building Research Establishment provides advice on particular foundation problems through its advisory service. The Geotechnics Division undertake investigation and research in connection with soil mechanics and foundation engineering, and the results of some of this work are published in current papers obtainable from the Establishment.

4 Bibliography

1 BUILDING RESEARCH STATION Current paper 8/72. Observation of retaining wall movements associated with a large excavation. K. W. Cole and J. B. Burland. April 1972 [(16.2)]

2 INSTITUTION OF STRUCTURAL ENGINEERS Civil Engineering Code of Practice no 2 1951: Earth retaining structures [(16)]

3 REYNOLDS, C. E. Reinforced concrete designer's handbook. 8th edition 1976, Cement & Concrete Association [Eq (A3)] £8

4 TERZAGHI, K. and PECK, R. B. Soil mechanics in engineering practice. 1968, J. Wiley & Sons [(L4)] 2nd edition. £7.25 (paperback)

5 HODGKINSON, A. Diaphragm walls. AJ, 1971, September 15, p593 [(16·2)]

6 AJ Handbook of Building enclosure. London 1974 Architectural Press. £6.95 (*paperback*). Section 3: EXTERNAL ENVELOPE: Lowest floor and basement. 8.9.72 and 15.9.72 [(9–)]

7 INSTITUTION OF CIVIL ENGINEERS. Civil Engineering CP 4: 1954 Foundations [(16)]. (*See 35 below*)

8 BS CP 2003: 1959 Earthworks [(11)] £3.00

9 BS CP 102: 1972 Protection of buildings against water from the ground [(12)] £2.00

10 BS CP 2007: Part 2 1970 Design and construction of reinforced and prestressed concrete structures for the storage of water and other aqueous liquids [184 (53)] £2.00

11 TOMLINSON, M. J. Foundation design and construction. London, 1975, Pitman Publishing, 3rd edition [(16) (A3)] £9.90 (*paperback*)

12 LEE, D. H. An introduction to deep foundations and sheet-piling 1961, Cement & Concrete Association [(16)] £1

13 TRANSPORT AND ROAD RESEARCH LABORATORY Road note 29. A guide to the structural design of pavements for new roads 1977, HMSO [12 (90·22)] £0.70p

14 Specification 1974, Architectural Press [Yy (A3)] £16

15 BS CP 2001: 1957 Site investigations [(11) (A3s)] £3.00

16 CAPO-BIANCO, A. J. Behaviour of complex foundations. Proceedings of the symposium on interaction of foundations and soil. 1971, Birmingham University [(16)]

17 INSTITUTION OF CIVIL ENGINEERS. Report on mining subsidence. 1959, [(L4)] o/p

18 The Coal Mining (Subsidence) Act 1957 [(L4)]

19 HEWITT, R. Guide to site surveying. London, Architectural Press, [(A3s)] o/p

20 BS 1377: 1967 Methods of testing soils for civil engineering purposes [(L4) (Aq)] £4.40

21 BRS Digest 90. Concrete in sulphate-bearing soils and ground waters. 1968, HMSO [Yq (S)]

22 BS CP 101: 1972 Foundations and substructures for non-industrial buildings of not more than four storeys [(1–)] £1.35

23 BAKER, A. L. L. Raft foundations, 1957, Cement and Concrete Association [(16)] o/p

24 TERZAGHI, K. Evaluation of coefficients of subgrade reaction. *Geotechnique*, 1955, December [(L4)]

25 WHITAKER, T. and COOKE, R. W. Proceedings of the symposium on large bored piles. 1966, Institution of Civil Engineers [(17·2)] £5.50 o/p

26 WHITAKER, T. The design of piled foundations. Oxford, second edition, 1975, Pergamon Press [(17)] (A3) £7.25 (*paperback* £4.25)

27 BS CP 2011: 1969. Safety precautions in the construction of large diameter boreholes for piling and other purposes [(17·2) (E2g)] £1.10 *metric units*

28 The Building Regulations 1976 SI 1676, HMSO £3.30; The Building (First Amendment) Regulations 1978 SI 723. HMSO £0.60p [(A3j)]

29 BS CP 114: Part 2: 1969 Structural use of reinforced concrete in buildings [(2–) Eq4 (K)] £3.45

30 The Building Standards (Scotland) (Consolidation) Regulations 1971 SI 2052 (S218), HMSO £1.30; The Building Standards (Scotland) Amendment Regulations 1975 SI 404 (S51) £0.29p; The Building Standards (Scotland) Amendment Regulations 1973 SI 794 (S65) £0.21p [(A3j)]

31 Greater London Council, London Building (Constructional) Bylaws 1972 GLC £1.20 *paperback*; *See also* PITT, P. H. and DUFTON, J. Building in inner London, 1976, London, Architectural Press £5.50 *paperback*

32 TRANSPORT AND ROAD RESEARCH LABORATORY. Soil mechanics for road engineers. 1952, HMSO [(L4)] £4.50

33 BRE Digests 63, 64 and 67 (Second Series) Soils and foundations, parts 1, 2 and 3. HMSO [(16) (L4)]

34 BRE Digest 95 Choosing a type of pile. HMSO [(17)]

35 BS CP 2004: 1972 Foundations [(16)] £6.90

Information sheet Foundations 1

Foundation design: general procedure

This sheet should be read with Technical study FOUNDATIONS 2. *Design procedure is set out in a flow chart*

Section 4 **Foundations and retaining structures**

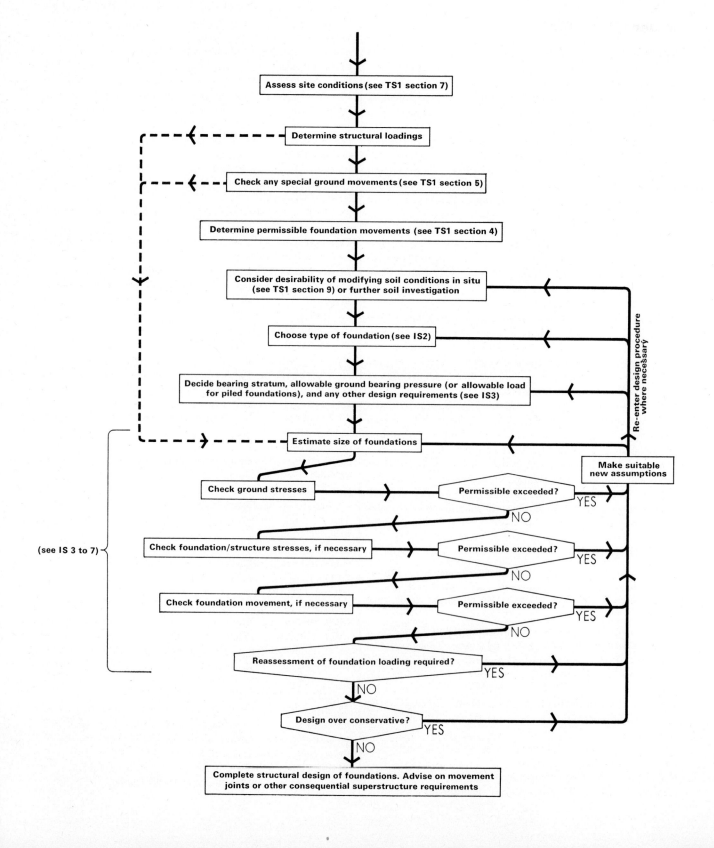

Information sheet
Foundations 2

Choice of foundation type

This sheet, to be read with Technical study FOUNDATIONS 2, *summarises in a table the general principles of foundation selection in normal conditions. Further investigation of possible choices may be required. For choice between different types of pile foundations, see Information sheet* FOUNDATIONS 7

Table I Selecting a suitable foundation

Type of soil[1]	Suitable foundations[5]	Remarks
Rock (including 'semi-rocks' eg chalk and clay shale)	Pads for individual columns. Strips for walls or columns closely spaced in rows	May be risk of swallow holes in limestone or chalk
Sands and gravels generally	As for rock. Rafts for very heavy buildings	Found above water table if possible. May settle if subject to vibration (then found deeper, or use raft)
Loose sands and gravels (especially if near water table)	Strips or raft[4]	Depends on bearing capacity and settlement requirements. Consider improving soil by vibroflotation (Technical study Foundations 1 section 9)
Clays and silts (except as below)	Pads, strips or raft for individual columns. Strips or raft for walls or closely spaced columns	Strips or raft may be required for bearing capacity or settlement requirements especially with silt or soft clay. Short bored piles (Information sheet Foundations 7) may have advantages for light buildings on clay
Clays where shrinkage or swelling likely to appreciable depth (eg existing or future trees or shrubs near building)	Deep foundations[2]	See Technical study Foundations 1 section 5
Peat	Unsuitable as bearing stratum	Unless peat can be removed, or is only thin layer at considerable depth below founding level, take foundations down to firm stratum below
Fill (existing or new)	Strips or raft	Depends on bearing capacity and settlement requirements; needs careful investigation. Foundations may have to be taken down to a firm stratum below very poor or variable fill[3]

Notes

1 It is assumed that there is a sufficiently deep and thick bearing stratum for 'shallow' foundations (ie up to about 2 m deep). If the strata are thin or mixed (ie between the main types mentioned), or if the proposed bearing stratum is underlain by softer material, further consideration will be required, see Information sheet Foundations 3.

2 If the soil at normal foundation level is soft or otherwise unsuitable it is necessary to provide deep foundations. Deep pads (piers) may sometimes be used, but more frequently piling will be economical (Information sheet Foundations 7). In both cases ground beams are required to support walls, and it may be desirable to design the lowest floor as suspended. Short bored piles are often used in shrinkable clay. Piling may also be necessary in very restricted or other unfavourable site conditions, particularly in the case of a high ground water table in non-cohesive soil. A basement may automatically provide a sufficiently deep foundation, and 'buoyancy foundations' (Technical study Foundations 2 section 5) are useful for heavy buildings.
Foundation depth may be governed by adjoining construction (Technical study Foundations 1 section 10). Piles can be used under other foundation types, but a mixture of high and low level foundations should be avoided.

3 Avoid fill containing appreciable refuse, combustible material or chemical waste. See also Technical study Foundations 1 section 6.

4 Strips and rafts may be designed to span over possible 'soft spots' or cavities in the ground.

5 Type and design of building superstructure may sometimes be adjusted to suit likely foundation settlements.

Information sheet
Foundations 3

Foundation design : common items

This sheet, which should be read with TS FOUNDATIONS 1 and 2, discusses factors common to any foundation design. It covers preliminary work, estimation of allowable bearing pressure, ground stress and settlement calculations, the interpretation of plate bearing tests and applications of standard penetration tests

1 Checklist of preliminary work

1 Assemble all available information about the structure, loading and site conditions, including special design requirements or foundation report.

2 Consider suitable types of foundation (Information sheet FOUNDATIONS 2) and bearing strata, taking into account:
● safe bearing capacity and likely settlements
● minimum foundation depth. In UK normally $0 \cdot 5$ m where frost protection is required; 1 to $1 \cdot 25$ m in shrinkable clay. Also check bearing pressure requirement, and relation to any adjacent existing or future construction or services (Information sheet FOUNDATIONS 6). Effect of vegetation
● risk of breaking into underlying softer strata. (For approximate pressure check use load spread, but note that a higher stratum, if harder and sufficiently thick, will act as a raft and prevent failure. Possibility of slips on sloping ground requires special consideration)
● precautions to minimise differential settlement. (Avoid mixing foundation types, bearing strata or pressures within one building where possible).
● effect of any overlap of soil stresses from new and adjacent foundations
● possible deterioration of bearing capacity eg by water seepage, changes in water table, vibration, climatic effects, constructional disturbance or underground cavitation
● constructional problems.

3 Decide if any geotechnical work is desirable, eg ground water lowering, vibroflotation or soil grouting, or if further soil investigation is required.

4 Knowing soil conditions (Technical study FOUNDATIONS 1 section 8), determine concrete mix and reinforcement cover. If no special requirements, mass concrete should contain not less than 180 kg of cement per m³ of concrete (approximately 1:8 by volume nominal mix). Reinforced concrete generally should be in accordance with CP 114[1], and the quality will normally be at least $21 \cdot 0$ N/mm² 28 day cube strength containing not less than 280 kg of cement per m³ of concrete.

2 Estimation of allowable bearing pressure for vertical loading

2.01 Terzaghi formula for shallow foundations ($z \geqslant B$, see **1**) in $c - \phi$ soils; mostly used for clays and silts. Net ultimate bearing capacity = increase in loading intensity over pre-existing at which soil fails in shear ie
$$q_{nu} = f_c.N_c.c + N_q.p_o + f_\gamma.N_\gamma.\gamma.B - p_o$$

or in terms of total pressures on soil, ultimate bearing capacity $q_u = q_{nu} + p$. [Where c = apparent cohesion; ϕ = angle of shearing resistance; f_c and f_γ are coefficients depending on the shape of foundation; N_c, N_q and N_γ depend on the value of ϕ; γ = average soil density below the underside of foundation (submerged density if below water table); p_o = effective overburden pressure; p = total overburden pressure at underside (see Technical study FOUNDATIONS 1 section 4); p_o = p if water table below.]

For a strip foundation, $f_c = 1$, $f_\gamma = 0 \cdot 5$
For a circle, $f_c = 1 \cdot 3$, $f_\gamma = 0 \cdot 3$
For a square, $f_c = 1 \cdot 3$, $f_\gamma = 0 \cdot 4$
For a rectangle ($B \times L$), $f_c = 1 + 0 \cdot 3 \left(\dfrac{B}{L}\right)$,
$$f_\gamma = 0 \cdot 5 - 0 \cdot 1 \left(\dfrac{B}{L}\right)$$
Values of N_c, N_q and N_γ are given in **2**.
For $\phi = 0$ (clay), $N_c = 5 \cdot 7$, $N_q = 1$, $N_\gamma = 0$.

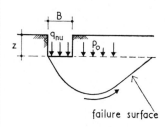

1 *Shear failure of shallow foundation*

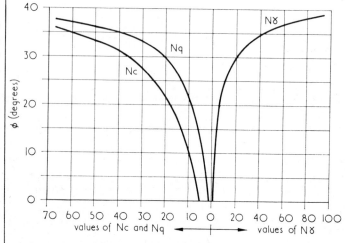

2 *Terzaghi's bearing capacity coefficients*

2.02 For *deep foundations*, the skin friction on the foundation perimeter may become a significant addition to the bearing capacity; suggested values are given in Information sheet FOUNDATIONS 7. There is also an important increase in N_c; **3** shows values proposed by Skempton.

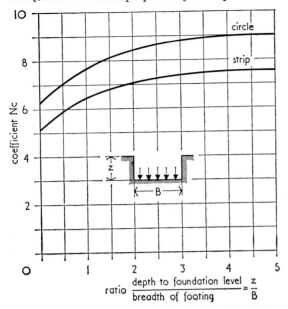

3 *Bearing capacity, coefficient Nc for deep foundations (including f_c factor)*

2.03 *Safe bearing capacity* is usually based on net ultimate bearing capacity (ie compared with the increase (q_n) in total pressure on the ground). It is q_{nu}/F, where F is a factor of safety (usually between 2 and 3), depending on certainty about soil conditions and likelihood of the maximum design loading being realised.

2.04 The allowable bearing pressure is the safe bearing capacity reduced if necessary to limit settlement to an acceptable value. It is usually increased by 25 per cent where the corresponding increase in actual pressure is due solely to wind. Allowable pressures on non-cohesive soils are normally based on penetration tests (see section 6), but the minimum foundation depth for a given pressure may be checked by the above formula, using a value of ϕ determined by penetration methods. Mixed soils may require loading tests (described later).

Bearing pressures in practice

2.05 Typical bearing pressures for different types of soil under normal conditions are given in table I, reproduced from CP 101[2]. These should be supplemented by soil tests, unless accurate bearing capacity is not essential (eg for small works), or is well known from local experience.

3 Ground stress calculations

3.01 Stress distribution below foundation may be approximately calculated assuming spread at constant side slope **4**: Pressure at depth z = σ_z which will be equal to

$$q_n \cdot \left(\cfrac{1}{1 + \cfrac{2z}{sB}} \right) \text{ for a strip foundation,}$$

and $q_n \cdot \left[\cfrac{1}{\left(1 + \cfrac{2z}{sB}\right)\left(1 + \cfrac{2z}{sL}\right)} \right]$ for a rectangular foundation B × L.

's' ($\tan^{-1}a$) may normally be taken as 3 (ie $a = 71 \cdot 6°$), but for depths greater than 2B or when checking overlap of stress from adjacent foundations, it is preferable to decrease this to 2 ($a = 63 \cdot 5°$). Stress increase of 10 per cent over net overburden pressure p_o may be taken as significant.

Table I Typical bearing capacities (from CP 101:1972)

Group		Types of rocks and soils	Bearing capacity kN/m²		Remarks
I Rocks	1	Igneous and gneissic rocks in sound condition	10 000		These values are based on the assumption
	2	Massively-bedded limestones and hard sandstones	4 000		that the foundations are
	3	Schists and slates	3 000		carried down to
	4	Hard shales, mud-stones and soft sandstones	2 000		unweathered rock
	5	Clay shales	1 000		
	6	Hard solid chalk	600		
	7	Thinly-bedded limestones and sandstones			To be assessed
	8	Heavily-shattered rocks and the softer chalks			after inspection

Group		Types	Dry	Submerged	Remarks
II Non-cohesive soils	9	Compact gravel or compact sand and gravel	> 600	> 300	Width of foundation not less than 1 m.
	10	Medium dense gravel or medium dense gravel and sand	200 to 600	100 to 300	Ground water level assumed to be a depth not
	11	Loose gravel or loose sand and gravel	< 200	< 100	less than the width of
	12	Compact sand	> 300	> 150	foundation
	13	Medium dense sand	100 to 300	50 to 150	below the base
	14	Loose sand	< 100	< 50	of the foundation

Group		Types	Bearing capacity	Remarks
III	15	Very stiff boulder clays and hard clays	300 to 600	This group is susceptible to
	16	Stiff clays	150 to 300	long-term
	17	Firm clays	75 to 150	consolidation
	18	Soft clays and silts	75	settlement
	19	Very soft clays and silts	< 75	
IV	20	Peat		Foundations should be carried down through peat and organic soil to a reliable bearing stratum below
V	21	'Made' ground		It should be investigated with extreme care

Note Due care should be paid to ensuring an adequate depth of the given soil, and in certain cases, in order to limit the amount of settlement, consideration may have to be given to restricting the allowable bearing pressure to a lower value than the bearing capacity.

Sand and gravel Table I assumes that the width of the foundation is of the order of 1 m. For narrower foundations on sand and gravel, the permissible bearing capacity decreases as the width decreases. In such cases the permissible bearing capacity should be the value given in table I multiplied by the width of the foundation in metres. In such soils the permissible bearing capacity can be increased by 12·5 kN/m² for every 0·3 m depth of the loaded area below the lowest ground surface immediately adjacent. If the ground water level in sand or gravel is likely to be at a depth of less than the foundation width below the base of the foundation, then the submerged value given in table 1 should be used.

Clay soils For the types of building considered in this code, the width and depth of the foundations do not have an appreciable influence on the permissible bearing capacity on clay soils. If a clay is examined under dry summer conditions the probable deterioration under winter conditions should be borne in mind. Mud stones and clay shales may deteriorate very rapidly if exposed to the weather or to ground water.

Mixed soils Soils intermediate between the main types given in table 1 may need to be assessed by test.

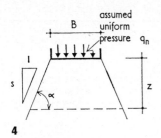
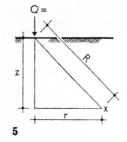

4 *Approximation to foundation load spread in ground*
5 *Boussinesq co-ordinates for stress at depth z. With a concentrated load on ground surface (Q) there is a vertical stress at X of σ_z*

3.02 Alternatively, if the foundation load can be taken as concentrated at a point near the surface, the ground stress may sometimes be determined by the Boussinesq formula for a uniform elastic medium of indefinite depth **5**.

Vertical pressure at $X = \sigma_z = \dfrac{3Qz^3}{2\pi R^5}$

or, in terms of horizontal distance r, $= \dfrac{3}{2\pi z^2} \cdot \dfrac{Q}{\left[1 + \left(\dfrac{r}{z}\right)^2 \right]^{5/2}}$

3.03 More accurate methods of calculating stresses under loaded areas are given by Tomlinson[3] and in *Soil mechanics for road engineers*[4] (Newmark's method).

4 Settlement calculations

4.01 The long term consolidation of a layer of clay may be estimated from $\rho = \sigma_z (\mu . m_v) H$ (where ρ = settlement; σ_z = average vertical pressure in the layer under the settlement point, calculated from the net foundation pressure (q_n) due to long term loads; μ = geological factor for particular clay; m_v = coefficient of volume decrease; H = layer thickness).

m_v is normally determined from soil tests as corresponding to a pressure of $p_o + \dfrac{\sigma_z}{2}$, but typical values are given in table II with the usual ranges of μ values.

Table II Typical clay consolidation properties

Type of clay	$m_v \left(m^2/MN \right)$	μ
Very soft alluvial	> 1·0	1·0 to 1·2
Normally consolidated alluvial	0·25 to 1·0	0·7 to 1·2, depending
Normally consolidated at depth	0·1 to 0·25	on sensitivity
Weathered and less stiff fissured	0·1 to 0·25	0·5 to 0·7
Boulder and very stiff fissured	0·05 to 0·1	0·2 to 0·7, depending on degree of over-consolidation
Heavily over-consolidated boulder and hard fissured at depth	< 0·05	0·2 to 0·5

4.02 The summation of all layer settlements under a foundation may be multiplied where appropriate by one reduction factor for depth (see Tomlinson[3]) and another for rigidity (usually 0·8 if completely rigid) to obtain the ultimate long term settlement of the foundation.

4.03 Methods of calculating 'immediate' settlements of clay and cohesionless soil are given by Tomlinson, eg those due to Steinbrenner and de Beer. The immediate settlement of clay usually lies in the range of 0·4 to 1·0 times the long term consolidation. If Young's modulus E is known, an approximate estimate may be made using $\dfrac{0·75}{E}$ instead of $(\mu . m_v)$ in the consolidation formula above. Excavation heave should be allowed for in the long term calculation, but the immediate heave will not affect the building.

5 Interpretation of plate bearing tests

5.01 A general procedure for testing is described in CP 2001[5]. The ultimate pressure is that at which settlement increases without appreciable increase in pressure, or it may be taken as that at which the settlement is 20 per cent of the plate width, if this value is attained first. For clay (which is not usually tested in this way), the ultimate pressure for a large foundation is the same as that for a test plate of similar shape. On frictional soils it is roughly proportional to the width of loaded area, and the full size value can be estimated from those for two or three different sized plates. But in such cases settlement usually governs the design.

5.02 To estimate settlement of the actual foundation at working pressure, multiply the corresponding settlement of the test plate by the ratio of actual to test widths for clay soils, but for frictional soil extrapolate settlements of several different sized plates. Note: for tests on sand Terzaghi and Peck[6] give the approximate relationship:

$$S = S_1 \left(\frac{2B}{B + 0·305} \right)^2$$

[where S_1 = settlement of a plate 0·305 m (1ft) square, S = settlement of a larger square or strip foundation of width Bm under the same ground pressure. (S = $4S_1$ when B becomes very large)].

5.03 In all cases beware the presence of soft strata or water within the zone effectively stressed by the actual foundation but not by the smaller test plate. Long term clay settlements cannot be thus determined; nor does the method allow for interaction of closely spaced foundations.

6 Applications of standard penetration test (SPT)

6.01 This is used mainly for non-cohesive soils; the blow counts per 305 mm penetration are measured preferably at not more than 1 m intervals over a depth equal to the width of the largest foundation, and averaged to give an N-value[7]. The lowest N found from the various bore holes is used in design. The corresponding relative density and allowable bearing pressure (chosen to limit total settlement to 25 mm and maximum differential to 19 mm), proposed by Terzaghi and Peck, may be determined from **6**.

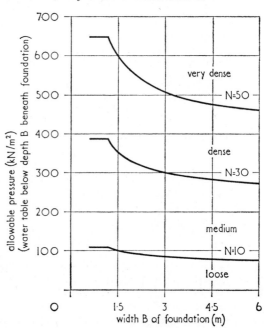

6 *Chart for estimating allowable bearing pressure for foundations on sand, on the basis of standard penetration tests. (Interpolate for N values between those shown)*

a *Dry or moist sand* (water table at least B below foundation). Use **6** directly. Pressure (kN/m²) is very roughly equal to 10 N.

b *Saturated sand* If very loose (N ≯ 5) shock may cause liquefaction, hence use piles or compact sand. If N > 5 use **6**. But, taking depth of foundation underside below adjacent ground surface as Df, if $\left(\dfrac{Df}{B}\right)$ is small, reduce pressure by 50 per cent, if $\left(\dfrac{Df}{B}\right)$ approaches unity, reduce pressure by $33\frac{1}{3}$ per cent.

Important: If N < 10 and Df < B < 2 m, check pressure for shear failure.

c *Fine or silty sand* If N > 15, use modified value $15 + \frac{1}{2}(N - 15)$.

d *Silt* (excluding loess). If N < 10, even less suitable than soft clay. If N > 10, treat 'rock flour' type as very fine sand but 'plastic' type as clay.

e *Gravel* Generally treat as sand, but results may be unreliable unless relative density checked by another method, eg by excavation. An 'equivalent cone' penetration tool may give greater accuracy.

Notes on SPT

● This applies to normal shallow foundations. Terzaghi and Peck suggest that the allowable pressure should be doubled in the case of rafts (because double total settlement may be tolerated) and piers with a depth/width ratio > 4 (because of reduced settlements at depth).

● The allowable pressure is usually determined for the largest foundation in a building and applied to all bases.

● At the pressures proposed, the factor of safety against shear failure may be expected to be at least 3, except perhaps for the case (n < 10) noted under b.

● The method should not be applied to sand or gravel subject to high frequency vibration (Technical study FOUNDATIONS 1 section 5), which required specialist advice.

● **6** appears to be based on penetrations measured at an effective overburden pressure in excess of 250 kN/m². Experts have proposed that where tests are carried out at lower pressures the measured N-values should be increased (before using **6**)—see table III. But large increases thus obtained at low overburden should be treated with some caution.

● SPTS are not normally used in clay soil, but if done give a rough indication of shear strength as table IV.

● For approximate checks on bearing capacity of cohesionless soil by the Terzaghi formula (eg for minimum foundation depth), it may be assumed that the angle of shearing resistance⁶ $\phi = 27 + 0\cdot3N$ (in degrees).

Table III Correction factors for standard penetration test (after Gibbs and Holtz)

Effective overburden pressure p_O (kN/m²)	Corrected N / Measured N
10	3·5
25	2·9
50	2·5
100	1·9
150	1·5
200	1·25

Table IV Standard penetration tests in clay

N	Very approximate apparent cohesion (kN/m²)	Usual description of clay
< 3	< 18	very soft
3 to 6	18 to 35	soft
6 to 12	35 to 70	firm
12 to 23	70 to 140	stiff
> 23	> 140	very stiff or hard

References

1 BS CP 114: Part 2: 1969 Structural use of reinforced concrete in buildings. Metric units. BSI [(2–) Eq4 (K)] £3.45

2 BS CP 101: 1972 Foundations and substructures for non-industrial buildings of not more than four storeys. BSI [(1–)] £1.35

3 TOMLINSON, M. J. Foundation design and construction. Third edition, 1975, Pitman [(16) (A3)] £9.90 (*paperback*)

4 TRANSPORT AND ROAD RESEARCH LABORATORY Soil mechanics for road engineers, 1952 HMSO. [(14)] £4.50

5 BS CP 2001: 1957 Site investigations. BSI [(11) (A3S)] £3.00

6 TERZAGHI, K. and PECK, R. B. Soil mechanics in engineering practice, 1968, J. Wiley & Sons Inc, second edition [(L4)] £17.50 (*hardback*) £7.25 (*paperback*)

7 Symbol N throughout sheet refers to this factor and must be carefully distinguished from Newton

Information sheet
Foundations 4

Section 4 **Foundations and retaining structures**

Pad and strip foundations

The basic design procedure for pads and strips outlined here should be read with Technical study FOUNDATIONS 2 *section 4*

1 Ground pressure calculations

1.01 It is normally assumed that pressure on the ground immediately under a simple pad foundation is either uniform or varies linearly in the horizontal direction. With a single load P at eccentricity e_1 from the base centroid **1**, the foundation is considered subject to centroidal loading P and moment Pe_1. The moment induced by the structure should be added to this moment Pe_1; the centroidal load will be increased by the weight of foundation (less soil displaced) when considering net ground pressures.

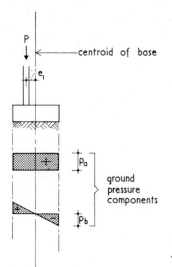

1 *Eccentrically loaded pad foundation*

1.02 If the total load is W and moment M, ground pressure is made up of uniform and 'bending' components:

$$p_a = \frac{W}{A} \text{ and } p_b = \frac{M}{Z}$$

where A and Z are plan area and section modulus of the base. For the usual case of a foundation symmetrical about its centroid, the total edge pressures will be:

$$\frac{W}{A} + \frac{M}{Z} \text{ and } \frac{W}{A} - \frac{M}{Z}$$

1.03 If $\frac{W}{A} \geqslant \frac{M}{Z}$ the final pressure diagram will be as **a**, **b** or **c** in **2**.

For a rectangular base L $\times$ B, $Z = \frac{BL^2}{6}$ and the edge pressures are $\frac{W}{BL} \pm \frac{6M}{BL^2} = \frac{W}{BL}\left(1 \pm \frac{6e}{L}\right)$

where e (total load eccentricity) $= \frac{M}{W}$

In case **c**, $\frac{6e}{L} = 1$, ie the eccentricity is at the edge of the mid-third of the base.

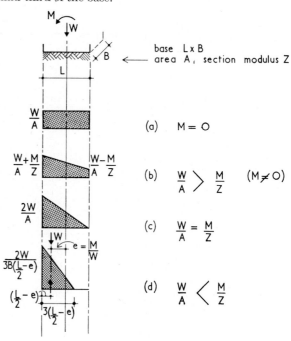

2 *Ground pressure under a rectangular base*

1.04 But if $\frac{W}{A} < \frac{M}{Z}$, (**2d**), the approach must be slightly different, because tension cannot be developed between the foundation and the ground. As centroids of total loading and ground pressure must coincide, maximum edge pressure for a rectangular base can be shown to be

$$\frac{2W}{3B(\frac{1}{2}L - e)}$$

The factor of safety against complete overturning should normally not be less than 1·5, which means the toe pressure should not exceed $\frac{4W}{A}$

1.05 If the foundation is asymmetrical or if bending occurs about both axes of the plan area, the same principles may be applied but the expressions will be more complex. A similar procedure may be used for pads or strips carrying two or more loads, if it is assumed that the base is rigid and that a linear pressure distribution is applicable. This is usually true for close loads (as with a wall) which are uniform or vary linearly along the strip.

1.06 In general the longitudinal pressure distribution may be very different from this simple assumption. The problem is then rather similar to the complex analysis of a raft, and cannot be dealt with here.

2 Design

2.01 Once pressure distribution on the ground is determined, bending moments and shears in the foundation may be calculated by simple statics, and structural design follows normally. Critical sections for shear and bending moments adjacent to structural columns or walls are shown in **3**. (Reinforcement details are discussed in Technical study FOUNDATIONS 2.)

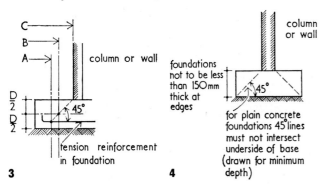

3

4

3 *Critical sections for shear and bending adjacent to loads on pads and strips. A is critical section for shear in cantilever projections* (BS CP 114); *B is critical section for shear in continuous foundation; and C critical section for bending moment*

4 *Load spread in plain concrete foundations*

2.02 If load dispersal through concrete at 45° from the edge of columns or walls **4** is sufficient to distribute pressure over the base area, reinforcement may be omitted completely (unless needed for another reason). Design of a strip foundation might be governed by the requirement to span or cantilever over 'soft spots' or cavities in the ground **5**. The settlement of the foundation should be considered carefully (deflection and end bearing areas). Details of strip foundations on sloping ground should normally be as shown in **6**.

Simple wall footings
2.03 In many cases simple wall footings may be designed by rule-of-thumb methods as in deemed-to-satisfy clauses in Regulations[1].

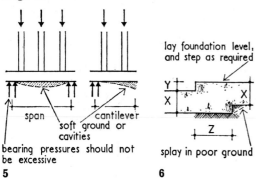

5

6

5 *Strip foundations in poor ground*
6 *Strip foundations in sloping ground, footing $X \not< 150$ mm, step $Y \not> X$, overlap $Z \not< 300$ mm or X or 2Y, whichever is largest*

1 THE BUILDING REGULATIONS 1972. Regulation D7. HMSO [A3j] £1

Information sheet
Foundations 5

Section 4 **Foundations and retaining structures**

Raft foundations

As mentioned earlier, design of foundation rafts is usually complicated and left to experts. This sheet, to be read with Technical study FOUNDATIONS 2 *section 5, covers simple design of small raft foundations including ground slabs*

1 Foundation raft design

1.01 Small rigid rafts may sometimes be designed in a similar way to rigid strip foundations (Information sheet FOUNDATIONS 4). As far as possible the centroids of loading and of the raft should coincide. Ground pressure at the edges should be checked for shear failure (Information sheet FOUNDATIONS 3).

1.02 Rafts, like strip footings, may be designed to span or cantilever over 'soft spots' or cavities. They should be made thick enough to obviate shear reinforcement, and a simple arrangement of main reinforcement should be provided in top and bottom. This main reinforcing should nowhere be less than $0 \cdot 12$ per cent of the concrete area in two directions at right angles in both top and bottom faces.

1.03 The best general method for manual design is probably that of Terzaghi[1]. See also the explanations of raft design and theory of subgrade reaction given by Terzaghi and Peck[2].

2 Ground slabs

2.01 Design proposals for ground bearing floor slabs where concentrated loads are not excessive are given in table I. Doubtful cases should be checked as below.

2.02 For heavy concentrated loads (eg fork lift trucks, heavy storage racks and plant) on uniform ground, three design methods are suggested:

a For nominally reinforced concrete slabs, check ground pressure in relation to settlement, assuming load dispersal through concrete at 45° and through any hardcore bed at 63° (2:1) to horizontal

b Alternatively Older's formula may be applied:

$$D = \sqrt{\frac{2400 \, W \times C}{S}} \quad \text{for wheel loading}$$

[where D = required slab thickness (mm), W = concentrated load (kN), C = coefficient of ground support from table II and S = design concrete tension, which may be taken as $1/14$ of the minimum 28 day cube strength (N/mm²)]

Table I Typical ground bearing floor slabs for normal loading

Soil type	Concrete thickness (mm)	
	Domestic and similar buildings	Industrial building where no special design requirements
Good (eg natural well graded sand or gravel)	100 (100)	150
Normal (eg indifferently graded sand, average clay)	100 (120)	150
Poor (eg soft clay, reasonable fill)	120*	180*
Very poor (eg very soft clay, indifferent fill, poorly graded loose sand with high water table)	150*	200*

* suspension of floor may have to be considered

Note
- All slabs reinforced except for alternative thicknesses shown in brackets.
- For reinforcement, concrete quality and hardcore bases see text.

Table II Values of coefficient in Older's slab formula

Nature of soil	C
Very soft and plastic	$1 \cdot 09$
Soft and plastic	$1 \cdot 00$
Fairly hard	$0 \cdot 90$
Hard	$0 \cdot 84$
Very hard	$0 \cdot 80$
Extremely hard gravel, rock etc	$0 \cdot 77$

c Where it is more economical to use a thinner slab and spread loads by reinforcement, this may be done by applying Terzaghi's method of subgrade reaction[1]. The type of ground pressure distribution to be expected, and the distribution from which slabs can be designed are shown in **1**. R is of the order of 7t (see **1**) on sand, but should be calculated for the actual soil conditions.

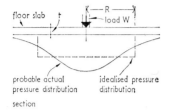

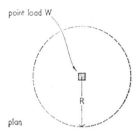

1 *Spread of loading in ground slab design, where R is radius of effective slab, t slab thickness, and W point load on slab (Terzaghi method)*

2.03 Ground bearing slabs may also have to span over potential soft spots or cavities. In such cases, and all slabs over 150 mm thick or using method c, top and bottom reinforcement should be provided. Otherwise it is usual to supply only top crack control steel. Types and positions of construction and movement joints must be considered in the design calculations (see AJ Handbook of Building enclosure section 3[3]).

2.04 Reinforcement is usually a square mesh fabric with 38 mm concrete cover. Unless otherwise required by the design it may be a nominal $2 \cdot 22$ kg/m² fabric, BS 4483 reference A142.

2.05 Concrete should generally contain not less than 280 kg of cement per m³ of concrete (28 day cube strength not less than $21 \cdot 0$ N/mm²) increased to 310 kg/m³ ($28 \cdot 0$ N/mm²) where it forms the finished surface in industrial buildings

2.06 Except on reasonably good sand, gravel or similar soil, there should be a consolidated hard core base of minimum thickness 100 mm for domestic and similar buildings, 150 mm for industrial work.

References

1 TERZAGHI, K. Evaluation of coefficients of subgrade reaction. *Geotechnique*, 1955, December [(L4)]

2 TERZAGHI, K. and PECK, R. B. Soil mechanics in engineering practice. 1968, J. Wiley and Sons Inc, second edition [(L4)] £17.50 (*hardback*) £7.25 (*paperback*)

3 AJ Handbook of Building enclosure. London, 1974, Architectural Press £6.95 (*paperback*). Section 3: EXTERNAL ENVELOPE: Lowest floor and basement [(9–)]

Information sheet Foundations 6

Underpinning

Underpinning design is outlined in the form of recommended specification clauses; this information should be read with Technical study FOUNDATIONS 2 *section 6*

1 Underpinning a brick wall

1.01 Concrete underpinning to an existing wall is a useful general example of underpinning technique. The method is illustrated in **1**. Work is done in a planned series of stages to preserve stability of the existing building. As shown in **1**, not more than 25 per cent of the total length is unsupported at any one time, but exact details must vary from job to job.

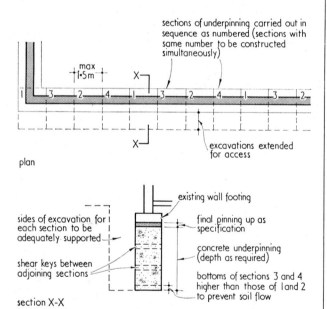

1 *Concrete underpinning details*

2 Suggested specification clauses

A The Contractor shall be responsible for ensuring that his operations do not in any way impair the safety or condition of the existing structure. He shall provide any temporary supports required for this purpose, and shall carefully inspect the condition of the structure both before and during execution of the work, and immediately inform the Engineer if he considers that any more stringent procedure than that specified is necessary.

B Underpinning is to be carried out to the satisfaction of the Engineer and Local Authority in short sections generally not exceeding 1·5 m in length, in such a manner that adequate support is at all times maintained to the underside of the wall for at least three-quarters of its length and that sections of work in progress at any one time are separated by a distance of at least 4 m.

C Projecting portions of the existing brick and/or concrete footings are to be carefully cut off where directed, and the undersides of the footings are to be cleaned and hacked free of any dirt, soil or loose material before underpinning.

D The body of the underpinning is to be constructed in mass concrete 1:2:4 nominal mix (19 mm maximum size aggregate) using rapid hardening Portland cement, and is to be cast to the widths and depths shown on the drawings. The bottoms of excavations are to be prepared as specified for foundations generally. As far as practicable excavation and concreting of any section of underpinning shall be carried out on the same day.

E The mass concrete is to be stopped off approximately 75 mm below the underside of the existing footing, and the final pinning up over the whole extent of the latter is to be carried out with a semi-dry fine concrete well rammed in as soon as possible after the foundation has set hard. The pinning-up concrete is to consist of 1 part by volume of rapid hardening Portland cement to 3 parts of aggregate (well graded from 10 mm maximum size down to fine sand) with a water/cement ratio by weight of 0·35.

F Excavation to any section of underpinning shall not be commenced until at least 48 hours after completion of any adjacent sections of the work.

G The Contractor shall keep a record on site of the sequence and dimensions of underpinning as actually executed, including the dates of starting excavation, casting concrete and pinning up for each section.

Information sheet Foundations 7

Piled foundations

For pile design, expert advice should always be taken, but the architect should be aware both of the possibilities and of reasons for technical decisions. This sheet explains various types of piling systems, and the method of pile design. The central sections give a checklist of information for piling tenders and a typical technical specification. The sheet should be read in conjunction with Information sheet FOUNDATIONS 2 *and* Technical study FOUNDATIONS 2 *section* 7

1 Characteristics of piling systems

1.01 The best type of pile for a particular job depends largely on the nature of the ground. Strata penetrated determine whether loads are to be transmitted to the soil by end bearing and/or skin friction. Shape and construction of the pile affects its load carrying and settlement characteristics. (Where more than one type of pile is considered suitable, the choice may be determined by competitive tendering, although allowance must be made for different pile caps, ground beams or other features required by each system.)

Displacement piles (driven)
General
1.02 Sizes do not usually exceed 600 mm. Normally driven by some sort of hammer, this type may cause unacceptable noise or vibration, sometimes alleviated by vibratory or jacking techniques. It is difficult to install where soil contains boulders or where very hard strata are to be penetrated. Ground heave can damage adjacent construction (eg newly concreted piles) and require redriving of end bearing piles, except with steel (small displacement) piles. Preformed piles may be installed by general contractor. Measurement of driving resistance can constitute a type of continuous load test.

Timber
1.03 This material is comparatively weak and subject to decay, but is occasionally used for temporary work or light loads.

Precast concrete
1.04 These may be prestressed. Unforeseen variations in length can be troublesome and uneconomic. Transport or a site casting yard will be required, therefore this type is not much used for urban work. It is useful where projection is required above ground level. Very high quality concrete is possible for aggressive soil conditions. There may be unseen damage under heavy driving. Precast concrete piles are not usually economical in small quantities.

Steel
1.05 Can easily be shortened or extended, and driven hard without damage. It is suitable where projection is required above ground; connections are easily made and have good resistance to horizontal loads. Steel may become corroded unless specially protected.

Driven and concrete cast-in-place
1.06 This type consists of in situ concrete within either a precast concrete shell or a steel tube. The latter may be withdrawn while concreting (shoe or end plug expendable) or left in the ground. Shells or tubes left in tend to be expensive, but are desirable in certain soil conditions (where other preformed piles may also be suitable) eg heavy ground water flow, soft squeezing ground or where redriving is necessary. If tubes are withdrawn, great skill may be required to avoid 'necking' or gaps in the concrete. Lengths may easily be varied to suit strata (shells can be in short lengths), and some enlargement of bases is frequently possible, also increase of load capacity by 'redrive' techniques.

Replacement piles (bored)
General
1.07 Formed by concrete cast in holes bored or excavated by various forms of tool according to diameter, depth and type of ground. Piles over 600 mm diameter are usually considered 'large diameter' (up to 3000 mm is possible). Temporary steel casings are needed through soft or non-cohesive or water-bearing strata,' but may usually be omitted in firm clay where bored piles can be very economical. Occasionally tubes may have to be left in because of adverse soil conditions (see para 1.06), and sometimes other preformed piles may be used in bored holes where driving is not practicable. Bored piles can be installed with minimal noise and vibration and where headroom is restricted. They may be very deep and lengths can easily be varied, but the concrete cannot be inspected after placing and is subject to the same problems as in driven cast-in-place piles. Installation in water bearing strata is difficult, in which case a compressed air concreting

technique may be desirable. Boring can loosen sands or gravels, and sometimes soft ground may be lost up the bore.

Large diameter bored piles
1.08 These can carry a relatively large load per pile, but the boring rig is heavy and movement may be costly in poor ground; also, if there are not many piles, the rig on-cost per pile will be high. In firm clay large diameter piles may be under-reamed or belled (see fig 17d, Technical study FOUNDATIONS 2) to form bases up to three times the shaft diameter, with a consequent decrease in the depth of the pile.

Short bored piles
1.09 If sunk in clay by hand or mechanical augering without lining, to depths not exceeding 6 m, these can be exceptionally economical. Often used for houses and other light building at about 2 to 3 m centres under ground beams **1**, especially in shrinkable clay (see Technical study FOUNDATIONS 1). Beams should be separated from the clay by soft fill such as ash, and ground bearing floor slabs should be isolated from the beams. But if the clay is liable to swell because of tree felling, the floors should be suspended over a void and the top 3 m of each pile reinforced and sleeved from the ground. Table 1 gives the load carrying capacity recommended in BRS Digest 67[1]. Length of the pile is the extent of contact between pile and clay.

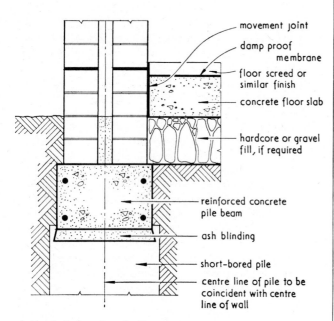

movement joint
damp proof membrane
floor screed or similar finish
concrete floor slab
hardcore or gravel fill, if required
reinforced concrete pile beam
ash blinding
short-bored pile
centre line of pile to be coincident with centre line of wall

1 *Typical short-bored pile construction*

Table 1 Load-carrying capacity of short-bored piles (in kN)

Strength classification	Diameter of pile (mm)	Length of pile (m)				
		2·5	3	3·5	4	4·5
		kN				
Stiff	250	40	48	56	64	72
(Unconfined shear	300	50	60	70	80	90
strength more than 70 kN/m²)	350	65	77	90	102	115
Hard	250	55	65	75	85	95
(Unconfined shear	300	70	82	94	106	118
strength more than 140 kN/m²)	350	95	108	120	132	145

Note The figures are for clay which increases in strength with depth to the 'stiff' and 'hard' classifications near the bottom of the piles. The figures should not be applied to piles in other situations

2 Information to be given when inviting piling tenders

The following is a checklist of the main items required; these should be prepared in detail by an experienced engineer.
1 Site plan with particulars of access
2 Condition of site when piling contractor starts work
3 Responsibility for maintenance of surface during his operations
4 Responsibility for setting-out, disposal of spoil, cutting away excess concrete deposited, removal of obstructions and provision of water and other facilities
5 All factual information on soil investigation (not recommendations)
6 Any restrictions on working, eg noise or vibration, or limitations on headroom
7 Information about underground obstructions or services on the site and adjoining properties
8 Proposed layout of piling
9 Working load on each pile. Settlement requirements of the structure
10 Extent of piling contractor's responsibility for design (see also TS FOUNDATIONS 2 para 7.04). Necessary technical specification clauses (see section 3 of this sheet)
11 Insurances required to be provided by piling contractor
12 Form of main contract and piling subcontract or other general conditions of piling contract, as appropriate
13 Dates for tender, start on site and completion of work
14 Provision to be made for testing piles
15 Any contingency sum to be allowed
16 List of the different rates to be provided in the tender (assuming that a complete bill of quantities is not issued)

3 Typical technical specification clauses (for concrete piles cast in situ)

3.01 The following clauses are taken from *Specification for cast-in-place piling* produced by the Federation of Piling Specialists (May 1971). They are given for general information, but the exact clauses to be used on a particular job must be decided by the engineer responsible.

Design
1 General design of piles shall be in accordance with the British Standard Code of Practice for Foundations CP 2004

2 Average compression stress in concrete piles under working load shall not exceed 25 per cent of the specified works cube strength at 28 days calculated on the total cross-sectional area of the pile shaft. Where the casing of the pile is permanent, the allowable compressive stress may be increased

3 Piles shall normally be designed to carry all compression loads in the concrete on the cross-sectional area of the nominal diameter

4 In the case of piles required to act in tension or bending, the stresses in the reinforcement shall be in accordance with CP 114

5 Ultimate bearing capacity of a pile shall be taken as defined in CP 2004

6 The factor of safety shall be taken as defined in CP 2004 as the ratio of the ultimate bearing capacity to the working load. Piles shall be designed to provide a factor of safety of not less than 2 (see note for guidance 3 below)

7 The cover on all reinforcement, where used, including binding wires, shall be not less than 40 mm

8 The piles shall be designed to carry the working loads shown on the drawings and, in addition, allowance shall be made for stated negative skin friction loads (see note for guidance 4)

Materials

9 Cement shall be ordinary or rapid hardening Portland cement complying with BS 12 or sulphate resisting cement complying with BS 4027

10 Aggregates shall comply with BS 882

11 Clean water free from acids and other impurities and in accordance with BS 3148 shall be used in the works

12 All steel shall be in accordance with the appropriate British Standard unless otherwise agreed

13 The slump of the concrete shall normally be in accordance with the following standards:

Piling mix	Slump Minimum	Range	Typical conditions of use
A	75 mm	75 to 125 mm	Poured into water-free un-lined bore. Widely spaced reinforcement leaving ample room for free movement between bars
B	100 mm	100 to 175 mm	Where reinforcement is not spaced widely enough to give free movement between bars. Where cut-off level of concrete is within casing. Where pile diameter is less than 600 mm
C	150 mm	150 mm or greater	Where concrete is to be placed by tremic under water or drilling mud

14 Any additive used in the concrete must be stated

15 Ready-mixed concrete may be used and shall comply with BS 1926

16 Test cubes shall be prepared and tested in accordance with BS 1881 (see note for guidance 6)

Driven piles

17 Piles shall be installed in such sequence that their construction does not damage any piles already constructed

18 Adequate measures shall be taken to overcome any detrimental effect of ground heave on the piles. When required by the engineer, levels shall be taken to determine the amount of any pile movement resulting from the driving process (see note for guidance 7)

19 When a significant change of driving characteristics is noted, a record shall be taken of the driving resistance over the full length of the next adjacent pile (measured as blows per 250 mm penetration)

20 In the case of end bearing piles, the final set of each pile shall be recorded either as the penetration in millimetres per 10 blows or as the number of blows required to produce a penetration of 25 mm

21 The temporary casing shall be dry after driving and before concreting commences

22 Where cut-off level is less than 1·5 m below working level concrete shall be cast to a minimum of 150 mm above cut-off level. For each additional 0·3 m below working level of the cut-off level an additional tolerance of 100 mm will be allowed. Cut-off shall be a maximum of 3·0 m below working level. (See also clause 35 and note for guidance 8)

Bored piles

23 A minimum length of 1·0 m of temporary casing shall be inserted in every borehole unless otherwise agreed

24 When boring through non-cohesive or very soft cohesive strata liable to collapse, temporary casing or another suitable technique shall be used to stabilise the hole. Temporary casing when used shall extend a sufficient depth below such strata adequately to seal off the unstable material

25 In dry non-cohesive strata water may be used to assist the advancement of the boring

26 When subsoil water which cannot be sealed off is encountered the water in the bore shall be maintained above the standing level of the subsoil water

27 When it is proposed to use a prepared drilling mud suspension the engineer must be advised

28 When under-reaming of the bore is carried out the slope of the under-ream must be a minimum of 55 deg to the horizontal

29 When it is not practicable to exclude ground water from the finished bore the concrete shall be placed by tremic tube

30 Where cut-off level is less than 1·5 m below working level concrete shall be cast to a minimum of 150 mm above cut-off level. For each additional 0·3 m below working level of the cut-off level an additional tolerance of 50 mm will be allowed. (See also clause 35 and note for guidance 8)

31 When concrete is placed by tremic tube the concrete shall be cast to piling platform level or to a minimum of 1·0 m above cut-off level with a tolerance from 1·0 m to 2·0 m

General

32 Piles shall be constructed within the following normal tolerances:

In plan	75 mm in any direction at piling platform level
Verticality	1 in 75
Raking up to 1:6	1 in 25

33 Each batch of concrete in a pile shaft shall be placed before the previous batch has lost its workability. Removal of temporary casings must be completed before concrete within casing loses its workability

34 In cold weather ice and snow shall be excluded from the material used in the manufacture of concrete for piles. Aggregates must not be heated to more than 38 deg C and the concrete when placed must have a minimum temperature of 5 deg C. The tops of the piles must be protected immediately casting is completed

35 When concreting dry pile holes through water-bearing strata the concrete must always be cast to a minimum of 0·3 m above the standing level of the subsoil water unless all water-bearing strata are effectively sealed off by permanent casing and this level of 0·3 m above standing water level shall be regarded as cut-off level for the purpose of calculating tolerances as defined in clause 30 and clause 22

36 Where concrete is not brought to piling platform level the empty pile holes shall be backfilled.

37 Safety procedures during piling operations shall comply with the recommendations of BS CP 2011[6] where applicable

38 The following records shall be kept of every pile:
pile number
piling platform level related to datum
nominal shaft/base diameter
date driven or bored
date concreted
depth from piling platform level to toe
depth from piling platform level to cut-off level
depth from piling platform level to top of concrete
final set (for driven piles), weight and drop of hammer
length of permanent casing
details of any obstructions encountered and obstruction time

Notes for guidance

1 *Preliminary test piles* Whenever possible a preliminary test pile or test piles should be installed to check the pile design. These should be constructed under the closest supervision in an area where the soil conditions are known and tested to a specified load of not less than twice the working load

2 *Tests on working piles* Working piles for testing are selected at random by the engineer and should be tested to 1½ times the working load

3 *Factor of safety* Pile design should ensure that (a) an adequate factor of safety is provided against reaching ultimate load of the pile or pile group, (b) required load settlement characteristics are achieved at and near to design working load

4 *Negative skin friction* The usual method of providing for this is to calculate working load from ultimate bearing capacity using the factor of safety and then add a net allowance for negative skin friction

5 *Concrete* Concrete for piles placed in the dry should contain not less than 300 kg/m³ of cement, and when placed under water by tremic tube a minimum cement content of 400 kg/m³ should be employed

6 *Concrete test cubes* Opinions vary as to the number of test cubes which should be required on a piling contract, but it is suggested that four cubes be taken for every 50 m³ of concrete used. The anticipated number of test cubes should always be included as a measured item in the bill of quantities

7 *Heave* The acceptable amount of heave depends upon whether piles are designed to carry the majority of their load by shaft friction. In cases where the pile carries the majority of its load in end bearing heave may be reduced by pre-boring or, alternatively, the contractor may elect to redrive piles where this is a practicable solution. Particular measures required will vary with each site and those adopted should be a matter for discussion and agreement with the piling specialist concerned

8 *Tolerances and cut-offs* When deep cut-offs are involved or long temporary casings have to be used in construction of piles, it is not possible to estimate the amount of concrete to form the finished level within normal tolerances. In such cases the tolerance should be a matter for discussion and agreement with the piling specialist concerned

9 *Data sheet* A piling inquiry data sheet should be enclosed with all piling inquiries
When piles are designed by the client, the inquiry data sheet will give all details including pile diameters, length or penetrations required, reinforcement and concrete specification
When piles are to be designed by the piling specialist or when alternatives are permitted, the inquiry data sheet will state required factor of safety (if different from that given in the specification), acceptable settlement of individual piles under test at working load and any other basic requirements which must be fulfilled by alternative pile designs
The inquiry data sheet will entail any variations from the standard specification which are required by the client

10 *In situ soil testing* Where in situ testing is required, the type and anticipated number of tests should be indicated in the inquiry data sheet and included as a measured item in the bill of quantities

4 Pile design and details

4.01 Piling design should always be carried out by an experienced engineer, but general notes are given here to help an architect appreciate the procedure and problems. Final design should be confirmed by load testing; where this is omitted an increased factor of safety is usually necessary. Preliminary designs are based on calculation and/or experience. Required driving resistance (penetration per blow) may be calculated, also load capacity from soil properties (especially in clay and for bored piles generally).

Calculation of load capacity
4.02 For a simple pile, net ultimate load $= Q_b + Q_s$
This $= q_{nu} A_b + f_u A_s$

where Q_b = net ultimate end bearing resistance; Q_s = ultimate shaft friction; q_{nu} = net ultimate end bearing stress; f_u = ultimate average skin friction stress; A_b = area of base $= \pi/4 \, (D^2)$ (D = base diameter); A_s = area of shaft surface in contact with soil, which $= \pi dL$ (d = shaft diameter and L = effective embedded length)

Notes
1 It is assumed that weight of pile equals weight of soil displaced, so this may be ignored if net ultimate load used
2 If markedly different strata are penetrated the skin friction for each stratum should be determined separately and added
3 In calculating L any soil near the surface subject to shrinkage or which is particularly loose or can develop 'negative friction' is neglected, also a depth of say 1·5D to 2D above the bottom of under-reamed piles is deducted to allow for dragdown. For method of dealing with 'negative friction' see technical specification clauses in section 3 above
4 For determination of q_{nu} see Information sheet FOUNDATIONS 3. Some usual values for q_{nu} and f_u are given below (straight-sided piles).

4.03 *Non-cohesive soil*
$q_{nu} = (N_q - 1) \, p_o$ (since N_γ term small)
For a driven pile use N_q corresponding to ϕ before driving, although dense soil will compact, and pile material may determine strength when driven to refusal.
For a bored pile the soil may be loosened and ϕ should not exceed 30°.
Normally f_u is assessed semi-empirically. Some values for preliminary guidance are given in table II. For bored piles it is wise to assume that the soil has been loosened.

Table II *Preliminary assessment of skin friction on driven piles in non-cohesive soil*

Density of soil	N value	Ultimate skin friction (kN/m²)
Very loose	0- 4	—
Loose	4-10	10-20
Medium	10-30	20-50
Dense	30-50	50-100
Very dense	over 50	100

Note N-value is the SPT (standard penetration test) blow count per 305 mm penetration. See TS FOUNDATIONS 1 para 7.13

4.04 *Cohesive soil*
$q_{nu} = N_c \, (\omega.C_b)$
(where $N_c = 9$ (deep foundation); C_b = cohesion at depth of base; ω = factor allowing for variation in soil strength between bulk and small sample (usually $0·75$ if C_b = mean triaxial strength, for fissured clays and for bases over 600 mm diameter, otherwise $1·0$)
$f_u = a \, \overline{c}$ (where $\overline{c}$ = average cohesion over depth L; a = adhesion factor depending on many circumstances (usually $1·0$ for soft clay ie $\overline{c}$ up to 35 kN/m², otherwise $0·45$)

Notes
1 f_u should not exceed 95 kN/m²
2 Sensitive clays may be softened by either driving or boring, and should be dealt with according to local experience

4.05 *Rock*
End bearing resistance may be assessed as for other rock foundations—see also BS CP 2004[2]. Friction in 'rock sockets' best assessed by load test.

Dynamic formulae
4.06 For a driven pile or tube the required penetration per blow (set) for a given ultimate load may be calculated approximately by a number of formulae[3], one of the best for concrete piles being Hiley's (see BS CP 2004). These formulae should be used only for piles mainly end bearing

in non-cohesive soil, and accuracy is improved by 'calibration' against load tests.

Factors of safety

4.07 Safe load on a single pile $= \dfrac{\text{net ultimate load}}{\text{F}}$

F, factor of safety, is usually 2 where there is a load test. For under-reamed piles the safe load should also be checked with $F = 3$ on base load and $F = 1 \cdot 5$ on shaft friction. The allowable load (taking into account permissible settlement) may be equal to or less than the safe load. For single piles F may have to be increased to about $2 \cdot 5$ to limit settlement. Large diameter and under-reamed piles require special consideration.

Settlement

4.08 Settlement of a single pile may be assessed by load test, or by calculation (Information sheet FOUNDATIONS 3) considering compression of the pile material and of the soil under the base, having allowed for development of skin friction up to the ultimate at working load. Another method for large diameter piles in London clay is given by Whitaker and Cooke[4].

Pile groups

4.09 Ultimate load per pile may be reduced and settlement increased where piles are closely spaced—they may in fact approximate to a single or block foundation. This is most important in cohesive soil (capacity per pile ranges from 60 per cent of that of a single pile at 2 diameters spacing to 100 per cent at 8 diameters in some cases). For details see Tomlinson[5] and Whitaker[3]. Settlement of a pile group is sometimes calculated as that of an equivalent raft **2**.

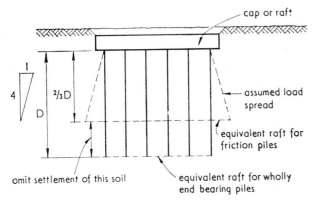

2 *Equivalent rafts for calculation of pile group settlement*

Loading test

4.10 General principles are discussed in BS CP 2004. See also technical specification clauses above. Whitaker[3] describes the constant rate of penetration (CRP) test.

Miscellaneous piling details

Concrete mix

4.11 The mix must be adequate for strength requirement (pile designed as a 'short' column when embedded in soil), including driving stresses (BS CP·2004); also for durability in given soil conditions. See also technical specification clauses (above) and Information sheet FOUNDATIONS 3.

Reinforcement

4.12 For precast concrete piles this is usually determined by handling and driving stresses. In situ piles are normally reinforced only for the top 5 m or so and through non-cohesive soil, with nominal steel (say $0 \cdot 4$ to $0 \cdot 8$ per cent of

3 *Typical walking type rig for driven and concrete cast-in-place piles (Frankipile Ltd)*

4 *In situ concrete pile exposed by excavation (John Gill Contractors Ltd)*

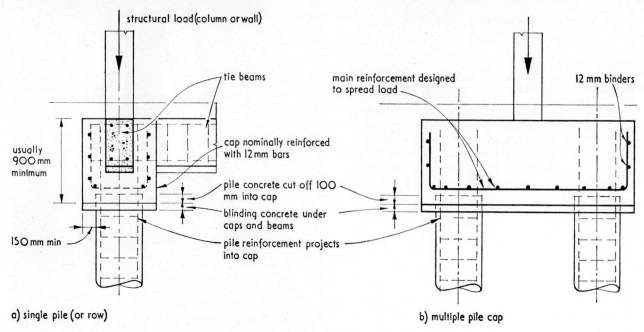

5 *Typical reinforced concrete pile cap and beam details:*
a *single (or row) piles;* **b** *multiple pile cap*

cross-sectional area with binder at 200 mm to 300 mm pitch). But special reinforcement may be required for high loads, lateral bending or uplift, including clay heave.

Minimum pile spacing (centres)

4.13 This depends on conditions but generally should be: for friction piles, not less than 3 diameters or 1000 mm, whichever is greater;
for end bearing piles, not less than 2 diameters or 750 mm, whichever is greater (effect of driving should also be considered).

Pile caps and tie beams

4.14 These are usually reinforced concrete. Caps must transmit the superstructure loads to the piles, and the latter are preferably arranged with their centroid coincident with that of the loading **5**. Single piles should be tied laterally at the top, normally by beams to other caps, unless lightly loaded. The system must resist any eccentric loading including that caused by inaccurate positioning.

References

1 BRE Digests 63, 64 and 67 (second series) Soils and foundations, parts 1, 2 and 3. HMSO [(16) (L4)]

2 BS CP 2004: 1972 Foundations [(16)] £6.90

3 WHITAKER, T. The design of piled foundations. Oxford, 2nd edition 1975, Pergamon Press [(17)] £7.25 (*paperback*)

4 INSTITUTION OF CIVIL ENGINEERS Proceedings of symposium on large bored piles. 1966, Reprinted 1968 [(17·2)] £10.00

5 TOMLINSON, M. J. Foundation design and construction. London, 1969, Pitman, 3rd edition [(16) (A3)] £9.90 (*paperback*)

6 BS CP 2011: 1969 (Metric) Safety precautions in the construction of large diameter bore holes for piling and other purposes [(17) (E2g)] £1.10

Information sheet
Foundations 8

Section 4 **Foundations and retaining structures**

Design of simple freestanding retaining walls

This sheet outlines a method of designing simple mass concrete and masonry, or rc retaining walls. It should be read in conjunction with Technical study FOUNDATIONS 3

1 Design method

1.01 This information sheet sets out a method of designing the simple cantilever wall shown in **1a**. It is applicable to a rc wall or to a mass wall having a brick, stone or plain concrete stem on a plain or rc base. The method may easily be adapted to a wall with a stepped or battered stem, or having a heel or toe 'beam' to improve resistance to sliding, also to a wall which supports applied vertical loading.

1.02 The wall is subject to active earth pressure and it is assumed that the soil and backfill are both non-cohesive, having similar properties:

angle of internal friction $= \phi$;

density above water table $= \gamma$;

submerged density $= \gamma_{sub} = \gamma_{sat} - \gamma_w$

where $\gamma_{sat} =$ saturated density and $\gamma_w =$ density of water (10 kN/m^3) (see Technical study FOUNDATIONS 1 section 4). Note: any consistent system of units may be used in calculation. Friction between fill and wall is ignored in this simple calculation.

Let $K_a =$ active pressure coefficient corresponding to ϕ. Civil Engineering (CE) CP 2* gives guidance on suitable values for the above design parameters.

1.03 Then at depth x from the top of the wall:

● If $x \leqslant h$:

effective vertical pressure $p_0 = w_s + \gamma \cdot x$

active pressure $p_a = K_a (w_s + \gamma x)$

where $w_s =$ intensity of surcharge

● if $x > h$:

$p_0 = w_s + \gamma \cdot h + \gamma_{sub} (x-h)$

$p_a = K_a \{w_s + \gamma \cdot h + \gamma_{sub} (x-h)\} + p_w$ (see **1b**)

where $p_w =$ water pressure $= \gamma_w (x-h)$

$\therefore p_a = K_a (w_s + \gamma \cdot x) + (K_a \cdot \gamma_{sub} + \gamma_w) (x-h) - K_a \cdot \gamma (x-h)$ *equation* 1 (see **2**).

1.04 Common values in practice are:

$\phi = 35°$; $K_a = 0 \cdot 27$; $\gamma = 17 \cdot 5 \text{ kN/m}^3$; $\gamma_{sub} = 10 \text{ kN/m}^3 = \gamma_w$; so that $K_a \cdot \gamma = 4 \cdot 7 \text{ kN/m}^2$ (a) (compare Technical study FOUNDATIONS 3 para **1.09**)

$K_a \cdot \gamma_{sub} = 2 \cdot 7 \text{ kN/m}^2$

ie $K_a \cdot \gamma_{sub} + \gamma_w = 12 \cdot 7 \text{ kN/m}^2$ (b)

where (a) is the lateral pressure per metre of depth above the water table, neglecting surcharge; (b) is the pressure below this to be added to the pressure at water table level.

*See bibliography at end of Technical study FOUNDATIONS 3

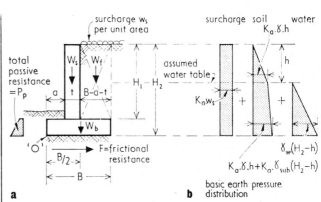

a   **b** basic earth pressure distribution

1a *Simple cantilever retaining wall for which a design method is given;* **1b** *earth pressure distribution*

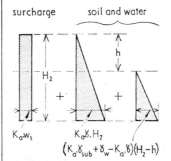

2 *Soil and water pressure distribution on cantilever wall (this is more convenient arrangement for calculation of wall shown in* **1**) *All pressures and loads are per unit length*

1.05 Putting $x = H_1$ in equation 1 and calculating total pressures, the shear (total horizontal force) per unit length of wall at the bottom of the stem for which this should be designed $= P_s$.

$$P_s = K_a \cdot w_s H_1 + K_a \cdot \gamma \frac{H_1^2}{2} + (K_a \cdot \gamma_{sub} + \gamma_w - K_a \cdot \gamma) \frac{(H_1 - h)^2}{2}$$ and the

corresponding bending moment per unit length $= M_s$.

$$M_s = K_a \cdot w_s \cdot \frac{H_1^2}{2} + K_a \cdot \gamma \cdot \frac{H_1^3}{6} + (K_a \cdot \gamma_{sub} + \gamma_w - K_a \cdot \gamma) \frac{(H_1 - h)^3}{6}$$

1.06 An rc stem is designed in the normal way for these forces and the vertical load.

For a mass wall stem thickness t, weight of stem W_s per unit length:

$$\text{maximum shear stress} = \frac{1 \cdot 5\, P_s}{t}$$

combined direct and bending stresses $= \dfrac{W_s}{t} + \dfrac{M_s}{t^2/6}$ (maximum compression) and $\dfrac{W_s}{t} - \dfrac{M_s}{t^2/6}$ (tension if negative—this may not be permissible)

1.07 If H_2 is written instead of H_1, in the expressions for P_s and M_s we obtain P_b and M_b, the total horizontal force and overturning moment on the base from the earth pressure.

Let total weight of fill and water above the base $= W_f$ per unit length (including surcharge).

Let total weight of base $= W_b$ per unit length.

Then taking moments of the vertical loads about the toe O,

$$\text{stabilising moment} = W_s\left(a + \frac{t}{2}\right) + W_f\left(B - \frac{B-a-t}{2}\right)$$
$$+ W_b \cdot \frac{B}{2} = M_r$$

(it is usual to ignore the passive earth pressure in front of the wall except for resistance to sliding)

Distance of centroid of vertical loads from $O = \bar{x}$

$$\bar{x} = \frac{M_r}{W_s + W_f + W_b}$$

and eccentricity of these on the base $= e = \dfrac{B}{2} - \bar{x}$

1.08 Also let total passive resistance against sliding $= P_p$ (see CE CP2) and total frictional resistance against sliding $=$ F.

According to CE CP 2,

$$F = [W_s + W_f + W_b - \gamma_w(H_2 - h)\, B] \tan \phi.$$

Then factor of safety against overturning $= \dfrac{M_r}{M_b}$

and factor of safety against sliding $= \dfrac{P_p + F}{P_b}$

According to CE CP 2 both these should be not less than 2.

1.09 The foundation base should be designed in the normal way for a gross vertical load $= W_s + W_f + W_b$ at eccentricity e, together with the overturning moment M_b (see Information sheet FOUNDATIONS 4).

1.10 If the soil is clay, the wall may be designed for the numerical values of pressure given above (ignoring water), provided that the actual clay pressures calculated in accordance with CE CP 2 do not give a worse case. The possibility of a rotational slip may have to be checked in this event, and the value of F must be reassessed as required by CE CP 2.

1.11 Trial sizing of wall before calculation might be
$t = 0 \cdot 25\, H_1$ for mass wall,
 $\quad 0 \cdot 085\, H_1$ for rc (230 mm minimum)
$B = 0 \cdot 5\, H_2$ to $0 \cdot 67\, H_2$
$a = 0 \cdot 33B$
Thickness of base $= 0 \cdot 15\, B$ (300 mm minimum)
Depth to underside to suit soil conditions.

AJ Handbook
**Building
structure**
CI/SfB (2-)

Section 5
**Structural
material:
Reinforced
concrete**

Section 5

Structural material: Reinforced concrete

Scope

Sections 1 to 4 of the handbook reviewed design generally and gave a résumé of constructional materials followed by some basic, analytical theory. Safety, statutory requirements and foundations were examined. The succeeding sections deal with each major structural material in turn so that the architect, having digested the properties of each material and the variety of structural form in use, can relate the information logically to his own design problem.

The perfect structure does not exist in building. It would need a form with minimum internal forces, most efficient structural shape and using the minimum, cheapest material giving at once durability, insulation, resistance to fire and occasionally to noise, and not requiring maintenance. Obviously good design achieves a compromise of all these aspects to satisfy the clients' brief. While some aspects of the design process may be a reasonably exact science, the whole is very much a question of general intelligence and intuition, sometimes referred to as 'thinking with the hips'. The handbook concludes with a comprehensive design guide which attempts to both pose the problems and summarise the solutions.

Author

The consultant editor of the handbook and author of this section is Allan Hodgkinson MEng, FICE, FIStructE, MConsE, a consulting engineer.

Allan Hodgkinson

Illustration over page shows board-marked concrete lift shafts

Technical study
Concrete 1

Section 5 **Structural material: reinforced concrete**

Concrete: the material and its properties

Having covered the theory and practice of structural design, the handbook now turns to a more detailed inspection of the various structural materials, their history, their properties and their present effects on design. This is the first of three studies on concrete by ALLAN HODGKINSON. *Here the various types of concrete are described with their basic constituents and relative properties. The next study will look at the uses of concrete as a structural form*

1 Materials

1.01 Reinforced concrete is now so commonly used that it tends to be compared with other structural materials as if it were, like them, homogeneous. In fact, not only is the reinforced concrete member a composite construction with steel but the concrete itself is a composition of cement, fine aggregate and coarse aggregate. Table I shows the range of concretes used structurally in this country. Most of the concrete used in building is composed of ordinary Portland cement, and sand with gravel or crushed stone, producing a density of from 2300 to 2400 kg/m³ and cube strength of from 21 to 27 MN/m². Aerated concrete is an additional structural material but is only available in the precast form owing to the method of manufacture.

Cement
1.02 Cement is a material with adhesive and cohesive properties which enable it to bond mineral fragments. The cements used in building have the property of setting and hardening under water and are usually described as hydraulic cements, with the sub-classifications natural, Portland and aluminous. They consist mainly of silicates and aluminates. Portland cement, the most commonly used, is manufactured from limestone or chalk; and the alumina and silica from clay or shale. Manufacture consists of grinding and mixing the raw materials and burning them in a rotary kiln at about 1400°C until the material fuses into

balls of 'clinker'. The clinker is cooled and ground to a powder, some gypsum added, and the final product stored in silos at the works. Distribution is usually in 50 kg bags or in bulk to silos on site. If delivered too early 'hot' cement may reach the site with adverse effects on the concrete.
1.03 *Ordinary Portland cement*, specification to BS 12, comprises 90 per cent of the cement used in Britain. British Standards relating to chemical composition, control the lime saturation factor and the magnesium and gypsum content. Strengths are consistently higher than prescribed.
1.04 *Rapid hardening Portland cement*, specification to BS 12, is similar to ordinary Portland. The acceleration is achieved by finer grinding and an increase in the tricalcium silicate content. Assuming the same water/cement ratio, rapid hardening produces 7 day strength equal to that produced by ordinary cement at 28 days. Extra rapid hardening Portland cement is produced by intergrinding rapid and calcium chloride. Strength is about 25 per cent higher than rapid at 1 to 2 days, and 10 to 20 per cent higher at 7 days. Setting time depends on temperature and can be as little as 5 to 30 minutes after mixing. This makes early placing vital.
1.05 *Low heat Portland cement*, to specification BS 1370, produces an ultimate strength equivalent to ordinary Portland but is less finely ground and is low on the more rapidly hydrating compounds, resulting in slower development of strength and lower rate of heat development. It is used primarily for bulk placing of concrete in dams or

Table I Range of concretes used in building structures in the UK

Coarse aggregate	Fine aggregate	Coarse aggregate bulk density kg/m³	Concrete density kg/m³	Compressive strength MN/m²	Conductivity 'K' W/m²	Particular use
	Sand or pulverised fly ash		540–800	3·5–5·5	3–5	Aerated concrete in building blocks and reinforced precast slabs
Crushed stone, mainly limestone and granite, (or) gravel, blast furnace slag	Sand or crushed stone	1380–1620	2000–2400	14–100	30–32	General structural use in the range 20-30. Mass and reinforced and prestressed concrete in situ or precast concrete. High strength building blocks
Foamed slag	Crushed	550–800	1850–1980	12–48		General structural use in the range 20-30
Expanded clay	Crushed		1500–1600	13–33		Reinforced and prestressed concrete.
Expanded shale	Crushed		1450–1650	15–42	10–13	Building blocks
Expanded slate	Crushed		1600–1800	13–50		
Sintered PFA	Crushed		1550–1710	18–56		
	If sand is used extra strength may be obtained with extra density					
Barytes		2500–3000 depends on shape, usually spheres	3400–3600	21		Special applications such as radiation shielding
Steel }						
Lead }				70		

large foundations where the heat of hydration might damage the setting concrete.

1.06 *Portland blast furnace cement,* to specification BS 146, is made by intergrinding Portland cement clinker and granulated b̶ furnace slag. The slag, a waste product from pig iron m̶ nfacture, is a mixture of lime, silica and alumina, the sa̶me oxides which make up Portland cement but in different proportions. The resulting cement is similar to ordinary Portland but has a lower heat of hydration and hardening rate.

1.07 *Sulphate resisting Portland cement,* to specification BS 4027, has a lower tricalcium silicate and tetracalcium aluminoferrite content than ordinary Portland, thus increasing the resistance to sulphate attack. Early strength is low and ultimate strength high. It develops only a little more heat than low heat Portland.

1.08 *Supersulphated cement,* to specification BS 4248, is not a Portland cement. It is made by intergrinding a mixture of granulated slag, calcium sulphate and Portland cement clinker. It is highly resistant to sea water, to the highest concentrations of sulphates normally found in soil or ground water, to oils and to the acids normally found with peat. At normal temperatures it has low heat of hydration and slow gain of strength. It must not be steam cured or mixed with other cements or additives.

1.09 *High alumina cement,* to specification BS 915, consists of about 40 per cent each of alumina and lime, with some ferrous and ferric oxides and about 5 per cent of silica. Raw materials are limestone or chalk and bauxite. Unlike Portland cement the materials are completely fused in the kiln, emerging as a molten material which is solidified, fragmented in a rotary cooler and ground. The result is a dark grey powder more in the nature of a chemical than a Portland cement. Although the cement is slow setting, it develops about 80 per cent of its ultimate strength within 24 hours. Resistance to chemical attack is high and, as with supersulphated, it must not be mixed with other cements or additives. The greatest care is required in its application and its use has been banned in structures.

1.10 *Coloured Portland cements* are available to specification BS 12. White cement is made from raw materials containing very little iron and manganese oxide. China clay is usually employed together with limestone or chalk free from impurities. Other colours are obtained by intergrinding white cement with various pigments. About 10 per cent more coloured cement is usually required to achieve the concrete strength provided by ordinary Portland.

1.11 *Hydrophobic cement* is obtained by intergrinding Portland cement with either oleic acid, stearic acid or pentachlorophenol. The hydrophobic properties are due to the formation of a water repellent film around each cement particle. The film, which remains intact until mixing, protects the concrete even during long storage in unfavourable conditions.

Table II indicates the relative costs of a nominal 1:2:4 mix concrete between a variety of cements and an average sand and gravel aggregate.

Table II Approximate cost of 1 m³ 1:2:4 gravel concrete with various cements, placed in an rc beam. Assumes batches of over 10-tonnes and delivered within the London area (March 1979 prices)

Cement	Concrete cost £/m³
Ordinary Portland	37·55
Rapid Hardening	38·00
Extra rapid	38·75
Sulphate resisting	40·30
Super sulphated	43·50
Aluminous	46·80
White	53·70

Aggregates

1.12 The most commonly used aggregates are described in table I. The lightweight variety produce concrete with a density of 1500-1800 kg/m³ while the normal weight produce 2000-2400 kg/m³. Lightweight aggregates are manufactured in a limited number of areas and transportation costs must be borne in mind before opting to use them.

1.13 Natural gravels and crushed rocks are still plentiful throughout the country and in certain areas sea-dredged aggregates are available. Aggregates from natural sources are normally specified to comply with BS 882 and to be sampled and tested in accordance with BS 812.

1.14 Consideration should be given to chemical and mineral composition, specific gravity, hardness, strength and physical and chemical stability. All of these may be regarded as properties of the parent rock. Properties particular to the aggregate are the particle shape and size, surface texture and absorption. All these properties may influence the concrete quality in both the fresh or hardened state. A good aggregate will usually produce good concrete.

1.15 Grading is achieved either by mixing screened single sizes or by stockpiling the various sizes on site and mixing by weight according to the requirements of a designed mix. Sea-dredged aggregates should be specially checked for shells and chloride salt content. A quite high shell content is permissible, but the inclusion of large shells containing soft sand can be a problem, particularly in fairfaced work.

1.16 Lightweight aggregates must comply with BS 877, 3797 and 3681. These include foamed blast furnace slag, expanded clay and shale, expanded slate and sintered pulverised fly ash. The fine aggregate obtained by crushing tends to be angular while only two types of coarse aggregate are rounded. The strength of the aggregate places a maximum value on the strength of the concrete in which it is employed.

Water

1.17 This essential ingredient should comply with BS 3148 and be clean and free from impurities. If it can be drunk, it is acceptable.

Admixtures

1.18 Admixtures are not yet subject to a British Standard. Calcium Chloride at restricted percentages was the main basis of acceleration mainly for use in cold weather but its use with reinforcement is banned. Proprietary materials are marketed to improve the nature of both wet and hardened concrete. Additives improving the hardened concrete may give an increase in control of heat hydration and expansion action, increased resistance to water penetration and chemical or fungicide attack and greater corrosion inhibition. Air entrainment in the wet concrete can give weight reduction and greater frost resistance in the hardened concrete. If high early strength is not essential, pulverised fly ash may be added in place of some of the cement. Admixtures should not be used indiscriminately. A well designed concrete mix with good site control and placing may well provide the answer.

2 Mix design

2.01 In the case of ready mixed concrete plants and precast concrete works, a range of mixes will be established on the basis of supplies from particular sources. For the on-site supply of concrete, there is the possibility of the use of nominal volume mixes, standard CP 114 mixes by weight, or designed mixes. A nominal mix could vary in strength from place to place and might not necessarily fulfil the design requirements of, for example, density, strength, water tightness, minimum shrinkage or abrasion.

2.02 In building work, strength is the usual criterion and is usually in accordance with Road Note 4 (HMSO, 1970). Table III shows the headings under which mix proportions should be considered.

Table III Factors in mix design

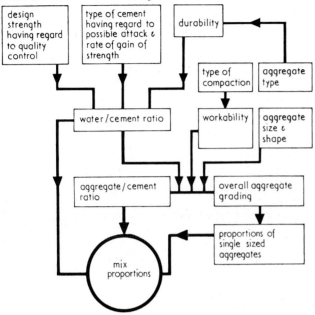

2.03 The materials are variable and properties cannot be assessed precisely so that the end product is not so much a design as an intelligent guess at something near the answer. Hopefully the number of trial mixes leading to the right solution can be kept to a minimum. Such trials are usually at laboratory level so it is still necessary to verify on site the results from the bulk materials and the particular mixing plant and placing methods.

2.04 Some basic facts guide the designer:

1 A specific amount of water is needed to hydrate the cement. Extra water is needed only to provide workability.

2 The strength of the cement paste relates directly to the amount of water used. Excess water reduces the strength and detracts from a number of other properties.

3 The cement paste covers the surface of the aggregate and fills the voids between the particles. The largest aggregate size with appropriate subsidiary gradings will therefore use the minimum paste. In foundation work it is common to use aggregate of maximum size 40 mm, but in some reinforced concrete members it may reduce to as little as 9 mm in order to get the concrete around the reinforcement.

4 Workability may be improved by plasticisers.

2.05 The Code of Practice CP 110 tabulates a number of mixes by 28 day cube compressive strength. Of these grades 20 and 25 suffice for the greater part of structural concrete used in buildings. Table IV shows the cost for varying strengths of concrete using ordinary Portland cement with average sand and gravel aggregate. These basic costs would likely be higher using a lightweight aggregate but the other properties and the consequential effects on the whole

Table IV Cost of concretes with varying strength, placed in bulk (March 1979 prices)

Strength at 28 days MN/m²	Concrete cost £/m³
20	37·30
25 30	38·60
40 50	40·30
60	45·35

structure and fabric of the building would have to be assessed in order to make a fair comparison.

2.06 For a designed mix for normal structural concrete, the specification should define strength grade, type of cement, minimum cement content in kg/m³ of finished concrete, nominal maximum size of aggregate, maximum water/cement ratio and the required workability. It is impossible to define the precise proportioning and then expect to hold the contractor responsible for the end result. Should special requirements be necessary, as mentioned earlier, the entire mix should be specified on a performance basis or special tests introduced.

3 Testing and acceptance

3.01 A variety of tests specified in BS 812 and BS 3681 check the aggregates as received on site; BS 1881 defines the testing procedures for both fresh and hardened concrete. These checks relate primarily to aggregates and concrete workability as cement in this country is sufficiently reliable to be covered by works certificates. The only likely sources of trouble here are overlong storage or too early delivery from the works.

3.02 Aggregate tests include: sieving to check the grading; a silt test to determine the extent of clay, silt or fine dust; an organic impurities test, primarily for sands; tests for bulking of the sand and bulk density of the aggregates; and a check on moisture content. Also to be checked are the storage condition for materials, the cement store or silo, the aggregate stock-piles, the water supply and the cleanliness and efficiency of the concrete batching plant and mixer.

3.03 The two basic checks on mixed concrete are on the workability using slump cone or compacting factor apparatus, and the making of 150 mm or 100 mm cubes. The samples for making cubes should always be taken at the point of placing. Modern specification should require compaction by vibration, in which case a slump of say 25 to 50 mm is appropriate to mass or lightly reinforced sections, and 50 mm to 100 mm for more heavily reinforced sections. The specification may demand a minimum 28 day or perhaps 7 day strength. Required values and test methods are given in CP 110 and CP 114.

3.04 Failure to comply does not necessarily mean that concrete placed previously has to be cut out of the structure, provided that consistent results have been obtained before and that tests have shown the cement and aggregates to be satisfactory. The cubes and their method of making and storage should be examined in case the test sample itself was at fault. There may also be ample reserve in the design to accommodate a slight understrength in a structural member. Non-destructive testing may be required on a part of the finished structure. This is dealt with in BS 4408 under the headings of 'gamma radiography', 'surface hardness methods' and 'ultrasonic pulse velocity'.

4 Distribution and placing

4.01 Distribution on site is the contractors' problem. It may be by pumping, by blowing, by skip and crane, or simply by dumper to concrete hoist, barrow and chute. Workability will vary to allow for the method employed. Satisfactory concrete once produced has still to be properly placed and compacted. A small reduction in compaction will mean a considerable loss of strength. Distribution and placing methods must be checked and further care taken as the concrete hardens and matures. This is particularly important in extremes of hot or cold weather. In the former there is a curing problem and in the latter the concrete could, at worst, be reduced to its basic constituents. Provided the concrete can be placed at about 12°C and then kept a few degrees above freezing it will eventually achieve full strength.

5 Properties of concrete

5.01 Table 1 summarised a range of concretes in terms of density, strength and insulation value. In specifying a particular concrete, the designer would consider strength, density, durability, and water tightness. Some of these are determined by aggregate choice but the major considerations are cement, quality of production, placing and curing.

Strength

5.02 Concrete has a high compressive strength and a low direct tensile strength (about 10 per cent the compressive). This last point is ignored in design. Shear strength is assigned a value, but this is meant to imply the principal tensile stress due to shear action in bending. A much higher value can be attributed to pure shear, for example where a castellation is used to interlock two members. Bond, the essential connection between the reinforcement and the concrete, is the combination of adhesion, friction and shrinkage of the concrete on to the steel. This again is a fraction of the compressive strength. Portland cement concrete which has not been subject to acceleration curing will gain strength over several years. The Code of Practice acknowledges this by allowing direct and flexural stress characteristics after one year 24 per cent higher than those after 28 days. Concretes containing pozzolanic material such as fly ash also tend to gain strength with time and many of the other properties, themselves related to strength, will become more pronounced also.

Young's modulus

5.03 Young's modulus cannot be employed in the usual way as the stress/strain graph is curved, not linear. The tangent or secant modulus are therefore considered. The value of the modulus for use in calculations is the relaxed value, in other words the value of the concrete under load; this may be half the value for prolonged loading as compared with instantaneous loading. Shear, tension, bond and modulus vary with the compressive strength but assume limiting high values and the CP 114 Code of Practice accordingly shows a method for deriving the allowable stresses for use in design.

Movement

5.04 Drying shrinkage results from the loss of water to surrounding, unsaturated air and to the thermal effect after cement hydration. Carbonation shrinkage results from CO_2 in the atmosphere reacting with the hydrated cement minerals when moisture is present. Shrinkage is of little consequence if unrestrained, but the differential between reinforcement and concrete can produce cracking in the member. Some aggregates give higher concrete shrinkage characteristics than others—a point for consideration in aggregate choice. Moisture movement may occur if the concrete is alternately wetted and dried, but initial drying shrinkage is never fully recovered even if the concrete is saturated for a prolonged period.

Creep

5.05 Creep is the increase in strain under a sustained stress. It can exceed strain under applied loading and its consideration is therefore important to design, particularly in pre-stressed concrete where a considerable portion of the preload could be lost through relaxation in the concrete as it 'creeps' under load.

Durability

5.06 Concrete must resist abrasion, fire, water penetration, thermal effects, erosion and chemical attack. Abrasion resistance and permeability are both achieved with a well compacted concrete of low water/cement ratio and a minimum of fines. Suitable additives may be used to prevent any small voids remaining in the mix. Abrasion resistance can be measured by a wearing test in which the sample is measured before and after treatment. Good concrete should be of such density as to be watertight but of course shrinkage cracks and construction joints are the danger points.

5.07 The coefficient of thermal expansion of concrete varies with aggregate and moisture content. It lies in the range $7—12 \times 10^{-6}$ per °C. Thermal problems usually relate to differential conditions, for example of constraints between members, and can normally be handled by adequate steel reinforcement. Fire effects have been described in section 3 of the handbook.

5.08 Frost damage is unlikely to occur in a dense concrete of low water/cement ratio but further resistance can be achieved by air entrainment in the mix. This in effect creates voids into which the frost expansion can take place. Erosion problems being similar to abrasion require a similar concrete. A dense concrete will have high resistance to normal atmospheric erosion. It will also withstand a high velocity flow of water but not the continual attrition of stones carried by water or waves.

6 Surface protection

6.01 Ground sulphates and acid or alkali effects have been considered in section 4 FOUNDATIONS. An enormous number of chemicals used in industry are injurious to even the best of concretes and some form of surface protection may be required. Certain surface treatments increase the resistance of Portland cement concretes to aggressive agents. The pores can be blocked to reduce permeability, chemical reaction induced with the free lime of the cement or a thick surface coating applied to prevent the aggressive attack registering. The former two treatments are of little value at depth. Severe attacks require use of surface coatings in the form of resins, chlorinated rubber, bitumen or coal tar, bituminous mastic, asphalt or special rubber compositions, acid- or alkali-resisting bricks or tiles, and plastic or glass wrapping.

7 Aerated concrete

7.01 This is more an expanded lime or cement mortar than a concrete. One method produces air bubbles in the mortar by using a foaming agent, usually with resin base, another uses a chemical agent generating a gas. The latter employs an autoclave, in other words a pressure-cooker curing system, which is essential to stability in structural aerated products. The aggregate is finely ground sand or pulverised fly ash. Compressive strengths are about $3 \cdot 5$ to $5 \cdot 5$ MN/m² in the density range 540 to 800 kg/m³. Production must be under close control. The aerated concrete emerges in large blocks and is sawn to close tolerances. Building blocks or reinforced slabs are provided to agreed standards, the latter being employed as large partition units, roof panels, or in lightly loaded floors. The reinforcement must be protected against corrosion and is usually in welded mesh form to give the necessary mechanical bond—all the strength properties being lower than with normal concrete.

8 Steel reinforcement

8.01 The manufacture of steel is dealt with in section 6 of the handbook. Steel reinforcement is only one type of steel fabricated by the mills. Shaped bars may be taken from the mills and their properties transformed by the pulling and twisting effects of cold working. British Standards cover different types of reinforcement such as

1 Hot rolled steel bars (BS 4449)
2 Cold worked steel bars (BS 4461)
3 Steel mesh fabric (BS 1221)
4 Plain hard-drawn steel wire for prestressed concrete (BS 2691)
5 Indented or crimped steel wire for prestressed concrete (BS 2691)

8.02 Steel for ordinary reinforced concrete has an ultimate strength range from 250 MN/m² to 410 MN/m², though proprietary steels with special bond control features go from 550 to 680 MN/m². Elongation for 250 steel is 22 per cent and for 410 steel is 14 per cent. Ultimate tensile strength must be at least 15 per cent greater than yield. Preferred bar sizes in mm diameters are:
6 8 10 12 16 20 25 32 40. 16 and 20 mm give the basic price with bars of greater or smaller diameter becoming increasingly expensive. Lengths greater than 12 m are also more expensive. Generally the cold worked steel is economic for structural application and tends to be used even in positions where mild steel would suffice, if only to avoid confusion on site. Hot rolled steel to BS 449 is used as an alternative.   British Standards require 460 grade bars up to 16 mm to have a minimum 12 per cent elongation and the 425 grade bars over 16 mm to have a minimum 14 per cent elongation.

8.03 Steel wire for prestressed concrete is supplied in plain or indented form with either normal or low relaxation—a reference to the creep characteristic—in an ultimate strength range of 1470-1720 MN/m² with minimum 0·2 per cent proof stress (in this case the equivalent of yield point) of 1250-1550 MN/m². Preferred sizes are 4, 5 and 7 mm diameter. Proprietary alloy steel round bars may be used as an alternative to steel wire in cable or strand form.

8.04 The majority of building structures are designed in normal reinforced concrete and the reinforcement is largely used in tension. As the hot rolled or cold worked high tensile bar costs only about 8 per cent more per tonne than mild steel, and has only two-thirds the equivalent weight, it is far more economic, so much so that it still tends to be used in the few cases where it is not the cheapest answer.

9 Advances in concrete

9.01 Radical innovation is unlikely to occur in a material dating from the Roman era, but refinement is possible. The last 30 years have seen improvement in materials and manufacture without any change of the basic ingredients, leading to stronger, cheaper, better concrete. The precast factory can produce extremely high strength concrete on account of the closer controls possible and the greater scope for applying techniques in placing, compaction and curing.

9.02 Strengths of 100 MN/m² are feasible using carefully selected materials, but only at higher cost. This would be of doubtful advantage in in situ work, even for localised areas of high stress, on account of the need for varying grades of concrete on site. It is in the long span structures, for example bridges, that such practice will bring overall economy.

Resins

9.03 Resins are used in three ways. They may act as a base for the concrete instead of cement and water. Resins lose strength at high temperatures so this type of concrete cannot be used in a fully structural capacity, only in semi-structural applications such as machine bedding and bolt setting. Resin impregnation is the filling of the voids in a normal mix with a polymer and then curing the polymer. This is expensive and has practical difficulties, but is being actively researched in the US. Resin additive concrete is produced by adding a liquid epoxide or polyester which cures by drying out or by chemical reaction. About 10 per cent of the weight of the cement is required before the resins can give worthwhile improvements. This virtually restricts their use to thin screeds.

Fibre reinforcements

9.04 The success of asbestos cement and glass fibre products has prompted research into fibre-reinforced concrete. This is in an effort to achieve a near homogeneous member instead of the compressive-tensile composite that is reinforced concrete, but it is unlikely to be economic. It would require about ten times the reinforcement weight of the conventional method in order to give the equivalent flexural and tensile strength. Small amounts may be used to increase impact strength and crack control and to affect the actual mode of failure in a member. Fibres used include metal, glass, plastic and carbon. The first two provide adequate tensile strength with the proviso that glass, with its low alkali resistance, must only be used with aluminous, rather than Portland cement. Plastic, such as polypropylene and nylon, has a low elastic modulus, confining its use to semi-structural members, for example cladding. Carbon fibres are currently too expensive to be considered. Generally fibres are difficult to handle. Both mixing and vibration pose problems so this again is a process for factory, not site.

For the time being it would appear that the well-tried methods and materials will continue in use, particularly for work on site.

Flint-faced Roman concrete, Burgh Castle, Suffolk

Structural concrete spans, Seaton Bridge, Devon (built 1876)

Technical study
Concrete 2

Structural applications
of reinforced concrete

In his first study on concrete ALLAN HODGKINSON *described the material, its constituents and properties. This study describes its structural applications*

1 Introduction

1.01 Because of its versatility, design flexibility, and inherent fire resistance, reinforced concrete, particularly in situ, can be used for a wide range of structures. Members can vary in shape and size to suit special design and layout requirements, and floors can be punctured to accommodate services, but once cast the structure is unalterable. The development of precast concrete has simplified the design and construction of skeletal structures. Prestressing has extended the spans to which concrete structures can be carried and structural steelwork has been partly deposed even in bridge building.

1.02 In the last two decades experience has proved that concrete can assume almost every structural form. However, the relationships between costs of materials, labour and plant vary with time and cause like variations in the cost-competitiveness of one material over another, and within a material of one structural form over another.

1.03 The designer must constantly bear in mind that the ultimate economy to be achieved is that of the building, not of the structure alone; so an in situ floor slab which has a ready-for-decoration soffit and a power-floated top surface ready to receive a carpet finish may show an overall saving compared with an apparently cheaper precast assembly which requires ceiling finish and screed to attain the same standard.

1.04 Superstructure and foundations must be considered together. A lightweight construction of longer span may provide a better solution than the apparently cheaper normal weight structure of shorter span. Thus there is no simple formula for deciding on a structural type and experience and intuition play a decisive part in the final choice. The purpose of this study is to illustrate, and comment on, the many structural forms which can be used.

2 The typical building frame

2.01 Over the past 20 years reinforced concrete has been used for most building frames, primarily because the light-weight casing of structural steelwork was slow to develop and the solid concrete casing required for fire protection precluded structural steelwork as a competitor. Building firms geared themselves to in situ concrete construction of entire frames and floors and though the margin between the two materials has lessened, the all-concrete structure is still the most common. Assuming a building of repeating bays 10 m × 7 m, the following solutions have been employed.

Column, main beam, secondary beam and slab
2.02 This was acceptable in the 1930s, but is only used nowadays for very heavy loading **1**. The construction se-

quence is to cast the columns and strip the formwork, erect the beam formwork and cast the beams up to the underside of the proposed slab, then strip the beam sides, erect slab formwork and finish the slab. Sometimes it may be convenient to erect shuttering for beams and slabs and place both in one operation. Then the formwork can be arranged so that the beam sides can be removed before the slab formwork (re-use of formwork and speed of re-use are essential to economy). Ways of leaving support until the concrete is capable of carrying its dead and construction load must be found; removing all the formwork and then replacing the props is a last resort but sometimes the only way. Time until stripping relates to span and loading as well as to strength maturance of the concrete. There will be occasions when construction loads are higher than live loads and so may require support for the ensuing wet concrete through two or more finished levels—a point which should be made to the contractor at the time of pricing. Part precast solutions can be employed in this type of structure in an attempt to cut out the more difficult aspects of the form-work process. However, the more bays there are in each direction, the more difficult does the cranage of heavy precast members become—placing dimensional restrictions on the use of this system.

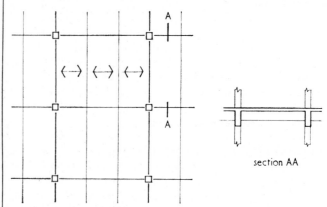

1 *Building frame: column, main beam, secondary beam and slab*

Column, beam, slab—one-way span
2.03 This is cheaper and now more commonly used **2**; but dispensing with the secondary beam leaves the structure dependent on some other way of resisting wind forces at right angles to the main beam line—usually some stiff lift or stair arrangement. The slab has to be thicker to span the extra distance, so to limit its weight without loss of strength, many types of void formers are used and formwork is then required to support the void-forming devices. Those requiring a full slab area formwork include hollow blocks, lightweight solid blocks, expanded metal boxes, cardboard

boxes, expanded polystyrene blocks, and either cylindrical, cardboard or inflatable-duct tubes **3, 4**.

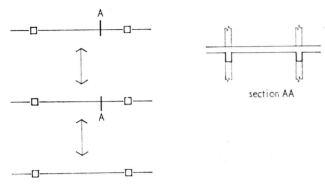

2 *Building frame: column, beam, slab—one-way span*

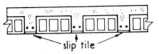

3 *Hollow pot floor slab*

4 *Void formers buried in slab depth to reduce weight*

2.04 The hollow pot **3** has done yeoman service for many years but its high labour content and breakage rate now tend to make the equivalent solid slab cheaper. Load to be carried to foundation must always be kept in mind but the labour content of the solid slab is less and it gives more flexibility where lift or lightwells or variable heavy loading occur. A cautionary word about the use of horizontal slip tiles between pots as permanent shuttering for the undersides of the beams—there is a danger of poor compaction between the underside of the reinforcement and the top surface of the tile and this can remain hidden. If slip tiles are used a deep cover should be allowed. Preferably expose the rib soffit by not having a slip tile and make clear to the contractor the problem of spacing the blocks when he tenders. Where spacing is a deflection constraint, perhaps owing to plaster finishes below or to sensitive machinery requiring level support, remember that a ribbed slab normally has 30/26 deflection of the solid slab. Thus allow an equivalent depth increase where minimum slabs are being compared. Where heavy point loads occur, cross ribs should be introduced. These act as load spreaders by forcing the main ribs to deflect together. Void formers buried in the depth of the slab give a performance almost equivalent to the solid slab, but locating the formers is difficult, for in addition to the problems of placing and vibrating, the former tends to float in the wet concrete **4**.

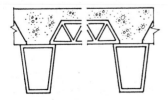

5 *Pressed concrete blocks as permanent soffit can be temporarily supported by triangular section telescopic formers spanning between beam edges or brick walls*

2.05 Pressed concrete blocks can be supported either by pro-

prietary telescopic formers which are adjustable to span between beam edges or brick walls, or on a complete soffit shutter **5**.

2.06 The larger void-forming devices such as plywood steel, fibreglass or polypropylene troughs can produce a very pleasant soffit appearance which the architect may be able to use to advantage. A flat soffit may be obtained by pinning plasterboard to the ribs. The trough can be standard, or fixed, or if the size of contract makes it worthwhile, purpose-made. One solution is shown where minimum formwork is employed and the trough can be removed after one or two days **6**.

2.07 Formwork can be kept to a minimum and part removed after one to two days allowing the troughs to drop but leaving props as long as is necessary to develop the strength of the rib. Cross ribs result from leaving gaps between the trough ends, their centring dependent on trough length.

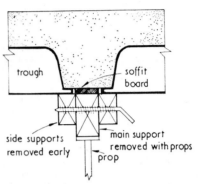

6 *Trough formwork can be struck leaving only temporary support as beams attain full strength*

Column, beam, slab—two-way span

2.08 With beams on each grid line the slab spans in both directions **7** and therefore can be thinner. The beam reinforcements are of the same order, and intersect at the column head. The top and bottom steel layers of adjacent slabs also converge at this point, and the resulting congestion places a limit on minimum slab depth. This form of construction has taken on a new lease of life since the fifth amendment of the Building Regulations came into force, because the two-way system provides enough reinforcement at enhanced allowable stresses to span the structure one way were a beam to be removed.

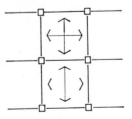

7 *Building frame: column, beam slab—two-way span*

2.09 As the spans increase, void formers can be employed to turn the slab into a two-way ribbed construction. The voids can be formed by steel, glass fibre, or polypropylene waffle pans and by the use of cardboard boxes, expanded metal boxes or expanded polystyrene blocks.

2.10 Waffle pans have recently been standardised and can be hired, though as with troughs, a large contract may permit use of purpose-made units Again depending on the type of former the soffit formwork may be an all over or partly removable system. With the waffle pan supported on two sides only, it is possible to use the method employed with the trough, provided the unsupported, open edge of the pan is taped.

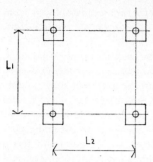

8 *Plan form for empirical design method described in para 2.11. See reference 3, p180.*

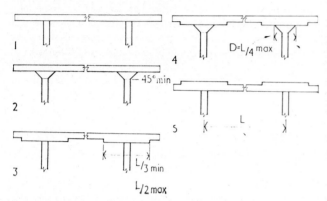

9 *Flat and drop slabs: column capitals and slab drops (2 to 5) reduce effective span of slab and transmit shear into column head*

Flat and drop slabs

2.11 An empirical method can be applied to flat slab design provided the plan form lies within certain limits **8, 9**. L_1 must not exceed $1·33 \times L_2$. Adjacent panels L_1 and L_2 dimensions must not differ by more than 10 per cent. End spans may be shorter but not longer than interior spans. The thickness of the slab should be constant. There should be at least three rows in each of two directions at right angles. Generally, a nearly square grid analysed by the empirical approach produces an economical answer for quite heavy loading conditions. This method does assume that a fairly high arbitrary moment is taken into the column. However, in the case of car parking floors, which work quite well on a 8 m square grid, the ideal column section is oblong to facilitate parking. In its weaker direction it could not accept the empirical moment, but given an alternative analysis and allowing only for the out of balance live load moment between adjacent spans it can be made to work.

2.12 Column capitals and dropped slabs **10** are used for the dual purpose of getting the shear into the column head and

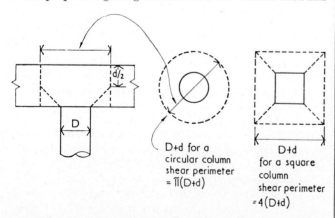

10 *Shear perimeter in slab using square and circular section columns*

reducing the effective span of the slab. The flat slab and simple column provides the cheapest formwork arrangement but the available shear perimeter is quite small.

2.13 The drop slab produced two critical shear conditions **11**. At 6 the total panel shear is carried on a perimeter determined by D and d_1, and at 7 by L_{drop} and d_2. In the same way the capital increases the slab shear perimeter relative to the maximum diameter of the capital head, and where the capital and drop are used together the critical section within the drop is raised yet again and very heavy slab loads can be carried. As in the two-way span supported by beams, the floor weight can be reduced by the use of waffle pans provided that there is sufficient solid concrete around the column head to deal with shear forces.

2.14 It is possible to reinforce the area around the column head by the use of welded crossheads in structural steelwork or by using a system of bent-up bars—a method to be used with caution and as a last resort, owing to practical problems of placing the reinforcing steel.

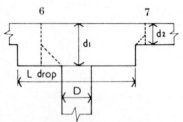

11 *Drop slab: shear perimeter shown dotted*

Flat slab and banded plate

2.15 The idea of abandoning the beam and making a wide strip of slab do the work evolved from the mid-1940s. First came the wide shallow beam with slightly simpler formwork, easier concrete placing and less overall height to the building, then in the early 1950s the beam disappeared altogether.

2.16 To consider the overall costs of the three schemes illustrated in **12** it is necessary to look at finishes and the building cube as well as at the structure itself. In the structure the amount of reinforcement will vary almost linearly with d_1, d_2 and d_3. Items 1 and 2 have extra concrete compared with slab 3, extra formwork costs in the sides and soffit, and extra plaster and decoration costs.

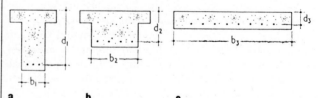

12a *Beam and slab;* **b** *drop slab;* **c** *flat slab*

2.17 If headroom beneath the beams is to be the minimum headroom in the building, as is usually the case, then the extra beam/stem depth must be added for every floor—in a 10-storey building this could be 5 m. The extra effect on the building cube is considerable and may result in the loss of a floor where overall building height is under planning control. An added advantage of the flat ceiling in office buildings is that there can be complete flexibility of partitions. However, the span of the solid slab is limited, especially when it has to double as a beam. The beamless floor deflects more and brittle partitions such as lightweight blocks and plaster must be used with care (see Technical study SAFETY 2, AJ 7.6.72).

2.18 By using void-forming devices spans can be improved without increase in weight and with the greater depth little increase in reinforcement. Making use of the trough or waffle pan can strike a balance between floor types **12** 1 and 3 and provide quite long spans while still giving a level soffit **13**.

2.19 A banded plate is a solid slab, with concentrated bands of reinforcement in place of beams. Whatever form it takes, the load has to be got into the column head in the same way as with the flat slab. Also, unless a trough-type void-forming system is used, a grid ratio in the 1:2 range will mean that the slab is spanning two ways and should be reinforced. Many structures have been incorporated in buildings with arbitrary one-way reinforcement and this is not good practice, particularly when the slab is also carrying the wind load in frame action.

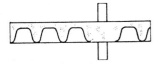

13 *Waffle pan shuttering, giving slab combining properties of* **12a** *and* **b**

3 Precasting

3.01 Precasting is a technique which has been in use for very nearly as long as reinforced concrete itself. It can be applied as a temporary site process or a permanent factory process, to a standard or a one-off product.

3.02 The decision to precast is generally complex and inevitably requires a degree of intuition and experience if the choice is to be successful. The factors to be considered are:

a, the ease or difficulty of formwork on site

b, the degree of repetition

c, handling or haulage to site or on site

d, plant, eg availability of cranes on the site

e, speed of erection required

Only once the project has been scheduled to start, and in a decided form, can a decision be taken on whether or not to precast. Provided the contract is to be negotiated with a chosen contractor, the best solution can be based on the contractor's existing resources or on what he can easily obtain. The problems are greater if the contract is open for competition. The designer cannot know how the successful contractor will choose to do the work. An intelligent guess has to be made in the design about the size of the members and their distance from the probable crane position. This may favour one tenderer while another would have submitted a lower price to a different arrangement. Unfortunately designs are usually too far advanced at this stage to be revised at a contractor's suggestion. On the other hand, where time permits, contractors can often submit a precast alternative for an in situ scheme at tender stage. However, the more difficult the in situ formwork problem, the easier is the decision to precast. Some projects are obvious targets. The simplest decision is that of the floor or roof in a steel frame or masonry structure. A quite different reason for precasting in certain parts of the UK is the local attitude of carpenters and steel fixers which may cause the contractor to limit absolutely the number of employees on site.

Factory precasting

3.03 In theory the precast factory is ideal. It should offer regular employment to an organised staff, minimum disruption by labour dispute and guaranteed delivery on time, ease of construction and therefore maximum control of quality and costs. In practice these are all variables and when the building industry as a whole is at low ebb the precast factory seems to suffer worst.

3.04 The standard product from the factory contributes considerably to modern building mainly through the supply of flooring units **15**. These can range from a few centimetres to more than 2 m in width, and up to 10 m in span. Most units are shaped to give maximum efficiency for weight—important both in transportation and working.

15 *Typical precast floor sections*

3.05 The units illustrated **15** can be placed side by side, requiring only a filling between them followed by a screed ready to receive a floor finish. The keying between the unit edges appears capable of spreading a concentrated load. When the units are contained by reinforced concrete, by main cross beams and edge beams, they can act as a structural diaphragm, imparting stability or carrying wind load laterally to some stiff point in the structure.

3.06 Units can be erected by builder's hoisting wheel or tower crane either individually or in batch and, depending on size, manhandled or craned into final position. Since the advent of prestressing in the late 1940s most manufacturers have equipped themselves with prestressing beds up to 100 m long and techniques have been introduced in which a machine moves along the bed, moulding the work and leaving finished concrete which can be sawn into required lengths.

3.07 A variety of frames and members have been introduced to capture the general building frame market, but success has been limited. A standard system The Public Building Frame, was promoted by the Ministry of Works in the mid-1960s but this found little favour outside the ministry itself. The main reasons for the limited use of the factory precast standard frame is that it imposes an immediate discipline in a material which for years has been advertised for its design flexibility; also it demands a design crystallisation

14 *Concrete Society award: North Thames Gas Board, Southend-on-Sea*

much earlier in the programme than with the in situ structure. In development and industrial work decisions to proceed are usually given too late relative to the required completion date and, even when given, are liable to variation as the structure proceeds. These circumstances favour the more adaptable in situ construction.

3.08 Industrial single storey frames are a different proposition. They provide a consistent cheap answer to the shed type building as discussed later under skeletal structures (para 9.09). Other structural factory products include sections of shell roofs, a complete range of standard beams for bridges, box sections for culverts and manhole rings, chimneys, staircases and cladding panels.

3.09 Apart from standard products, modern road and rail haulage has enabled transportation of enormous members over long distances. Although there is a range of standard beams, motorway development has led to the production of purpose-made whole beams, or components of hollow girder bridges and arches.

Site precasting

3.10 For a multi-storey project with design for table formwork there is no reason why in situ concrete should not be the fastest mode of construction. Such schemes still have edge beams, lintels, stairs and cladding panels and these members have often been produced from a factory on site, either by one central casting yard or a number of small yards within crane reach of the building in question. Precasting on site or nearby permits use of much larger members since normal traffic requirements will no longer apply. Hammersmith flyover **16** shows the scale of precasting which can be achieved.

3.11 Shell construction becomes more feasible when the complete shell can be cast in a site factory and lifted bodily into position. This usually applies to northlight shells spanning between precast frames, for example the Gallacher factory in Ireland. A similar process was used on the Hull Technical College where one stressing bed on site provided the entire precast roof system for the workshops. At the main drainage outfall buildings, Beckton, part rings of a barrel roof were precast, propped in position and stressed together.

16 *Hammersmith flyover;: 16-span prestressed structure; superstructure of precast units is prestressed to act as one continuous member*

3.12 Precast units may act compositely as permanent formwork for in situ concrete. In one method **17** columns are constructed in situ.

3.13 Precast beams are positioned on top and use either local supports down the column face or odd supports along the length of the beam. Precast floors are spanned from beam to beam, and in situ concrete placed to the beam/column connection and the whole area of the slab. Depending on the form of the precast slab, virtually the whole value of an equivalent in situ construction can be achieved. Variations on this theme include a variety of steel-to-steel connections concreted into precast members. In one range of precast products **18** a concrete soffit is provided containing the main floor reinforcement such that the in situ concrete once placed can work compositely to provide a finished structural floor with many of the advantages of the fully in situ slab. The soffit unit may be capable of spanning between supports or may have to be propped temporarily to enable it to carry the wet concrete and working construction load. With the aid of hollow pots, concrete blocks, etc a wide variety of the flooring units described previously (paras 2.01 to 2.07) can be used in composite construction either with or without temporary propping.

3.14 A recent addition to the precast standard range is the double-T spanning up to about 15 m **19**. Such systems can be economic because of ease and speed of erection without necessarily developing the full strength of an in situ structure of the same dimensions.

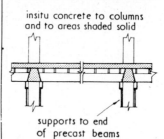

insitu concrete to columns
and to areas shaded solid

supports to end
of precast beams

17 *Composite structure: precast beams and floor units act as permanent framework to in situ concrete*

18 *Composite structure: precast units with exposed reinforcement act as permanent soffit to reinforced concrete placed above*

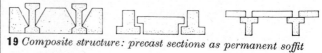

19 *Composite structure: precast sections as permanent soffit*

4 Jointing

4.01 The weak links in any structure are the site joints, so joint control in both precast and in situ work is important. The earliest attempts at precasting relied on a full dead bearing with beams landing on corbels. In an effort to improve appearance, beams were scarfed at their ends and took the profile of the corbel. However, corbels are not easy to cast and today beam-to-column connections tend to rely on steel inserts cast in opposing faces and bolted or welded together. If the connection is to be neither too costly nor too critical, the all precast frame must not rely on continuity to the extent of the in situ construction, but can compensate by using a higher strength concrete from the factory.

4.02 The column-to-column joint in multi-storey work has always been a problem. One solution is to make the lower half of the upper column hollow to receive the vertical bars from the lower column. The void has to be successfully grouted for the joint to transmit direct load and relies on stability from some strong point such as a staircase enclosure or crosswall. This led, before the Ronan Point collapse, to some rather dubious constructions. A more positive connection is the threaded bar, with right and left hand threaded couplings capable of developing most of the strength of the bars so joined. Up to four storeys it is sometimes possible to cast the columns to full height and then use beam-to-column steel connections to complete the frame. Up to four storeys in narrow width buildings it may be possible to precast a section of the elevation to full height and one and a half bays width, thus including two columns within the section with horizontal beam infills to complete the elevational frame.

4.03 A final method of connecting columns is to weld the steel at each column end to a steel plate, mate the plates and weld them round the exposed perimeter. The final joint stability in any arrangement of members can be achieved by post-stressing, though this is only likely to be economic in large span structures and is seldom used in normal building work.

5 Prestressed concrete

5.01 Prestressing is the act of compressing a concrete member by pretensioning its reinforcement. This enables it to carry an amount of direct or bending tension without the concrete going into tension or at least keeping the tension within the allowable tensile stress. Three methods are commonly used. Concrete may be cast around stressed wires and the wires released from their anchorages after the concrete has reached a strength of about 35 N/mm². The concrete grips the wires and is compressed as the wires tend to return to their original length. Alternatively the wires or tendons may be housed in a sheath and the sheath concreted into the member. At a concrete stress of about 40 N/mm² the tendon can be tensioned and retained by an end anchorage. Finally an external jack may be used, operating between the member and a buttress, but this is most likely to be employed to adjust a reaction in a redundant structure.

5.02 Stressed wires are commonly used in precasting factories. A permanent bed is constructed on which a variety of members can be cast in line, member size being restricted by crane coverage. An early disadvantage was that the wires could lie in only one position along the bed and therefore the stresses at the end of the member were similar to those at mid-span, thus losing some of the design advantage in both bending and shear. Developments include deflecting the wires or coating them to cause disbonding at the ends of a member.

5.03 The post-tensioning method is perhaps the most flexible, assuming the tendon can be made to take a desired shape and thus work to maximum efficiency. On the other hand members have to be stressed individually and the expensive anchorage units at both ends are permanently lost within the concrete. Also the cable has to be grouted and this can be a source of trouble if badly done. The main advantage in the normal range of structures is that the concrete is not cracked at the working stress level. This protects the steel and gives a better deflection characteristic because the full section of the member remains available (normal rc will be in tension below the neutral axis). Prestressed concrete has come to be regarded as another material but it is just the same concrete with another means of reinforcement. With higher yield steels certain shapes of member can be made more economically in ordinary reinforced concrete, even with long spans. As ultimate loading is reached there is not much difference between the stressed member and the unstressed member containing the same steel. Nearly all the notable successes in precast, or in situ, prestressed concrete are in long span structures eg bridges.

6 Sliding formwork

6.01 Originally developed for construction of silos and bunkers, it is now used in building the service, lift and staircase cores for multi-storey buildings **20**. Formwork about 1·3 m deep is erected at foundation level on the line of the structure walls. Concrete is placed in the forms which are then raised slowly by a screw or hydraulic mechanism arranged to climb vertical steel rods cast in the foundations and extended upwards by couplers as the work proceeds. If the weight of concrete in the forms is below a certain optimum the concrete may be lifted by friction with the form; so 150 mm is about the thinnest section to use. Success depends on uniform and consistent supply of concrete and construction is usually carried on non-stop for the full lift. Progress is in the order of 150 mm to 300 mm per hour. Checks on verticality are made as the height increases and the jacking is varied to compensate for any tendency to lean. Holes for windows or doors, or pockets for beams, can be cast as work proceeds.

20 *Sliding formwork: services tower, Addenbrooke's Hospital, Cambridge*

7 Lifting and jacking

7.01 One answer to the formwork problem is to cut out staging by doing the work close to the ground and then lifting or jacking it into its final position. The most straightforward system is 'Lift slab' as applied to the column and flat slab structure. Columns are first cast to part height, or, if not too high, to the full height of the proposed structure. Slabs are cast in their plan position at ground level, one on top of the other, using a separating medium between the layers for connection to the columns when the slabs have been raised to the correct level. But because lifting is by jacks at the column heads, care must be taken that the columns are stable throughout. Slabs can be constructed as plate floors by any of the design methods described earlier (paras 2.01 to 2.19). Jacking from below is usually applied to whole units of beam and slab, or shell roofs. This still requires formwork of the shape of the member, the object being to achieve this on the ground rather than on staging in the air, with resulting economies in formwork support and concrete placing.

8 Tolerance

8.01 The greatest disservice to concrete is to expect it to be exactly as drawn. First there are the moulds—if the number of members is less than 100 these are usually timber—which will not be perfect in the mould shop let alone after deterioration with use. Further inaccuracies may occur during assembly. The entire formwork in an in situ scheme must be erected to plumb line and level, and so allowances must be made for inherent inaccuracies of measuring in space. The formwork has to be held in position against construction loads incurred in placing the concrete. The concrete then deflects on striking and in time shrinkage and creep change the deflection again.

8.02 A sensible tolerance has to be allowed. The tighter that tolerance is, the higher the cost. An example of ill-considered detailing would be to place a window unit between two concrete columns and horizontal members without the minimal 15 mm clearance all round the theoretical concrete inner face. There may be greater accuracy in the shape and size of precast members, but the actual assembly may demand even greater control to validate the method of jointing.

8.03 Moulds used in precast work are likely to be accurate. However, subsequent reinforcing and handling, as well as the results of over early demoulding and ill-considered stacking and transportation, can play even greater dimensional havoc than that found with in situ. Individual, small-section, pretensioned members are the worst, and tolerances should be carefully specified. Length, cross-section, straightness or bow, squareness, twist or flatness should be defined specially for the job in hand. In post-tensioned units particular reference should be made to the camber and the variation of camber between adjacent members. There have been some unsatisfactory results in projects where the only control of tolerance has been the manufacturer's statement that the work would comply with CP 116 *The structural use of precast concrete.*

9 Building and structural types

The narrow depth building
9.01 Speculative office building has led to many buildings being planned on two office depths plus a central corridor —an overall structure width of from 10 m to 15 m. Framing has usually been provided with one or two internal columns related to the corridor line and an elevation column module to satisfy the window arrangements. All the options of column, beam and slab structure described earlier are open. The first solutions were conventional **21** but gave way to the banded plate floor **22** and, in cases of particular occupation, to the clear span based on the deep ribbed floor or precast, prestressed beams across the building carried by a deep beam set within the solid part of the wall elevation **23**.

The single span building
9.02 This follows logically from the narrow depth type. In the early '50s secondary schools and technical colleges demanded a two-to-four storey structure of one classroom and corridor width, usually about 10 m. This led to multi-

storey single bay portals, spanned with a variety of floor types, usually ribbed for lightness **24**. The portal frame is sensitive to movement and may be open to differential settlement and so needs careful consideration. One solution in a difficult foundation situation involving deep piling was to cantilever from a single central base. The external cladding frame was hung from the roof-level beam **25**.

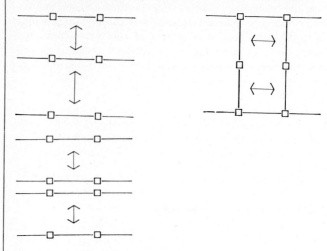

21 *Narrow depth building: frame plan*

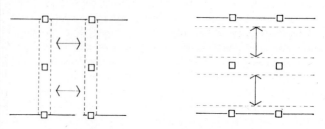

22 *Narrow depth building: banded plate floor*

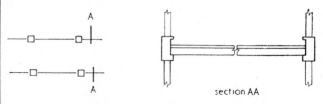

section AA

23 *Narrow depth building: frame plan. Section shows precast floor units spanning to deep beam within wall elevation*

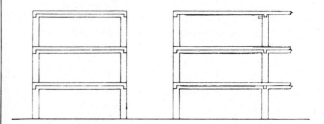

24 *Single span building. Typical sections showing multi-storey portal construction*

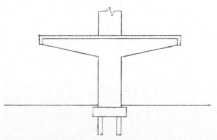

25 *In a difficult foundation situation, frame may cantilever from single central base*

The hull core building

9.03 Problems of vertical circulation, escape, services and storage in office block developments with from 20 to 30 storeys led to the grouping of these facilities in one or two areas on plan which could be protected by reinforced concrete walls through the full height of the building and able to cantilever from the base for stability **26**.

9.04 The structure outside the cores could be designed with primary concern for vertical loading, the columns taking only local bending from wind-load could be of minimum section. The conventional vertical wall formwork, constructed floor by floor, first gave way to larger panels. The most recent development is construction of the inner core by sliding formwork, enabling the whole of the inner cores to be erected as towers at the rate of 150 to 300 mm per hour. Deep cantilever frames coincide with each service floor. Suspended hangers provide tensional support to the perimeter of the building **27**. It is early to predict whether these latest developments are economic. Fashion in structure as elsewhere is not always well-founded.

26 *Hull core building. Stability is from vertical reinforced concrete circulation and service cores cantilevering upwards from base*

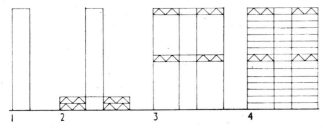

27 *Hull core building. Cantilevered horizontal frames at each service floor. Suspended hangers tension building perimeter*

High-rise dwellings

9.05 Replacement of war damage, slum clearance, and the desire to provide open landscaped areas around housing sites helped to promote tower and slab dwelling blocks. Starting in the late 1940s on the column, beam, slab basis, these hives of flats and maisonettes rapidly progressed into cellular and crosswall structures rising to 20 or more storeys. The introduction of the housing yardstick put a premium on minimum superstructure cost, with little regard to foundation cost. Many techniques were employed to rationalise the combined construction and finishing processes. In in situ work, shuttering developments—including the development of table top formwork—resulted in a sufficiently high standard of finish to both wall and ceiling to allow decoration with minimum preparation. Getting the best from this type of construction requires several disciplines. There should be no downstand beams within the building and preferably no edge beams. This allows maximum flexibility for the table formwork. If edge beams are required they should be upstands cast one floor later, so that the tables can easily move out to the crane and upwards to the next floor. If an edge upstand or a combined cladding and structural member has to be concreted in at the same time as the floor, the table legs can be arranged to hinge and clear the sill. Lintels should be avoided, allowing door units to go to the underside of the slab above. Selected walls may be linked by heavily reinforced lintels to provide stability under wind load.

9.06 A particular example of this is the Borough of Wandsworth's Surrey Lane Project **28** of four 21-storey towers. For economy, the entire structure including elevations is in lightweight concrete. Precast edge members were hoisted into position on the table formwork and made continuous with the slab. Walls 175 mm thick, even in lightweight aggregate concrete, give adequate sound reduction and fire protection between tenancies and also the fire protection required around lifts and stairs.

28 *Surrey Lane project, Wandsworth. Precast edge beams occur on two opposite elevations, enabling in situ edge beam shuttering of other two elevations to span between*

9.07 A large estate at Milton, Glasgow, shows the ultimate in rationalised tradition. Here a purpose-made steel table formwork was designed for the complete floor of a tower. Walls were shuttered internally in large steel panel formwork, and externally by precast, exposed aggregate, permanent formwork which could be placed by crane and strutted from the inside of the building. The elevations were designed with full height slits enabling the table formwork, with side leaves hinged down, to be withdrawn on a special platform, and lifted by the crane. This slit was later clad in glass and timber applied from inside the building. No scaffolding was required.

9.08 Industrialised building was a step beyond rationalised tradition in an attempt to provide an entirely precast structure incorporating the maximum of finishes and services. Although composed largely of reinforced concrete, this type of construction is classed as masonry and will be dealt with in section 8 of the handbook.

Skeletal structures

9.09 Where the formwork and its supports are large compared with the quantity of concrete contained, and where the structure is composed of single members of skeletal type, in situ concrete is rarely economic. Here precast work comes into its own. Skeletal structures are primarily of the industrial or shed type and cheap buildings can be provided in the pitched roof form up to about 35 m in span **29**. Joints can be made at convenient points and the structure analysed accordingly, taking account of hinges or placing them at the points of contraflexure. Light roof spans up to 16 m can be achieved with trusses of traditional steel shape, while longer spans and heavier loading require heavier, probably prestressed, trusses **30**. Purlins for these skeletal roofs can be made in a variety of shapes **31**. The rectangle is not the best structural section but, LS and TS present problems both during manufacture and in the deflection of sloping roofs. Plugs can be cast-in for cladding connections or, in the case of lightweight concrete, masonry nails will suffice.

9.10 Space frames in concrete are essentially single layer grids and can be in situ, part precast or wholly precast and then post-tensioned together **32**. Although some structures have been built in this form, developments in single, double and triple structural-steel layer grids make concrete uneconomical. The fire resistance value must be of great importance to justify concrete in this form **30** as structural steelwork, with its low weight/strength ratio, is the basic economic solution.

Surface structures

9.11 The primary purpose of the shell form of construction is to use the cladding surface as a structural member **33**. Suitable curving or corrugation of the surface can produce efficient structural form, so concrete with its mouldability immediately suggests itself. Some exciting forms of shell have been constructed, particularly in South America and Spain. While in Britain the story follows the familiar pattern, fashion is often the only reason for the adoption of an innovation and so misuse quickly leads to disrepute.

9.12 In the late 1940s 'barrel' and northlight cylindrical shells were represented as an economic solution to all roofing problems and in early work exaggerated claims of low cost led to inappropriate applications **34**. Certainly, one or two projects with adequate repetition resulted in satisfactory costs, but single shells have high formwork, setting-up, and distribution costs (remembering the small quantities of concrete involved). Certainly, the minimum support roof could be visually exciting, but in the industrial field it tended to restrict the flexibility of buildings.

9.13 A 'barrel' roof spanning an area of 50 m by 25 m would have a dead load of 320 kg/m², about four times more than that of an equivalent steel-framed metal-clad roof where both are designed to carry 70 kg/m² superimposed load. Foundations must obviously be urgently considered in the overall costing. The shaping of the shell from which strength is derived may well increase the volume of the building and therefore the heating load.

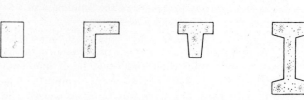

30 *Precast lightweight trusses for roof spans up to 16 m. Greater spans require prestressed units*

31 *Purlin sections commonly used in skeletal roofs*

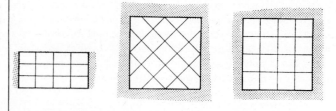

32 *Concrete space frames on plan: these are single layer grids, in situ, part of wholly precast and often post tensioned together*

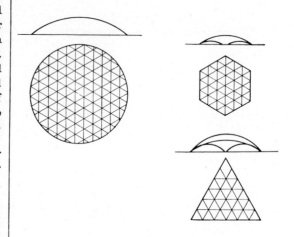

33 *Plan and elevation of common shell types*

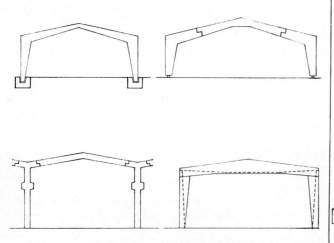

29 *Precast building frames*

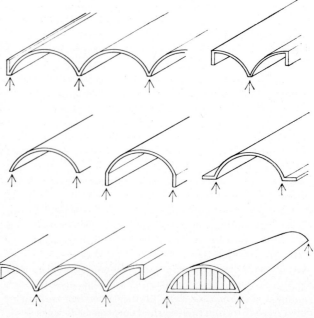

34 *Barrel and northlight cylindrical shells*

9.14 With rapidly rising labour rates, formwork costs for individual shells are likely to make them uncompetitive, but cost is not always the final consideration. If all the requirements of the project indicate a shell, namely the need for an enclosed envelope or the use of large unloaded spans, then a wide variety of shapes are available.

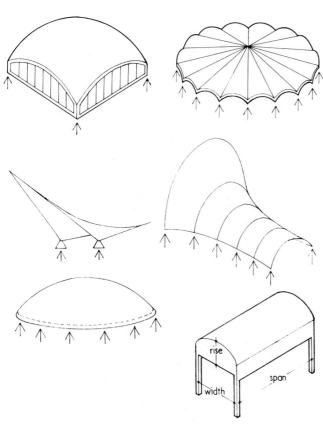

35 *Surface structure: square dome, warped and barrel*

9.15 Meaningful calculation may only be possible with certain shapes but testing of models in micro-climate with fine wire reinforcements can demonstrate to an engineer the manner in which the shell behaves under load. Considering first the simpler form of cylindrical shells, the 'barrel' and northlight were investigated in depth in the early 1950s **34**.
9.16 Whether an edge beam is used or not, the rise to span ratio should be about 1:10 **35**. To obtain this depth an edge beam is necessary with a narrow barrel but as the width is increased the appropriate depth may be obtained in the rise of the shell itself. Compromise between these requirements and the cost of the columns suggests a span to width ratio of 2:1. Typical proportions in this range are shown **36**. End diaphragms must be provided and can take a variety of forms **37**.
9.17 As in structural steelwork, northlight shells provide certain advantages but at extra cost. The quantities of concrete and steel are greater than in the 'barrel' of similar span; also unit construction costs are higher. It is usually difficult to obtain as large an effective depth (see ANALYSIS 2 AJ 24.5.72 p1171), and the beam below the glazing, which forms the gutter, tends to carry a high proportion of the load. To increase the depth of construction the width must be not less than half the span, relating as shown **38**.

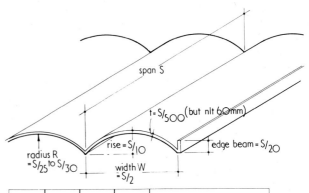

S (m)	W (mm)	R (m)	t (mm)	approximate weight / m² (including min live load (kg)
20	10	7	60	270
30	15	11	70	300
40	20	15	80	350
50	25	18	100	390

36 *Barrel structure properties*

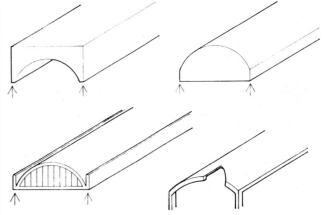

37 *End diaphragms commonly used with barrel structures*

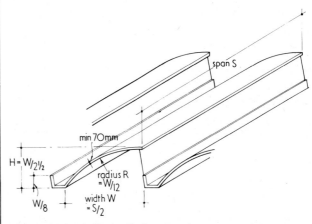

38 *Northlight shells: typical proportions*

9.18 Square areas can be covered by a square 'barrel', though this is not likely to be economic. The alternatives are two or three 'barrels' of less rise carried on a common end beam, or a square dome—ie a barrel curved in two directions **35**. Such a shell can theoretically span up to 100 m with little more than 100 mm thickness. The more recent fashionable shell is the hyperbolic paraboloid—effectively a warped parallelogram or a surface of translation **39**. In the latter case the surface is obtained by moving a vertical parabola having upward curvature over another parabola with downward curvature, the parabola of translation lying in a plane perpendicular to the first but moving parallel to it. This is shown graphically where the saddle shaped surface is formed by moving parabola ABC over parabola BOF.

9.19 A variety of roof forms may be developed either by use of the entire warped surface or by combining parts of it in various ways. The supports shown, if too flexible, would be tied across to avoid the spread of the arc **39**. The complete warped surface has been used with striking effect in churches, restaurants and roofs of garage. Structures formed by combining four stiff quadrants at angles to one another are suitable for covering the large rectangular areas common to industrial plans. An alternative grouping produces the umbrella structure. This can be horizontal or tilted—if tilted the units may be combined to produce a northlight roof effect **40**. The span/rise ratio is usually taken as not more than 9:1. Typical sizing for an umbrella is shown.

9.20 Though the double-curved surface appears a difficult formwork problem, it in fact requires only straight line generator wood joists on which the warped face can be achieved by nailing flexible plywood sheathing. Stresses in a properly proportioned roof are low and 75 mm thickness suffices for the normal span. The University of Mexico has one roof only 16 mm thick but in the UK 60 mm is about the right answer for steel reinforcement mesh plus cover.

9.21 The folded plate structure differs from the shell in that it is built up from slabs, and not necessarily thin as in the membrane. These carry bending as well as direct and shear forces. Prismatic structures consist of rectangular plates which are immovably restrained in relation to one another by means of transverse stiffening diaphragms or rigid frames. The roofs illustrated fall into this category **41**. Pyramidal structures occur as silos, cooling towers and peaked roofs **42**, while the prismoidal structure falls between the two, being a sawn off version of the pyramid **43**.

9.22 However, as every slab-to-slab junction can be regarded as a beam support almost any combination of folded plates will provide a stable structure. The scale of the plates should be such that the so-called 'beams' have span to depth ratios of less than 2:1 and are designed as 'deep beams'—as distinct from normal beam theory. Formwork is conventional and can be made up from panels as large as can be handled on site.

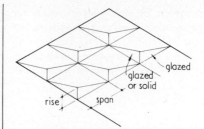

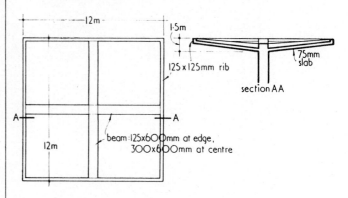

40 *Umbrella structure: if tilted, individual quadrants may combine to form northlight roof effect*

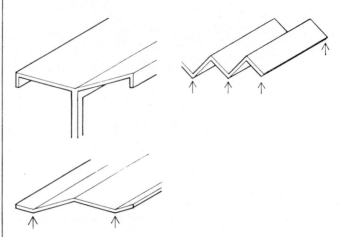

41 *Folded plate structures*

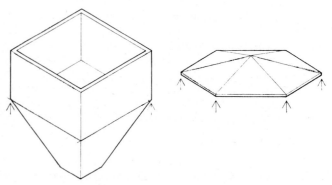

42 *Prismatic structures: these comprise rectangular plates immovably restrained in relation to one another*

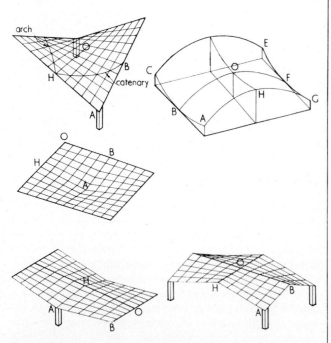

39 *Roofs by hyperbolic paraboloids*

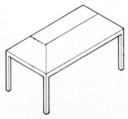

43 *Prismoidal structure*

Technical study
Concrete 3

Section 5 **Structural material: Reinforced concrete**

Simple design in reinforced concrete

ALLAN HODGKINSON'S *thesis in this final study on structural reinforced concrete is that an architect need not be afraid of designing simple concrete members himself, and that certainly he should be aware of what such design involves. Assuming an understanding of the analysis (section 2 of the handbook), this study explains the stress conditions in loaded rc sections and tabulates those permitted; it touches on cover, bar spacing and explains reinforcement detailing. Finally examples of typical structural situations are worked through in detail to illustrate the earlier sections of the study. Cover and bar spacing, with all code requirements, are detailed in Information sheet* CONCRETE I *which follows this study*

1 Introduction

1.01 Design variables in reinforced concrete are many; and although attempts have been made to produce design charts for reinforced concrete sections similar to those for structural steelwork, design is still apparently done each time from first principles. Perhaps it is this which causes architects to regard all concrete design as the province of the engineer—but there is no reason why an architect should not be as prepared to do simple design in concrete as he is, say, in timber.

1.02 On projects where no engineer is employed, simple solid floor slabs, stairs, simple beams, lintels and small retaining walls could well be dealt with in an architect's office. Apart from his ability to do this work, an architect often has a responsibility in site supervision and should be aware of good practice in steel detailing and fixing, as well as legislation governing the general use of concrete.

1.03 In Technical study SAFETY 1, the modern approach to design was discussed; for concrete, the new unified code of practice CP 110 will apply in 1973 but this does not concern architects' design—indeed many practical engineers consider it is not even a code of good practice. However the existing Code (CP 114[1]) will remain in use for some years and, provided no condition of the unified code is more onerous, there is no reason why CP 114 should not be used indefinitely for simple designs.

2 Stress and strain

2.01 Section 2 of the handbook (Structural analysis) showed how structures are analysed and how the direct forces, bending movements and shear forces induced in the structural members by the applied loading are determined. In designing a reinforced concrete member it is necessary to place the reinforcement in the section so that it anticipates the tensile stresses which the concrete is not able to accept. The following paragraphs (2.02 to 2.21) explain the stress conditions in the concrete beam or column and define the allowable stresses permitted in CP 114.

Beam or slab in bending

2.02 Early design was based on the assumption that concrete was elastic in the working stress range and the ratio of the modulus of elasticity of the steel to that of concrete was 15. In other words the approach was that an area of steel contributed 15 times that of the same area of concrete. This had the effect of penalising higher strength steels and preventing their economic use.

2.03 However tests proved that as failure was approached, the compressive stresses in a member, instead of being a maximum at the edge farthest from the neutral axis, adjusted themselves to a nearly even value from neutral axis to extreme fibre—giving a total compression greater than that assumed in the elastic theory. This *stress block* resulted from the concrete assuming a 'plastic' rather than an 'elastic' state.

2.04 The 'plastic' design permitted in CP 114 accepts the concept of a rectangular stress block, as in **1**, provided that its depth is not greater than half the depth from extreme compression fibre to the centre of the tensile reinforcement.

2.05 Forces at working load are shown in **1** and this assumes steel reinforcement in compression as well as in tension. (Compression steel is ignored for the time being.) The tensile

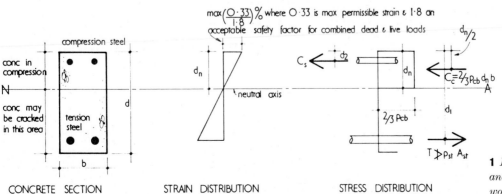

1 *Load factor method; stress and strain distribution of working load*

force (T) is the product of the permissible tensile steel stress (p_{st}) and the area of tensile steel (A_{st}), i.e:

$$T = p_{st} A_{st}$$

and the compressive force (c) is two thirds of the permissible concrete stress (p_{cb}) multiplied by the stressed area (b breadth of section and d_n depth of the compressive stress block); ie:

$$C = \tfrac{2}{3} p_{cb} \, b \, d_n$$

The code limit of $d_n \not> \dfrac{d_1}{2}$ means that the maximum moment

of resistance of the concrete in compression is $\dfrac{p_{cb}}{4}(b \, d_1{}^2)$

at which point the lever arm is $\left(d_1 - \dfrac{d_n}{2}\right)$ which equals $\dfrac{3d_1}{4}$

2.06 If this maximum is more than that required to resist the external moment, then a stress block of less depth than $\dfrac{d_1}{2}$ is needed; if the concrete is still stressed to the allowable value the lever arm is therefore greater and less steel will be required. A graph **2** can be plotted to express this condition.

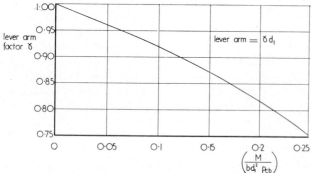

2 *Lever arm curve for load factor design. Graph of lever arm factor (γ) against* $\left(\dfrac{M}{p_{cb}(bd_1{}^2)}\right)$ *factor. When the latter factor exceeds $0 \cdot 25$, compression reinforcement is required*

2.07 In order to calculate the tension steel required it is therefore necessary to derive the appropriate value of $\dfrac{M}{p_{cb}(bd_1{}^2)}$ from the applied moment (M) section properties (b and d_1) and the permissible concrete stress (p_{cb}). The lever arm factor (γ) can then be read off the graph. The tension steel can then be calculated from $\dfrac{M}{p_{st} \;\; \gamma d_1}$

2.08 If the value of $\dfrac{p_{cb}}{4}(b \, d_1{}^2)$ is less than the applied moment (or in other words $\dfrac{M}{p_{cb}(b \, d_1{}^2)} > 0 \cdot 25$) the stress block would be deeper than the permitted half-section. In this case it may be possible to provide the extra resistance by using compression steel. The lever arm to calculate the amount of steel is the distance between the compression steel and the tension steel ($d_1 - d_2$) and

the amount of steel is $\dfrac{M - \dfrac{p_{cb}\,(bd_1{}^2)}{4}}{(d_1 - d_2)\,p_{sc}}$

That is, the required compressive resistance divided by the product of lever arm and permissible steel compressive stress (p_{sc}). This amount must not exceed 4 per cent of the area of the member section; if it does, then a deeper section is required.

Rectangular sections under direct compression

2.09 The plastic properties of concrete under stress have, for a considerable time, been used in calculating axial load

resistance in columns. The safe load can be easily assessed as $A_c \, p_{cc} + A_{st} \, p_{sc}$ where A_c is area of concrete, p_{cc} the allowable direct stress in compression in concrete; A_{st} is area of steel and p_{sc} allowable compression stress in the steel.

2.10 Although the term 'crushing' is used in concrete testing the concrete cube fails on diagonal planes and if this tendency can be resisted by some binding force the concrete can carry enormous compressive stress. Thus while rectangular steel links are usually employed to prevent a column bar from buckling, the use of helical binding would allow a higher safe load to be calculated.

Rectangular section under direct and bending stress

2.11 The plastic design (load factor method) for bending and direct stress is an extension of the process defined earlier for beams. When the magnitude of the bending moment relative to the direct load is such that primary failure of the member would be by the yielding of tensile steel, the assumed stress distribution for resisting the working loads is as shown in **3**.

2.12 The formulae presented in the code are far too complicated to use each time a section has to be designed; but making the assumption that equal reinforcing steel will be used in each of the two bending faces of the member, graphs can be produced which show a series of curves plotted to relate direct load, moment, section width and depth, and cover to main reinforcement and allowable steel stress (all as related to any concrete quality). A typical graph is shown in **4** for the particular (d_1/d) ratio shown inset. With so many variations, it is necessary to have a family of graphs, so that each one expresses a different ratio of effective depth of tensile steel to total section depth; as steel has been chosen to be symmetrical this means that the curves are constructed in direct relationship to the width and depth of section.

Having calculated $\left(\dfrac{P}{p_{cc}\,bd}\right)$ and $\left(\dfrac{M}{p_{cc}\,bd^2}\right)$, a point on the graph can be plotted. This lies between curves 4 and 5, say 4.4. When this is multiplied by a constant for the graph, it gives r, the total percentage of reinforcement in the column, therefore $A_{st}(= A_{sc})$ is given as $A_{st} = \dfrac{(bd)r}{2}$.

2.13 Other curves can be drawn for differing configurations of reinforcement with $d_1/d = 0 \cdot 7, 0 \cdot 85, 0 \cdot 9$ and $0 \cdot 95$ (see **5**). Curves can also be drawn for circular columns and for columns subjected to bending on two axes at right angles.

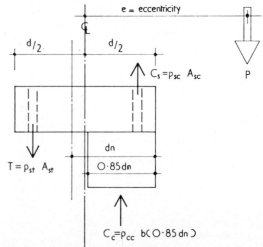

3 *Load factor method: stress distribution in rectangular section under direct load and bending at working load (figures are all defined in text)* $M = Pe$

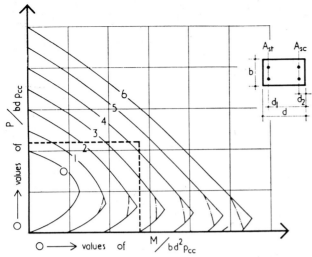

4 *Typical graph for determining steel reinforcement required in the rectangular column shown inset* $(d_1/d = 0.8)$

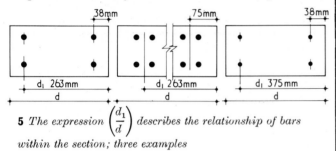

5 *The expression* $\left(\dfrac{d_1}{d}\right)$ *describes the relationship of bars within the section; three examples*

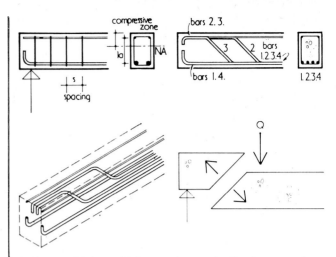

6 *A normal beam with bent up bars or inclined compression bands to counteract shear stress acting as in* **7**
7 *Shear failure of concrete*

Shear stress

2.14 A considerable number of research workers have contributed to the subject of failure in shear and the Institution of Structural Engineers issued a report in 1970 on the assessment of recent work[2]. The still unpublished unified code approaches the subject in a similar way and provides some economy in design compared with CP 114. However, as was said earlier, CP 114 remains adequate for simple designs.

2.15 Failure in concrete with no vertical steel will occur as shown in **7** when the principal stress developed across a plane at approximately 45° exceeds the tensile strength of the concrete. Provided the average shear stress q does not exceed the allowable shear stress the section is assumed to be adequate and q is given by:

q = Q/b la,

where Q is applied shear, b width of section and la the lever arm.

2.16 If the allowable stress is exceeded then links or bent up bars must be added to the member to provide a lattice beam action with vertical tension members and inclined compression bands **6**. If bent up bars are used they lie more or less in the direction of the principal tensile stress. Links should be provided at a spacing of

$$s = \frac{p_{st}\ A_w la}{Q}$$

where A_w is the cross sectional area of the link and p_{st} the permissible stress in tension for shear. For given permissible steel stresses tables can be found in any concrete designer's handbook with bar diameter plotted against spacings to show the resistance value of the two legs of the link where this is designated as $\left(\dfrac{Q}{la}\right)$.

Bond stress

2.17 Only resistance of concrete to horizontal shear stresses and the adhesion of concrete to steel reinforcement allow the compression and tension zones to co-operate in produc-

ing the resistance moment of a beam.

The vital bond is difficult to assess numerically as it is a combination of adhesion, the gripping effect of shrinkage, the shape of the bar section and the form of anchorage.

2.18 The grip of the concrete on the steel preventing slippage equals the difference in total steel force at any two adjacent sections. This force (δ_{st}) equals the product of shear stress (q), breadth of beam (b) and length (δl), ie $\delta_{st} = qb\delta l$.

2.19 Assuming local bond stress of uniform intensity, the total steel force also equals the product of bond stress, perimeter of all the bars (o), and δl.

Therefore local bond stress $= \dfrac{q\ b\ \delta l}{o\ \delta l} = \dfrac{q\ b.}{o}$

But, as q was defined in para 2.15 as q = Q/la.b, finally local bond stress = Q/la.o. A permissible stress is given in CP 114 for each quality of concrete.

2.20 In addition, for anchorage or lapping, the bar must extend from its particular point of stress a distance in tension of:

bar diameter $\times \dfrac{\text{tensile stress}}{4 \times \text{permissible average bond stress}}$;

and in compression of:

bar diameter $\times \dfrac{\text{compression stress}}{5 \times \text{permissible average bond stress}}$.

In practice particular combinations of concrete and type of steel bar will require a certain number of diameters of length for anchorage when using the full permissible stress in the steel. It is then easy enough with this figure in mind to design the correct bond length or reduce it proportionately for lesser stresses. A reduction in the length is made for L-hooks or U-hooks at the end of bars as in **9**.

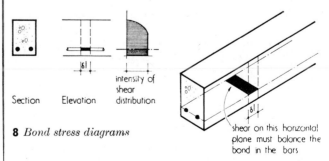

8 *Bond stress diagrams*

Permissible stresses

2.21 The actual stresses allowed for concrete and steel in structural reinforced concrete are given in CP 114 and exemplified in tables I and II (considering for this purpose only the two grades of concrete most likely to be used

1:2:4 concrete at a 28-day cube strength equal to 21 MN/m² and 1:1½:3 equal to 25·5 MN/m²).

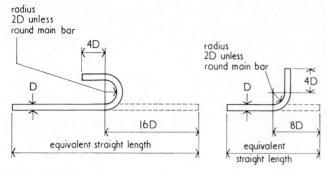

9 *Reduction in length for bond—permissible with* L- *or* U-*hooks at end of bars* (D = *bar diameter*)

Table I *Permissible concrete stresses* (*from* CP·114)

Mix proportions	Permissible stresses (in MN/m² or N/mm²)				
	Compressive		shear	Bond	
	direct	bending		average	local
1:2:4	5·3	7	0·7	0·83	1·25
1:1½:3	6·5	8·5	0·8	0·93	1·40

Table II *Permissible steel stresses* (*from* CP 114)

Type of stress	Permissible stresses (MN/m²)		
	Mild steel bars		High yield steel having a guaranteed yield or proof stress (fy)
	diameter <40 mm	diameter >40 mm	
Tensile stress other than in shear reinforcement	140	125	0·55 fy but not more than 230 for bars not exceeding 20 mm dia 0·55 fy but not more than 210 for bars not exceeding 20 mm dia
Tensile in shear reinforcement	140	125	0·55 fy but not more than 175
Compressive stress	125	110	0·55 fy but not more than 175

where fy = guaranteed yield or proof stress or specified characteristic strength.

3 Cover and bar spacing

3.01 Earlier in this handbook (Technical study SAFETY 3) the requirements to protect reinforcement in the event of fire attack were defined. Reinforcement must have adequate cover both to protect it from weather and to give it adequate bond, and it must be distributed in such a way that cracking due to shrinkage and temperature movement of the concrete is controlled. Good *detailing* of the reinforcement is just as essential as calculating correctly the right *amount* of reinforcement.

3.02 The maximum size of the aggregate from which the concrete is made demands some control over the cover and spacing of bars. Information sheet CONCRETE 1, which follows this study, illustrates the code requirements and relates to 20 mm maximum size aggregate.

4 Detailing and scheduling the bars

4.01 The main principles of good detailing are as follows: a, to depict in the drawings and schedules the number and shapes of bars so that the designer's intention is properly interpreted and the steel-fixer can fix the steel in the same manner

b, to put in just sufficient reinforcement to guard against shrinkage and movement cracking, and to allow the bars to be caged together when only the steel required to support

the structure is indicated in the calculations

c, to detail the bars in member sections so that there are no sudden changes from heavy to light reinforcement and no areas of concrete completely unreinforced.

4.02 Detailing should be carried out in accordance with BS 4466. In general, this involves the use of standard shapes (each of which is given a reference number known as *shape code*) and a list of dimensions A, B, C, D, E or R which are written into the schedule in place of the bar sketch.

4.03 Two tables of standard bar shapes are given in BS 4466 *Table 6—preferred shapes* and *Table 7—other shapes;* those shown in table 6 should be used wherever possible. If a bar is required which is not covered by a standard shape, '99' is entered in the shape code column and a sketch of the bar is made in place of the standard dimension A, B, C etc **10**.

4.04 The dimensions of standard shapes to be entered in the bar schedule are restricted to those shown in the fourth column of (BS 4466) tables 6 and 7. This means that the closing dimension is not given. Note that non-standard radii are not always shown as such but may be indicated as an overall dimension (eg shape 36).

10 *Bar schedule example*

Tolerances

4.05 Cutting and bending tolerances, shown in table I of BS 4466, should be noted carefully, and when detailing the bars on drawings, account should be taken of them. In particular note that the tolerance on the length of straight bars is ±25 mm, and that on normal links (up to 1000 mm, maximum side dimension) is ±5 mm.

4.06 In special cases, where the tolerances in table I (BS 4466) could lead to an unacceptable reduction in cover and where this is critical (eg precast concrete cladding panels), stricter tolerances can be adopted for particular bars; this must be stated on the bar schedule. As this will undoubtedly increase the cost of the reinforcement, it is to be used only where absolutely necessary.

Bar marks

4.07 Bar marks are retained as plain serial numbers 1, 2, 3 etc, starting at 1 on each drawing. Type and size are entered in the adjacent column on the schedule, the size being the number of mms in nominal size commonly used in the round and square ranges of areas (see clause 2.2 of BS 4466), and the types most commonly used being R (mild steel) and Y (high yield steel), both in the round range. Square twisted bars or bars not covered by R or Y are denoted X.

5 Examples

Floor slab

5.01 The first of three examples covers the design of a four-span continuous floor slab carrying a 50 mm screed, lightweight partitions, plaster to the soffit and a superimposed load of $2 \cdot 5$ KN/m² **11**.

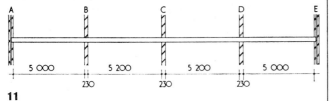

11

12

The span dimensions used in design is the lesser of a, centre to centre of supports or b, clear dimension between supports plus effective depth of slab; the depth of slab must not be less than $\left(\dfrac{\text{span}}{31 \cdot 5}\right)$ if high yield steel is used as reinforcement.

In this case,

$$\frac{\text{span}}{31 \cdot 5} = \frac{5200 + \text{say } 200}{31 \cdot 5} = 172 \text{ mm; use slab depth of } 175 \text{ mm.}$$

Effective span for end bay $= 5000 + 175 = 5175$;
effective span for inner bay $= 5200 + 175 = 5375$.

Loading: 175 mm slab $4 \cdot 2$ KN/m²
finishes $1 \cdot 5$
partitions $1 \cdot 0$
superimposed $2 \cdot 5$
giving a total of $9 \cdot 2$ KN/m².

5.02 The code (CP 114) gives an approximate analysis for continuous slabs of three spans or more when the spans do not differ by more than 15 per cent in length. This analysis is quite adequate with spread loading and nearly equal spans, and there is no need to do any more complicated analysis. The moments (assuming Wd = total dead load distributed; Ws = total superimposed load distributed; and 1 = effective span) are given in table III.

Table III Moment formulae

	Middle of end span	Penultimate support	Middle of internal span	Internal support
Moment due to dead load	$+\dfrac{Wdl}{12}$	$-\dfrac{Wdl}{10}$	$+\dfrac{Wdl}{24}$	$\dfrac{Wdl}{12}$
Moment due to superimposed load	$+\dfrac{Wsl}{10}$	$-\dfrac{Wsl}{9}$	$+\dfrac{Wsl}{12}$	$-\dfrac{Wsl}{9}$

In this particular case they give the figures in table IV (where all moments are in KNm per m width).

Table IV Actual moments for this example

	Mid AB	Support B	Mid BC	Support C
Moment due to dead load	$+14 \cdot 84$	$-17 \cdot 80$	$+ 8 \cdot 05$	$-16 \cdot 10$
Moment due to superimposed load	$+ 6 \cdot 68$	$- 7 \cdot 42$	$+ 6 \cdot 01$	$- 8 \cdot 00$
Total	$+21 \cdot 52$	$-25 \cdot 22$	$+14 \cdot 06$	$-24 \cdot 10$

5.03 Use 1:2:4 concrete, 21 MN/m² at 28 days, $p_{cb} = 7$ MN/m².

$$\frac{M}{p_{cb}(bd_1)} = \frac{M}{7 \times 10^3 \times 1 \times 0 \cdot 15^2} = 0 \cdot 00637M. \text{ (b and } d_1,$$

as in **12**).

At each of the four points at which the moment has been calculated, it is necessary to read from the graph of lever arm

factor (γ) against $\dfrac{M}{p_{cb}(bd_1{}^2)}$, the correct γ factor, and thus the correct lever arm for each case. This, taking permissible steel stress (p_{st}) to be 230 MN/m², the area of steel (Ast) is calculated, as in table V.

Table V Steel required

Moment KN/mm run	$\dfrac{M}{p_{cb}(bd_1{}^2)}$ $= 0 \cdot 00637M$	Lever arm factor (γ)	$A_{st} = \dfrac{\text{moment}}{p_{st}(\gamma)d_1}$ in mm²/m width	Use steel
$21 \cdot 52$	$0 \cdot 137$	$0 \cdot 88$	$\dfrac{21 \cdot 52 \times 1000}{230 \times 0 \cdot 88 \times 0 \cdot 15}$ $= 712$	12 mm bars at 160 mm
$14 \cdot 06$	$0 \cdot 09$	$0 \cdot 93$	$\dfrac{14 \cdot 06 \times 1000}{230 \times 0 \cdot 93 \times 0 \cdot 15}$ $= 437$	12 mm bars at 260 mm
$25 \cdot 22$	$0 \cdot 161$	$0 \cdot 86$	$\dfrac{25 \cdot 22 \times 1000}{230 \times 0 \cdot 86 \times 0 \cdot 15}$ $= 845$	12 mm bars at 135 mm

5.04 Distribution steel must be placed across main reinforcement partly to distribute the load evenly, but also to avoid shrinkage cracking. With high yield steel, the distribution steel area should be at least $0 \cdot 12$ per cent of the cross sectional area of slab,

ie $\dfrac{0 \cdot 12 \times 1000 \times 175}{100} = 210$ mm² per metre width.

Therefore use 10 mm at 300 mm centres for both top and bottom steel.

5.05 The shear can be tabulated as in table VI. Resistance to shear $= 0 \cdot 7 \times 0 \cdot 15 \times 0 \cdot 87 \times 1000 = 91$.

Table VI Shear

Shear	Support A	Support B		Support C
		span AB	span BC	
Static load	$2 \cdot 5 \times 9 \cdot 2 = +23 \cdot 0$	$+23 \cdot 0$	$+23 \cdot 9$	$2 \cdot 6 \times 9 \cdot 2 = +23 \cdot 9$
Change in load due to moment	$-\dfrac{25 \cdot 22}{5 \cdot 075} = -4 \cdot 9$	$+ 4 \cdot 9$	$+ 0 \cdot 2$	$-\dfrac{25 \cdot 22 + 24 \cdot 1}{5 \cdot 375} = -0 \cdot 2$
Total	$+18 \cdot 1$	$+27 \cdot 9$	$+24 \cdot 1$	$+23 \cdot 7$

5.06 This calculation was carried out to show the procedure; but with solid slabs, unless subjected to superimposed loads much higher than the dead load, shear stress is unlikely to exceed permissible stress.

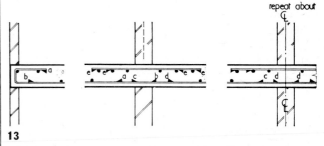

13

5.07 A section through the floor would be detailed as in **13**. At A, alternate bars from the bottom are taken round and into the top of the slab to provide a nominal resisting moment to counter the fixing moment induced by clamping action of the wall above (though this end was treated as simply supported in the design analysis). Chairs, as **14**, are required at internal supports to keep the top steel in its correct position and should be detailed on bar schedule.

5.08 Letters on **13** are bar marks, and these define the bars on the bar schedule. Dimensions are required to indicate the amount by which the top steel (e) over the supports is staggered in alternate bars. The bars must develop their anchorage bond length from the point of maximum stress, but as the bending moment falls sharply **15** the first bar can be curtailed quickly, and the alternate bar then carried on to just beyond the quarter span point.

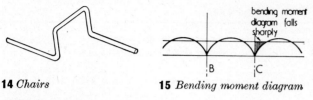

14 *Chairs*   **15** *Bending moment diagram*

Continuous footing

5.09 The second example is design of a continuous footing. Assume that the wall carries 260 KN/m² run and the soil has a permissible bearing pressure of 216 KN/m²; use high yield steel and 21·0 MN/m² concrete **16**.

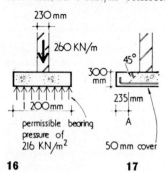

16   **17**

Width of footing required say 1200 mm. The wall load spreads down to the level of the reinforcement as shown in **17**, so the critical position for shear failure is along line A. (Depth to reinforcement = 240 mm.)
Load to left of A = $0·235 \times 216 = 51$ KN per m width. This is obviously quite safe with a slab depth of 300 mm.

5.10 Considering the ground pressure as a load on a cantilever springing from the centre line of the wall, the bending moment at the centre line of the wall

$$M = \frac{Wl}{2} = 0·6 \times 216 \times 0·3 = 38·88 \text{ KN m}^2.$$

$$\frac{M}{bd_1^2 p_{cb}} = \frac{38·88}{1 \times 0·24^2 \times 7 \times 10^3} = 0·095;$$

lever arm factor = 0·92.

$$A_{st} = \frac{38·88 \times 1000}{230 \times 0·92 \times 0·24} = 766 \text{ mm}.$$

5.11 Use 12 mm bars at 150 mm centres and distribution steel of 12 mm bars spaced evenly in the width of the base. In foundations bond length should always be checked for anchorage from maximum stress position to end of bar and if large bars are used the local bond should also be checked. In this particular case the bar length measured from the centre line of the wall is about 50 times the diameter of the bar and therefore acceptable. The design would be as **18**. Note that a 60 mm blinding concrete has been shown for two purposes. It both protects the trench bottom from weather after excavation and also allows a clean surface on which to place the steel.

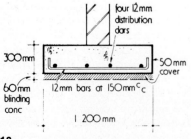

18

Lintel

5.12 The third example is the design of a lintel. Assume that the lintel, carrying brickwork and roof trusses, spans 2·3 m and supports a spread load of 35 000 N/m including its own weight **19**. Use high yield steel and 21·0 MN/m² concrete.

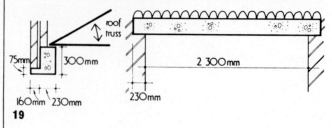

19

Effective span = 2300 + 230 = 2530 mm = 2·53 m

Using the formula for the bending moment $M = \dfrac{Wl}{8}$,

$$M = \frac{35\,000 \times 2·53 \times 2·53}{8} = 28\,100 \text{ Nm}$$

$$\frac{M}{bd_1^2 p_{cb}} = \frac{28\,100}{0·230 \times 0·267^2 \times 7 \times 10^6} = 0·243;$$

la factor = 0·75

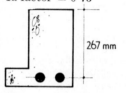

20

$$A_{st} = \frac{28\,100 \times 10^3}{230 \times 0·75 \times 267} = 615 \text{ mm. Use two 20 mm bars.}$$

Shear force at support = $1·15 \times 35\,000 = 40\,250$ N
Shear resistance = $0·7 \times 1000^2 \times 0·267 \times 0·85 \times 0·230$
= 36 540 N
Shear reinforcement is therefore required.
The spacing for 2 legs 10 mm links is

$$\frac{175 \times 157 \times 267 \times 0·85}{40\,250} = 155 \text{ mm}$$

Say 100 mm links at 150 mm centres for 500 mm then the concrete will be able to carry the shear; so use the same links but spaced out to 300 mm.

5.13 In practice, by the use of designer handbook tables, shear links could have been found by calculating the value

$$V = \frac{Q}{la} = \frac{40\,250}{267 \times 0·85} = 177.$$

This value would be read off from the table as requiring the 10 mm at 150 mm centres as already calculated. The beam would be detailed as in **21**.

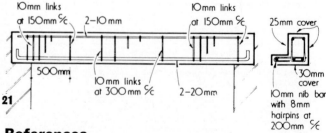

21

References

1 BS CP 114: Part 2: 1969 The structural use of reinforced concrete in buildings. Part 2 metric units. £3.00
2 INSTITUTION OF STRUCTURAL ENGINEERS The shear strength of reinforced concrete beams. 1969. o/p
3 The empirical method to CP 114 is no longer applicable.
4 Concrete admixtures, use and applicating. M. R. DIXON. The Construction Press

Information sheet
Concrete 1

Cover, bar spacing and laps

This sheet, to be read with Technical study CONCRETE *3, gives the code requirements from* BS CP 114, *and relates to 20 mm maximum size aggregate.* A *and* B *show typical requirements,* C *to* G *show special situations which an architect may encounter*

A Typical cover and spacing requirements for beams and slabs

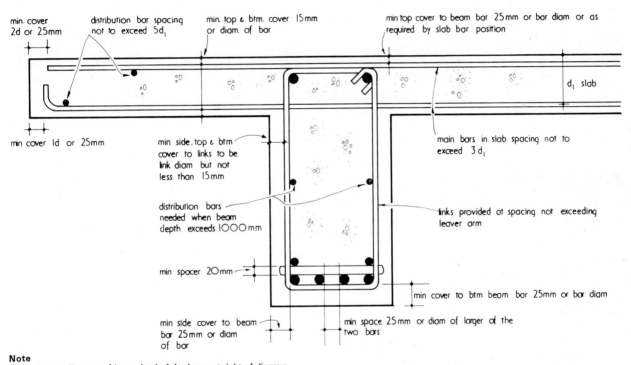

min. cover
2d or 25mm

distribution bar spacing
not to exceed 5d₁

min. top & btm. cover 15mm
or diam. of bar

min top cover to beam bar 25mm or bar diam or as
required by slab bar position

d₁ slab

min cover ld or 25mm

min side, top & btm
cover to links to be
link diam but not
less than 15mm

main bars in slab spacing not to
exceed 3 d₁

distribution bars
needed when beam
depth exceeds 1000mm

links provided at spacing not exceeding
leaver arm

min spacer 20mm

min cover to btm beam bar 25mm or bar diam

min side cover to beam
bar 25mm or diam
of bar

min space 25mm or diam of larger of the
two bars

Note
● d refers to diameter of bars; depth d₁ is shown at right of diagram.

B Typical cover and spacing requirements for columns

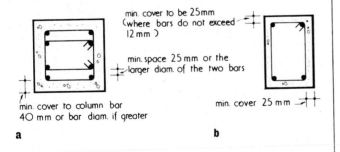

min. cover to be 25mm
(where bars do not exceed
12 mm)

min. space 25mm or the
larger diam. of the two bars

min. cover to column bar
40 mm or bar diam. if greater

min. cover 25 mm

a

b

Notes
● a is general case, with minimum main bar size of 12 mm; b is special case, with columns of 200 mm or less width and where bars do not exceed 12 mm.
● The total main reinforcement area should generally be at least 0·8 per cent of concrete area; preferably it should not exceed 4 per cent of concrete area, and it definitely must not exceed 8 per cent. No lap between upper and lower bars to exceed 8 per cent of concrete area.
● Links are to be spaced at the least of the following dimensions:
Least lateral dimension of the column, 12 × diameter of the bar linked, or 300 mm.
● Link diameter to be not less than 5 mm or less than ¼ of the main column bar dimension. Every main bar to be linked in two directions unless otherwise directed.

C Concrete members exposed to moisture: open platform

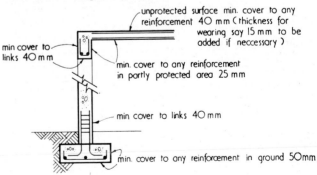

unprotected surface min. cover to any
reinforcement 40 mm (thickness for
wearing say 15 mm to be
added if neccessary)

min cover to
links 40 mm

min. cover to any reinforcement
in partly protected area 25 mm

min cover to links 40 mm

min. cover to any reinforcement in ground 50mm

Note: With asphalt tanking, waterproofing membrane or surfacing covers **may** be reduced to normal as in A or B

D Concrete members exposed to moisture: exposed column and edge beam

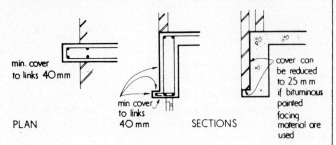

min. cover to links 40mm

PLAN

min cover to links 40 mm

SECTIONS

cover can be reduced to 25 mm if bituminous painted facing material are used

Note: for nibs less than 75 mm depth, special consideration is required.

E Concrete members exposed to moisture: exposed walls and stairs

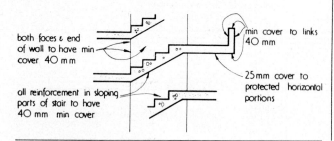

both faces & end of wall to have min cover 40 mm

all reinforcement in sloping parts of stair to have 40 mm min cover

min cover to links 40 mm

25mm cover to protected horizontal portions

F Cladding panels

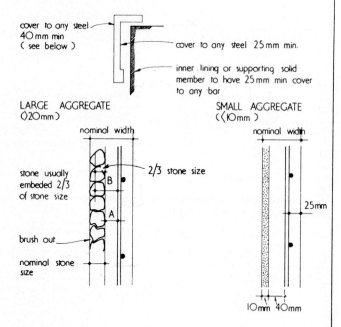

cover to any steel 40 mm min (see below)

cover to any steel 25 mm min.

inner lining or supporting solid member to have 25 mm min cover to any bar

LARGE AGGREGATE (>20mm)

nominal width

stone usually embeded 2/3 of stone size

2/3 stone size

B

A

brush out

nominal stone size

SMALL AGGREGATE (<10mm)

nominal width

25mm

10mm 40mm

Notes
● Aggregate between 10 mm and 20 mm if used to be treated as large.
● The use of heavily galvanised or stainless steel reinforcement can reduce the cover to as little as 20 mm
● If stones are not well bonded A must be 40 mm minimum, otherwise B must be 40 mm minimum.

G Special considerations of cover

Notes
● Lightweight aggregate concrete when exposed to moisture requires 10 mm more cover than any cover dimension given in this sheet.
● Add 10 mm extra cover to exposed aggregate faces, or where retarders (Redalon etc), bush hammering or grit blasting has been used.
● With water-retaining structures, walls and slabs above and below ground, 40 mm cover is required for all reinforcements in faces against water or earth. Also 25 mm minimum cover is required for any bar in a slab soffit over the tank water.

AJ Handbook   Section 6
Building structure   Structural material:
CI/SfB (2-)   Steelwork

Section 6

Structural material: Steelwork

Scope

This, the second section of the handbook devoted to the structural use of a material, discusses steel. To expect a simple guide (or the idea that a specific situation demands the unilateral choice of a specific construction and material) is unrealistic. Only comparative analysis, including variables not essentially structural, will reveal the best choice for a particular job.

The first technical study discusses properties of the material and forms of structural section available, while the second study carries this into a major discussion of structural forms, from single-storey large span to high rise, suitable for steel construction.

The third part assumes that a choice of steel as structural material has been made. It discusses the design of simple steel members and argues, as the handbook did earlier with reinforced concrete, that the architect need not be too shy to design his own simple steelwork, and he should certainly be able to understand how his engineer works.

Author

The author of section 6 is Gordon M. Rose CEng, FICE, FIStructE, FASCE. Against further specialisation among engineers, he is not only a leading advocate of tension structures but, for example, was consulting engineer on the award winning office building used as frontispiece to the concrete section of the handbook. He has worked in Canada and US and is now with British Steel Corporation, for whom he regularly lectures on tubular structures.

Technical study
Steel 1

Section 6 **Structural material: Steelwork**

Production, properties and shapes of structural steel

In his first study GORDON ROSE *briefly describes steel production, discusses the available forms of structural steel section, and notes how it can be protected from weather and fire*

1 Production

1.01 An important starting point for an understanding of the structural possibilities of any material is a knowledge of its origin and properties. With no material is this more important than modern structural steel.

1.02 Steel is refined iron plus measured amounts of other elements. The original Bessemer process takes ore from various parts of the world, grades it, crushes it and then charges it into a blast-furnace with coke and limestone; 90 per cent pure iron is continuously tapped off from a white-hot pool of molten metal, on top of which impurities float as 'slag'. The metal contains some carbon, phosphorous, manganese and silicon **1.**

1.03 From this iron, steel is then produced by one of three major processes: *open hearth, electric arc* and *basic oxygen;* the open hearth process is declining and the basic oxygen process **2** is fast becoming the major method. Here steel scrap and molten iron are fed into a furnace and oxygen is blown through; impurities are eliminated as a slag, and the refined steel is cast either into ingots, or into slabs by a continuous casting process.

1.04 As the ingots or slabs cool, they take up a brittle structure. Subsequent shaping processes modify this structure and confer the greatly increased tensile strength and other properties.

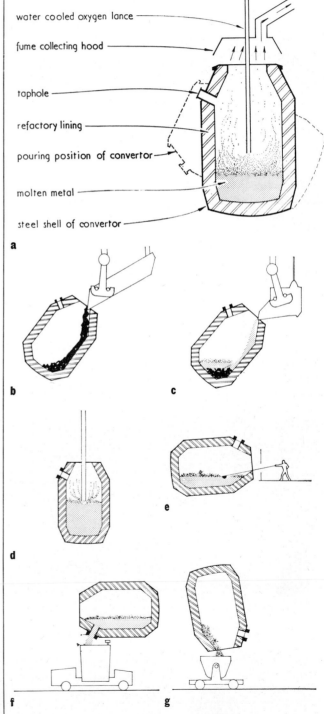

water cooled oxygen lance
fume collecting hood
taphole
refractory lining
pouring position of convertor
molten metal
steel shell of convertor

a

b

c

d

e

f

g

2a *Basic oxygen furnace production of steel;* **b** *charging scrap;* **c** *charging hot metal;* **d** *'blow';* **e** *sampling;* **f** *pouring and* **g** *slagging*

1 *Molten steel production*

2 Shaping process

2.01 Rolling, forging and extruding are the three primary processes for shaping hot steel. Rolling is generally used for structural steel shapes, the other methods for specialist applications.

Rolling

2.02 The ingots or slabs are reheated in a 'soaking pit' at the first mill. Here they are rolled down (to become *blooms*) and the rough ends are cropped off. The bloom then enters another mill, is cut into 10 m lengths and allowed to cool. After external and internal examination for defects, the steel is reheated before passing into either the *heavy section rolling plant* or the *light section plant*. The former may use either a 'Cross country mill', with no adjustment during rolling, or a 'Universal beam mill', in which both horizontal and vertical adjustments are used to create the profile. The light section plant produces profiled sections, smaller beam sizes and rails, and also solid rounds, squares, hexagons and flat plate. Flat plate is often further processed to produce wire for reinforcement.

2.03 About half the steel rolled in the UK is turned into flat plate, sheet or strip. Flat plate, called *heavy plate* if from 10 mm to 50 mm thick and *light plate* if from 3 mm to 10 mm thick, may be up to 3·5 m wide. *Sheet* is between 1·2 mm and 3 mm thick and up to 2 m wide, although some plate, finished to the closer sheet tolerances, may be up to 12 mm thick. *Strip* is from 0·75 mm to 12 mm thick, but is only from 20 mm to 600 mm wide. For distribution most light plate, sheet and strip is rolled into *coils*; a coil weighs up to 12 tonnes.

2.04 The most advanced steel rolling technology uses wide hot strip. Vast mills (kilometres long), house long production lines that are continuous from reheating furnace to coiling plant. The strip, although emerging at 65 km/h, is still controlled in thickness to within $\pm 0\cdot 05$ mm. Narrow strip is produced either by slitting wide strip, or in special 'skelp' mills. For bulk production of very thin sheet (down to 0·15 mm thick) a final reduction is made without the usual re-heating in another mill; this is *cold-rolled steel* **3**.

2.05 When steel is coldworked it becomes stiffer and harder, and so has to be *annealed*, that is heated and then gradually cooled under controlled conditions.

3 *Cold-rolled steel*

Hollow sections

2.06 Smaller sizes of structural hollow sections (SHS), up to 150 mm, are made from cold-rolled strip that is formed into a circle and then electrically welded in a continuous process that produces 120 m lengths at one time. This tube is then re-heated and passed hot through rolls where it is stretched and reduced to smaller diameters, squares and rectangles.

2.07 The re-heating replaces annealing, and so the properties of the finished hollow section are equal to those of a hot-rolled section. But care must be taken to determine the origin of sections because some producers of circular section—mostly outside UK—offer the cold-formed product without hot-finishing. The large sizes of SHS, up to 450 mm overall and 12 mm thickness, are similarly formed, but without stretching, from plate of the appropriate thickness; and these have the hot-formed properties.

2.08 Extra thick-walled hollow sections (up to 30 mm thick) are now being produced in the UK. An ingot of steel is pierced, creating a hollow bloom which is then forged over a mandrel and rotated until it becomes a seamless tube, that may then be re-rolled if desired into square or rectangular sections.

Wire

2.09 Wire is not finished to size in a rolling mill. Wire-rod, produced from a heated billet in a high-speed rod mill, is drawn, in a cold process, through a tapered die that reduces the diameter by as much as 90 per cent. This treatment greatly increases both tensile strength and hardness. Heat treating and controlled cooling enables larger rods to be used; and this can further enhance final properties. Finally the wire is coiled, and passed to other manufacturers for processing into mesh, wire ropes, prestressing strand, etc, or it may be spun into larger diameter cables for use in major tension structures.

2.10 So the primary processes for structural steel are massive operations, directly geared to the overall national economy; the user can do much to help himself by taking this into account when considering any project. For example, a preliminary order or letter of intent placed with a supplier can ensure prompt delivery at the time of the contract; discussions with the supplier might help considerably towards a sensible programming of production. If this were done on a wide scale it would be to the advantage of the whole industry; the savings could be well worth the additional contractual effort.

3 Specification of strength

3.01 Tables I and II, which come from BS 4360[1], are reprinted to give examples of the composition and strength of various grades of structural steel, classified from 40A to 55E. Not all of these grades are readily available in every size and shape, and manufacturers should be consulted as to availability.

3.02 The terms 'ladle' and 'product' in table I refer to two analyses done on the steel; the first during production (ladle) and the second during manufacture of the sections (product). All grades in these tables are of weldable quality steel, provided that BS 1856 procedures are followed and appropriate electrodes and control methods used.

Table I *Chemical composition of steel (from* BS 4360)

Grade		Carbon maximum	Silicon	Manganese maximum	Niobium maximum	Sulphur maximum	Phosphorus maximum
40A	Ladle	0.22	—	—	—	0.050	0.050
	Product	0.27	—	—	—	0.060	0.060
40B	Ladle	0.20	—	1.5	—	0.050	0.050
	Product	0.25	—	1.6	—	0.060	0.060
40C	Ladle	0.18	—	1.5	—	0.050	0.050
	Product	0.22	—	1.6	—	0.060	0.060
40D	Ladle	0.18	—	1.5	—	0.050	0.050
	Product	0.21	—	1.6	—	0.060	0.060
40E	Ladle	0.16	0.10/0.50	1.5	—	0.040	0.040
	Product	0.19	0.10/0.55	1.6	—	0.050	0.050
43A1	Ladle	0.25	—	—	—	0.050	0.050
	Product	0.30	—	—	—	0.060	0.060
43A	Ladle	0.25	—	—	—	0.050	0.050
	Product	0.30	—	—	—	0.060	0.060
43B	Ladle	0.22	—	1.5	—	0.050	0.050
	Product	0.26	—	1.6	—	0.060	0.060
43C	Ladle	0.18	—	1.5	—	0.050	0.050
	Product	0.22	—	1.6	—	0.060	0.060
43D	Ladle	0.18	—	1.5	—	0.040	0.040
	Product	0.21	—	1.6	—	0.050	0.050
43E	Ladle	0.16	0.10/0.50	1.5	—	0.040	0.040
	Product	0.19	0.10/0.55	1.6	—	0.050	0.050
50A	Ladle	0.23	—	1.6	—	0.050	0.050
	Product	0.27	—	1.7	—	0.060	0.060
50B	Ladle	0.20	—/0.50	1.5	0.10	0.050	0.050
	Product	0.24	—/0.55	1.6	—	0.060	0.060
50C	Ladle	0.20	—/0.50	1.5	0.10	0.050	0.050
	Product	0.24	—/0.55	1.6	—	0.060	0.060
50D	Ladle	0.18	0.10/0.50	1.5	0.10	0.040	0.040
	Product	0.22	0.10/0.55	1.6	—	0.050	0.050
55C	Ladle	0.22	—/0.60	1.6	0.10	0.040	0.040
	Product	0.26	—/0.65	1.7	—	0.050	0.050
55E	Ladle	0.22	—/0.60	1.6	0.10	0.040	0.040
	Product	0.26	—/0.65	1.7	—	0.050	0.050

Table II *Strength and yield stress of steel (from* BS 4360)

Grade	Tensile strength	Up to and including 16 mm	Over 16 mm up to and including 25 mm	Over 25 mm up to and including 40 mm	Over 40 mm up to and including 63 mm	Over 63 mm up to and including 100 mm
	hbar	hbar	hbar	hbar	hbar	hbar
40A	40/48	—	—	—	—	—
40B	40/48	24.0	23.0	22.5	22.0	21.0
40C	40/48	24.0	23.0	22.5	22.0	21.0
40D	40/48	24.0	23.0	22.5	22.0	21.0
40E	40/48	25.5	24.5	24.0	23.0	22.5
43A1	43/51	—	—	—	—	—
43A	43/51	25.5	24.5	24.0	23.0	22.5
43B	43/51	25.5	24.5	24.0	23.0	22.5
43C	43/51	25.5	24.5	24.0	23.0	22.5
43D	43/51	25.5	24.5	24.0	23.0	22.5
43E	43/51	27.0	26.0	25.5	24.5	24.0
50A	50/62	—	—	—	—	—
50B	50/62	35.5	34.5	34.5	34.0	32.5
50C	50/62	35.5	34.5	34.5	34.0	32.5
50D	50/62	35.5	34.5	34.5	34.0	By agreement
55C	55/70	45.0	43.0	41.5	—	—
55E	55/70	45.0	43.0	41.5	40.0	—

Notes to tables I and II

● 1 hbar = $10^7 N/m^2$

● Hollow sections, bars, rounds and squares are excepted; these are covered elsewhere in BS 4360

● Some grades containing either 0.20-0.35 per cent or 0.35-0.50 per cent copper may also be supplied if specified on the order

● For sections and flat bars over 16 mm thick and for round and square bars over 25 mm thick, a maximum carbon content of 0.22 per cent for ladle analysis, and 0.26 per cent for product analysis is permitted

● Where niobium is indicated, other grain refining elements may be used, each up to the maximum percentage indicated. If micro-elements other than niobium are to be used the manufacturer shall inform the purchaser at the time of the inquiry or order

● Facilities for normalising sections are limited

● Where normalised sections and flat bars can be supplied, impact test values equivalent to those specified for plates of the same grade and thickness can be provided

● Minimum yield stress values for flat bars over 100 mm thick to be agreed between the manufacturer and the purchaser

● For sections and flat bars under 6 mm thick, only bend tests required

● Minimum tensile strength 48 hbar for material over 50 mm thick

● Minimum yield stress 24.5 hbar for universal beams, columns and bearing piles to BS 4 *Structural steel sections* Part 1 'Hot-rolled sections' with flange thicknesses not exceeding 40 mm

● For universal beams, columns and bearing piles the maximum tensile strength may be increased to 54 hbar

● For universal beams, columns and bearing piles the yield stress obtained on test pieces taken from the web shall be not less than 1.5 hbar greater than the specified minimum value

4 Basic structural sections

Hot-rolled sections

4.01 The traditional hot-rolled sections **4** are available in three main grades to BS 4360 ie grade 50, grade 43 and grade 55. All are weldable, but it is necessary to specify with a suffix eg grade 50c, when material with a particular set of properties for using in 'impact' loaded situations is required (see tables I and II).

4.02 Manufacturers usually give the size range of sections available in the form of a handbook but all sizes shown are not necessarily immediately in stock, so the delivery situation at the time of intended construction should be checked. There is usually a range of weights of section within any one serial size eg 305 × 165 UB 40, 305 × 165 UB 46 or 305 × 165 UB 54. This extra is achieved by raising the rolls as each section is manufactured, with the result that depths and widths of section also alter, hence the dimensions of the above sections are 304 × 165, 307 × 166 and 311 × 167 respectively, although all are known as 305 × 165 UB. (Dimensions are in millimetres; the figure following the term UB, eg UB 40 is the weight in kilogrammes per metre run.)

Hot-rolled hollow sections

4.03 Hollow sections, as illustrated in **5**, are available in three main grades, but are generally suffix c, ie with controlled impact qualities. Suffix D is available to order where precise checking of the qualities is called for. The tolerances of these sections are generally better than those of traditional shapes, but they are usually more expensive.

4.04 As with traditional sections, a range is available within any serial size, but here the additional material is added internally, increasing the thickness while maintaining the same outside dimension.

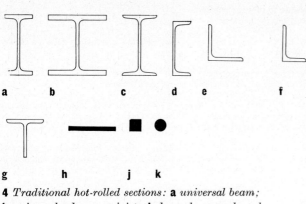

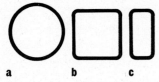

4 *Traditional hot-rolled sections:* **a** *universal beam;*
b *universal column;* **c** *joist;* **d** *channel;* **e** *equal angle;*
f *unequal angle;* **g** T-*bar;* **h** *flat;* **j** *square;* **k** *round*

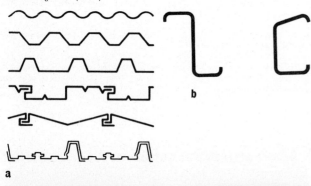

5 *Hot-rolled hollow sections:* **a** *circular* (CHS); **b** *square;*
c *rectangular* (RHS)

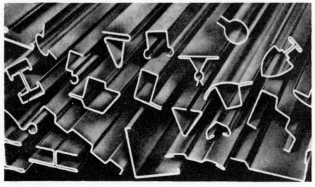

6 *Cold-formed sections*

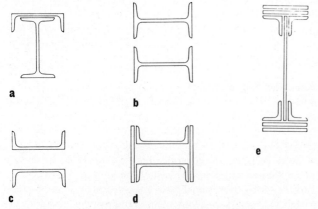

7 *Made up sections:* **a** *gantry girder;* **b, c** *and* **d** *compound*
stanchions; **e** *plate girder*

Cold-formed sections

4.05 Cold-formed steel products are generally associated
with sheet and cladding materials **6a** although they may be
made in thousands of shapes **6b**. Their structural use is well
established in the z-purlin **6c** to which it is well suited and
is a most economic form. Many patent floor and permanent
shuttering systems employ shaped sheets, and they may
be used as the flanges of open-web standard joists.

Made up sections

4.06 Many additional shapes can be made from basic rolled
sections by combining two or more shapes. Manufacturers
supply the components; fabricators make up the section,
but some of these have been so well-used over the years
that their combined properties have found a place in manufacturers' lists **7**. Other shapes and sections are produced
for special purposes, usually under patent or trade names.

4.07 These sections are automatically welded, and with full
supervision can be entirely acceptable. Open-web joists **8c**
come in many forms, made up of any combination of shapes
for top and bottom chords, either hot-rolled or cold-formed
section, usually held by a continuous braced web member
of one or two rods. They are particularly suitable in lightly
loaded situations; for heavier loads the castellated member
8a is preferable, and the welded plate girder (Autofab or
otherwise made up **8b**) can carry the very heaviest and
dynamic industrial loading.

Cased universal sections

4.08 Addition of a concrete casing enhances the compression strength of any steel section; it also improves fire
resistance. Unfortunately there is no direct mathematical
relationship between the capacity of the cased and uncased
sections, so all work here is experimental. Figures are
available for cased UC and UB sections, but it cannot be
assumed that other shapes of equal steel area would produce
the same results.

4.09 Only 38 mm concrete casing is required for fire protection, but a 50 mm minimum is needed for the casing to
become loadbearing, and furthermore the steel must be
wrapped in additional reinforcement. Published figures
enhance only the direct compression on the member,
bending must be carried by the steel section alone.

Composite sections

4.10 Steel and concrete may be combined in ways other
than complete encasement **9**. Heavy loading on shallow
beams can be economically dealt with by composite sections
but in this situation deflection may become critical and this
must be checked. A T-beam is improved by welding steel
connectors to the upper flange of the steel section, and then
casting a reinforced concrete slab onto it. Because the

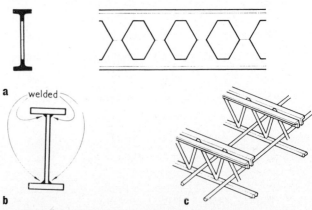

8 *Other compound sections:* **a** *castellated universal beam;*
b *Autofab beam;* **c** *open-web joist*

slab may be an integral part of the structure the extra expense of studs may be small compared with the load carrying benefit. Spacing of studs and propping of beams are covered by BS CP 117[2] and may be carried out by any competent contractor. Preflex beams, although a patent process, may also finally be incorporated with studs into an in situ slab **9c**. Preflex beams are put into reverse bending in the factory, and cased on one flange to hold this shape during erection. The working load is designed against this initial prestress bending, giving long spans and shallow depths.

4.11 Under certain circumstances concrete filling of structural hollow sections enhances their fire resistance and at all times it increases load-carrying capacity **9d**. Published figures give loads for both direct compression and combined compression and bending. The filling can be carried out by a main contractor on site, but the history of this section is short; still it is established practice in heavy engineering and its real value may only be realised when factory-filled sections become available to order.

Cables

4.12 Any hot-rolled, cold-formed or made-up section can be used as a tension member; but the lightest and slimmest will always be a cable, since it is made of a vastly higher strength steel than any of the other normal shapes **10**.

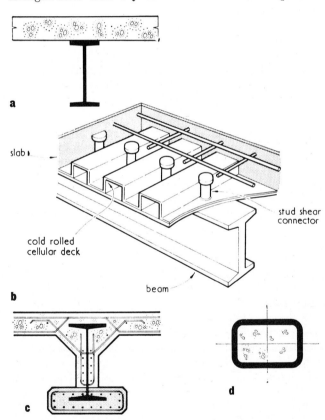

a

slab

stud shear connector

cold rolled cellular deck

beam

b

c

d

9 Composite section: **a** *beam (friction only);* **b** *composite beam (with studs etc);* **c** *Preflex beam (typical section with in situ floor and web casing);* **d** RHS *filled with concrete*

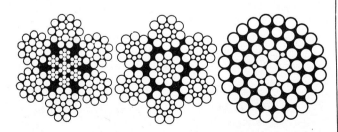

10 Cables: three typical cross sections

a

b

c

d

e

11 Cable structures: **a, b, c** *circular structures are particularly suited to single or double layer tension solutions, since the circle balances out the holding down forces and returns them through a tension ring or similar device;* **d** *and* **e** *are rectangular structures using ropes as inclined suspension ties to supplement other forms; in fact* **d** *is basic space frame and* **e** *basic portal*

Strangely, the relatively few tension structures—other than bridges—in the world seldom use the cable as a vertical hanger, even though its size enables it to be hidden in the slenderest of casings. Lower grade steel sections are preferred possibly because, due to the higher stress, the cable has a greater extension and this could distress the finishes of a building. But this is probably just conservative thinking, since by judicious prestressing cables could well be used for vertical members.

4.13 Today, however, the cable is only used as either a catenary rope, the main tie of a simple truss system, or as a net **12**. Traditional cable fixings develop from those used by mechanical engineers, tent makers, sailors etc but do not necessarily lend themselves to developing full structural strength, or acting as a structural analysis would require. Hence connections when non-standard tend to be expensive and often need proof-testing. New opportunities in this area would be opened if cast shaped connections, as on the Munich Stadium roof **12**, were capable of small-scale production.

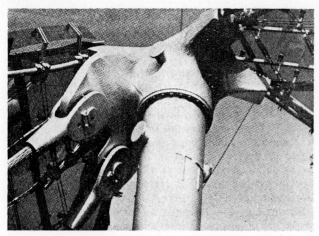

12 *Cast shaped connections on roof of athletics arena for Munich Olympic Games (Behnisch and Otto):* **a** *cast steel point connecting two cable nets and supported on mast;* **b** *roof during erection showing net details, edge cables, system points and parallel strand main cables*

5 Corrosion protection

5.01 Finishes may be decorative, anti-corrosive or for fire-proofing and often fulfil a combination of these requirements. For decorative purposes a single coat of red oxide followed by primer and finishing coat of paint is often adequate.

Corrosion

5.02 Corrosion is not inherent in steel, but depends largely upon the environment in which it is used; however, the economics of corrosion protection are complex and can fundamentally affect the architect's choice of structure.

5.03 Corrosion is an electro-chemical process, and it can be halted by the following methods:

Coating: putting a physical barrier around the steel to insulate it from the environment

Cathodic protection: application of an electric potential to prevent electrolytic action

Use of alloy steels: adjusting the chemistry of the steel to form a protective surface film

Atmospheric control: treating and controlling the surrounding atmosphere to render it harmless to steel

Structural design factors: designing the structure to reduce locally aggressive situations.

Coatings

5.04 All coatings, other than concrete or asbestos casing, depend for adhesion on surface preparation. This may be done by hand, using wire brushes (possibly powered by compressed air), or flame guns. This can be done by a subcontractor, but galvanising or metal spraying are usually carried out by specialists, who quote for preparation and coating. This sub-subcontract may require additional transportation, possibly even the breaking down of a fabricated assembly and re-making afterwards. An architect should beware letting such operations detract from the work, either in finished specification or progress.

Coating methods

5.05 Coatings can be applied as paint, hot-dip zinc galvanising, metal spray, vitreous enamel coating, plastic coating, hot coal tar pitch, or protective tapes. The metal face can be prepared using wire brush or flame, a sulphuric acid bath, or shot-blasting.

5.06 Paints may be any colour; hot-dip galvanising gives a zinc colour that may dull after initial shine; metal spray is usually grey matt; coal tar pitch is black; tapes may be black or dark green; and plastic coatings variously coloured for sheets, steel mesh and structural hollow sections.

5.07 In galvanised work, welded connections can be made good if painted with a zinc-rich paint while still warm, and a cold plastic compound may be used for joints in plastic coated work. Shot-blasting and metal spray techniques do not lend themselves to making good and this should be allowed for by treating the largest possible units.

Alloy steels

5.08 Stainless steel is available as sheets, rods and certain sizes of tube. Generally it is too expensive for use as a structural member; but pressed panels could be used in surface active structures. However, the current price structure could make stainless steel a reasonable structural material in the not too distant future.

5.09 All structural sections are available in weathering steels and, used wisely, these can be economic. They are copper bearing steels and fully weldable provided that the correct electrodes are used. However, care is needed during fabrication to ensure that unwanted markings or initial unbalanced weathering are not built into the finished work; the architect must evaluate site exposure conditions precisely. It is very easy to overlook simple everyday details when working in weathering steel; recourse to literature[3] and the faults of others can prevent difficult problems.

5.10 Most suppliers of alloy steels require minimum orders of a certain tonnage of a certain section; this should be determined at an early planning stage, for large stocks are unlikely to be held.

6 Fire protection

6.01 A favourite selling point in the post-war boom in reinforced concrete was said to be the elimination of casing costs necessary with steel. True, but the real test is: does the cost of the final fire-proofed frame exceed that of the other material? Further, have factors of speed, accuracy, flexibility, etc been adequately weighed?

6.02 For many years these questions were not asked because structural steelwork was not available in the UK anyway. Still the idea that fireproofing of steel involved a completely extra cost on a contract was born. But this was never wholly true, and using modern concepts of fire safety, real comparisons are worth investigation.

6.03 Yet perseverence is needed, for the relevant regulations were founded in dim history, where superstition, surmise, imagination and not a little hysteria were cheaper and quicker than real research. Unfortunately, today some research is still coloured by an instinctive fear of fire, but despite the ever-present spectre of a major disaster (so easily mis-used to represent statistical fact) design is becoming more rational.

6.04 The basic factor in fire protection is that at temperatures exceeding 550°C the yield strength of steel is halved (see Technical study SAFETY 3 earlier in this handbook). The capacity of the section to carry stress therefore reduces to a figure equal to the design load and if it is being applied at that time, failure occurs.

6.05 The protection required by a building relates to its occupancy and size—regulations give notional time periods for various forms of construction and particular categories of building. It is generally assumed that these give personnel time to evacuate before the structure fails; only secondary significance is given to the loss of contents. But with increasing costs of buildings, cost of lost production, and increasingly valuable building contents (eg computers and electronics) fire periods laid down in regulations may not be sufficient to protect an architect's client from loss. Where alarms, sprinklers, or fire walls are involved, blind application of structural fireproofing regulations is not adequate. With a finished, adequately protected building, structural steelwork protection is only a small part of the cost.

6.06 For most single-storey buildings, steelwork does not need to be protected, although supporting columns should have the same fire resistance as an enclosing wall. Multistorey structures do require protection, but dispensation may be obtained for car parks, where it has been adequately demonstrated that the fire loading in the building, even fully occupied, cannot produce the required critical temperatures **13**.

6.07 Placing columns outside the external cladding increases their life under fire conditions; up to half an hour can be added to a column's life when fire has to first penetrate the cladding **15**. However, more usual fireproofing methods are casing, coating and, sometimes, hydraulic **14**.

Casing

6.08 Casing can be concrete (preformed or in situ), non-structural asbestos (preformed), loadbearing concrete (in situ, external with wire mesh or internal for SHS), or block or brickwork.

6.09 Since the objective of fire resistance is to keep the steel below 550°C for a certain required period, any means of accomplishing this is worthy of inspection. The heavier the steel section, the longer it will take to absorb the heat from a given fire load, consequently a thicker section has a better fire rating than a thinner one carrying equal load **16**. Here the recent introduction of very thick-walled

13 *Exposed steelwork in multi-storey car park*

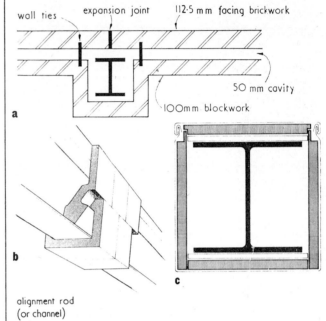

14 *Fire protection of steel;* **a** *blockwork encasement on external wall;* **b** *precast lightweight concrete interlocking blocks;* **c** *expanded metal lath and plaster encasement to stanchion;* **d** U-*shaped body and lid used for free-standing column weighing 54 kg/m gives 2 hour fire resistance when lined with 25 mm low density Asbestolux*

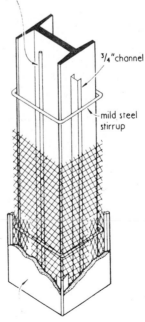

structural hollow sections of uniform thickness can prove an advantage.

6.10 Hollow sections also suggest water filled protection systems, and these are now being attempted in various parts of the world. For example, the US Steel Building in Pittsburgh, **17**, exploits all these concepts in one integrated solution. The columns are outside the structure, hence of weathering steel, and are box section, water filled.

References

1 BS 4360: Part 2: 1972 Specification for weldable structural steels [(2-) Yh2 (D4)] £3.00

2 BS CP 117: Part 1: 1965 Simply supported beams in buildings [(2-) Gy (K)] £1.35

3 AJ series on weathering steels, AJ 25.10.72 to 29.11.72

15 *External unclad steel columns*

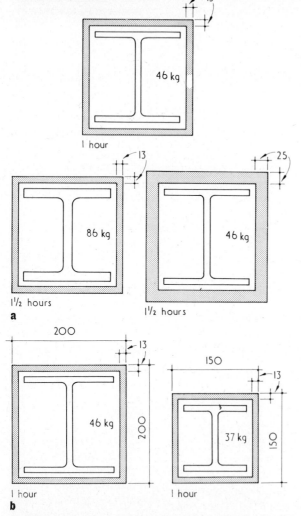

16 *Effect of steel weight, size of section and thickness of protection on fire resistance: **a** with section measuring 200 mm × 200 mm thickness of required protection varies with steel weight; **b** with 13 mm protection, section size can vary with steel weight*

a

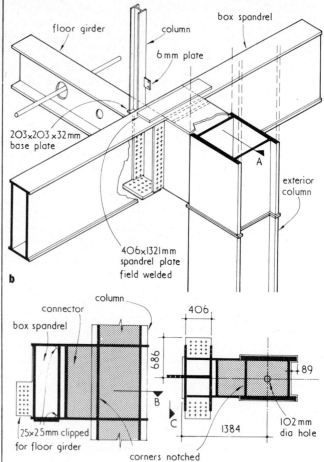

b

c vertical section at A   **d** plan section at B

e   end view at C

17 *External structure frame of US Steel Building, Pittsburgh. Exterior column is tied to structural frame at every third floor. Stub connector, box shaped, links box column and box spandrel. Liquid for fireproofing fills column and stub, but not spandrel, which is conventionally fireproofed*

Technical study
Steel 2

Section 6 **Structural material: steelwork**

Structural forms of steelwork

GORDON ROSE's *second study leads from the basic structural elements into a discussion of suitable forms for single storey, low-rise and high-rise building. He concludes with a look at the constraints of actual construction—site, type of contract, fabrication etc*

1 Basic structural elements

1.01 Beam and column are fundamental structural elements; and although both have evolved into complex units they are still capable of simple consideration—indeed a quick intuitive check can often prevent much unnecessary or inaccurate sophistication.

Beams
1.02 Beams may be designated by their support condition, and this is usually more critical than the shape of the section. Any steel section may be used as a beam, but its efficiency will depend upon the suitability of the section to withstand the loading pattern. The best condition is when most material is as far from the neutral axis as possible. (At the neutral axis of a section, as discussed in Section 2 ANALYSIS, stress is zero as it changes from tension to compression.) For a simple section, simply supported, it is easy to decide the optimum size; but with continuous beams (perhaps of differing spans) a symmetrical section can be wasteful, and the choice of built up or composite shape can require long investigation. With moving loads, and loads in more than one direction, a real design problem emerges **1**.

1.03 Although it is not important with massive construction, with steel-beam construction the compression flange must be adequately restrained; not only at the completion of the structure, but during all stages of construction. Restraint may be achieved by friction between a flange and a concrete slab, and shear connectors can be used here to assist very heavily loaded sections. Where adequate restraint cannot be provided, the allowable compressive bending stress in the beam must be reduced.

1.04 When flange restraint is provided by secondary beams, or purlins, the resisting force must be $2\frac{1}{2}$ per cent of the maximum flange force in the beam. (Timber floor joists should not generally be regarded as capable of stabilising steel beams.) Thus, for example, rectangular hollow sections are capable of spanning greater unrestrained distance than open sections, simply because the 'flange' is effectively restrained by two 'webs'.

1.05 As materials improve and permissible stresses increase, the ability to carry loads on beams of less and less depth increases. If flange restraint is provided, the critical condition will not be bending failure, but unacceptable deflection.

Camber
1.06 Deflection of beams under live load should not exceed one three-hundred-and-sixtieth of their span. To compensate for deflection long beams can be *cambered* (cold curved at the works up to 100 mm on a 20 m length). Preflex and other built-up sections can be cambered during fabrication.

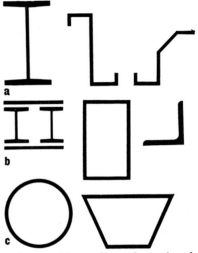

1 *Preferred beam shapes for various bending conditions:* **a** *bending primarily in one direction;* **b** *in two directions, and* **c** *bending in two directions and with torsion*

Columns
1.07 Columns and struts may be any shape; the final choice will be influenced by the means of fixing and continuity at either end and the desirability of intermediate lateral restraint. For pure axial compression the most efficient column section must be a circle. In older structures this was recognised by using solid round sections (still available) but in recent times these have been largely superseded by the circular hollow sections. A hollow section of regular shape is better than an open section because of its better stress flow, geometric properties and torsional rigidity.

1.08 For axial compression combined with bending in one direction, an I-section is best, although the less efficient rectangular hollow section may have other advantages. For axial compression plus bending in two directions, built-up sections of plates can be prepared. Two angles together are usually inefficient, unless formed into a rectangle when, like the rectangular hollow section, they will carry greater storey heights without lateral restraint.

Arch
1.09 Arches may be formed of single steel sections, or made up in a variety of combinations **2**. They carry load in pure compression, and so shapes discussed for columns apply to them also. Asymmetrical loading (due to wind or snow etc) will produce bending, and when significant will determine the section, and even the validity of using the arch-form.

1.10 For larger spans the arch may be the cheapest structural form, but it produces a great deal of 'dead space', and with ever increasing demands for maximum coverage (and of course heating and air-conditioning), it may lose ground as a big-span solution. Arches can be hinged or fixed at each end, or pinned one end and fixed the other; they may also have an additional pin at the crown. Whichever form is chosen the joints must be made to work strictly according to the design. An arch must produce a vertical reaction at the ground plus a thrust; often the deciding factor in the choice of construction is the way in which the thrust is carried.

2 *Arches:* **a** *three pin arch from single curved rhs;* **b** *three pin tubular truss*

Truss
1.11 In trusses the bending effect seen with beams is translated into compression and tension forces (bending stresses become relatively small secondary effects). Still probably the most economical form of modern structure, trusses may be constructed from any available shape of section, and are very simple fabricating shapes. But ever-increasing wind load requirements lead to more bracing in wall and roof planes and produce reversal of stresses in members. This increases the weight of material and so possibilities for mass production of standard trusses are reduced and fabrication costs increased **3**.

1.12 A truss cannot economically resist any but the lightest nominal service loads. Heavier loads are usually carried by a separate structure, even if using combined columns.

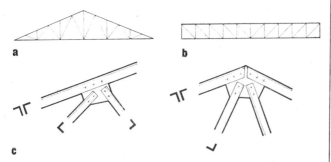

3a, b *pitched and horizontal trusses using rolled angles;* **c** *details of typical junction on trusses made from angles*

1.13 The basic elements—column, beam, arch and truss—may be combined with other materials—brick, block, asbestos, timber etc—for support and cladding. A balanced combination of these may well form the most economic space enclosure, even when compared with more exotic envelopes.

2 Single storey structures

2.01 Structures can be classified in many ways; here the use of steel is considered under the headings—single storey structures, low-rise structures, and high-rise structures. Many factors overlap these arbitrary divisions and the best results come from combining a knowledge of up-to-date techniques in each.

2.02 The main single storey forms—skeletal structures and surface active structures—can often be mixed in one project. Skeletal structures have a self-supporting steel framework that is clad with another material. Surface active structures rely for stability-in-use on the full structural action of the cladding which can be taken into account during the analysis. (But a steel shed with columns stabilised by brick panels, for example, is not known as a surface active structure—for structures to be so classified an intuitive sense of the cladding's behaviour must be replaced by a full analytical understanding.)

Skeletal structures
2.03 Until recently definition of skeletal structure was simple because tension members and suitable connections were neither cheap, easily available nor well understood. Even in 1960 tension members were little used in the UK, but within the last ten years attitudes have been more enlightened (but unfortunately not yet exploited by designers) so that skeletal structures can be discussed as compression structures, composite structures, or tension structures. With these sub-divisions established, other considerations, such as single bay or multi-bay, one-way spanning, two-way spanning, and multi-directional spans of single layer or double layer apply.

One-way spanning
2.04 The cheapest single storey structure is the truss supported on posts (all traditionally shaped sections). This owes little to sophisticated analysis, inspired intuitive design, consideration of cost in use, aesthetics etc but is based on the market structure—a fabricating industry geared to the production of traditional structures, giving the purchaser freedom to shop around. So where capital cost is the only yardstick, the simple traditional structure must win—and win handsomely **4**.

4 *One-way spanning truss*

2.05 Maintenance costs, aesthetics, mass production and even better conditions of work help to cut this traditional superiority, but it is the increased wind loading requirement —generated by a completely unconnected and irrelevant event—which will be the real spur towards more sophisticated structure. Under the 1972 Code (BS CP 3 Ch v Pt 2) the pitched roof suffers: loadings are increased (as for all structures) so that they often result in complete reversal of dead load stresses. Bottom ties become compression members, resulting in uneconomic members and connections. The structure shown in **4**, for example, may not now be viable in exposed locations. Particular attention should be paid to renovation of existing structures because modern requirements may lead to much larger members.

2.06 The second and vital change due to the new code is the far greater time an engineer needs to analyse all the required loading combinations. Costs are higher, and the use of standard answers greatly limited. Increasingly the engineer will have to look to the desk computer (and its memory); if he must do this for a 'simple' structure, then why not design a 'complex' structure, which his computer program remembers just as well. Thus a wild application of revised loading may press the designer to consider alternatives—many of which will continue to be used even if the Code of Practice loadings one day return to sanity. The designer will need to look elsewhere for economy.

2.07 Engineering changes have resulted from new sections or new methods of analysis becoming available, eg an alternative truss design using hollow sections is shown in **5**. Traditional truss geometry does not exploit the compressive or repetitive abilities of hollow sections, but this easily-jigged alternative, with its long compressive members and easily welded joints lends itself to the single span roof.

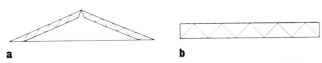

a　　　　　　　　　　　　　　　**b**

5 *Trusses using structural hollow sections—compare with* **3***:* **a** *roof truss;* **b** *trussed beam*

Rigid frame

2.08 With welded *knee braces*, partial rigidity has always been possible; but the mass-produced welded *portal frame* **6**, using single members at shallow pitch, has become the most usual application of modern moment distribution analysis. This is also the major application of *plastic theory* in structural steelwork. Elsewhere *elastic analysis* appears to give adequate results, but here plastic analysis gives very realistic results, because deflection is significant with slender members and can be most accurately forecast by plastic analysis.

6 *Welded portal frame*

2.09 Built as frames, cross-braced for stability in the plane of the walls and the roof-line, welded rigid frames may be used to form any of the known shed buildings of particular shape **7**. They are generally constructed as skeletal buildings, the roof covering, purlins, and side sheeting rails (added later) serve no structural purpose except where purlins and rails are deemed to restrain main members from buckling. This greatly influences the design and shape of purlins and rails, and there is growing awareness that lightweight structures are adequate for working loads, but expensive, complicated and sometimes impossible to erect. Increasingly, ease of fabrication and erection rather than the weight of steel contained determine costs of lightweight buildings.

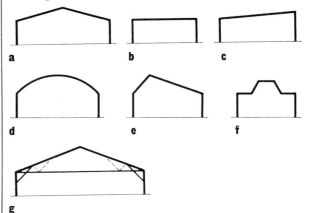

a　　　　　　　　**b**　　　　　　　　**c**

d　　　　　　　　**e**　　　　　　　　**f**

g

7 *Welded rigid frames:* **a** *pitches;* **b** *flat;* **c** *sloping;* **d** *curved;* **e** *northlight;* **f** *monitor;* **g** *knee-braced truss*

2.10 This is particularly true of the recent investigation of 'surface active' sheds **8**. Here frame, purlins, sheeting rails and the sheeting itself should act as one unit; results with steel sheets are encouraging, asbestos sheets less so. The sheeting fixings are more elaborate than simple 'hook bolts' and many details await refinement **9**. But the surface active shed will be a breakthrough and worthy of consideration.

2.11 Both the simple and rigid frames lend themselves to multi-bay construction, and are competitive in span for a range 7 m to 20 m, and for welded portals even up to 35 m span.

8 *Surface-active portal frame shed under full scale load tests at Salford University. Note shear connector, out of length of Z-purlin, between rafter and sheeting*

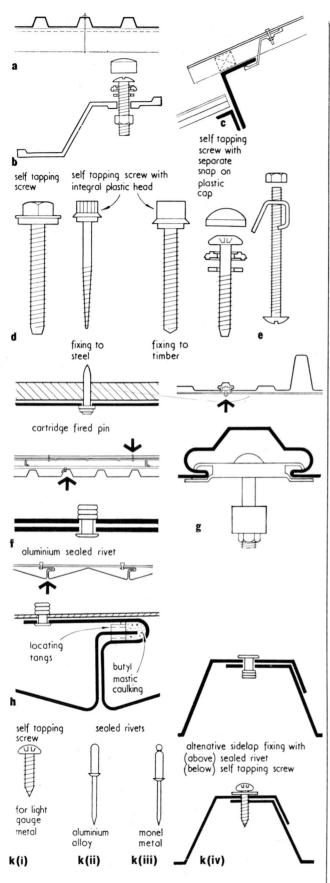

Long spans

2.12 Despite growing interest in 'long spans', the rule of thumb that 'the more columns in a building the cheaper the structure' will remain valid for some time. But as demand for clearer longer spans increases, and more methods become available, this price difference is reducing.

Arch

2.13 The arch is the oldest long span structural form; its commercial use is limited because of the finished shape and 'dead space'. Where this can be overcome, an economic structure may use a lattice system to divide the whole into segments and so simplify transportation and allow repetitive production. Arch up to 300 m have been proposed using such methods **10**.

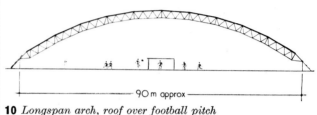

10 *Longspan arch, roof over football pitch*

Beam and column

2.14 In applications where deflection is not critical and the depth of beam can be at least one-twentieth of span, up to 100 m spans are now being economically achieved with beam and column **11**.

Two-dimensional tension structures

2.15 Just as steel ropes lend themselves to suspension bridges so too they are suitable for long span linear structures. The large span capabilities of any traditional structural form may be enhanced by the use of steel tendons **12**. Applications are at present limited to structures where deflections are not critical, because even a stiff framework will deflect if supported by ropes; here, attempts to eliminate rather than accept and treat deflections can result in costly details and poor appearance. It would seem preferable to develop true 'tension structures' rather than use ropes to support traditional forms.

9 *A selection of modern sheeting fixings:* **a** *hookbolt fixing through 'crown' of flutes;* **b** *sheet clip;* **c** *sheet clip used above glazing;* **d** *self-tapping screws;* **e** *Oakly clip;* **f** *wall construction incorporating steel liner tray and box rib steel outer section, with enlarged details of aluminium sealed rivet fastening side-laps of box-rib sheet, and cartridge-fired pin fastening steel tray;* **g**, **h** *two secret fix units;* **k** *self tapping screws and sealed rivets in position*

11 *The world's longest beam and column (beam depth 6 m, span 100 m). Because they are lighter, cheaper and need not interrupt the field of view of the spectator such spans can make the cantilever construction obsolete for such applications. Football stand at Celtic Park, Glasgow*

12 *Steel tendons to allow increased span*

Two-way spanning

2.16 The demand for large clear spans combined with shallow construction depth has increased the use of two-way structures. During the past decade the structural engineer, aided by the computer, has been able to open up a whole new range of structures which had previously (because no one could pursue them) been thought impossible. This is now possible with great accuracy; for example, the double layer dome **13** may be analysed as single trusses spanning from centre hub to outer tension ring—an established big span solution. It can even be made to open up in segments **14**; analysis is still traditional. Breakthroughs in costs will now be with single layer big span domes, effectively using only half the quantity of steel.

13 *Double layer dome*

14 *Dome which can be opened up in segments (Pittsburg seen also in* STEEL 1, **17a**

2.17 Unfortunately such structural forms have no intuitive 'correctness', and the designer has to rely, to a great extent, on his structural analysis. But this phase is now passing and checking by alternative methods of analysis and models is available. Yet the age old statutes of stability still rule: the sums of all vertical and horizontal forces and moments must be zero, ($\Sigma V = 0$, $\Sigma H = 0$, $\Sigma M = 0$); while M, of course is still often given by WL/8. The application of these simple truths can save much time and trouble; a surprising number of sophisticated three-dimensional structures can be reasonably checked by these formulae.

2.18 The truss on posts is the simplest one-way structure, but the interpretation can be infinitely varied; in the same way the interacting trusses of the *space frame* have become the general form of the two-way structure. The advantages of the space frame are:

a, it gives a shallower depth than the equivalent one-way system

b, it is more efficient at carrying asymmetric or moving loads

c, supports can be fewer and more widely spaced

d, it lends itself to prefabrication and mass production

It is seldom economic to use a space frame where these are not overriding considerations.

Double-layer systems

2.19 At its best when square on plan, the efficiency of the double-layer space frame decreases as the length/width ratio increases; when this exceeds 2:1 a one-way system is probably a better solution. Plan dimensions of over 25 m are usually required for economy. Generally the best column spacing for a square plan form is four columns inset **15a**. This is seldom acceptable in commercial structures (except very large multiple areas) so the next best solution is two or more columns on opposite sides, or fully supported all round. The least efficient, if most spectacular, is one column in each corner.

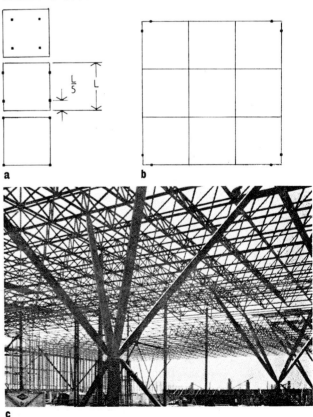

a   **b**

c

15 *Space-frame column spacing;* **a** *for single space frames;* **b** *for multi-span frames;* **c** *space frame under construction* (RAI *Amsterdam architects and engineers Drexhage Sterkenburg Bordon Uenstra*)

2.20 The space frame roof with its great torsional rigidity is particularly suited to buildings on poor foundations; span/depth ratios can be as high as 30:1, **16**. Shallow depth has perhaps reached a limit; it is often more valuable to use an 'inefficient' depth, and be able to carry service loads within this depth. As well as providing extra 'lettable area' at little extra cost, monorail cranes can be carried, eliminating a need for traditional infrastructure of crane gantries, columns etc.

16 *Lifting a 25 m square space grid into place. Two cranes would probably have been adequate as the whole frame has great stiffness (*BSC *Nodus frame later used at Surrey University, see* AJ *1.11.72 p1001. Joint detail as* **17a**)

2.21 Steel two-way structures are very light, therefore the form of joint chosen has a greater effect on final cost than with other forms. Though an all-welded joint is apparently best, it requires controlled site techniques; for welding is best done at the works, leaving site connections to be bolted. For analytical purposes welding provides full fixity of a member at the joint but many patent forms of connection give lesser degrees of continuity.

2.22 Such systems are often based on a particular geometric pattern, some even restrict the module, but by using them space frame construction can be competitive with traditional one-way forms **17**. Most double-layer space frames can be cambered to falls for drainage without greatly affecting design, but real curves—with the exception of domes—are better done as a single layer.

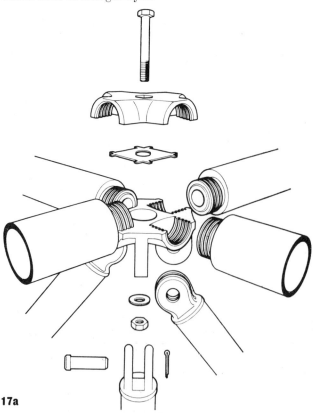

17a

17 *Fourteen commercial survivors from the vast variety of space frame joints. Which is best? Much depends on the job. By welding, or designing a purpose-made joint, the best geometrical layout will be achieved. By using a patent joint, the benefits of mass production, erection skill, availability of parts etc, may be offset by the inefficient module and pattern. The designer should always compare at least one 'system' with at least one 'purpose designed' scheme*

Single-layer systems

2.23 Insufficient knowledge of action under asymmetrical loading held back the single-layer dome for some years, but this is being overcome, and single-layer construction—perhaps with surface active cladding—has obvious advantages for the middle range of spans.

2.24 Restrictions imposed by length/width ratios may be relieved by curving or cranking the space frame in one or more directions. This is more common with single layer construction but would impossibly complicate the geometry and joints of double layer frames. Methods of calculation and fabrication now allow more exotic shapes (previously restricted to timber, plastic or reinforced concrete construction) to be built in steel with lightweight cladding **18**. Since most of these structures carry roof loads only, the reduction in self-weight represents an appreciable saving in structure and foundations; the resulting grid patterns lend themselves to interesting decorative treatment.

2.25 Here again, choice of joint is critical to cost—it can be as much as 50 per cent of the cost of the space frame. Patent joints are available but the welded joint is most efficient, particularly with shallow curves on big spans, where asymmetrical loading is often the critical criterion, and care must be taken to avoid *snap-through*. This is a form of local failure peculiar to two-way curved structures that, if uncontrolled, can lead to full failure. A rise of at least one-eighth of the span seems to be necessary before such considerations become secondary; much research is being done on this form of construction.

3 Multi-storey structures

3.01 A significant outcome of legislation following the Ronan Point collapse was the requirement to investigate the validity of structures of more than four storeys. Though arbitrary, this has created a demarcation between the approach to low-rise (two to four storey structures) and high-rise structures. Despite marginal exceptions, low-rise buildings and their methods of construction can be shown to be quite different from high rise.

Low-rise structures

3.02 Well documented over many years, the merits of competitive components are well known. Stability under asymmetrical loading must of course be shown to satisfy the latest requirements, but at these heights the traditional forms of construction usually remain competitive **19**.

a

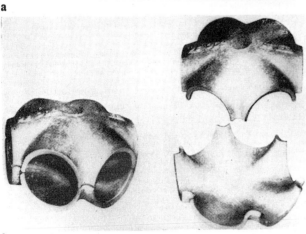

b

c

18a *Barrel vault, single layer space grid (PLA warehouse engineers Harris & Sutherland); b single layer joint (patent du Chateau); c joint made by site bolting and welding the mitred corners of two square panels in a shallow single layer dome (BSC)*

19 *Low rise steel frame*

3.03 Any material or combination of materials may be used for the floor system but its composite action with the steel structure is seldom worth considering. The frame comprises a secondary and main beam system. The secondary beam, depending largely on the fire-proofing requirements, could use lattice steel trussed beams as close as 5 m centres. Larger centres may be gained using deeper slabs on universal beams and castellated beams, and steel may be used as a permanent floor shuttering/reinforcement **20**. (The main beam system here is universal beams.)

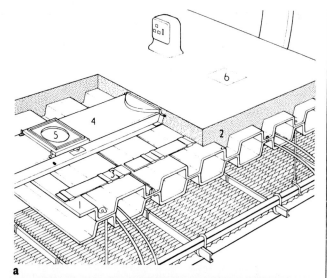

a

b

20 *Steel as permanent shuttering:* **a** *cutaway diagram,*
1 steel flooring, 2 concrete topping, 3 shear straps, 4 header
duct, 5 junction, 6 flush finish junction box; **b** *Holorib*
permanent shuttering

3.04 With steel frame construction a great advantage is the
possibility of a shallow floor depth, but this is not always an
efficient use of the material. The growing need for heavy
floor loadings and service runs makes the use of a false
ceiling at soffit level well worth considering. The traditional
rule of thumb of keeping storey heights to a minimum can
no longer be arbitrarily applied. A deep construction, with
sensible exterior treatment, may well be justified, or a
deep space frame floor, with big clear spans and full facilities for trunking of services could be appropriate.

High-rise structures
3.05 For this discussion, high-rise multi-storey structures
must be sub-divided. Buildings of traditional *proportions*, no
matter how high (ie limited only by the site) lend themselves
(perhaps with greater use of composite construction) to the
treatments outlined for low-rise. Slender buildings are open
to detailed comparative methods and these vary only with
height **21**.
3.06 An arbitrary limit seems to occur at over 100 m; the
UK has few structures higher than this and there is a
tendency to use some of the methods developed for the
US, 'ultra high-rise' **22** on buildings that need not be
considered as really high buildings. Using a sophisticated
frame technique for a building whose height is impressive
only by parochial standards, is as wasteful and self-defeating
as using a space frame to cover a garden shed. In fact it is
negative, because the resulting cost statistics, must be
misleading and make potential users wary.
3.07 Choice of suitable high-rise frame layout will be
determined by:
a, function of large open areas and small compartments

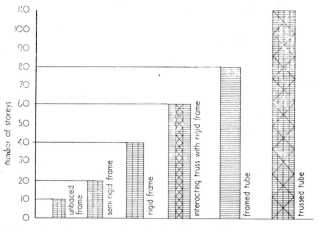

21 *Structural systems related to heights of buildings*

22 *412 m high World Trade Center, New York (architect*
Minoru Yamasaki & Associates, engineers Worthington
Skilling Helle & Jackson)

b, height
c, plan shape, dimensions and height ratio
d, access and distribution of personnel and services,
e, ground conditions and basement layout
f, loading—normal, exceptional, dynamic, earthquake.
(Loading conditions are given in BS CP 3, Chapter V 1967:
Part 1 deals with gravity loads, Part 2 with wind loads.)
3.08 Although there is much data on wind loading, the
design of high rise structures is little influenced by margin-
ally more or less wind load; if common sense is used, little
can be gained by refinement of wind loading. More attention
should be paid to local environmental effects on people of
wind velocities, wind noises and weather.

Beam and column portal frame
3.09 The most common design for single bay structures is
the portal frame. They may be either *simple design* or *fully
rigid design*. Generally the fully rigid system gives more
slender members, but greater workmanship at the joints,
and improved site operations are required **23**.

23 *Single bay portal, with cantilever beams. Now that analysis has reduced steel content to an absolute minimum thought must be given to methods of erection. With this (Dutch) method care must be taken to brace such slender frames during erection. Permanent bracing, a considerable proportion of frame costs, must be allowed for (Architects Groosman NV, Engineer Corsmit)*

Beam and column frame—braced panels/centre core

3.10 A semi-rigid analysis which simplifies fabrication can be applied to this common form with its braced panels and central core. Side forces are taken through the floor system either to vertical shear walls at the ends of the structure, or to a central core. To reduce tension on the whole building symmetry of plan form is desirable; the floor system must be substantial enough to take these additional secondary stresses which can be difficult to evaluate. In the UK the centre core is often reinforced concrete using rapid methods of construction. A steel frame 'core' could lead to even faster construction, and might also simplify services installation **24**.

Staggered truss system

3.11 The staggered truss is an increasingly used framework, and is particularly suited to flats and hotels where partitions are in known locations. Lattice beams can eliminate costly and unsightly junctions between floors and walls; deep, long span joists and slender columns may be completely hidden (and fireproofed) in the thickness of 150 mm partitions.

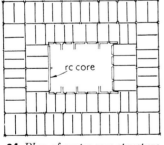
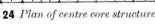

24 *Plan of centre core structure*

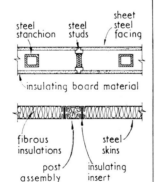

25 *Sections through partition showing structure hidden*

3.12 Using a storey-height girder, with a Vierendeel panel in the centre for an access corridor, the full width of a building up to 20 m can be spanned. But it is not necessary to have such deep beams on all column lines; alternate floors may be carried on top and bottom boom **26** giving a dramatic reduction in steel content.

27 *Systems for very high buildings by F. R. Khan* SOM *chief structural engineer;* **a** *rigid frame with shear truss interacting;* **b** *same with horizontal stiffening truss;* **c** *framed tube structure;* **d** *diagonal truss tube system;* **e** *column diagonal truss tube system*

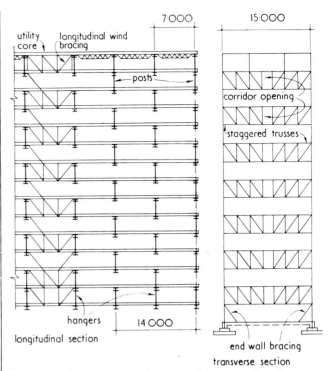

26 *Staggered arrangement of trusses: this means that floor plank need span only 7 m, yet 14·5 m clear space is provided between trusses;* **a** *longitudinal section,* **b** *transverse section*

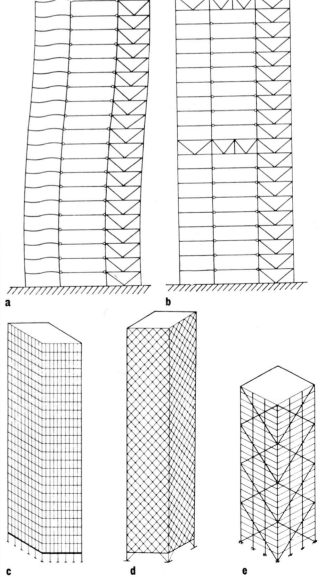

Ultra high-rise structures

3.13 Various structural layouts exist for buildings over 100 m high. It is said that the UK cannot sustain such structures, even that soil conditions are inadequate; but such arguments have been shown invalid in the past. A brief look at the ultra-high rise situation may be useful against the day when UK development requires (and socially accepts) these greater heights **27**.

3.14 The recently developed composite framed tube system fits somewhere in this category, probably at about 40 to 50 storeys **28**. At such heights a frame clad by surface-acting steel panels (stainless or plastic coated) might have some merit **29** but this is at the very limits of present engineering understanding.

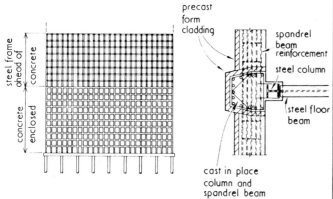

28 *Framed tube composite system*

29 *Not 'surface active', but it might be* (GLC)

3.15 Suspended structures **30** come into their own on truly monumental buildings, perhaps larger than yet built anywhere in the world. Meanwhile 'total' suspension may remain a novelty, but lessons being learned from US suspended buildings already built show the skilful designer various ways of incorporating the benefits of suspension into otherwise normal construction. These include the ability to hide vertical members in very small spaces **25**, and the subsequent ability to produce non-linear shapes for very little extra cost. An optimum and distinctive shape and style of suspended multi-storey building has yet to evolve; this is unlikely while the rectangular box prevails **32**.

3.16 The value of suspension from a central core is limited, for there is a penalty to be paid in the cost of the roof level works. In 'ultra high rise' work, a planning demand for two, three or more service and access cores can be foreseen; here the suspended form could prove its worth.

a

b

c

30a *Suspended structure—is this pointing the way? Although only 12 suspended floors it demonstrates that cables may adequately be used as suspension members replacing exterior ties of solid steel;* **b** *cables (no more than 300 mm diameter each) are easily hidden in fascia or internal partitions;* **c** *stiffness of the curtain is obtained by forming simple trusses within the opaque depth of the wall. (Westcoast Building, Vancouver, Canada; engineer Bogue Babicki & Associates)*

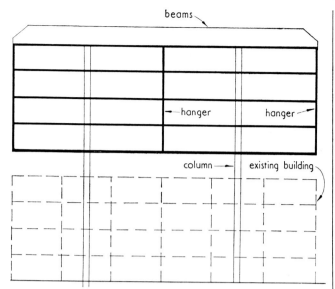

31 *One suggested method of increasing the height of buildings without adding to existing loads on the members*

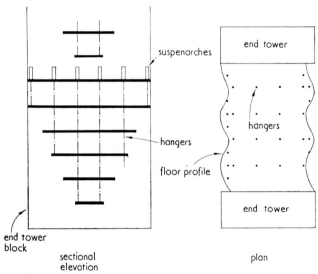

32 *Use of hidden suspension members to provide variable floor areas and freedom for wall shapes*

4 Construction

The steel erector

4.01 The selection of the best construction technique for any project is vital before the final stage of design. Just as the brief description of the steel making processes in Technical study STEEL 1 provided a useful background against which to make design decisions, so a consideration of the form of contract is appropriate here. Contractual and commercial interaction between main contractors, sub-contractors, suppliers, and stockholders has played a significant part in the evolution of steel structures. Historically in the UK the steel frame was generally a sub-contract. As the frame cost element grew, so did the sub-contractors; but (unlike their counterparts in the US) they did not aspire to main contracting. This worked well as long as the design and erection processes held some magic. But with the post-war move towards reinforced concrete structures main contractors could erect (rc) structures themselves. This gave them the double advantage of being in control of events on site, and not having to give away profit on a large fraction of the work.

4.02 Such steel, sub-contractor methods and mentality persist, but they are changing as new structural steel techniques become available. Meanwhile main contractors,

offering management and design packages, are starting to consider structural steel frames, and thus give steel the use and development it has lacked to date because of the structure of the industry.

4.03 Steel structures can still, as in the beginning, be the lightest and fastest form of construction, imposing the least load upon foundations. Also they can now be produced in exotic and dramatic forms as well as everyday structures. However, market forces—particularly in skeletal structures—may easily overwhelm technological considerations, and an architect must consider the following constraints.

Site conditions

4.04 Site influences structural choice in two ways: geologically and geographically. Steel structures—capable of being fully rigid, fully flexible or anywhere between these extremes—enable the designer to produce the structural form most suited to specific ground properties. Early investigation of the interaction between structure and foundations can always show an overall benefit to the project.

4.05 Location is critical. A steel structure is particularly suited in in-fill sites in city centres; members may be fabricated and transported to site in off-peak hours, if there is no site storage, only enough for one shift at a time need be delivered; costs are greatly influenced by transportation between fabricating shop and site, less so by distance between rolling mill, or stockist, and the fabricator. Architect and engineer should carefully consider available routes—road junctions, sharp corners, overhead hazards, provision of police escorts, maximum load signs and bridge weight restriction and even lay-bys for overnight stops. All such significant details should be known to the sub-contractor, for these might have a determining effect on the detail design.

Contractor

4.06 The contractual method should be considered at an early stage. The steelwork contractor must have adequate facilities available at contract date to handle the work. Ample setting out space, cranes, machines, cutting, welding, supervision, transportation, stockyards and experience are essential; sophisticated equipment is no substitute for good production layout and management expertise.*

Proprietary systems

4.07 Although simple beams, columns or purlins may be taken from catalogues and safe load tables, with made-up fabrications, trusses or portal frames, an architect must require calculations to be submitted for local authority approval before acceptance. A number of proprietary systems are available for houses, schools, or hospitals and here an architect should endeavour to evaluate past experience, bearing in mind his own site and loading conditions. It is seldom competitive to adapt a system designed for one particular application to another, especially with differing loadings; smallest variations can have widely disproportionate effects; purpose-designed structures are usually more suitable. This is equally true where clients intend to provide for change of use; foresight can protect an inexperienced client from future problems and expense.

4.08 An architect should ascertain that his chosen system was designed by a chartered engineer, and that subsequent modifications have kept within the spirit and letter of the original concept. With a proprietary structural system for the superstructure, an independant engineer is often still

* It may be to the client's advantage, when looking for a steel contractor to go a little further (than required by contract or tradition) toward the European approach of interim payment, whereby the sub-contractor does not have to carry such heavy risk and on-costs as in present UK practice.

needed to design foundations and ancillary works, and this engineer must comprehend the system used; cheap superstructure needing expensive foundations is a bad buy. For overall safety and economy an independent engineer should be retained to check any system-structure. Even architects in multi-discipline consortia could usefully ask their engineer colleagues to consult a specialist engineer with particular experience before embarking upon any of the more elaborate structural forms discussed.

Package deal

4.09 Design and build contracts, package deals, 'Turnkey projects', etc are all amenable to structural steel projects—provided always that a chartered engineer is in a position to influence safety and stability. Whether employed by the main contractor or as an independent consultant, the engineer's services are only available to the architect, and his client, through the contract; and this can limit preparation of work in its formative stages. (With these forms of contract the architect should be satisfied as to the indemnity insurance of engineer and contractor for the particular contract under consideration.)

4.10 Traditionally a main contractor is selected either by nomination or by competitive tendering. Competitive tenders may be based on Bills of Quantities (based on detail drawings) or a lump sum based on outline proposals only. The latter should not be too sketchy and an attempt should be made to include rates for the type of work involved in the steel sub-contract, even if actual tonnages and details are not known. Equally, engineers details should not be too rigidly defined, and should leave open the possibility of the sub-contractor suggesting alternative methods that may suit him better.

4.11 The design engineer is today becoming more closely involved in erection procedures, previously the domain of the steelwork contractor (with engineers content for it to be so) but with greater design sophistication the erection process has become more vital to economic success and to stability itself, and its importance will grow as structures become more refined. An erector can no longer look at a detail and know from experience that it is acceptable.

4.12 The erector must be educated in the idiosyncrasies of new methods, and until this training is common the design engineer plays a vital part in erection methods. (This may mean an increase in engineer's fee to draw up or approve methods, and a greater application of the present additional fee permitted for site supervision.) Only the design engineer will be familiar enough with it to supervise and guide erection; reliance on statutory authorities or other qualified engineers cannot replace this knowledge.

4.13 The designers' overall concept—description of building-function, site and ground conditions, programme of erection and bracing, provision of joints, methods of fixing, methods of protection, description of the structural analysis and loading, materials, special to tolerances, and services—must be passed to all concerned with the projects. Perhaps this should be bound and stored for the future use of demolition contractors.

Loading conditions

4.14 Immediately after erection and bracing a steel frame is loadbearing. Thus it can support a crane capable of lifting very heavy loads to help construct secondary stages. The steel skeleton fits easily around temporary raking or flying shores and facilitates their early removal. Difficult shapes of load and structure are easily catered for, and steel structures are particularly suited to carrying the heaviest industrial loads, both active and passive.

Services

4.15 In these days of extensive and costly services, architect and engineer must finely balance the requirements of services and structure; steel is a very flexible medium, lending itself to modification in those cases where pre-planning goes adrift. Ease of fixing and rectifying manufacturing errors at site allows variations to a member, although a designer's normal objective may be for repetition and mass production.

4.16 The steelwork sub-contractor, normally controls fabrication; he prepares programmes, drawings and such calculations as are needed for individual connections of members. He also chooses erection methods, geared closely to the overall contract schedule.

Construction techniques

4.17 Riveting—other than light fixing by various 'blind' rivet systems—is no longer used; even these are in competition with shot fired fixings, self-tapping screws, stud welding, spot-welding etc. Riveting has been replaced by automatic welding processes in the shop, and high strength bolts on the site. Site welding can be used for many applications, particularly with hollow sections, since uniformity of wall thickness of metal and greater accuracy of rolling tolerances lend themselves to good welding techniques, even in the highest grade steels.

Welding

4.18 The general reluctance to use site welding is related to the need for inspection—often of a joint inaccessible when finished. Site welding will become more competitive if such problems can be overcome by pre-planning. Generally small jobs should not be site welded since contract price cannot sustain an adequate welding or inspection system. Larger contracts with provision for periodic testing of welders' capabilities, protection of operative and welding in temporary heated enclosures, pre-heating of the parts to be joined, and even site facilities for annealing heavy welded sections, are well suited to site welding.

4.19 Many organisations site test finished work by non-destructive methods such as ultra-sonics and gamma-rays. Provided that this is part of the initial contract, and not regarded as extras, welded site connections with their vastly superior appearance need present no problems, and the occasional sample weld, cut out of the work and tested in the laboratory, will be satisfactory.

Bolts

4.20 High strength friction grip bolts have largely replaced all others for both shop and site fixings—they have reduced the size of connection and the number of bolts. A torque wrench applies the load to the bolt, and the strength of the connection is enhanced by added induced shear strength between members. While overtightening must be avoided, a number of devices are available to ensure this, and provide simple checks. In welded structures, a shop-welded connection plus site bolting can often give a more expensive (and cumbersome) finished joint than would be provided by full site welding. Bolts, holes, flanges are all expensive items and for economy should be kept to a minimum.

Proprietary members

4.21 Many fabricating shops have specialist capability and these are worth uncovering. It may, for example, be castellated beams—produced by separating the top and bottom flanges of a universal beam and rewelding—or some portal frames whose tapered sections are produced by similar techniques.

Technical study
Steel 3

Section 6 **Structural material: steelwork**

Simple steelwork design

The steelwork section concludes with an account of simple design which an architect can undertake—a specific beam and column example is worked through. GORDON ROSE *adds an analysis of a horizontal truss which an architect should at least understand and the increasing limits on his scope for structural design without engineering assistance are briefly discussed*

1 Introduction

1.01 Many of the techniques described in Technical studies STEEL 1 and STEEL 2 need the full knowledge, experience and attention of the practising structural engineer, and his specialist consultants, to realise. The amount of 'design' that can be attempted by the non-specialist is rapidly reducing, and even where safe, it may today be grossly inefficient. There are nevertheless occasions when an architect, builder, or surveyor, will wish to design for himself; the following notes are for him, in the hope that he will be then able to satisfy his local authority.

1.02 Before going further a word of warning must be given to all those who over the years have come to rely on a favourite 'steel handbook', issued by one of many old steel companies whose names are now history. Even if full of invaluable tit-bits, well-thumbed and falling open of themselves at just the right place, their time has passed; designers must move to the most up-to-date replacements possible. Of these the most important are the *Handbook on structural steelwork* and its *Supplement*, published jointly by BCSA and CONSTRADO[1], and *Structural hollow sections. Properties and safe load tables*[2], published by BSC tubes division. Metrication, improved shapes of section and recent drastic alteration of loading and material standards has completely invalidated older handbooks; their use will result in disappointment at best and structural failure at worst.

Metrication
1.03 For some years tables and documents for structural steelwork will be metric equivalents of imperial sizes, while the steel rolled sections themselves remain imperial; thus a 406 × 178 UB 54 (406 mm deep, 178 mm wide, and weighing 54 kg per metre) is still the well known 16 × 7 UB 40.

1.04 Similarly many BS stresses and loads are conversions from imperial. All this can at times lead to very cumbersome numbers; these may be no trouble when using computers, but are difficult to handle manually, eg the moment of inertia (I), of the above section is now given as 21 520 cm⁴. Sensible rounding-off at an early stage will prevent undue work and lessen the possibility of error. (Approximate metric conversions for use in steel calculations are given in a brief appendix to this study.)

2 Examples

2.01 Design procedure for the simplest situation — a simple beam and the column which supports it **1**.
Dead load = $2 \cdot 0$ kN/m²
Live load = $1 \cdot 5$ kN/m²
Calculate load on beam A:
Dead load = $8 \cdot 5 \times 5 \cdot 0 \times 2 \cdot 0$ = $85 \cdot 0$ kN
Estimate self weight of beam at 60 kg/m, plus
casing at 130 kg/m = $8 \cdot 5 \times 190$ = $16 \cdot 2$ kN
Live load = $8 \cdot 5 \times 5 \cdot 0 \times 1 \cdot 5$ = $63 \cdot 8$ kN
                                                $\overline{165 \cdot 0}$ kN

1 *Plan showing beam* A *and column* B

2.02 Reference to p197 of the BCSA/CONSTRADO *Handbook* **2** shows that a 406 × 178 UB 60, grade 43, will carry a safe uniformly distributed load of 155 kN (including its own uncased weight) over a span of $9 \cdot 0$ m.
Check to see whether it is adequate in this actual lesser span of $8 \cdot 5$ m.

Maximum bending moment = $\dfrac{WL}{8} = \dfrac{165 \cdot 0 \times 8 \cdot 5}{8} = 175$ kNm

2.03 The properties of 406 × 175 × UB 60, grade 43 steel are given in the *Handbook* p43 **3** as:
$Z = 1059$ cm³
$I = 21\,520$ cm⁴ (= say $21 \cdot 5 \times 10^7$ mm⁴)

Actual $f = \dfrac{M}{Z} = \dfrac{175}{1059} \times 1000 = 165$ N/mm²

2.04 Allowable bending stress p_{bc}, grade 43 steel = 165 N/mm² provided adequate lateral restraint of the top flange is available. A cased section built into a reinforced concrete floor, for example, is obviously well restrained; a similar section carrying, say, timber joists, may not be. BS 449[3] gives examples of forms of restraint, and where a beam flange is unsupported, the permitted working stress is

UNIVERSAL BEAMS

SAFE LOADS FOR GRADE 43 STEEL

Serial size	Mass per metre	Safe distributed loads in kilonewtons for spans in metres and deflection coefficients													Critical span Lc
		2.00	2.50	3.00	3.50	4.00	4.50	5.00	5.50	6.00	7.00	8.00	9.00	10.00	
mm	kg	112.0	71.68	49.78	36.57	28.00	22.12	17.92	14.81	12.44	9.143	7.000	5.531	4.480	m
457 × 191	98	†1066	1033	861	738	645	574	516	469	430	369	323	287	258	4.00
	89	†983	935	779	668	584	519	467	425	389	334	292	260	234	3.86
	82	†911	851	709	608	532	473	426	387	355	304	266	236	213	3.74
	74	†832	771	643	551	482	429	386	351	321	276	241	214	193	3.65
	67	†771	684	570	489	428	380	342	311	285	244	214	190	171	3.53
457 × 152	82	†995	822	685	587	514	457	411	374	343	294	257	228	206	3.03
	74	†913	742	619	530	464	412	371	337	309	265	232	206	186	2.92
	67	†825	660	550	471	413	367	330	300	275	236	206	183	165	2.81
	60	†728	591	493	422	370	329	296	269	246	211	185	164	148	2.78
	52	626	501	418	358	313	278	251	228	209	179	157	139	125	2.62
406 × 178	74	†801	699	583	499	437	388	350	318	291	250	218	194	175	3.65
	67	†721	627	523	448	392	348	314	285	261	224	196	174	157	3.54
	60	†634	559	466	399	349	310	279	254	233	200	175	155	140	3.45
	54	611	489	407	349	305	271	244	222	204	174	153	136	122	3.28
406 × 140	46	513	411	342	293	257	228	205	187	171	147	128	114	103	2.58
	39	414	331	276	236	207	184	165	150	138	118	103	92	83	2.41
356 × 171	67	†662	567	472	405	354	315	283	258	236	202	177	157	142	3.72
	57	†574	473	394	338	296	263	237	215	197	169	148	131	118	3.50
	51	†519	420	350	300	263	234	210	191	175	150	131	117	105	3.38
	45	453	363	302	259	227	201	181	165	151	130	113	101	91	3.23
356 × 127	39	377	302	252	216	189	168	151	137	126	108	94	84	75	2.33
	33	311	248	207	177	155	138	124	113	104	89	78	69	62	2.18
305 × 165	54	†479	398	331	284	249	221	199	181	166	142	124	110	99	3.69
	46	†412	342	285	244	214	190	171	155	143	122	107	95	86	3.53
	40	370	296	247	212	185	165	148	135	123	106	93	82	74	3.38
305 × 127	48	404	323	269	231	202	180	162	147	135	115	101	90	81	2.59
	42	351	280	234	200	175	156	140	127	117	100	88	78	70	2.45
	37	311	249	207	178	156	138	124	113	104	89	78	69	62	2.37
305 × 102	33	274	219	183	156	137	122	110	100	91	78	68	61	55	1.90
	28	232	185	154	132	116	103	93	84	77	66	58	51	46	1.79
	25	190	152	127	109	95	84	76	69	63	54	47	42	38	1.64
254 × 146	43	333	267	222	191	167	148	133	121	111	95	83			3.41
	37	286	229	191	164	143	127	115	104	95	82	72			3.22
	31	233	186	155	133	117	104	93	85	78	67	58			2.96
254 × 102	28	203	163	135	116	102	90	81	74	68	58	51			2.01
	25	175	140	117	100	88	78	70	64	58	50	37			1.87
	22	149	119	99	85	74	66	66	54	50	43	37			1.75
203 × 133	30	184	147	123	105	92	82	74	67	61	53				3.03
	25	153	122	102	87	77	68	61	56	51					2.80

Loads printed in italic type do not cause overloading of the unstiffened web, and do not cause deflection exceeding span/360.

† Load is based on allowable shear of web and is less than allowable load in bending. See also footnotes to page 196.

Loads printed in ordinary type should be checked for deflection See page 15.

2 Handbook on structural steel Supplement *p197, with relevant section circled. The two top rows (bold type) are: upper, span and lower, deflection coefficient*

UNIVERSAL BEAMS

DIMENSIONS AND PROPERTIES

Serial size	Moment of inertia			Radius of gyration		Elastic modulus		Ratio D/T
	Axis x–x Gross	Net	Axis y–y	Axis x–x	Axis y–y	Axis x–x	Axis y–y	
mm	cm⁴	cm⁴	cm⁴	cm	cm	cm³	cm³	
457 × 191	45717	40615	2343	19.10	4.33	1956	243.0	23.9
	41021	36456	2086	18.98	4.28	1770	217.4	26.3
	37103	32996	1871	18.84	4.23	1612	195.6	28.8
	33388	29698	1671	18.75	4.19	1461	175.5	31.6
	29401	26190	1452	18.58	4.12	1296	152.9	35.7
457 × 152	36215	32074	1143	18.62	3.31	1557	149.0	24.6
	32435	28744	1012	18.48	3.26	1406	132.5	27.1
	28577	25357	878	18.29	3.21	1250	115.5	30.6
	25464	22611	794	18.31	3.23	1120	103.9	34.2
	21345	19035	645	17.92	3.11	949.0	84.6	41.3
406 × 178	27329	24062	1545	16.96	4.03	1324	172.0	25.9
	24329	21425	1365	16.87	4.00	1188	152.7	28.6
	21508	18934	1199	16.82	3.97	1058	134.8	31.8
	18626	16457	1017	16.50	3.85	925.3	114.5	37.0
406 × 140	15647	13765	539	16.29	3.02	777.8	75.7	36.0
	12452	11017	411	15.88	2.89	626.9	58.0	46.0
356 × 171	19522	17045	1362	15.12	3.99	1073	157.3	23.2
	16077	14053	1109	14.92	3.92	896.5	128.9	27.5
	14156	12384	968	14.80	3.87	796.2	112.9	30.9
	12091	10609	812	14.57	3.78	686.9	95.0	36.2
356 × 127	10087	9213	357	14.29	2.69	571.8	56.6	33.1
	8200	7511	280	14.00	2.59	470.6	44.7	41.0
305 × 165	11710	10134	1061	13.09	3.94	753.3	127.3	22.7
	9948	8609	897	13.00	3.90	647.9	108.3	26.0
	8523	7384	763	12.86	3.85	561.2	92.4	29.9
305 × 127	9504	8643	460	12.50	2.75	612.4	73.5	22.2
	8143	7409	388	12.37	2.70	531.2	62.5	25.4
	7162	6519	337	12.28	2.67	471.5	54.6	28.4
305 × 102	6487	5800	193	12.46	2.15	415.0	37.8	29.0
	5421	4862	157	12.22	2.08	351.0	30.8	34.8
	4387	3962	120	11.82	1.96	287.9	23.6	44.6
254 × 146	6558	5706	677	10.91	3.51	505.3	92.0	20.4
	5556	4834	571	10.82	3.47	434.0	78.1	23.4
	4439	3879	449	10.53	3.35	353.1	61.5	29.1
254 × 102	4008	3569	178	10.52	2.22	307.9	34.9	26.0
	3408	3046	148	10.29	2.14	265.2	29.0	30.8
	2867	2575	120	10.04	2.05	225.7	23.6	37.2
203 × 133	2887	2476	384	8.72	3.18	279.3	57.4	21.5
	2356	2027	310	8.54	3.10	231.9	46.4	26.0

In calculating the net moment of inertia, each flange of 300 mm or greater width is reduced by two holes, and each flange less than 300 mm wide by one hole. See page 10.

3 Handbook Supplement *p43*

reduced by a percentage dependent on the radius of gyration of the section, span, depth of section, effective flange thickness and end restraint conditions of that flange. The radius of gyration (r), a geometric property of a section (see Technical study ANALYSIS 2 para 2.06) is given in the *Handbook* 3 as 3·97 cm for a 406 × 178 UB 60.

2.05 In 3, $\dfrac{D}{T}$ is given as 31·8 (where the section depth is D and effective flange thickness T). The actual length of flange is 8·5 m but if partial restraint of the ends is assumed, BS 449 (clause 26a) allows 0·85 × span, therefore effective flange length = 8·5 × 0·85 = 7·2 m

Slenderness ratio $= \dfrac{l}{r_y}$; effective flange length $l = 7·2$ m

radius of gyration $r_y = 3·97$ cm $\therefore l/r_y = \dfrac{7·2 \times 100}{3·82} = 181$

Were the beam flange deemed to be unsupported then permissible bending stress units p_{bc} (from table 3a, BS 449, **4**) would be 76 N/mm². The top flange must therefore be effectively restrained.

2.06 Note that there is no top flange support requirement on structural hollow sections used as beams, provided the depth-to-breadth ratio of the whole section is not greater than 4. Shear considerations therefore rule, and safe load tables are available for RHS beams up to 254 × 152·4 and 7·0 m spans. The treatment of combined stresses in long span and larger size RHS is too complex to discuss here.

Deflection (δ)

2.07 Deflection must not exceed $\dfrac{\text{Span}}{360} = \dfrac{8·5 \times 1000 \text{ mm}}{360} = 23$ mm

With a live load of 63·8 kN (63·8 × 10³ N) and E, Young's modulus, of 2·1 × 10⁵ N/mm²

$$\delta = \frac{5WL^3}{384EI} = \frac{5\,(63·8 \times 10^3) \times (8·5 \times 10^3)^3}{384 \times (2·1 \times 10^5) \times (21·52 \times 10^7)}$$
$$= \frac{5 \times 63·8 \times 8·5^3 \times 10^{12}}{384 \times 2·1 \times 21·52 \times 10^{12}}$$
$$= \frac{5 \times 8·5^3 \times 63·8}{384 \times 2·1 \times 21·52} = 11·3 \text{ mm, therefore acceptable.}$$

TABLE 3a. ALLOWABLE STRESS p_{bc} IN BENDING (N/mm²) FOR BEAMS OF GRADE 43 STEEL

l/ry	D/T							
	10	15	20	25	30	35	40	50
90	165	165	165	165	165	165	165	165
95	165	165	165	163	163	163	163	163
100	165	165	165	157	157	157	157	157
105	165	165	160	152	152	152	152	152
110	165	165	156	147	147	147	147	147
115	165	165	152	141	141	141	141	141
120	165	162	148	136	136	136	136	136
130	165	155	139	126	126	126	126	126
140	165	149	130	115	115	115	115	115
150	165	143	122	104	104	104	104	104
160	163	136	113	95	91	91	94	94
170	159	130	104	91	85	82	82	82
180	155	124	96	87	80	76	72	71
190	151	118	93	83	77	72	68	62
200	147	111	89	80	77	72	64	59
210	143	105	87	77	70	65	61	55
220	139	99	84	74	67	62	58	52
230	134	95	81	71	64	59	55	49
240	130	92	78	69	61	56	52	47
250	126	90	76	66	59	54	50	44
260	122	88	74	64	57	52	48	42
270	118	86	72	62	55	50	46	40
280	114	84	70	60	53	48	44	39
290	110	82	68	58	51	46	42	37
300	106	80	66	56	49	44	41	36

Intermediate values may be obtained by linear interpolation.

NOTE. For materials over 40 mm thick the stress shall not exceed 150 N/mm².

4 *Table 3a*, BS 449. *Allowable stress in bending for beams. Relevant section circled*

2.08 A preliminary guide to the deflection may have been gained by use of the deflection coefficient C (see **2**), in this case 5·531 for a span of 9·00 m.
The load required to produce a maximum deflection of

$$\frac{\text{span}}{360} = \frac{C \times I \text{ (in cm}^4\text{)}}{1000} \text{ kN}$$

$CI = 5\cdot531 \times 21\cdot52 = 119$ kN

It would have been reasonable to expect a load of 63·8 kN on a somewhat shorter span to be well within the allowable deflection. Under normal conditions of loading and end support, this 406 × 178 UB 60 will be acceptable. Occasionally these matters will necessitate investigation, especially for point loading or narrow supports, when the following checks should be made.

Shear

2.09 The allowable shear stress in an unstiffened web of a rolled section of grade 43 steel is 100 N/mm². If reaction

$= \dfrac{165}{2} = 82\cdot5$ kN; depth of web $= 406$ mm, and web

thickness $= 7\cdot8$ mm, then shear stress $p_{q'} = \dfrac{82\cdot5 \times 1000}{406 \times 7\cdot8} =$

26 N/mm², therefore acceptable.

Web buckling

2.10 The web of a joist may buckle **5b** under the action of a concentrated load or support; this must be restricted to
$W = p_c t B$
where p_c is the axial allowable stress as given in clause 30a (BS 449); W is permissible reaction under concentrated load on unstiffened web, and t is web thickness.

$B = \dfrac{D}{2} + t_p + l_b$ for end bearings of simply supported

beams or $D + 2t_p + l_b$ for intermediate bearings or concentrated loads (where D is overall depth, t_p is thickness of bearing and/or flange plate if any, and l_b is length of stiff portion of bearing, as defined in clause 28a (I) of BS 449). If beam A were resting on a wall rather than being jointed into a beam, web buckling would be a possible mode of failure. If the length of bearing were, say, 100 mm, the buckling load of the 406 × 178 UB 60, using a 20 mm thick flange plate, would be:
$= C_j + t_p C_p + l_b C_b$ where C_j is beam component, C_p flange plate component and C_b stiff bearing component. The figures for a 406 × 178 UB 60 are given on p260 of the *Handbook*. Therefore buckling value (W)
$= 165\cdot33 + 20 \times 0\cdot75 + 100 \times 0\cdot75$
$= 165\cdot33 + 15 + 75 = 255$ kN

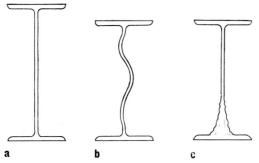

5 a *Beam; **b** buckled; **c** crushed in*

Web crushing

2.11 The web may crush **5c**, and hence the load W must also be less than $C_j + t_p C_p + l_b C_b$. The possibility of combined bending and shear, or combined bending, shear, and bearing must also be considered, although it is unlikely to be critical for simply supported flanged beams under uniform loads. BS 449 gives allowable stresses for such cases, and experimental data is available for structural hollow sections. In beam A therefore, the permissible direct bearing value to prevent web crushing from p261 of the *Handbook*,
$= 62\cdot89 + 20 \times 2\cdot57 + 100 \times 1\cdot48$
$= 61\cdot87 + 51\cdot4 + 148 = 272$ kN
Thus the maximum allowable load on the web is 255 kN. The beam reaction is 82·5 kN, thus a 20 mm thick flange plate with a stiff length of bearing of 100 mm is more than adequate, and—from this point of view—could be reduced **6**.

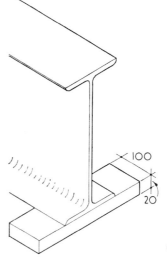

6 *Flange plate*

Column design

2.11 The action of a column is greatly influenced by the degree of fixity provided at the top and bottom. BS 449 devotes some 15 pages to drawings of various arrangements that give different 'effective lengths'. This effective length may vary from 0·7L for a completely fixed condition, to 2L for a virtual cantilever. The maximum allowable stress is related to the ratio L/r (r = radius of gyration; L = effective length), r is given in the handbooks, and for all sections care must be taken to relate the particular r about one axis to the correct effectiveness length for that same axis.

2.12 The *Handbook on structural steelwork* gives values for column loading under direct compression for cased and uncased sections, and separate tables are available for uncased structural hollow sections. Neither of these give values for combined bending with axial compression, but a new handbook gives such values for concrete filled hollow sections. Combined loading is dealt with as follows:

If bending is on one axis only, then $\dfrac{f_c}{P_c} + \dfrac{f_{bc}}{P_{bc}} \not> 1$.

Or if bending occurs on both xx and yy axes at the same time,

$\dfrac{f_c}{P_c} + \dfrac{f_{bcx}}{P_{bcx}} + \dfrac{f_{bcy}}{P_{bcy}} \not> 1$.

To design the column it is therefore necessary to determine the working stresses f_c and f_{bc} and the allowable stress p_c and p_{bc}.

2.13 Working direct and bending stresses f_c and f_{bc} are determined quite easily: f_c is simply the direct load on the column (W) divided by the area of the proposed section (A); and f_{bc} is the bending moment (M) applied by the method of connection of the beam or truss divided by the modulus (Z) of the proposed section obtained from the section table. M is found by multiplying the load on the column by the eccentricity resulting from the method of connection.

$f_c = \dfrac{W}{A}$

$f_{bc} = \dfrac{M}{Z}$ where $M = W \times$ eccentricity (e).

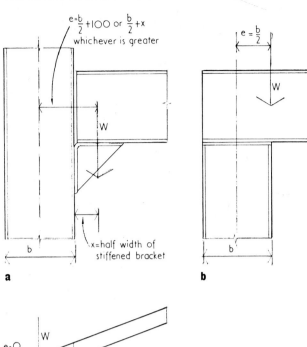

a

b

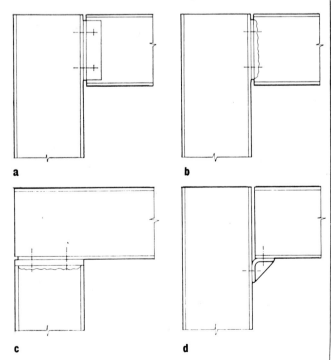

c

5 *Eccentricity:* **a** BS 449 *here defines eccentricity as* $\left(\dfrac{b}{2} + 100\right)$
or
$\left(\dfrac{b}{2} + x\right)$ *whichever is the greater;* **b** *here it is defined as* $\dfrac{b}{2}$.
In the case of **c** *no eccentricity needs to be taken*

a

b

c

d

6 *Connections capable of transmitting only shear from beam to column:* **a** *angle web cleat;* **b** *welded web plate cleat;* **c** *welded plate cap;* **d** *seating cleat*

The eccentricity is defined in BS 449 as the greater of two dimensions, 100 mm outside the face of the section or at the centre of the bearing of the beam. In cap connections the load may be assumed to be at the face of the section and in simple roof truss bearings no eccentricity needs to be considered **5**.

Encased columns
2.14 If an I-section column is bound with 150 mm mesh and cased in 1:2:4 concrete with 50 mm cover to the flanges, the radius of gyration about the axis in the plane of the web may be taken as: 2(b + 100), **7**.

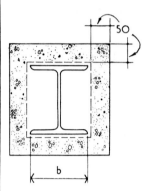

7 *Encased column*

Simple connections from beam to column
Types of connection
2.15 The connections shown in **6** are suitable for transmitting shear only from the beam to the column.

Angle web and welded web plate cleats
2.16 The welded web plate is a neater version of the bolted angle web cleat. The latter, although not dealt with in *Handbook on structural steelwork*, is detailed in standard form to suit a variety of beam depths with appropriate loads in the publications of the main steel suppliers and fabricators. These usually also include tables from the *Handbook on structural steelwork*. The welded plate detail can be interpreted from these same standard forms.

Seating cleat
2.17 The seating cleat **6d** may be required by certain authorities and in this case the web buckling and web crushing effect on the beam must be considered (see beam design example above). The load cannot be shared between the seating cleat and the web unless the web is secured to the column by close tolerance bolts.

Welded plate cap
2.18 The welded plate cap **6c** presents a neat solution where the beam can be carried over the column. The load can be transmitted to the column by fillet welding of the cap plate to the inner face of the flange and web.

Column connection to foundation
The slab base
2.19 For axial loads the simplest connection is the slab base fillet welded or butt welded to the column **8**. The steel

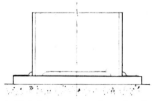

8 *Slab base connection of column to foundation*

section handbooks mentioned above tabulate safe axial loads on various sizes and thicknesses of base slabs for universal columns. The base slab can be chosen to spread the load to the allowable intensity appropriate to the grade of concrete under the slab. (See section 4 FOUNDATIONS of this handbook.)

Comparisons of column methods for direct load only

2.20 Let total direct load on column B be; say, 450 kN; if storey height = 3·0 m and effective length (height) = 0·85 × 3·0 = 2·55 m. Using grade 43 steel there is a choice of:

a) UC section = 152 × 152 I 30 (from p272 of *Handbook of structural steelwork supplement* **9**)

b) UB section = 254 × 146 I 37 or 305 × 127 I 42

c) Cased UC section = 260 × 260 (152 × 152 I 23)

d) Cased UB section = 310 × 240 (203 × 133 I 25)

or   = 370 × 210 (254 × 102 I 28)

e) Circular (CHS) = 139·7 × 32·0

or   = 168·3 × 25·2

(from p34 of *Constrado Handbook*, reference **2**)

Serial Size in mm	Mass per metre in kg	SAFE CONCENTRIC LOADS IN KILONEWTONS FOR EFFECTIVE LENGTHS IN METRES													
		1.0	1.5	2.0	2.5	3.0	3.5	4.0	4.5	5.0	6.0	7.0	8.0	9.0	10.0	11.0
254 × 254	167	3164	3100	3036	2980	2905	2809	2686	2537	2363	1977	1610	1307	1070	886	744
	132	2497	2445	2394	2348	2286	2206	2104	1981	1838	1526	1236	1000	817	676	567
	107	2033	1991	1949	1910	1858	1790	1704	1600	1480	1222	986	796	650	537	450
	89	1695	1659	1624	1592	1547	1490	1417	1328	1227	1010	813	656	535	442	370
	73	1381	1351	1323	1295	1259	1210	1149	1075	991	813	653	525	428	354	296
203 × 203	86	1622	1580	1541	1488	1416	1321	1205	1077	950	728	562	443	356		
	71	1341	1306	1273	1229	1168	1087	990	883	777	594	458	360	290		
	60	1116	1086	1058	1020	966	896	812	721	632	481	370	291	234		
	52	977	951	927	892	845	783	708	628	550	418	321	252	203		
	46	865	842	820	789	745	689	622	550	481	365	280	220	176		
152 × 152	37	685	661	6..		429	357	298	249	180						
	30	552	533	4	460	40	341	283	235	197	142					
	23	429	413	3..	.3	253	208	172	144	103						

9 Handbook *p272, with relevant section circled*

Designation						Maximum axial compression in kN for effective length in metres						
Outside diameter D	Thickness t	Mass per metre	Area of section	Radius of gyration	Elastic modulus	0.5	1.0	1.5	2.0	2.5	3.0	3.5
mm	mm	kg	cm²	cm	cm³							
21.3	3.2	1.43	1.82	0.650	0.72	20	7.1	3.3	1.9			
26.9	3.2	1.87	2.38	0.846	1.27	31	15	7.1	4.1	2.6		
33.7	2.6	1.99	2.54	1.10	1.84	35	23	12	7.2	4.7	3.3	2.4
	3.2	2.41	3.07	1.08	2.14	42	27	14	8.4	5.5	3.8	2.8
	4.0	2.93	3.73	1.06	2.49	51	32	17	9.9	6.4	4.5	3.3
42.4	2.6	2.55	3.25	1.41	3.05	47	38	24	15	9.7	6.8	5.1
	3.2	3.09	3.94	1.39	3.59	57	45	28	17	11	8.0	6.0
	4.0	3.79	4.83	1.36	4.24	69	55	33	20	13	9.5	7.0
48.3	3.2	3.56	4.53	1.60	4.80	66	57	40	25	17	12	9.0
	4.0	4.37	5.57	1.57	5.70	81	69	48	30	20	14	11
	5.0	5.34	6.80	1.54	6.69	98	84	57	36	24	17	13
60.3	3.2	4.51	5.74	2.02	7.78	85	78	64	47	33	24	18
	4.0	5.55	7.07	2.00	9.34	104	96	79	57	40	29	22
	5.0	6.82	8.69	1.96	11.1	128	117	95	68	47	34	25
76.1	3.2	5.75	7.33	2.58	12.8	109	104	95	79	61	46	36
	4.0	7.11	9.06	2.55	15.5	135	129	117	97	75	56	43
	5.0	8.77	11.2	2.52	18.6	167	159	144	119	91	68	52
88.9	3.2	6.76	8.62	3.03	17.8	129	125	117	105	88	70	55
	4.0	8.38	10.7	3.00	21.7	160	155	145	130	108	86	67
	5.0	10.3	13.2	2.97	26.2	198	191	179	159	132	104	82
114.3	3.6	9.83	12.5	3.92	33.6	189	184	178	169	155	137	116
	5.0	13.5	17.2	3.87	45.0	260	253	245	232	212	186	158
	6.3	16.8	21.4	3.82	54.7	323	314	304	287	262	229	193
139.7	5.0	16.6	21.2	4.77	68.8	322	315	308	298	284	265	240
	6.3	20.7	26.4	4.72	84.3	400	392	383	371	353	328	297
	8.0	26.0	33.1	4.66	103	502	491	480	464	441	409	368
	10.0	32.0	40.7	4.60	123	617	603	589	569	540	499	448
168.3	5.0	20.1	25.7	5.78	102	391	384	377	369	360	346	327
	6.3	25.2	32.1	5.73	125	489	480	471	462	449	431	407
	8.0	31.6	40.3	5.67	154	614	602	591	579	562	539	509
	10.0	39.0	49.7	5.61	186	756	743	729	714	692	663	624
193.7	5.4	25.1	31.9	6.66	146	487	479	472	464	455	443	428
	6.3	29.1	37.1	6.63	168	566	558	549	540	529	515	497
	8.0	36.6	46.7	6.57	208	713	702	690	679	666	647	623
	10.0	45.3	57.7	6.50	252	880	867	853	839	821	798	768
	12.5	55.9	71.2	6.42	303	1086	1069	1051	1034	1012	983	944
	16.0	70.1	89.3	6.31	367	1362	1340	1318	1295	1267	1228	1177

10 Structural hollow section properties and safe load tables, *p34, with relevant section circled*

f) Rectangular (RHS) = 120 × 120 × 27·9

g) Concrete filled CHS = 114 × 16·8 (from *Design manual concrete filled hollow section steel columns*, published by CIDECT[4]).

The final choice from this wide selection will evolve from the factors discussed earlier in this section of the handbook. Use of higher grades of steel will enable smaller sections to be used, but cost must be checked. The final chosen section may then be checked for combined compression and bending.

3 Design other than simple column and beam

Simple trusses

3.01 As stated in Technical study STEEL 2, it is probable that the simple pitched roof truss that has given decades of useful service will, in the near future, be little used except in standard prefabricated products. Recent loading requirements especially against wind—make it impossible to apply the old simple forms of analysis. Despite the volume of work now attached to pitched roof design, it is still the cheapest form to construct.

3.02 Therefore it will no longer be particularly useful for the architect to know how to design a simple pitched truss. Even in the smaller 'one-off' construction, where a seemingly uneconomic design (up to 100 per cent) could be accepted, the local authority may now demand calculations beyond the architect's ability. In the following example of a flat truss **11**, where the wind loading requirements are less complicated to handle, the loadings nevertheless need lengthy calculations—involving plan size, height, location, and amount of size of wall opening. They are here assumed to be already determined.

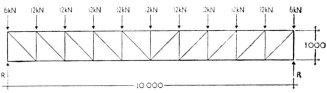

11 *Horizontal truss example*

3.03 Span = 10 m; depth = 1 m; panel width = 1 m. Load per panel point = 12 kN; total load = 120 kN; therefore reaction R = 60 kN.

Main member forces can be quickly estimated:

Maximum mid-span bending moment $= \dfrac{120 \times 10}{8} = 150$ kNm

Maximum chord force $= \dfrac{M}{\text{depth}} = \dfrac{150}{1} = 150$ kN

Maximum force in vertical bracing = 60 − 6 = 54 kN

Maximum force in diagonal bracing $= (60 - 6)\sqrt{2} = 76 \cdot 5$ kN

3.04 Various text books give the classical mathematical and graphical derivation for finding loads in members. These are based on the assumption that all ends are pin jointed, and are used even for welded girders. Tables of coefficients are also published enabling forces to be directly calculated, knowing the loads and physical dimensions. Table I (with **12**) is such a table, derived from the *Steel designers' manual.* Hence:

force in top or bottom chord = coefficient × $\dfrac{\text{length of panel}}{\text{depth of girder}}$ × panel load;

force in vertical bracing = coefficient × panel load

force in diagonal bracing = coefficient × $\dfrac{\sqrt{d^2 + p^2}}{\text{depth of girder}}$ ×

panel load, (where d = depth of girder and p = length of panel).

Table I *Coefficients for truss illustrated in* **12** *below*

10-panel girder		Case			
		1	2	3	4
Chords	TC1	−12·5	−12·5	+12·5	+12·5
	TC2	−12·0	−12·0	+12·0	+12·0
	TC3	−10·5	−10·5	+10·5	+10·5
	TC4	−8·0	−8·0	+8·0	+8·0
	TC5	−4·5	−4·5	+4·5	+4·5
	BC1	+12·0	+12·0	−12·0	−12·0
	BC2	+10·5	+10·5	−10·5	−10·5
	BC3	+8·0	+8·0	−8·0	−8·0
	BC4	+4·5	+4·5	−4·5	−4·5
	BC5	nil	nil	nil	nil
Verticals	V0	−1·0	nil	nil	+1·0
	V1	−1·5	−0·5	+0·5	+1·5
	V2	−2·5	−1·5	+1·5	+2·5
	V3	−3·5	−2·5	+2·5	+3·5
	V4	−4·5	−3·5	+3·5	+4·5
	V5	−5·0	−4·5	−0·5	nil
Diagonals	D1	+0·5	+0·5	−0·5	−0·5
	D2	+1·5	+1·5	−1·5	−1·5
	D3	+2·5	+2·5	−2·5	−2·5
	D4	+3·5	+3·5	−3·5	−3·5
	D5	+4·5	+4·5	−4·5	−4·5

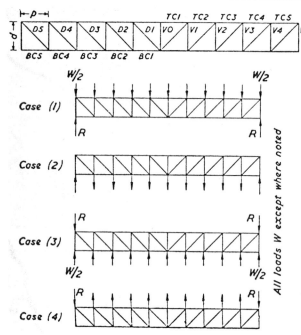

All loads W except where noted

12 *Explanatory diagram from* Steel designers' manual, *accompanying table* I

3.05 Therefore:

maximum top chord force TC1 =

$$-12 \cdot 5 \times \frac{3}{3} \times 12 = 150 \text{ kN}$$

maximum bottom chord force BC1

$$= 12 \times \frac{3}{3} \times 12 = +144 \text{ kN}$$

maximum vertical bracing force V4 = 4·5 × 12 = 54 kN

maximum diagonal bracing force D5 =

$$4 \cdot 5 \times \frac{\sqrt{3^2 + 3^2}}{3} \times 12 = 76 \cdot 5 \text{ kN}$$

Size of members

3.06 *Top chord* Using bolted construction, an effective length of 0·85 × 1 m = 0·85 m may be used. From the *Structural Steelwork Handbook* p98, a 100 × 65 × 8 angle weighing 9·94 kg/m will be adequate to carry 150 kN at this length. Alternatively, using welded construction, and an effective length of 0·7 × 1 m = 0·70 m, from *Safe load tables of structural hollow sections*, p38 a 60 × 60 × 5 □ weighing 8·54 kg/m could be used.

□ weighing 8·9 kg/m could be used.

3.07 *Bottom chord* Using bolted angle construction, allowing one hole deducted from area of cross section (*Handbook* p129) a 100 × 75 × 8 angle 10·6 kg/m would be adequate. Using welded construction (*Tables* p11) a circular hollow section 76 mm diameter, 4·5 mm thick, weighing 7·9 kg/m would be suitable. Internal bracing members may also be sized from safe load tables, and a variety of sections to suit particular requirements are possible.

General comment

3.08 The choice of a flat truss to demonstrate the analysis and selection of member sizes was deliberate, because an architect may still attempt this himself. But the range of structures which he can tackle decreases rapidly, and even if he is capable, his design may well be uneconomic. Not only might his choice of member sizes be too conservative, but he may really have selected the wrong section. An occasional item such as a steel truss in a house or factory extension, may not be condemned if twice as heavy as required, but 20 such trusses in a new complex should and would be.

3.09 In demanding the best architectural solution, a client calls unwittingly for a very high degree of sophistication. The architect using the latest materials and methods, exposes himself to the need for an understanding of their action that he cannot hope to have. In fact he must involve a specialist engineer at an early stage; and consultant computer facilities which most engineers now have are particularly useful.

3.10 One advantage of structural steel construction may yet be that it is the only common material for which a handbook, giving a wide range of the simple members, is available. However, whenever loading or form are complex many authorities will require detailed calculations and a structural specialist will be required to endorse it.

References

1 BRITISH CONSTRUCTIONAL STEELWORK ASSOCIATION LTD/ CONSTRADO (Constructional steel research and development organisation). Handbook on structural steelwork. Revised edition 1978. [(2–) Yh2 (K)] £4.50
2 BRITISH STEEL CORPORATION TUBES DIVISION Structural hollow sections. Three leaflets. 1978. [(2–) Hh2 (K4)]
3 BS 449: Part 2: 1969 Metric units. Specification for the use of structural steel in building [(L) (2– (L) Yh2 (K)] £3.00
4 CIDECT (Comité international pour le développement et l'étude de la construction tubulaire) Concrete filled hollow section columns design manual. British edition. London, 1970 [(2–) Hh2 (K)] £2.50 (obtainable from BSC)
5 CONSTRADO Angle Handbook

Appendix

Approximate metric conversions for use in calculations:

4in = 100 mm   16 sq in = 1000 mm²

10ft = 3·05 m

$$10\text{lb} = 4 \cdot 5 \text{ kg} = \frac{4 \cdot 5}{100} \text{ kilonewton, say } 0 \cdot 05 \text{ kN}$$

$$10 \text{ lb per ft run} = 15 \text{ kg/m} = \frac{95 \text{ kN}}{100} / \text{m}$$

10 lb per sq ft = 50 kg/m² = 0·5 kN /m²

10 ton per sq in = 155 kN/mm², or 10 kN = 1 ton

10 kg/m = 6·5 lb per ft run

1 kN/m² = 20 lb per sq ft

100 N/mm² = 6·5 tons per sq in

AJ Handbook
Building structure
CI/SfB (2-)

Section 7
Structural material: Timber

Section 7

Structural material: Timber

Scope

This section of the handbook is somewhat longer and more detailed than those dealing with other structural materials. There are two reasons for this. First, timber is both more common and more varied in its uses and in the forms in which it is available today. Second, although it can be used successfully for certain types of large buildings, it is pre-eminently a structural material for domestic and other small- and medium-sized buildings which the architect may feel sufficiently competent to deal with himself without calling in a structural consultant. Adequate guidance is given to cope with such structures in detail. For larger and more complex structures, it provides sufficient insight to enable the architect to select suitable materials and forms and to instruct his consultant engineer as well as advise his client with skill and discernment.

The first technical study briefly discusses the development of structural timber and its position today. Technical study TIMBER 2 examines the material properties both of wood itself and its structural forms (from poles, sawn-woods and laminates to woodwool slabs) and Information sheet 1 summarises these properties. This will be followed by a major article (Technical study TIMBER 3), which will discuss structural forms in timber. The specific forms, or at least the simpler ones, are discussed in further detail with methods of their calculation in a series of eleven information sheets later in the section. Before these, there will be two studies, one relating structural timber to building regulations (Technical study TIMBER 4) and another on jointing and connections (Technical study TIMBER 5). The design information sheets will cover column, beam and portal frame selection, suspended domestic floors and flat roofs, laminated, plywood and stressed skin construction. Finally technical study TIMBER 6 will discuss production and erection factors, production procedure and inspection, etc.

Levin

Burgess

Foster

Johnson

Masters

Brown

Herman

Mettem

Definitions

A few properties or characteristics which appear in the first studies and information sheets could usefully be defined at the start.

Texture This is defined in BS 565 as being the structural character of *wood* as revealed by touch or reaction to cutting tools. This is largely determined by the distribution and size (relative or absolute) of the various wood elements. It is distinct from both *figure* and *grain*. Common descriptive terms are:

Coarse texture Texture of wood with relatively large elements or unusually wide growth rings for the species.

Fine texture Wood with relatively small elements or narrow growth rings.

Medium texture Has been used here as intermediate between coarse and fine.

Durability The natural durability of heartwood in conditions most conducive to decay (ie in contact with ground) is defined as: *low*, 5-10 years; *medium*, 10-15 years; *high*, 15-25 years. Sapwood of all species is non-durable in these conditions.

Moisture movement Degree of shrinking or swelling with changes in moisture content. Given for sawnwood in tangential direction (ie plain sawn) and measured in percentage of dimensional change between air-dry condition (20 per cent moisture content) and equilibrium moisture content at around 10 to 12 per cent. *Small:* 1 to 2 per cent (ie 3 mm to 6 mm per 300 mm), *medium:* 2 to 2·8 per cent (ie 6 mm to 8 mm per 300 mm), *large:* over 2·8 per cent (ie more than 8 mm per 300 mm). Movement in radial direction is smaller and longitudinal movement is negligible for most species.

Price Owing to inflation and other market forces actual price ranges to define 'low, medium and high' could soon be out of date and misleading. These descriptions should therefore be taken only as indicating relative values between species for similar grades, sizes and degree of seasoning and processing.

Author

The timber section of this book has been produced largely at the Timber Research and Development Association under Ezra Levin, RIBA, TPDip, deputy director and chief architect. Ezra Levin, after working with the Ministry of Works, BRS and an inter-departmental committee on house construction spent three years as city architect of Haifa, Israel. He has been at TRADA since 1957, and is assisted here by his colleagues H. J. Burgess, BSc, MIEE, CEng, MIMechE; W. L. Foster, BSc, CEng, MICE; V. C. Johnson, CEng, MIStructE; M. A. Masters, AMIStructE; W. H. Brown, AINSc, AMIWM; P. R. Herman, and C. J. Mettem Grad 1 Mech E. There was also a contribution by R. W. Wands of Rainham Timber Engineering.

Technical study
Timber 1

Section 7 **Structural material: Timber**

Introduction to structural timber

The timber section of the handbook begins with a brief survey by EZRA LEVIN *of the structural development of timber, and its position today as a structural material.*

1 History

1.01 The story of the development of wood as a building material spans the history of mankind, going back many thousands of years into prehistoric times. Early uses by prehistoric man already contained the germ of structural ideas—the use of timber as beams or rafters. However, for the development of true timber structures three separate but interrelated elements were required: tools for cutting and shaping, methods of connections and concepts of structural form for space enclosure (or for other purposes such as bridges).

1.02 Once the means of cutting and splitting wood had been acquired, probably in the early Stone Age (flints, stone axes), the way was open for development of cut joints, permitting different kinds of connection for load transfer at member junctions and more sophisticated framed structures. Developing needs and means demanded higher space and structure concepts, and these in turn stimulated the advance of tools and building techniques. Although timber has, through history, undergone more profound changes than any other material, until quite recently basic innovations in wood technology were few.

1.03 While the Bronze and Iron Ages improved cutting tools and craftsmanship and later natural forces were harnessed and engines invented to drive the cutting edges, in essence these were but refinements on the basic invention of the Stone Age: the physical means of separating wood fibres longitudinally and transversely.

1.04 But structural forms did change gradually and were perfected within the available means of processing and connecting wood, and later, by the use of metal in conjunction with wood or by borrowing structural forms first developed with other materials (eg arch forms from masonry construction). Timber was soon eclipsed in the 19th century as the main material for framed structures, and forms which combined timber and iron were evolved. Recent innovations and improvements in timber technology have been far reaching: eg development of glued lamination, wood-based sheet materials and efficient means of connection and preservation. These in turn opened up possibilities of new structural concepts for timber construction and a new era in timber engineering.

1.05 In an attempt to show the relationship between the structural forms and the technical means, stages in the historical development of timber housing, roofing and beam construction are shown in tables I, II and III.

Housing
1.06 In all forested areas, including most of Europe and the UK, timber was the primary, and often the only, structural material for small houses from the Stone Age to the 18th century. Elsewhere, eg North America and Japan, it remains the most important domestic structural material. Historically, the development is from flimsy but statically stable huts to sturdy and massive cruck-framed structures which produced the great framed house from the Middle Ages to the 17th century. Light-braced framing with small sections of sawn timber came into use only in the 19th century, first in balloon framing and later in platform framing. The development of exterior grade plywoods led from the sheathed frameworks to stressed skin panel construction and eventually, in recent years, to three-dimensional 'module' houses capable of full prefabrication and factory finish.

Roof construction
1.07 In Europe, two basic types of pitched roof structure (distinct in statical concept and in appearance) developed, at first independently and later intermingled. The rafter roof, a direct descendant of the primeval stake hut, was developed particularly in North Western Europe and in England. Purlin-type roofs, generally of slacker pitch, derived from Graeco-Roman culture, flourished early in the Mediterranean basin. In late medieval times roof frameworks, with cut carpentry joints designed to transfer all types of stress, reached a peak of excellence; but the Renaissance contributed little to the art of roof construction. Iron tie-rods were introduced in the 18th century. In France, as early as the 16th and 17th centuries, vertically laminated bolted arches were introduced to cope with long spans and the growing scarcity of long timbers. Horizontally laminated and bolted or dowelled arches came into use early in the 19th century in Germany, France and England. Structural glued lamination was also introduced early but did not become an adequately durable industrial product until the advent of casein and resin adhesives in the 20th century. Timber-shell roofs of multi-layer construction of planed boarding, and folded-plate roofs (usually of plywood) are also of recent design and owe their origin to the development of shuttering for concrete shells and the development of plywood as a structural material in recent decades.

Beams
1.08 The development of beams from simply supported logs to the sophisticated production of light plywood web beams and lattice girders went through various stages: strutted or bolstered construction to reduce the unsupported clear span; methods of building up larger beams capable of resisting shear; combination with iron and steel in flitched beams and trussed beams and glued lamination. It was not until the early Renaissance, due to the growing shortage of timber, that the structural significance of placing members on the narrower edges rather than the wider faces was appreciated. Cantilevered construction was already widely practised in medieval times, eg jettied houses, but the use of points of contraflexure in multi-span construction to form unstressed joints was derived from Gerber's work on steel bridges in the 19th century.

Table 1 Development of timber house

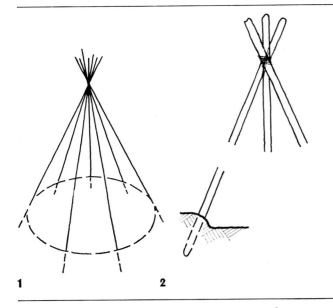

1 *Circular hut* **2** *Top and bottom joints*

Material: branches, split stakes, etc
Joints: tied at apex (withies, bark strips, etc) feet stuck in ground
Tools: none required, use of broken off branches, later burned off

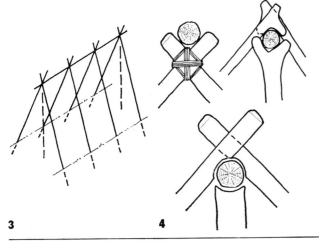

3 *Elongated ridge pole hut*
4 *Joints at ridge support*

Material: as above, and small logs
Joints: first tied, later using natural forks, later still crudely cut
Tools: flints, stone axes

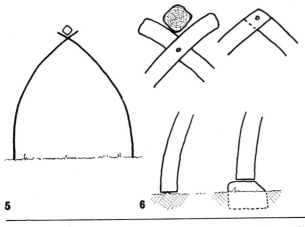

5 *Cruck frame* **6** *Joints at apex and feet*

Material: halved trunks of naturally bent trees
Joints: top halved or lapped and pinned; bottom: resting on ground, later on stone pads or dwarf walls
Tools: iron saws, axes, adzes, augers

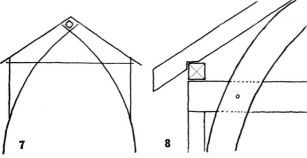

7 *Cruck frame with rafter roof supported on tie beam and posts*
8 *Detail of rafter supported on extended collar tie beam*

Material: as above and roughly squared members
Joints: halved and pinned or mortise and tenon bird's mouthed rafters
Tools: as above

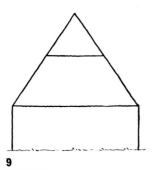

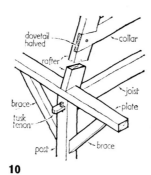

9 *Separate framing of roof and walls*
10 *Detail of connections at junction of wall and roof*

Material: squared timbers, generally sawn and adzed.
Construction and joints: tie beams variously joined to posts or over wall plates. Collar halved and dovetailed to rafters
Tools: as above

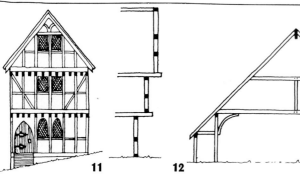

11 *Multi-storey town house, jettied*
12 *Diagram of framing of large house (hall), barn or church*

Material: as above
Construction and joints: as above; wall framing either close timbered or widely spaced with braced panels
Tools: as above

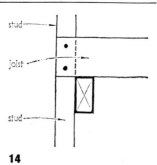

13 *Balloon framing*
14 *Detail of nailed joists to studs*

Material: sawnwood in small sections and long lengths; planed boards for sheathing
Construction and joints: nailed joints to studs; minimum cutting of members; closely spaced studs suitable for clapboard cladding and lath and plaster interior lining
Tools: mechanised sawmills producing small section material; hammer and nails and handsaw on site

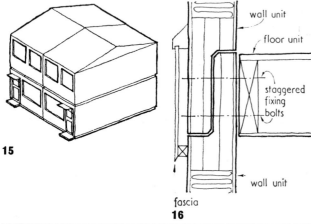

15 *Stressed skin panel house*
16 *Junction of floors and walls*

Material: sawn and planed small section timbers and exterior grade plywood
Construction and joints: plywood sheathed panels with light framework glued, insulated and bolted at connections of panels. Walls, floors and roofs may be of similar construction
Tools: modern sawmilling and planing; press gluing of plywood panels. Semifactory finishes optional

17 *'Module' home; requires stiffness of panel connections to resist transport racking*

Material: as above
Construction and joints: three-dimensional units providing factory finished house or section of house for wheeled transport and crane erection on site. Joints glued, nailed, stapled etc
Tools: large presses and handling equipment in shop and on site. Equipment for mechanised installations and finishes in shop

Table **II** *Developments in timber roofs*

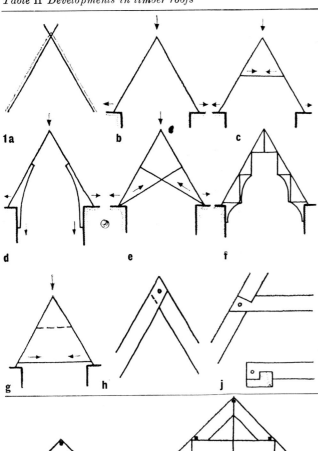

1 *Medieval rafter roofs (open) derived from primitive stake hut with various methods of relieving thrust:* **a** *original stake hut;* **b** *couple roof;* **c** *collar tie;* **d** *arch braces taking part load to walls;* **e** *scissor beams;* **f** *by single or double hammerbeams and posts—often with arch reinforcement, spans attained up to 21 m in most cases, culminating in triangulated frame;* **g** *relieving walls of thrust;* **h** *ridge junction;* **j** *collar tie beam*

1 Material: round wood, roughly squared then well adzed; later cut by saw and adzed finish
Joints: wide variety of cut joints to suit type of stress—halved and pinned, or tenoned and pinned, or pinned or nailed to ridge and piece at top junction. Collars tenoned or dovetailed and pinned; tie beams fixed (tenoned) to posts or cogged to wall plate

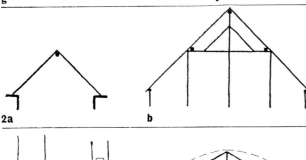

2 *Purlin and ridge-pole type roofs. Rafters act as simply supported beams; roof pitch generally lower than with rafter framing types. In late medieval, Renaissance and later, the two types were mingled*

Material and **jointing:** generally as above
a *simple ridge pole roof;*
b *purlins carrying rafters*

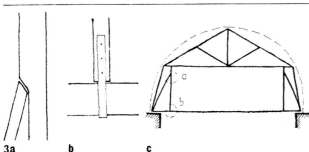

3 *Mansard roof. Idea— utilise roof space and reduce pitch. (Also flat topped roofs developed in late Renaissance and Baroque period.)* **a** *and* **b** *details of strutted posts and hanger straps supporting tie beam;* **c** *section*

Material and **jointing:** generally as before; ceiling beam supported by metal straps

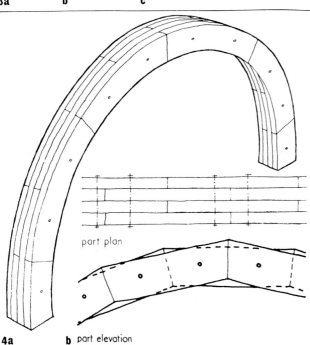

4a *Philibert de l'Orme's (16th century) method of vertical bolted laminations for arches and domes. Idea— wood in compression, and to overcome shortage of great lengths. Used extensively in France and Germany 17th and 18th century for arches and domes.* **b** *Detail elevation and plan showing alternate layers*

Material: sawn planks in short lengths, bolted together, joints staggered
Joints: butt joints in place of laminates straggered in alternative layers, bolted

part plan

4a   **b** part elevation

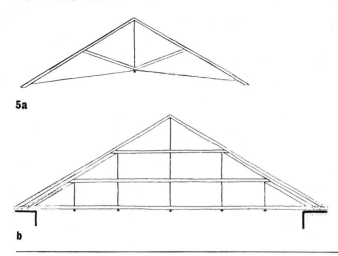

5a

b

5a *From early 18th century on, combined timber and iron trusses. Idea—timber in compression, wrought iron in tension;* **b** *Longer frame with iron hangers but wood tie beams, rafters, collars and struts*

Material: sawn wood, sometimes planed; wrought iron rods with threaded ends, washers
Joints: timber to timber—bird's mouthed, tenoned, etc. Iron rods passing through timber members and anchored by plates

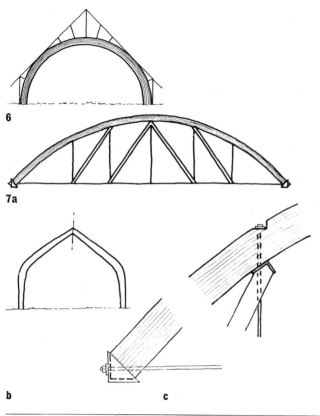

6

7a

b   **c**

6 *Horizontally laminated arches and introduced by Emy in France in late 1820, but preceded by Wiebeking, Bavaria, for bridges, up to 60 m clear span. Used also in England for railway bridges and stations 1830-1850*
7 *Modern glue laminated timber:* **a** *bow string truss (steel or timber tie);* **b** *portals—single- or three-hinge types, etc (arches of mid-19th century construction found in England);* **c** *details of timber and iron rod connections with shoes and anchor plates*

Material: planed timber boarding generally 25-50 mm thick; casein (within) or durable resin adhesives (exterior) used for gluing when forming to required shape
Joints: usually steel plates and bolts at top and pin hinged base sockets. Shear plates, ring connectors, etc also used when required

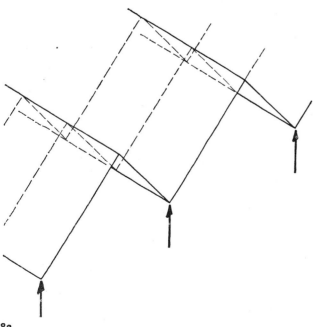

8a

8 *Plywood uses in roof construction;* **a** *folded plates;*

Material: exterior grade plywoods. Small section dressed timber for frame members and stiffness (in folded roofs)
Joints: glued (pinned, nailed or stapled) plates bolted together on site

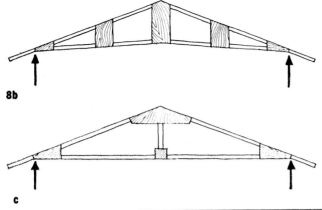

b and **c** *small span roof trusses with plywood glued or glue nailed gussets (mid-20th century)*

Note: plywood can be used in trusses as an alternative to other joint types such as nailed, press inserted, toothed plates, etc

8b

c

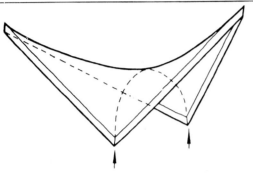

9 *Hyperbolic paraboloid shell roof; conoidal and other forms also used (mid-20th century*

Materials: planed tongued and grooved boarding in three or more layers (generally laid in cross directions). Durable adhesives, laminated edge beams

Joints: Glue-nailing or gluing of board layers over whole or part surface

Table III Development of the timber beam

Form concept		Technology
	1 *Log, simply supported*	Only cutting to length required
	2 *Strutted beam*	Span between posts increased by: struts, with housed or mortise and tenon joints (pinned with squared timbers)
	3 *Arch strut or bracket*	As above, but with naturally anchored strut or cut bracket
	4 *Bolstered beams*	Saddle piece or bolster to support ends of beams. Adjacent surfaces adzed, later pinned with trenails

Form concept

Technology

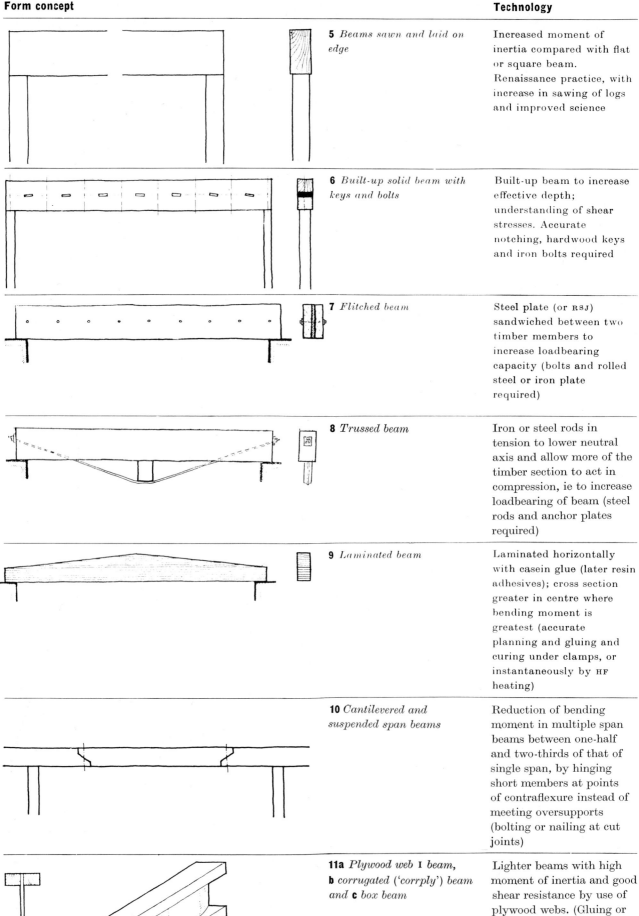

5 *Beams sawn and laid on edge*

Increased moment of inertia compared with flat or square beam. Renaissance practice, with increase in sawing of logs and improved science

6 *Built-up solid beam with keys and bolts*

Built-up beam to increase effective depth; understanding of shear stresses. Accurate notching, hardwood keys and iron bolts required

7 *Flitched beam*

Steel plate (or RSJ) sandwiched between two timber members to increase loadbearing capacity (bolts and rolled steel or iron plate required)

8 *Trussed beam*

Iron or steel rods in tension to lower neutral axis and allow more of the timber section to act in compression, ie to increase loadbearing of beam (steel rods and anchor plates required)

9 *Laminated beam*

Laminated horizontally with casein glue (later resin adhesives); cross section greater in centre where bending moment is greatest (accurate planning and gluing and curing under clamps, or instantaneously by HF heating)

10 *Cantilevered and suspended span beams*

Reduction of bending moment in multiple span beams between one-half and two-thirds of that of single span, by hinging short members at points of contraflexure instead of meeting oversupports (bolting or nailing at cut joints)

11a *Plywood web I beam,* **b** *corrugated ('corrply') beam and* **c** *box beam*

Lighter beams with high moment of inertia and good shear resistance by use of plywood webs. (Gluing or glue nailing and structural grade plywood)

Form concept		Technology
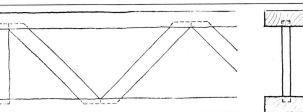	**12** I-*beam with diagonal cross-boarded web*	As **11**, but using planed wood boarding for web as well as flanges (glue nailing)
	13 *Open web girder*	Lightweight girders for floors and roofs of considerable spans; (several systems for . industrialised production with special machines for accurate grooving, end matching of struts gluing and pinning of joints)

2 Structural timber today

2.01 Since the middle of the 19th century, timber had been almost completely ousted as the major structural material from all but small domestic work, first by iron and steel and later by reinforced concrete. For forms beyond traditional sawnwood domestic floors and roofs, designers today are less at ease with timber. They either think in terms of other materials or leave design to the structural engineer who may be equally inhibited about novel uses of timber. Its deep roots in past traditions, the fragmented nature of its industries, increasing constraints motivated by fear of fires, and its slow technological development prevented timber from meeting the technical and economic challenge of new materials born of the industrial revolution.

2.02 In the last few decades, timber as a structural material has regained some of the ground lost during the previous century. Many changes have taken place. Swift advances have occurred in timber technology and engineering and adequate design methods, standards and codes of practice have been developed. The application of science, invention and engineering principles to an age-old material has greatly extended the range of suitable structural forms, producing more efficient and economical designs not only for small- and medium-sized buildings but also for large spans and a wide variety of purposes.

2.03 Important changes have been made in the kind and size of material available, and new sources of timber supplies have appeared—softwoods from the Baltic, Europe and North America, and many hardwoods of great strength and durability from the tropical forests of Africa and South-East Asia. Their structural properties are being systematically investigated in research laboratories in the UK and other countries and working stresses for many of these timbers are already included in our codes of practice.

2.04 British timbers for building are relatively scarce, with the bulk of our supplies being imported—an important element in the national balance of trade. Depletion of the primary forests in Europe and elsewhere and increasing reliance on younger plantation trees to the supply of softwoods has resulted in a dearth of timbers of great dimensions and has stimulated the development of built-up members and structures from smaller sections. Modern resin adhesives have given impetus to the development of laminated timber on an industrial scale and although here the UK lags behind North America and some West and

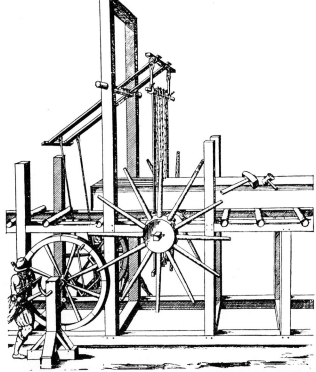

Through and through sawing of logs—16th century and 20th century

North European countries, many impressive and large buildings have been erected with glue-laminated arches, portals or domes. The Timber Research and Development Association (TRADA) has also pioneered the development of timber-shell roofs of various suitable forms and these have been more extensively used in the UK than elsewhere.

2.05 The mechanical and physical properties of major wood species used in building construction are now well known and much research has been carried out on the relationship of their strength properties to moisture content, density, knots, slope of grain and so on. Elastic constants and ultimate fibre stress values have been established and, in laying down working stresses, the statistical concepts of probability have been introduced. Grading rules enable timbers of the same species but which have different degrees of strength reducing features, to be distributed into well defined strength classes; species displaying similar basic strength characteristics have been grouped in the code of practice[1] to aid the designer and ease supply. Within the last few years stress grading by electro-mechanical graders has been introduced in the UK, opening up the possibility of selecting higher strength grades than are possible by visual inspection.

2.06 Laboratory tests have also enabled the establishment of differential values of stress for different kinds of load, such as dead loads and superimposed loads of long duration; short period loads such as wind or snow and loads of great intensity but very short duration such as wind gusts and impact. The refinement, now embodied in the code, is of particular value for design with timber which has very favourable characteristics in behaviour under short-term and impact loads.

2.07 Development of wood-based sheet materials, such as plywood and chipboard in grades suitable for structural purposes has had great significance for design in timber. Plywood in particular has been used for many components and building elements of great stiffness and high strength: weight ratios. Membrane structures with these very light materials can be prefabricated in large sections and erected with a minimum of mechanical plant.

2.08 Methods of jointing wood members to ensure more efficient load transfer or connections than was possible with traditional carpentry joints have recently improved. These range from better nails and nailing patterns through a variety of mechanical connectors designed for insertion by hand or press to glued joints with strong weather-resisting adhesives.

2.09 The durability of wood and wood-based materials has been enhanced by improved methods of preservation and, when suitably treated, even naturally non-durable timbers can be used with impunity in conditions of extreme exposure to decay hazards. Fire-retardant treatments have also been developed either for surface application or impregnation often together with decay inhibiting substances, thus adding low spread of flame rating to the naturally high fire resistance of timber.

2.10 Recent national building regulations, based to a large extent on functional requirements and performance standards, have within this decade advanced timber construction by introducing a number of important relaxations affecting its use and thereby making possible many structural applications previously forbidden under model by-laws. Partly due to their complexity and partly to the slow adaptability of the building industry, the new freedom is as yet far from being fully exploited. But the timber trade has been undergoing continual transformation. Concentration through amalgamations and take-overs has been accompanied by an increasing measure of integration with the processing of the materials and the manufacture of building components and prefabricated buildings, including in some cases the provision of specialist erection teams.

2.11 Many companies in the timber trade now possess considerable expertise not only in general properties of the materials and components they sell, but on their behaviour in building structures and their suitability for specific applications. Some of the larger companies engaged in timber engineering have their own staff of designers and others employ specialist consultants. Trade associations, grouping timber-engineering firms and manufacturers of industrially produced timber buildings and components, issue manufacturing codes and exercise some control on levels of competence, albeit on a voluntary basis.

2.12 These developments and changes offer the architect the choice of a broader range of timber materials, structural forms, suppliers and producers. The range of possibilities as well as the inherent limitations must however be fully grasped if this freedom is to be exercised. The wealth of variants can be a source of embarrassment and confusion: reliable and independent guidance is essential in a field which has suddenly become rich in innovation.

2.13 What of the future prospects of timber as a structural material? The recent revival in the use of wood in new and 'rationalised' old forms may well be a long-term trend. With growing concern over the rapid depletion of world resources, raw materials and fuels, wood is seen as a renewable resource, and modern reafforestation policies and improved silviculture are gradually increasing yields. Use of wood in building is also favoured by other environmental and anti-pollution factors.

2.14 The association of wood with other materials seems particularly promising. Already the combination with plastics has produced resin-bonded plywood, particle board, durable laminated wood and many types of sandwich materials, all with structural and other properties greatly modified from those of natural wood.

Technical study
Timber 2

Properties of timber and wood-based materials

This second study, by EZRA LEVIN *and* W. H. BROWN, *introduces the structural properties of timber- and wood-based materials. It discusses the physical properties of wood itself and then in detail describes the forms it can take; sawnwood is covered in detail and followed by sections on glue-laminated timber, plywood, blockboard, chipboard, fibre building boards and wood wool slabs. Characteristics and structural properties of all these materials are tabulated in the seven tables of Information sheet* TIMBER 1

1 Introduction

Scope

1.01 Structural timber, as defined in BS 565[1], is 'timber used in framing and loadbearing structures where strength is the major factor in its selection and use'. Timber is generally taken to imply 'natural' wood, sawn and sometimes planed, and this is the form in which it is mostly used in building construction, particularly for structural purposes. But there are today many kinds of modified or reconstituted wood, with or without the addition of other substances, whose importance in building is constantly growing, not only in substitution for sawnwood, but using their specific properties for new structural applications and forms.

1.02 The main forms discussed here are glued-laminated timber and the basic wood-based panel products—plywood and blockboard, fibre building board, chipboard and wood wool slab. Panels of composite construction—softwood core with hardboard facings, hollow-core panels, metal or plastic faces, etc—are too numerous and varied to be considered here. They are generally produced to obtain some desired specific properties for particular uses, not necessarily structural. When their use for any loadbearing purpose is required, their structural and other relevant properties should be ascertained from the manufacturers or by carrying out appropriate tests and performance evaluations.

Applications

1.03 At present the code of practice for the structural use of timber (CP 112[2]) contains design guidance only for solid timber—sawnwood and laminated wood and for plywood. Nevertheless, the other panel products, whether in sheet or slab form, are covered by British Standards which lay down dimensions and quality controls, minimum structural requirements relating to their major end-uses, as well as sampling and testing procedures. Over the years a good deal of practical experience has been amassed on the structural behaviour of these materials when properly applied in their normal uses.

1.04 When designing with these materials for unconventional structures the skilful designer may well achieve economy in cost; but he may face the following difficulties: problems in obtaining approval from the building control authority; lack of adequate field experience; absence for some materials of well established working stresses for particular strength properties of importance for the required use; profusion of brands for most of the materials with widely varying strength values and other characteristics.

Specialist advice, as well as an understanding of the fundamental characteristics of the different types of material, is essential. For works of considerable importance, prototype testing is recommended.

Suitable end-uses

1.05 All types of timber and wood-based materials discussed here are commonly applied to one or more structural or semi-structural end-uses in building. Some are interchangeable for certain uses, as shown in table I, and selection may be governed largely by availability or price. They have, however, distinct characteristics which for a particular purpose may render them more or less technically suitable than their competitors. This is true with regard to the panel products whose fields of application overlap considerably. The following is only a general introduction, and reference should be made to the subsections dealing with the various materials and their tabulated properties and, in individual cases, to the specific brands of the materials under comparative study.

1.06 The rheological properties, ie deformation under long-term loads, should be taken into consideration when designing for such uses as heavily loaded shelving or storage floors, etc. Plywood, blockboard and laminboard, with their elastic properties, are similar to natural wood in bending and less subject to creep than particleboard or fibre building boards. For the latter this implies a reduction in the permissible extreme fibre stresses under such conditions. Even for lightly loaded domestic floors (including dead and live loads), Lundgren[3] suggested the following comparative thicknesses of different panel products with joists at 600 mm spacing: plywood, 15 mm; hardboard faced lumbercore panel, 19 mm; particle board (650 kg/m³ density), 22 mm.

1.07 Fibreboards and particleboards are also more vulnerable to large or cyclical changes in moisture content (mc). For sarking or decking in roof construction, where a number of different types of boards or slabs may be used, fibreboards and particleboard should not be exposed to the risk of high humidity (eg condensation in unventilated roof spaces). The properties of wood wool slab are less affected by mc variations.

1.08 Panel products, especially the thicker boards, impart excellent racking resistance to wall systems when properly fixed to the framework as sheathing, lining or cladding. The effect of high humidity in reducing the mechanical properties of panels is generally of little consequence when no loadbearing other than racking resistance is considered in

Table 1 Timber and wood-based materials and their main structural uses

Available types of material	Main structural and semi-structural uses
1 *Poles (debarked logs)*	Piles, transmission line poles, posts (eg in pole barns), shoring, dock and harbour work, log cabins
2a *Baulk and* **b** *half-timber (sawn or hewn)*	Piling, shoring, beams and columns in heavy construction (mill construction), dock and harbour work
3 *Sawn sections (sawn, precision or planed)*	Carcasing in housing, skeletal frameworks generally, trusses, stud frames, built-up members, floor and roofs members generally
4 *Planed boards and planks (square edged or tongued and grooved)*	Flooring, roof decking, sheathing, lining, cladding, shell roofs
5 *Laminated timber:* **a** *'architectural' and 'industrial' classes;* **b** *'economy' class (unplaned sides)*	Beams, columns, arches, portals, bow string trusses (generally where large members and long spans are required). 'Economy' class for buildings where members are concealed or appearance not important: heavy-duty flooring (on edge)
6 *Plywoods:* **a** *threeply;* **b** *multiply (five, seven etc)*	Flooring, roof decking, sheathing, cladding, wall and partitions lining and panelling, webs in I- and box-beams, gussets for frameworks, stressed-skin panels, folded plates and shell roofs, formwork
7 *Solid core boards:* **a** *blockboard;* **b** *laminboard*	Shelving, flooring, panelling, partitions
8 *Particle board:* **a** *single layer;* **b** *three layer;* **c** *graded density;* **d** *extruded;* **e** *extruded, hollow cored*	**a, b** and **c** Flooring, roof decking, wall lining and panelling, ceilings, partitions, **d** and **e** partitioning (principally as core stock, faced with other materials)
9 *Fibre building boards:* **a** *hardboard and tempered hardboard;* **b** *medium board* **c** *insulating board (sheathing and sarking grade)*	**a, b** and **c** Sheathing, sarking, linings; **b** and **c** ceilings, **a** and **b** cladding; **a** stressed skin panels, folded plates, gussets etc. (Limited experience, protection required from high humidity and temperature rises)
10 *Woodwool slab (unreinforced, standard grades)*	Roof decking, linings, ceilings, partitions

design. An important criterion, particularly with cyclic exposure, is the lateral moisture movement of panels. This being greater with particleboard and hardboard requires proper restraint against buckling or differential movement at joints. Closed joints are possible with all panel products under conditions of small mc variations. Warping, owing to large humidity differentials of opposing faces, also requires restraint especially in the case of thin panels with high strength and moisture movement characteristics.

1.09 Glued stressed skin structural panels are usually plywood, but hardboard—which is used extensively in stressed skin applications such as flush doors—has been used in some structural stressed-cover applications. Protection from exposure to large mc variations and temperature rises is, however, an essential condition. Tempered hardboards, displaying the smallest reductions in strength properties and the greatest moisture stability, should be selected.

1.10 Exterior grade plywood is indicated for structural uses where high strength and stiffness are required even under long-term loads, combined with dimensional stability and resistance to moisture variations (eg webs in I-beams, gussets in trusses and stressed membrane structures under extreme exposure conditions).

1.11 Price comparisons of the materials themselves are a poor guide to choice, as a true cost comparison would require an estimate of all the factors necessary to achieve identical performance standards in every respect.

2 Properties of timber

2.01 Structural (and other) properties of timber are closely related to the physical and anatomical structure of wood and its chemical composition, to the growth characteristics of a particular species—even of the individual tree from which the timber was produced—to the mode of converting the log and to the methods of subsequent processing and treatment. Wood-based panel products conserve some of the properties of natural wood and modify others, some are improved and others diminished. Products made up of wood layers—glue laminated, plywood or even blockboard —have properties nearer to natural woods than products in which wood fibres are broken up and reconstituted under pressure with or without the admixture of binders and additives. Wood wool slab, which contains up to 50 per cent by weight of inorganic binders such as Portland cement, is furthest removed from natural wood in its properties.

Nomenclature of species
2.02 Commercial timbers are divided into 'softwoods' and 'hardwoods', the former being derived from coniferous trees (eg pines, spruces, firs) and the latter from broad leaved or deciduous species (eg oak, beech, teak). Common names are sometimes loosely applied to timbers of different species; the botanical names in Latin are more precise, and consist of two parts, the first indicating the genus and the second the species (both common and botanical names of construction timbers are shown in tables III, IV and V Information sheet TIMBER 1).

2.03 Deciduous species are more numerous than coniferous but the bulk of construction timbers used in the UK (and most other countries) are 'softwoods' produced from a few species, which are relatively light and cheap, easy to work (soft) and abundant. Most construction hardwoods are heavier and less easy to work but, as shown in these tables, have superior strength properties.

Composition and structure
2.04 Both the chemical composition and anatomical structure of wood are complex and vary to some extent between species. The mechanical strength of timbers is derived mainly from the fibrous tissue forming the thin walls of the long cells which run roughly in the direction of the trunk.

2.05 The main chemical substances of the fibres are cellulose, hemicellulose and lignin. Cellulose, a carbohydrate of long-chain molecules, is extremely strong in tension, hygroscopic, combustible, but resistant to ordinary solvents; it provides between 45 and 60 per cent of the total substance of the wood. Hemicellulose accounts for between 15 and 25 per cent, has shorter chain molecules and is chemically less stable. Lignin, of complex chemical composition, interpenetrates the cells and binds them together. It accounts for 25 to 35 per cent of the wood tissue.

2.06 Different species also contain small and varying amounts of other substances such as tannins, resins and oils, calcium, silica, starch, etc, which affect certain properties and uses of the timbers derived from them. Thus the tannin in oak contributes to its durability; sapwood is susceptible to fungal and insect attack because of the starch-food in it; certain oil infiltrates, eg in western red cedar, are toxic to attacking organisms and increase durability; gums and resins add to weight without increasing strength and adversely affect the working qualities and ability to take surface finishes of the various species in which they occur.

Rates of growth
2.07 Trees grown in climates with seasonal variations display in cross-section annual growth rings of lesser and greater density. The faster grown *springwood* zones tend to be more porous, lighter and weaker than the *summerwood* cells which are thick walled, denser and stronger **1** and **2.** The strength of timber is related to the proportion of summerwood to springwood. In softwoods and *ring porous* hardwoods (such as ash or oak) there is a connection between this proportion and the rate of growth (measured by the number of annual rings per 25 mm of cross-section). Extremes of fast and slow growth tend to produce weaker timbers. The optimum for softwoods is considered to lie between seven and 20 rings per 25 mm, and for ring porous hardwoods between seven and 15 rings. In *diffuse porous* hardwoods, eg beech, mahogany, keruing, afzelia, there is

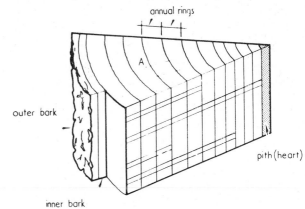

1 *Sector of tree trunk showing annual growth rings*

2 *Cross-section of pine (area* **A** *in* **1***) greatly magnified, showing denser cell formation in summerwood zones than in the faster growing springwood zones*

generally no distinction between summerwood and spring-wood, and variations in ring width bear no relationship to the strength of the timber. BS 4978: 1973 requires four rings per 25 mm for visually graded softwood of GS or SS grade. CP 112 requires six rings per 25 mm for 65 grade and eight rings for 75 grade. No limits are laid down for structural tropical hardwoods to BS 5756 or for machine graded timber generally.

Moisture content (mc)

2.08 Moisture in timber is an important natural constituent which affects its weight and strength, durability and shrinkage characteristics. Moisture content is measured as a percentage ratio of the weight of contained water to the oven-dry weight of the wood substance. In a freshly felled tree it may amount to 150 per cent or more. Most of this moisture is contained in the cell cavities, but when the timber is dried to around 30 per cent mc all the free water will have evaporated so that the timber is at 'fibre satura-tion point'.

2.09 On further drying, moisture will leave the cell walls resulting in drying shrinkage and increased strength and stiffness through hardening of cell walls. These processes continue until equilibrium is reached with the surrounding atmosphere, at a point which varies with the relative humidity of the air.

2.10 Imported softwoods from European sources are generally dried to a mc below 23 per cent before shipment to the UK. Softwoods from Canadian and US sources are generally unseasoned before shipment and their mc may exceed 'fibre saturation point' on arrival and require further seasoning. Imported hardwoods from all sources are usually dried to some extent before shipment, but their mc may vary over a wide range. Further seasoning is normally carried out in the UK.

2.11 All timber, whether imported or homegrown, which is thoroughly air-dried in this country will attain a moisture content between 17 and 23 per cent, depending on weather conditions at the end of the drying period. The average moisture content that construction timbers attain in a dried-out building in the UK may be generally taken at between 15 and 18 per cent, being lowest for roof timbers and highest for ground floor joists. There are some seasonal and regional variations and some differences that are caused by the ways in which the buildings are heated. BS CP 112[2] accepts carcassing timbers at maximum 22 per cent mc at the time of erection, but advocates 17 per cent for precision work in prefabricated timber buildings. It also recommends that during fabrication the mc of timbers should not exceed that expected in use by more than 3 per cent.

Moisture movement

2.12 Below fibre saturation point (about 30 per cent), timber shrinks or swells with changes in the relative humidity of the atmosphere. Owing to the natural structure of wood, this movement is about 40 to 70 times greater across the grain than along it and may be between 1·5 and 2 times greater tangentially (along the growth rings) than radially (across them). In commercial species (with the exception of Parana pine) movement along the grain may be ignored for practical purposes, but movement across grain must be considered. In drying from fibre saturation point to 18 per cent mc, a piece of Douglas fir will shrink about 1½ per cent of its width if quarter sawn (see **3.02**) and 2 per cent if flat sawn; and at 10 per cent mc by 3 per cent and 4·25 per cent respectively. Maximum shrinkage at the oven-dry state is 5 per cent and 6·6 per cent approximately.

2.13 The equilibrium mc attained by structural timber in buildings in the UK is generally between 15 and 18 per cent. Softwoods air-dried to around 22 per cent will not shrink more than 1 per cent in flat sawn members and less in quarter sawn. For precision work, timber may have to be kiln-dried so that it approximates the mc in use. In accord-ance with BS 4471[5] and appendix A to CP 112 timber sizes are measured at 20 per cent mc. For any higher mc up to 30 per cent, the size should be greater by 1 per cent for every 5 per cent of mc in excess of 20 per cent; for a mc lower than 20 per cent, the size may be reduced by 1 per cent for every 5 per cent mc below 20 per cent. No increase in dimensions is required for increases in mc above 30 per cent.

Measurement of mc

2.14 The mc of wood can be determined with accuracy by the oven-drying method laid down in BS 373[6]. This is a slow specialised lab method requiring samples to be cut out of the timbers. For checking mc on site, in the shop or the timber yard, electrical moisture meters of various types and makes are available. They are mostly based on the dielectric properties of wood and give instant readings to within ±2 per cent of true mc. Information on many types of moisture meters and practical guidance on aspects of use, including effective range, temperature effects, density, local wetting, etc may be obtained from TRADA or the Princes Risborough Laboratory of the Building Research Establish-ment.

2.15 Because the effect of mc on the strength and stiffness of timbers is considerable and because timbers are available and may be used at widely varying moisture contents, CP 112 gives different basic stresses, working stresses and moduli of elasticity for dry timbers (defined at 18 per cent mc or under) and for green timbers (defined as over 18 per cent mc). The values for the latter are considerably lower and are recommended for use when the timber used is appreciably above 18 per cent or likely to remain above that level subsequent to construction. The various values are shown in tables III to V Information sheet TIMBER 1.

Specific gravity and density

2.16 The specific gravity of wood tissue is about the same for all species: 1·50 (ie a weight of about 1500 kg/m³). However, the density of timber varies greatly between species and within species, even when pieces are dried to the same mc. This is due chiefly to the differing ratios of cell voids to solid tissue. The average weight of balsa wood may be 100 kg/m³ or less compared with 385 kg/m³ for western red cedar and about 1060 kg/m³ for green heart. Weight variations within species are less extensive and are chiefly related to rates of growth. Although gums and resins in certain species add appreciably to weight without contributing to strength, weight is a good general indicator of likely timber strength, particularly of different pieces within a given species. Tables III to V of Information sheet TIMBER 1 show the relationship of density to strength properties of the various species.

2.17 Density is also a good guide to the relative strengths of different boards within a type of wood-based panel. Plywoods of the denser wood species tend to be stronger than those of lighter woods, the more compressed and denser fibreboards or particleboards are usually stronger than the lighter ones. Mode of manufacture and additives do, however, modify this rule with some products.

Colour and texture

2.18 The colour and texture of timbers of different species have no bearing on their structural properties. They may, however, affect choice where the structure is visible and a certain decorative effect or type of finish is intended.

Durability

2.19 The durability of timber in use depends on: natural resistance of the particular species to decay; presence or absence of sapwood; and moisture content. The cross-section of a log from a living tree will reveal an annular outer band of sap-carrying tissue. It is often lighter in colour than the central core of heartwood, which gives only mechanical support to the tree trunk.

2.20 The sapwood band varies from about 25 mm to 175 mm in thickness and although its wood has much the same weight and strength properties as that of the heartwood, it is generally non-durable—ie liable to be attacked by fungi and insects—because of the presence of starches and other 'food' substances. On the other hand, it is more absorbent and easier to treat with preservatives than heartwood. Softwoods, timbers derived from the smaller, second growth European trees, are liable to contain a higher proportion of sapwood than those derived from Douglas fir or western hemlock from British Columbia. The natural durability of heartwood varies considerably between species, and is generally classified as being non-durable, moderately durable or very durable. Most softwoods are non-durable, but many structural hardwoods are either moderately durable or very durable (see tables III to V Information sheet TIMBER 1).

2.21 The equilibrium moisture content of timber, which is affected by its conditions of use, has a great bearing on durability. Green timbers of non-durable species, or containing sapwood of durable species, are more likely to be attacked by wood-destroying organisms than dry timber. At below 25 per cent mc the risk is greatly reduced and below 20 per cent mc all wood is reasonably immune from decay. When the mc cannot be kept below safe limits by design or protective coatings, non-durable timbers can be treated by a variety of preservative materials and methods suited to the degree of exposure and other factors.

2.22 The durability of wood-based panels depends on their type and manufacture. Plywood of durable species (see the fourth column in table VI, Information sheet TIMBER 1) and made with WBP adhesives to BS 1203 is suitable for external exposure or high internal humidities. Wood wool slab is generally considered immune to decay or insect attack. Particleboards and fibreboards are not naturally durable; however, they may be treated to inhibit decay and improve weather resistance to a certain extent. Non-durable plywoods can also be rendered durable by preservative treatment (see Information sheet TIMBER 16).

Fire properties

2.23 Being a cellulose-base material, all wood tissue is combustible. This property inhibits its use for certain structural elements under present legislation. However, the fire resistance of wood-built elements is high and compares favourably with those built of some other non-combustible material (particularly structural steel and aluminium) which suffer loss of strength at relatively low temperatures or spall or cause collapse through expansion. As already discussed in Technical study SAFETY 3, wood chars slowly as it burns and is reduced in cross-section (on the sides exposed to the fire) usually at the rate of 0·64 mm/min even in high temperatures of 900°C to 1200°C. The timber left uncharred is hardly affected in strength, stiffness or dimensional stability (beyond the oven-dry state). Consequently, large cross-section elements will withstand fire longer under full load, and 'sacrificial timber' may be added in design to obtain the requisite fire-resistance period. Heavy timber construction attains high fire ratings for this reason. In elements with small cross-section members (eg stud frame walls) only part of the required fire resistance is provided by the timbers, the remainder being supplied by protective sheathing and infilling materials.

2.24 By the spread-of-flame test criteria of BS 476[7], untreated wood surfaces of all species of a density not less than 400 kg/m³ fall in class 3. Where spread of flame requirements relating to structural surfaces may be higher (eg the underside of folded plates or shell roofs exposed as ceilings), structural timber may be impregnated or surface treated to attain class 1. (See Technical study TIMBER 4 for conditions of the requirements and methods of treatment.)

2.25 Wood wool slab is unique among wood-based panels in that it contains a high percentage of non-organic material which enrobes the wood strands. It is virtually non-combustible and attains class 0 without further treatment. In general, wood-based panels are at a disadvantage compared with solid timber because they are both combustible and relatively thin. In separating elements such as walls or floors, they may, however, perform extremely well compared either with timber boarding or non-combustible sheet materials. They are better than the former because of the relative absence of joints and they have superior thermal insulation to most of the latter. For example floor-ceiling panels, with timber joists, particleboard or plywood flooring and plasterboard ceilings, have attained fire ratings of up to one hour. Similarly, plywood box beams with 10 mm webs were shown in fire tests to have a longer endurance than steel bar joists of equal load capacity.

2.26 As with solid timbers, the spread of flame classification of wood-based panels depends largely on their density. Improved performance can be obtained by surface treatment (with intumescent paints), a method often employed with low-density fibreboards, or by deposition of fire-retardant salts on the wood fibres (or chips, or veneers) during manufacture. With plywood pressure impregnation is also possible after manufacture (see Information sheet TIMBER 4).

Other properties

2.27 Some other non-structural properties need to be mentioned briefly because they may affect the choice of timber- or wood-based materials in conditions where these properties are important.

Thermal properties

2.28 The thermal conductivity of wood is low compared with that of other structural materials due to its low density. It is therefore easy in framed timber structures to avoid 'cold bridges' and pattern staining. Timber membrane structures for roofs or walls have good inherent thermal insulation values. The thermal movement of timber is low and may be ignored for practical design purposes.

Sound insulation

2.29 CP 3: chapter III[4] lists numerous heavy constructions of separating walls and floors that attain grade I insulation, but those given for timber joisted or light framed constructions are generally rated grade II or worse; and where heavily pugged, joisted floors with 'floating' boarding are said to attain grade I only if supported on heavy walls. This information is somewhat out of date, for in recent years numerous double-frame timber stud walls and a variety of joisted floors (with and without pugging) have attained in laboratory and field tests grade I insulation.

Chemical resistance

2.30 Wood is highly resistant to attack by most chemicals, including organic materials and solutions of acid or neutral salts, but less to caustic solutions. This property makes it a particularly suitable material for processing or storage buildings of many chemical and allied industries and generally for building structures exposed to highly polluted and humid or salt-laden atmospheres.

Electrical properties

2.31 Wood has a high resistance to the passage of electrical currents. This dielectric property is exploited industrially in heat-curing glue-lines (eg in finger jointing or glue-lamination), as a dielectric surface heats up when subjected to a high-frequency electrical field. The rapid change in resistance due to mc variations is also used to determine the mc of wood by means of electrical moisture meters (see para 2.14).

3 Forms of timber and wood based materials

Wood poles

3.01 Timber can be used efficiently in round form, ie debarked logs, generally for poles in power and tele-communication lines, but also for posts (eg in pole barns) and for piling. BS 1990[8] specifies quality requirements, dimensions and minimum retention for creosoted softwood poles of redwood or Scots pine and various larch species which have been proved in practice to be suitable. The standard also applies to other timber species, including Douglas fir, western red cedar, European and Sitka spruce. Creosoting to a minimum retention level of 130 kg/m³ is to be carried out after reducing mc to 28 per cent or less.

3.02 BS 1990 divides poles into three classes and gives guidance tables for application of lateral load at a point 600 mm from the top. Calculations, depending on pole diameter, planting depth, and whether stayed or unstayed, can be made on the assumption of an ultimate extreme fibre stress of 53·8 N/mm² for redwood or Scots pine and 65·5 N/mm² for larch, with respective moduli of elasticity of 10480 N/mm² and 11380 N/mm².

Sawnwood

Conversion

3.02 Most timbers used in building construction are in the form of sawnwood sections, conversion being the term applied to sawing logs into marketable timbers. There are various ways of cutting up a log and some of the character-istics of the resulting timbers depend on the angle between the face of the timber and the cut growth rings. If the angle is not less than 45° in any part of the broader surface, the piece is called 'quarter sawn' (or rift sawn, edge grained or vertical grained) **3**. If less than 45° on at least half its broader surface, the piece is known as 'flat sawn' (or tangentially cut, plain sawn, slash or bastard grained) **4**. 'Through and through' sawing **5, 6** is the most economical method of conversion and produces about two-thirds of flat sawn pieces and one-third quarter sawn.

3.03 To increase the quantity of quarter sawn material other methods of conversion are sometimes employed **7**. Quarter sawn timbers have the advantage over flat sawn because they shrink less in width, have less tendency to cup or twist, are less likely to check or split in seasoning, can be kilned more rapidly and wear more evenly on face (eg in flooring). Because of the greater cost involved, selection of quarter sawn material is rarely justified for structural purposes. Softwood conversion is often designed to produce

the largest sections practicable from the log, with 'boxed heart' or cutting through the centre **8a** and **b**. Where pith is objectionable on the face, a centre board is cut out **9** (a method common in central European mills).

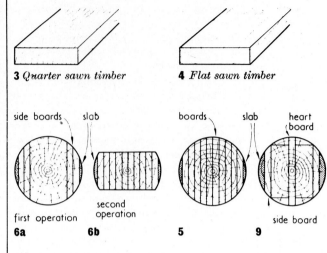

3 *Quarter sawn timber*

4 *Flat sawn timber*

6a   **6b**   **5**   **9**

6 *Two stage conversion of log to boards:* **a** *first; and* **b** *second operation*

5 *Through and through sawing to produce boards*
9 *Conversion to produce timbers free of pith on face*

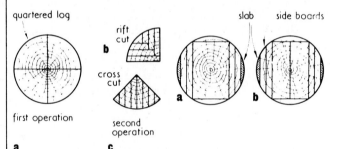

a   **c**

7 *Conversion in two operations to secure high percentage of quarter sawn boards:* **a** *first; and* **b** *or* **c** *second operation*

8 *Conversion to produce construction timbers from small logs:* **a** *with 'boxed heart';* **b** *cut through heart*

Sawn softwood dimensions and tolerances

3.04 BS 4471: Part 1[5] specifies dimensions and tolerances for sawn softwoods, the basic sizes for which, at 20 per cent mc, are given in Table II. Actual sizes are required to be 1 per cent greater for every 5 per cent increase in mc up to 30 per cent and may be smaller by a similar measure for mc below 20 per cent. 'Resawn ex larger' pieces are allowed 2 mm reduction in size and for regularising widths, a reduction of 3 mm for widths up to 150 mm and 5 mm over that width is permitted. Allowed reductions for planing two opposing faces are 3 mm up to 100 mm, 5 mm for 101–150 mm and 6 mm above that. For trim, joinery etc greater reductions are allowed. The standard includes an appendix on surfaced constructional timber imported mainly from Canada and produced to North American standards. Tables of geo-metrical properties of the standard sections are given in CP 112.

Standard hardwood sizes

3.05 BS 5450[17] gives a standard range for sawn hardwoods at 15 per cent mc. It also allows reductions for lower mc and increased sizes for mc up to 30 per cent but gives no rule in this respect, owing to the difficulty of predicting accurately the drying and shrinkage of most hardwoods. The allowances for processing are similar to those in the softwood standard.

3.06 Tables II and III below show the basic ranges of sections of softwood and hardwood. Actual sizes take into account permissible reductions or addition due to manufacturing tolerances, resawing, processing and mc, as laid down in the respective standards. The sizes and lengths of the different species which are available commercially vary considerably and are given in some detail in Tables III–V Information Sheet Timber 1.

Table II Basic sizes of sawn softwood (cross-sectional dimensions) at 20 per cent mc. All dimensions are in millimetres

| Thickness | Width | | | | | | | | |
	75	100	125	150	175	200	225	250	300
16	x	x	x	x					
19	x	x	x	x					
22	x	x	x	x					
25	x	x	x	x	x	x	x	x	x
32	x	x	x	x	x	x	x	x	x
36	x	x	x	x					
38	x	x	x	x	x	x	x		
44	x	x	x	x	x	x	x	x	x
47*	x	x	x	x	x	x	x	x	x
50	x	x	x	x	x	x	x	x	x
63		x	x	x	x	x	x		
75		x	x	x	x	x	x	x	x
100		x		x		x		x	x
150				x		x			x
200						x			
300									x

The smaller sizes contained within the dotted lines are normally but not exclusively of European origin. The larger sizes outside the dotted lines are generally but not exclusively of North and South American origin. Standard lengths in BS4471 are 1·8m—7·2m in 300mm increments.

* This range is usually available in constructional quality only.

Table III Basic cross-sectional sizes of sawn hardwoods in millimetres

| Thickness | Width | | | | | | | | | | |
	50	63	75	100	125	150	175	200	225	250	300
19			x	x	x	x	x				
25	x	x	x	x	x	x	x	x	x	x	x
32			x	x	x	x	x	x	x	x	x
38			x	x	x	x	x	x	x	x	x
50				x	x	x	x	x	x	x	x
63						x	x	x	x	x	x
75						x	x	x	x	x	x
100					x	x	x	x	x	x	x

Basic lengths are from 1m in multiples of 100mm. The availability of *required sizes and lengths in any particular species should be checked.*

Strength reducing characteristics

3.07 The basic stresses which can be assumed for given timber species (see tables III to V Information sheet TIMBER 1) relate to clear, near perfect specimens. In practice, conditions of growth, of conversion and of seasoning in commercially available timbers produce defects which reduce their strength and consequently the working stresses which can be safely applied to them. The limits of these defects and characteristics are defined in standards which specify stress grades for structural timbers.

Stress grading standards

3.08 BS 4978[22] defines two structural grades for visually graded timbers, GS (General Structural) and a better grade, SS (Special Structural). It also contains provisions for machine graded timbers, two of which, MGS and MSS, have similar strength characteristics to those of the corresponding visual grades and two additional 'numbered grades', M50 and M75, the numbers denoting the approximate percentage of the basic bending stress. CP 112[2], from which the latter grades were taken, also contains grading rules for equivalent visual grades, 50 and 75, as well as other grades, 40, 65 and

a composite 40/50 grade which was much used for carcassing timbers before the GS and SS grades became available a few years ago. The CP 112 visual grading rules are more complex than those of the more recent BS 4978 and are now virtually superseded.

3.09 The limits of defects permitted by BS 4978 for the visual grades are in respect of knots, wane, slope of grain, fissures, rate of growth, and distortion. Knots are measured by the 'knot area ratio' (KAR) at the worst cross section of the piece, a distinction being drawn between knots which occupy more than half the margin area of the projected cross section and those which do not (see 10). For the 'margin condition' a KAR of 1/3 is allowed for GS grade and 1/5 for MS. Where no margin condition exists, these are increased to 1/2 and 1/3 respectively for the two grades. Groups of knots, each exceeding 90 per cent of the permitted KAR must be separated along the piece by at least 1/2 its width. Knots under 5 mm diameter are ignored and no distinction is made between knot holes, dead or live knots.

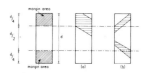

10 *Knot in piece (a) exceeds half a margin area so there is a 'margin condition'; but knots in piece (b) do not occupy half of either margin area so there is no ('margin condition'.) The 'knot area ratio' KAR depends on this condition.*

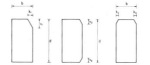

11a *Wane on face is* $\dfrac{v_1}{d}$

or $\dfrac{v_2 + v_3}{d}$.

Wane on edge is $\dfrac{k_1}{b}$ *or* $k_2 + k_3$

11b *Slope of grain (α) can be determined by means of a swivel-handled scribe.*

3.10 The ratio of wane to edge or face of a piece is limited for GS to 1/3 if within 300 mm from end but to 1/2 if further and not exceeding 300 mm in one continuous length. For SS it is limited to 1/4. **11a** shows method of wane measurement. Slope of grain of 1 in 6 is allowed for GS and 1 in 10 for SS. **11b** shows method of measurement. Fissures or pitch pockets not deeper than half the thickness of the piece are unlimited in number. If equal to the thickness they are only permitted at the ends of SS pieces and to a length not exceeding the width. For GS, end fissures are allowed to one and a half times the width and elsewhere not over 600 mm long. Fissures of intermediate depth are allowed up to 900 mm long for GS and 600 mm for SS, but in both cases not more than 1/4 the length of the piece. Sap stain, pin holes and worm holes are permitted to a limited extent, provided there is no active infestation, but fungal decay, brittle heart or other abnormal defects cause rejection. Pieces that are bowed, sprung or cupped excessively, having regard to the end use, may also be rejected.

3.11 Tropical hardwoods are now excluded from the provisions of BS 4978 and are visually graded in accordance with the recent BS 5756[23] which specifies the limits of characteristics for a single hardwood structural grade. Knots may not exceed 1/4 of edge or width to which they are related and if separated by less than twice the width of the piece their sum is the limiting dimension. Slope of grain is not to exceed 1 in 11 and interlocked grain 1 in 4. Fissures are not limited in length or position if not deeper than 1/3 the thickness, but if deeper they are limited in length to one and a half times the width or 1/5 of the length, whichever is smaller; fissures through the thickness are allowed only at ends and to a length not exceeding the width of the piece. Resin pockets are more restricted than in BS 4978. Excessive distortions are prohibited and spring, ie curvature on the strong axis, is limited to 7 mm per 2 m length.

3.12 The strength grading of timbers by electro-mechanical grading machines has been in operation in the UK for over a decade and there are now many machines in use. They generally operate by applying a constant load on the piece as it passes through and the measure of the deflection determines the grade, since there is a fairly good correlation between stiffness and strength properties. The machines can be programmed to select the required grade for a given species or group of species. Machine graded timbers do not have to be inspected for knots, slope of grain or rate of growth, but the remaining limitations in the standards have to be determined as for visually graded timbers. Grading machines are subject to approval and inspection by the British Standards Institution. Visual graders are trained and certificated by the Stress Grading Association. Stress graded material is required to be stamped, whether imported graded or graded in the UK.

Grade stresses

3.13 Since its amendment in 1973, CP 112: Part 2[2] contains many tables of working stresses for the various grades in BS 4978 in respect of bending, tension, compression parallel to grain and perpendicular to grain, shear parallel to grain and both mean and medium moduli of elasticity. There are separate tables for 'green' timbers, ie with a mc over 18 per cent and for 'dry' timbers, not exceeding 18 per cent. The tables cover a range of imported and home-grown softwood species both individually and grouped for related strength properties (s1, s2 and s3 groups). There are separate tables of values for Canadian timbers, imported graded to NLGA standards. Table III shows the 'dry' stresses and elastic moduli for European Redwood and Whitewood.

Grade stresses for the structural hardwoods, in CP 112 and tables III and v of Information sheet TIMBER 1, are given in the code for all the numbered grades (40, 50, 65 and 75) and separately for green and dry timbers. The permissible dry bending stresses for some common imported and home-grown hardwoods, in the medium or low price range and in good supply, are given in Table v below. Stress values corresponding to the structural grade for tropical hardwoods

Table IV Dry grade stresses and moduli of elasticity for European redwood and whitewood in N/mm²

Property	Grade GS	MGS	M50	SS	MSS	M75
Bending	5·1	5·1	6·6	7·3	7·3	10·0
Tension	3·5	3·5	4·6	5·1	5·1	7·0
Compression parallel to grain	5·6	5·6	7·1	8·0	8·0	10·8
Compression perpendicular to grain	1·38*	1·38*	1·55*	1·55*	1·55*	1·80*
Shear parallel to grain	0·86	0·86	0·86	0·86	0·86	1·28
Mean modulus of elasticity	8 600	8 800	9 000	10 000	10 200	10 700
Minimum modulus of elasticity	4 900	5 400	5 600	5 700	6 400	6 700

* These values may be increased by a factor of 1·1 if redwood is used separately.

Table v Dry grade stresses in bending for hardwoods in N/mm²

Grade	Species					
	Green-heart	Gurjun/Keruing Ash,* Beech*	Opepe	Sapele, Jarrah	Iroko	Abura
40	16·5	7·9	11·7	8·6	9·3	6·2
50	20·7	9·7	14·5	11·0	11·7	7·9
65	26·9	12·4	18·6	14·1	15·2	10·3
75	31·0	14·8	22·4	16·9	17·6	12·1

* Home-grown

in accordance with the grading rules of BS 5756: Part 1 will appear in the revised CP 112 as BS 5268: Part 2—Permissible stress design.

Glued-laminated timber

3.14 Glue lamination is a process of building-up timber members of large cross-sections and long lengths by gluing together under pressure small cross-section boards not exceeding 44 mm in thickness. The grain of the wood in adjacent laminations should be parallel to the axis of the member (as distinct from plywood, in which the grain of adjacent veneers is generally at right angles). Laminations in a built-up member may be vertical or, more commonly, horizontal.

3.15 Glued-laminated members can be made of any of the species in CP 112 or any other species with known strength properties. However, ease of gluing and securing a firm bond should be ensured and expert opinion sought when specifying unusual species. Members can be straight or formed to a curvature, of uniform or variable cross-section as required. Practically any cross-section or length can be produced, the limiting factors generally being transportation and erection facilities.

3.16 BS 4169[9] lays down control of production; maximum thickness, tolerances, moisture content and other conditions for laminations; types of adhesives for internal and external members; jointing methods; processes of manufacture, methods of preservative and fire retardant treatments. It also classifies laminated members by appearance into *architectural*—machined and sanded after manufacture and free of blemishes, *industrial*—machined after manufacture but blemishes permitted, and *economy*—without surface finishing (recommended for concealed members). There is no structural distinction between these classes if the members are manufactured from similar materials, provided the overlaps of adjacent boards in the 'economy class' do not exceed those allowed in the standard and which vary from 6 mm ($\frac{1}{4}$in) to 32 mm ($1\frac{1}{4}$in) depending on the width of the laminations. Design methods for glued-laminated members must be in accordance with CP 112 and they are briefly described in Technical study TIMBER 3 and Information sheet TIMBER 3.

Plywood

3.17 Plywood is the oldest wood-based panel product, and the most important in terms of volume of production and the range of its structural applications. It is defined in BS 565 as 'a product of balanced construction, made up of plies assembled by gluing, the chief characteristic being the crossing of alternate plies to improve the strength properties, and minimise movement in the plane of the board'. Boards formed of more than three plies are usually designated as 'multi-ply', but the number of plies is always uneven **12**. The plies (or veneers) are usually all rotary cut for constructional plywood[7] **13**, but may be sliced, usually for face veneers in decorative plywoods **14**.

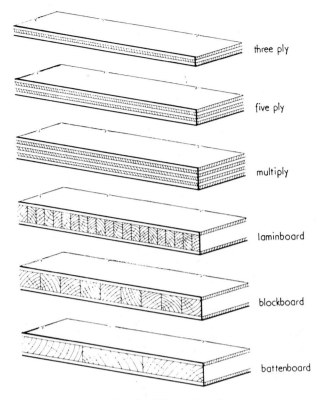

12 *Types of plywood and solid core boards*

three ply

five ply

multiply

laminboard

blockboard

battenboard

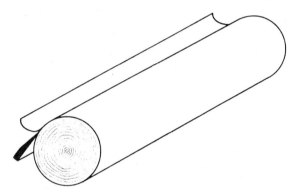

13 *Peeled or rotary cut plywood veneer*

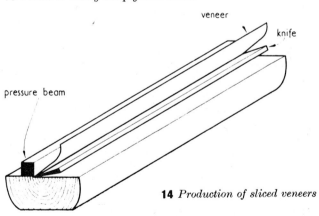

veneer

knife

pressure beam

14 *Production of sliced veneers*

Structural properties

3.18 Wood is 25 to 30 times stronger along the grain than across grain. Cross bonding of adjacent veneers in plywood tends to equalise the strength in both directions of the board as the number of plies increases. The cross bonding of plywood also gives it greater resistance to impact loads than can be sustained by solid timber boards of the same species and equal thickness. The strength properties of individual types of plywood depend on: the species (or strength group of species) of the veneers; the grade of each layer; the geometry of the section and the bonding agent.

3.19 The sanding of plywood surfaces to obtain smoother faces reduces the outer plies and modifies the section properties. CP 112 gives details of the various grades of structural plywoods of Douglas fir (Canadian), European birch (Finnish) and tropical hardwoods (British made), their dimensions and section properties, their grade stresses and their moduli of elasticity and rigidity. Table VI of Information sheet TIMBER 1 lists the characteristics and structural properties of a selected range of these types of plywood. For other types, whose properties and structural suitability are not included in the code, information should be sought from TRADA, PRL or the appropriate plywood association.

3.20 The construction of plywood and the resulting structural properties make it suitable for structures where its high panel shear values combined with flexual rigidity and light weight can be fully exploited, eg in webs of I- and box-beams, folded plate roofs of considerable span and stressed skin structures generally (see Information sheet TIMBER 4 and 5). Plywood has no line of cleavage and therefore cannot be split. Nails, screws or rivets may be placed in close proximity to edges of boards without risk of splitting, but edge distances in receiving timber members should generally comply with those of table 24 in CP 112.

Moisture movement

3.21 By crossing the grain of adjacent veneers in plywood, the moisture movement is restrained. Tests on three-ply construction have shown that from a saturated to oven-dry condition the panels shrunk by approximately 0·45 per cent in the grain direction of the face plies and 0·67 per cent in the grain direction of the core veneer. However, movement is reduced and equality in both directions approached with an increase in the number of plies and glue lines. In internal situations where equilibrium mc of wood is likely to vary from 9 to 12 per cent, movement of plywood is negligible. For external exposure, where three-ply plywood installed at 10 per cent mc might attain 20 per cent mc, the amount of expansion can be calculated by relating the difference in mc to the values given; ie 20-10 per cent = 10 per cent increase in mc, giving maximum potential swelling along face grain 10 per cent of 0·45 = 0·045 of length, and across grain 10 per cent of 0·67 = 0·067 of width. For multi-ply of five, seven, nine, etc, construction the values are reduced *pro rata*. The use of water-resisting WBP glues also reduces swelling in the thickness of plywoods. The balanced construction generally prevents distortions of the plane due to moisture movement, but extreme differentials of temperature and humidity on two opposing faces may require additional restraints, particularly in the case of thin three-ply constructions.

Durability and preservation

3.22 The durability of plywood depends on the natural durability of the timber species from which it is made (see table VI, Information sheet TIMBER 1), and on the weather-resistant properties of the adhesives used in its manufacture. Generally structural plywoods for which values of stress are given in CP 112 are bonded with WBP-type adhesives (usually phenolformaldehyde). BS 1455[10] specifies a number of tests and test conditions to determine the quality of bond.

3.23 Structural plywoods of non-durable species can be treated efficiently with various preservatives during manufacture or subsequently. Details are specified in BS 3842[11] (see Information sheet TIMBER 6). Plywoods can also be obtained impregnated or surface treated with fire retardants to class 1 or 2 spread of flame classification to BS 476: part 1.

Fire behaviour

3.24 The spread of flame classification of plywood is the same as that of the timber species of the face veneers, generally class 3 to BS 476. However, it can be treated to attain class 1 or 2 as required by impregnation or surface application of suitable paints or varnishes.

Blockboard and laminboard

3.25 Three- or five-ply constructions using a thick built-up core of wood and thin outer veneers have long been manufactured for internal uses such as manufacture of furniture, panelling, partitions and doors. The main objectives of such construction are to stabilise the panel and achieve economy using high quality decorative woods for face veneers and using lower quality and usually lower density woods for the core.

3.26 BS 3444[12] regulates the construction requirements, permissible defects, quality of veneers and cores, moisture content, standards of adhesion and surface stability requirements. It also specifies dimensional tolerances, marking requirements and procedures for sampling and testing. The distinction between blockboard and laminboard is essentially in the maximum width of the glued up strips— up to 7 mm for laminboard and up to 25 mm for blockboard. Battenboard, which is not included in the standard, is generally defined as having strips up to 75 mm wide **12**.

3.27 Principal imported and British made boards are listed in BS 3583[13] along with species used in cores and veneers, bonding agents and overall dimensions and thicknesses. (A more up-to-date list and other notes are available from TRADA[14].) The range of available sizes and thicknesses varies with countries of manufacture and individual mills. Widths are commonly 1220 mm, lengths are generally not less than 2440 mm, and thicknesses range from 12 to 30 mm. The sizes are in most cases a metric conversion from the existing manufacturing sizes in imperial measure. Modular metric sizes are usually cut from the next greater imperial sizes.

Structural properties

3.28 As with plywood, structural and other properties of particular boards will depend on many factors, including the species and grades of the cores and plies, the geometry of the section and the bonding. Little, if any, structural engineering applications have yet been made, but in practice these boards are often used under conditions of loading which require high flexural rigidity and strength. Recommended values for permissible bending stresses and moduli of elasticity for three- and four-ply blockboards and laminboards of Finnish manufacture* have been published by the Finnish Plywood Development Association[15] and are based on tests of short duration loading carried out at the State Institute for Technical Research, Helsinki. Those values and the conditions to which they apply are shown in table I Information sheet TIMBER 1.

Durability

3.29 Blockboard and laminboard are mostly made of species of low or moderate durability, and usually not with exterior type (WBP) adhesives. Consequently they are not recommended for external or internal use where the mc is likely to be above 18 per cent for prolonged periods.

Wood chipboard

3.30 The most recent development of wood-based panels, chipboard has been the fastest to expand in the building industry in recent years, its growth trends rivalling those of plastics. It is made in a wide variety of densities and

*Accounting for about 80 per cent of the blockboard and laminboard imported to the UK.

thicknesses of wood chippings and other particles, bonded with synthetic resins and/or other organic binders and cured under heat and pressure. Particleboards are also made, on a smaller scale, from other ligno-cellulosic materials (such as flax shives) and flaxboards have also been used extensively in the UK. BS 5669[16] specifies minimum requirements for various grades of wood chipboard (see table II Information sheet TIMBER 1) and test methods for all particle boards.

3.31 According to their method of manufacture, particleboards may be homogeneous or have surface layers of higher quality or a different texture from the core. They may be single layer, three layer, graded density (all platten pressed), or extruded boards, including hollow core boards.

Structural properties

3.32 Although the denser chipboards in general tend to be stronger than the lighter ones, there is no consistent correlation between strength and density. The recent BS 5669, unlike BS 2604 which it supersedes, lays down no density limitations for chipboards for structural use. It defines and gives the levels of quality, including basic structural properties of four types of board: Type I—standard, Type II—flooring, Type III—improved moisture resistance and Type II/III combining the strength levels of II with the moisture resistance of III. Table II of Information sheet TIMBER 1 gives a summary of the mean strength values required by the standard for the different types and thicknesses of board. The standard does not deal with extruded boards nor with particle boards of non-wood fibres.

3.33 The essential strength properties required for boards intended for structural use are mean bending stresses and moduli of elasticity and resistance to impact loading. Flooring grade boards are not normally available in thicknesses less than 18 mm and the standard recommends joist spacings which should not be exceeded for domestic loading. These are: 450 mm for 18 mm or 19 mm thickness and 610 mm for 22 mm thickness. For other loadings or other structural uses, design data should be obtained from manufacturers or their agents. Test methods covering a wide range of properties including shear in the plane of the board and compression parallel to the plane are also stipulated in the standard, as well as permissible deviations and tolerances and periodic quality control.

3.34 For fixing with mechanical fasteners, the permissible loading values, distances to edges, etc should be determined. BS 5669 lays down test methods for resistance to axial withdrawal of screws and the limiting values for standard boards. Lower density and layered boards are generally inferior in this respect than high density and homogeneous boards, but where this property is important information should be obtained on the basis of tests in accordance with the standard. Chipboards are more susceptible to creep deformation and permanent set than either solid timber or plywoods. For conditions of long duration high level loading it is important that acceptable limits are established by tests and measurements laid down in the standard.

Moisture effects

3.35 Chipboards in accordance with the standard are dispatched from the factory at a mc between 7 and 13 per cent. Considerable increases in moisture content during transport, site storage, erection and use will cause:

1 Expansion and swelling. Boards may not contract fully to the original dimensions when dried to the mc on dispatch. The standard does not lay down limits in this respect and where this property may be significant information should be obtained from manufacturers.

2 Reduction in strength properties and increase in creep and permanent set. Boards of Type III or II/III are manufactured

with moisture inhibiting characteristics, their swelling is limited by the standard to 8 per cent even after 24 hours immersion or 72 hours of thrice repeated cyclic tests including immersion, freezing and air exposure. They also recover appreciable tensile strength perpendicular to the plane of the board, indicative of their general strength recovery. These boards are therefore suitable for conditions of short term exposures to water and/or high humidity. However, they do not necessarily resist prolonged exposure to weather. In some countries certain chipboards with high moisture resistance are permitted for external use.

3.36 Deterioration of the board and fungal attack may also result from prolonged wetting or exposure to very damp conditions.

Fibre building boards

3.37 Fibre building boards are manufactured from the fibres of ligno-cellulosic materials (mainly wood), the primary bond being derived from the felting of the fibres and their inherent adhesive properties. Bonding, impregnating and other agents may be added during or after manufacture to modify particular properties of the boards. In production a wide variety of wood species and much forest residue is used, including tops, branches and thinnings, as well as sawmill and planing mill wastes. Various processes are used for chipping, pulping, felting, pressing and subsequent treatments (eg oil tempering of hardboards), and production in modern mills is increasingly rationalised and automated. Price trends of fibreboard, like those of particleboard, have in the past couple of decades compared favourably with those of other wood products and most building materials.

3.38 Early production of fibreboards was concentrated on highly compressed hardboards and on low-density insulating boards. In recent years, however, there has been a greater diversification to meet specific end-uses in building. New types of board, particularly in the medium-density range, have been produced with good dimensional stability and adequate strength for many end-uses.

3.39 BS 1142[18] for fibre building boards is in three parts. Part 1 lays down methods of test for a wide range of properties and quality control, including the crucial ones of moisture movement and bending strength. Part 2 contains specification and values for hardboards and medium boards, and part 3 for insulating boards including sheathing and sarking grades. Table VII of Information sheet TIMBER 1 contains a summary of the main characteristics of fibre building boards of various types, their minimum bending strength requirements and their maximum moisture movement in length, width and thickness.

3.40 The BS does not give limiting values for moduli of elasticity, tensile strength parallel and perpendicular to surface, resistance to withdrawal of nails or screws and other structural properties of practical importance. These have to be established by appropriate tests or data of performance tests obtained from manufacturers.

3.41 A recent series of tests carried out by the PRL[19] on four brands of 4·8 mm hardboard (three standard grade and one tempered) gave valuable information on the effects of board direction, face location, temperature and relative humidity on the mean modulus of rupture in bending. Bending strengths in normal temperature and humidity conditions were considerably higher than the minima in the standard and a safety factor of 3 in deriving working stresses was considered appropriate.

3.42 High moduli of elasticity were also established for the boards (6·28 kN/mm² for the tempered board and 3·78 kN/mm² for the worst of the standard hardboards tested).

Over a range of relative humidities from 30 to 90 per cent (at a constant temperature of 25°C) strength values dropped by nearly 30 per cent on average and elasticity moduli by up to 40 per cent. The effect of temperature rise to 40°C (with a slight rise in relative humidity from 60 to 65 per cent) on both strength and stiffness was limited. It was clear from these tests (and others reported in detail in a FIDOR technical bulletin[20]) that proper allowances can be made for various loading and exposure conditions as a basis for structural design with hardboards.

Moisture effects on stability

3.43 Like other wood-based products, fibreboards are subject to dimensional changes with variations in moisture content induced by changes in the relative humidity of the atmosphere. The limiting values for the various types of board are shown in table VII (Information sheet TIMBER 1). BS 1142 specifies tests and limits for water absorption for hardboards, medium boards and sheathing and sarking boards. Tempered hardboards (types TE and TN), high-density medium board (type HME) and low-density medium board (type LME, treated in manufacture to reduce absorption) are the most suitable for conditions of high relative humidity.

Durability

3.44 The denser boards, particularly the oil-tempered ones, have a higher natural resistance to decay under conditions of high humidity than the lower-density and insulating boards; but all boards can be treated to inhibit decay.

Wood wool slab

3.45 Mineral bonded wood wool slabs have been manufactured for more than half a century and were first produced industrially in Austria in 1914. They are different in appearance, manufacture and properties from other wood-based panel products. The inorganic binders approximate to half the weight of the slabs, which are moreover comparatively thick (25 mm to 102 mm), rendering them heavy and bulky in proportion to their value. They are also brittle and subject to damage during shipping and these characteristics restrict their use to areas or countries of manufacture.

3.46 Although other hydraulic binders are used in some countries in the manufacture of wood wool slabs, BS 1105[21], recently revised and metricated, defines them as 'consisting essentially of wood wool and Portland cement, mixed together, pressed and matured and containing not more than 2·5 per cent of anhydrous calcium chloride'. The quality of the slab depends on the properties of the wood strands which must have high strength and good felting qualities and be free of chemical substances which could be deleterious to the binder. Pine and spruce are mentioned in the BS as suitable in this respect.

3.47 BS 1105 specifies sizes, weights, strength, thermal conductivity and fire performance requirements of slabs of specified thickness and methods of measurement and testing of these characteristics as well as of sound absorption when acoustic properties are required. Two classes of slab are laid down, type A for non-loadbearing purposes (such as linings, ceilings, partitions, insulation and permanent shuttering) and type B of greater strength but not less than 51 mm thick and intended primarily for roof construction but suitable also for the purposes of type A. Type B slabs must be marked with a white line 25 mm wide on a short edge.

Sizes

3.48 BS 1105 specifies metric modular sizes of a standard co-ordinating width of 600 mm and lengths from 1800 mm to 3000 mm in increments of 300 mm and a further length of 4000 mm. These are also the work sizes and the maximum limits of manufacturing sizes, the minimum limits being set at 6 mm below these lengths and width. Work size thicknesses are 25, 38, 51, 64 and 76 mm with a manufacturing tolerance of 3 mm above and below each of these thicknesses. Metric equivalents are also given in the standard for slabs which will continue for a time to be manufactured to imperial sizes, of the same thicknesses as those made to the metric modular slabs in a width of 610 mm (2ft 0in) and lengths from 1829 mm (6ft 0in) to 3048 mm (10ft 0in), and 102 mm (4in) thick slabs for lengths of 3429 mm (11ft 3in) and 3810 mm (12ft 6in). Reductions to minimum manufacturing sizes and $\pm$ tolerances for thicknesses are the same as those for the metric modular slabs. Maximum weights of air-dry slabs are 18 kg/m² for 25 mm slab to 47 kg/m² for 102 mm slab, thus allowing for a reduction of density as the thickness of slab increases.

Strength
3.49 The BS does not specify a modulus of rupture for wood wool slabs but simply requires the boards to sustain a specified load at quarter points of their lengths for a short duration (1 minute); type A slabs in the dry state with the transverse bearers at 450 mm apart and type B slabs after steeping and draining, with more than double that load and with bearers at 600 mm apart. This is an empirical test simulating use conditions and the induced extreme fibre-bending stress values have been calculated for the A slabs from about $0 \cdot 2$ to $0 \cdot 9$ N/mm² and for B slabs from $0 \cdot 4$ to $0 \cdot 8$ N/mm².

Other properties
3.50 The enrobing of the wood wool with cement imparts excellent fire properties to the slabs, making them virtually non-combustible. Slabs conforming to the BS are required to have class 1 spread of flame rating and a fire propagation index not exceeding $12 \cdot 0$ with sub-index (i) not exceeding $6 \cdot 0$ when tested in accordance with BS 476: part 6 (ie conforming to class O of the Building Regulations).
3.51 Because of their higher density, wood wool slabs have a higher thermal conductance than wood-based products without mineral binders. This, however, is compensated by the greater thickness of the slabs in normal use. BS 1105 sets maximum limits of thermal conductivity for slabs of 25, 38 and 51 mm thickness—$0 \cdot 1$ W/m°C for 25 mm and 38 mm slabs and $0 \cdot 093$ W/m°C for 51 mm slabs.
3.52 The BS gives no information on the behaviour of wood wool slab in humid conditions or on its resistance to fungal or insect attack, but laboratory tests and field experience in countries with humid tropical climates and with termite risk have been satisfactory. Nevertheless, although occasional wetting and redrying does not affect the strength performance of the board, it is not generally recommended for use in persistent humid conditions.
3.53 The sound absorption (or noise reduction coefficient) is high but varies with thickness and density of the slab, its surface texture and mode of support. Wood wool slabs combine properties which render them attractive for use in a roofing system; eg the B type in particular has adequate strength, the thick slabs have good thermal insulation value and good sound absorption properties when exposed to the ceiling side. Wood wool slab can be laid in mortar or fixed by nailing to studs or joists and can be sawn and cut fairly easily.

References

1 BS 565: 1972 Glossary of terms relating to timber and woodwork
2 BS CP 112: Part 2: 1971 The structural use of timber (metric units)
3 Food and Agriculture Organisation, Plywood and other wood based panels, papers submitted to the International Consultation on Plywood and other Wood-based Panel Products, Rome 8–9 July 1963, FAO Rome 1965. (Paper by S. A. Lundgren—FAO/PPP CONS/Paper 5.11)
4 CP 3: Chapter III: 1972 Sound insulation and noise reduction
5 BS 4471: Dimensions for softwood, Part 1: 1978 Sizes of sawn and planed timber, Part 2: 1971 Small resawn sections
6 BS 373: 1957 Testing small clear specimens of timber
7 BS 476 Fire tests on building materials and structures: Part 6: 1975 Fire propagation test for materials; Part 7: 1971 Surface spread of flame tests for materials
8 BS 1990: 1971 Wood poles for overhead lines (power and telecommunication lines)
9 BS 4169: 1970 Glued laminated timber structural members
10 BS 1455: 1972 Plywood manufactured from tropical hardwoods
11 BS 3842: 1965 Treatment of plywood with preservatives
12 BS 3444: 1972 Blockboard and laminboard
13 BS 3583: 1963 Information about blockboard and laminboard
14 TRADA Plywood its manufacture and uses. Technical Brochure TBL 7, Hughenden Valley, the Association 1972 (Revision 1979)
15 FPDA Finnish birch blockboard and laminboard; Technical Publication no 13
16 BS 5669: 1979 Specification for chipboard and methods of test for particleboard (supersedes BS 2604 and BS 1811)
17 BS 5450: 1979 Sizes of hardwood and methods of measurement
18 BS 1142: Fibrebuilding boards, Part 1: 1971 Methods of test: Part 2: 1971 Medium board and hardboard; Part 3: 1972 Insulating board (softwood)
19 Building Research Establishment, Princes Risborough Laboratory study subsequently published by FIDOR in 20
20 FIDOR structural properties and applications; Technical bulletin TB9/72/2, London, FIDOR, 1972
21 BS 1105: 1972 Wood wool slabs up to 102 mm thick
22 BS 4978: 1973 Timber grades for structural use
23 BS 5756: 1979 Grading of structural hardwood for structural purposes
24 BS 5268: Part 2 Permissible design stress (to be published, superseding CP 112: Part 2)

Technical study
Timber 3

Structural forms and design

This study by EZRA LEVIN *covers the range of structural forms which are suitable for construction in timber and wood based materials. It covers skeletal structures and their component members, portal frames and arches as well as surface structures such as folded plates and shells. The degree of skill in design and manufacture required for the various forms and suitable methods of connection are indicated. Information sheets* TIMBER 5 *and* 6 *will amplify the detail or offer tabulated data for design with most of the forms described. (To ensure that timber construction is permitted, this should be read with Technical study* TIMBER 4 *'Timber and the Building Regulations 1972', and the tables in Information sheet* TIMBER 2.)*

1 Structural forms of timber

1.01 Wooden poles and sawn timbers from logs are naturally rectilinear materials; this renders them most suitable for skeletal structures such as post-and-beam construction and common trusses. Natural limitations on size and length of members are overcome by building up with mechanical fasteners or more effectively by glue lamination, a procedure which also permits the production of curved members such as arches or portals. Although surface structures are possible with sawn boards, especially if crossed diagonally, the development of structural plywoods has rendered them more widespread and economical. However, sawn boards remain pre-eminent in the construction of double curvature shell roofs.

1.02 Choice of form will depend on many factors—particularly the relative suitability for the specific function of the building or the structural element from both the functional and aesthetic points of view; availability of suitable materials; means of connection and manufacturing capabilities; transport and erection facilities and comparative costs. The relative ease of design may also affect choice, but will certainly be a factor in deciding if the design requires a specialist consultant or is to be entrusted to a manufacturer with an adequate design staff.

Design procedure

1.03 Structural design with timber, as with other materials consists essentially of two distinct phases:
1 Determining the forces acting on the structure and analysing their distribution through its component parts
2 Establishing the size of members and required connections between them to sustain loads safely and to ensure structural stability.
The first phase is largely independent of the nature of the structural material, except that the dead load of the structure may be an important component of total load.

1.04 In buildings or elements of light or moderate imposed loadings, materials with high strength and stiffness-to-weight ratios are more likely to be competitive in cost than in those subjected to heavy loadings. As a generalisation it may be said that timber is most suited for structures or components which are large in relation to the loads they have to carry, eg single-storey buildings of all kinds, particularly those of great height and span, roofs of all kinds, lightly loaded floors, footbridges, etc.

1.05 Loadings will generally be to the recommendations in BS CP 3: Chapter V[1] as modified by mandatory requirements in Building Regulations or by-laws and the determination of the actual loading conditions by the designer. An example of determination of loads for a simple timber building is given in Information sheet TIMBER 5.

1.06 In calculating the reaction of timber members to the forces acting upon them, timber is considered as an elastic material to which the classic elastic theories of strength of material apply. These are modified, however, in the light of experience and test data to suit timber's departures from ideal elastic bodies, particularly in respect of its anisotropic structure (ie different strength and elastic characteristics along and across the grain) and its response to different durations of load. These differences are generally expressed in BS CP 112, the code of practice for the structural use of timber[2], as *modification factors* of the permissible stresses.

1.07 In this study consideration is first given to the design of individual members subjected to different kinds of stress—bending, compression tension or combined stresses. Next, the assembly of members into planar elements or components is reviewed (eg trusses, portal frames, framed walls and floors). Finally, attention is given to spatial structures, both those assembled from planar elements and surface structures such as shells. Jointing is the most crucial aspect of successful design in timber. Technical study TIMBER 5 and Information sheets TIMBER 3 and 4 deals with various jointing techniques in detail. Here suitable connection methods for the various forms of structure described are indicated.

2 Solid structures

2.01 Timber can be used in solid construction for walls and floors, by building up from logs, planks or battens.

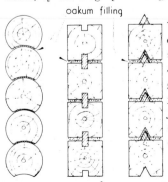

oakum filling

1 *Typical cross sections of log cabin walls:* **a** *partly machined logs;* **b** *and* **c** *squared logs, with* **b** *filled grooves with preservative paste and* **c** *nailed triangular fillets. Horizontal joints are caulked with oakum and courses dowelled together.*

a   b   c

Walls

Horizontal log construction

2.02 Solid walls of log construction, with logs laid horizontally in courses, are still popular in many forest areas for mountain chalets, huts, village schools and even large structures. They are structurally inefficient as the material is stressed across the grain in the direction of least strength. They make profligate use of forest resources (unless thinnings are used), but are easy to fabricate, transport and erect. People like them for their rusticity and robustness and, probably, for the smell of pine. They are designed according to tradition and attempts at engineering refinements (in reducing sections, etc) lead to a rather mean appearance. Sections still common in Russia are shown in **1**.

Vertical plank panels

2.03 Jointing is necessary to provide cohesive elements and distribute stresses evenly. Efficient methods take the form of tongued and grooved joints or splined joints. Tongues and grooves worked on each component are shown in **2**. Timber worked with grooves and small rectangular section splines of wood or metal to keep components together are shown in **3**. Timber can also be grooved longitudinally on both edges and fitted together as in **4**. This method has the advantage of increasing resistance to buckling.

2.04 Plank walls are designed as columns (see below), their maximum loading being determined by the ratio of effective length (height) to r (radius of gyration) or d (thickness of panel) and the elastic modulus of the wood used.

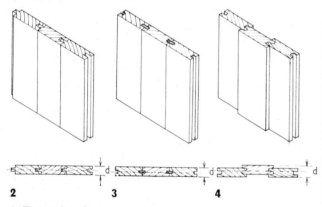

2 *Tongued and grooved jointing for wall panels*
3 *Spline jointing for wall panels*
4 *Staggered plank construction for wall panels*

Roof and floor slabs

2.05 These are designed to span between walls, beams, purlins or arches. They may be used in single or continuous spans or in random lengths (with splined ends). Design limits are usually set by deflection of end spans. These forms of construction are at their most efficient when slabs are continuous over at least three supports and least efficient in single spans. This can be deduced from the relevant formulae for deflection which are for:

$$\text{single span}\frac{5wl^4}{384EI} \text{ and for multiple span}\frac{wl^4}{185EI}$$

2.06 Three methods of construction are used: tongue and groove, spline jointing and mechanical lamination. Tongued and grooved is similar to the conventional method used for floor boarding **5** but double tongue and double groove edges are sometimes used. Spline joints are similar to wall panels **6**.

2.07 Mechanically laminated construction is essentially a series of rectangular section timber joists, or planks on edge packed together and connected by nails driven through the sides of successive members. Nails should be $2\frac{1}{2}$ × width of plank (w) in length, spaced 750 mm apart for spans of

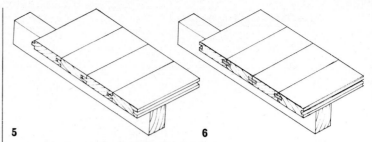

5 *Tongued and grooved joint for roof or floor slabs*
6 *Splined joint for roof or floor slabs*

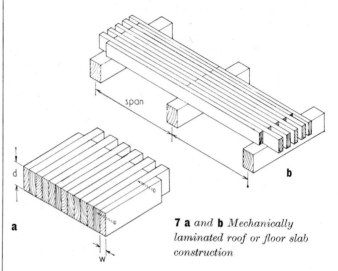

7 a *and* **b** *Mechanically laminated roof or floor slab construction*

about 600 mm and graded down to 450 mm apart for spans of 1200 mm and over. They should be driven staggered top and bottom of the middle third of the depth **7**. The limits of economic span are between 1·2 m and 4·8 m for this form of construction. Such floors are usually employed for heavy loading, often in conjunction with heavy timber framing (mill construction).

3 Skeleton structures

Columns and struts

Elastic behaviour

3.01 The general behaviour of elastic columns and struts was described in section 2 ANALYSIS. Timber columns follow this pattern of behaviour within certain limits set by the nature of the material. CP 112[2] gives some stress modification factors for use in relation to slenderness ratio and to type of loading. The factors cover low- and high-grade softwoods and hardwoods. Limits of permitted slenderness ratio relative to loading conditions are also given in the code: these are normally 180 for dead and imposed compressive loads but up to 250 for certain other conditions.

3.02 In compression members subject also to bending (as often happens, say in chords of trusses) the code limits the sum of the ratios of applied to permissible stresses in compression and bending to 1·0 in short members $\left(\frac{l}{r}\leqslant 20\right)$ and to 0·9 for longer members.

Effective length

3.03 For purposes of structural design effective length is governed by the conditions of restraint at the ends of actual column length. CP 112 lays down the relationships shown in **8**, **9**, **10**, **11** and **12**. Alternatively effective length may be taken as the distance between two adjacent points of zero bending moment on the column subjected to the worst loading conditions.

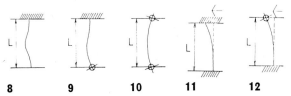

8 **9** **10** **11** **12**

8 *Effective length* $= 0 \cdot 7L$ *(actual length) where a column is restrained at both ends in position and direction.*
9 *Effective length* $= 0 \cdot 85L$ *where both ends are restrained in position and one in direction.*
10 *Effective length* $= L$ *where both ends are restrained in position but neither end in direction.*
11 *Effective length* $= 1 \cdot 5L$ *where one end is restrained in position and direction and the other partially in position but not in direction.*
12 *Effective length* $= 2L$ *where one end is restrained in position and direction and the other neither in position nor direction.*

Rectangular solid sections

3.04 For freestanding columns, square sections are equally efficient about both planes of buckling (x-x and y-y). Rectangular columns are more efficient if laterally supported in direction of least resistance to buckling **13**. Solid rectangular columns are economical in cost (particularly knee braced **14**) and limited only by available sizes and lengths of timber, which vary according to species (see Technical study TIMBER 2). Tables of permissible loads for low grade softwood columns of cross-sections up to 300 mm in depth and up to 6 m in height, are given in Information sheet TIMBER 6. Hardwoods of high compressive strength may prove economical for heavily loaded columns.

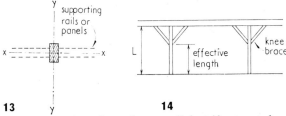

13 **14**

13 *A rectangular column is more efficient if supported laterally in the direction of least resistance to buckling.*
14 *Knee braces are a common method of reducing effective length of a column, and also spans of beams.*

Built-up columns

3.05 Where single-piece columns of adequate cross-section are not available they may be built up of several pieces, mechanically fastened together (nailing, bolting, and so on) to make up the required cross-section **15**; butt joints should be carefully machined and staggered. Metal plate inserts in joints improve efficiency. According to *Wood handbook*[3], *Timber engineering handbook*[4] and *Timber design and construction handbook*[5], such built-up columns have a reduced strength compared with similar section one-piece columns, depending on the effective length: least width ratio of the column as table I.

Table I Strength of built-up columns

Length: depth ratio	Strength (per cent; one-piece=100)
6	82
10	77
14	71
18	65
22	74
26	82

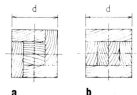

a **b**

15 *Solid built-up columns: a boxed solid core; b edges tied with cover plates*

3.06 Box-section hollow columns **16** can also be built up and are more efficient in relation to timber content, as radii of gyration are increased $\left(r = \sqrt{\dfrac{I}{A}} \right)$. For a given cross-sectional area A, the moment of inertia I increases with increased distance to axis.

3.07 I-section columns **17** are also efficient in relation to cross-sectional area, but if web or flanges are thin they may fail at a point below the Euler formula load through flange wrinkling or column torsion.

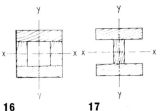

16 **17**

16 *Box-section hollow column*
17 *I-section column*

Spaced column

3.08 Spaced columns built up from two or more parallel members spaced apart by end and intermediate packing blocks **18**, glued, nailed, bolted, screwed or connected adequately, are covered by CP 112. Rules are given for length, spacing and fixing, also factors for increased effective length according to efficiency of connecting method and ratio of space to thickness of thinnest member (this varies from $1 \cdot 1$ for glued members side by side to $3 \cdot 5$ for nailed members with space three times the thickness of member.

3.09 In general, spaced columns are more efficient and economical than solid built-up columns. (More detailed design guidance on spaced columns and built-up members generally, subjected to a variety of loading conditions and end fixity conditions, is to be found in *Wooden structures* (pp187-207) edited by G. G. Karlsen[6].)

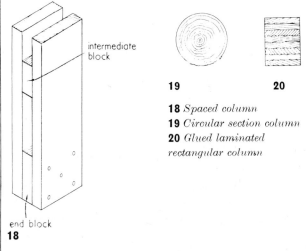

18 **19** **20**

18 *Spaced column*
19 *Circular section column*
20 *Glued laminated rectangular column*

Circular sections

3.10 Round columns **19** are generally used in pole barns, towers, and (in heavy sections) in warehouse construction. There are design formulae for tapering columns. A simple method is to design square columns and then determine a diameter which will give an equivalent cross section. In the case of a tapering column, this should be at $\frac{1}{3}$ from the top. For alternative fibre stresses, moduli of elasticity and strengths of poles of various dimensions see BS 1990[7]. Pressure treated poles are generally fixed directly in the ground to save foundation costs.

Glued laminated columns

3.11 These are generally rectangular in cross-section **20** but alternatively may be of I-section and/or tapering. Considerably more costly than solid or built-up columns, but have advantage of better appearance and freedom from checks or splits, as well as higher permissible stresses in accordance with the provisions of CP 112.

Loadbearing stud walls

3.12 In stud wall design, it is assumed that the studs are efficiently supported laterally either by sheathing or cladding materials or by nogging or diagonal bracing **21**. The studs can therefore be regarded as columns in a load-sharing system (spacing must not exceed 610 mm) and their strength is calculated about the plane parallel to the wall, using a 10 per cent increase in permissible stresses and the mean moduli of elasticity. (Examples of calculations are given in Information sheet TIMBER 5 including a table of permissible loading for studs 38 mm × 100 mm to 50 mm × 125 mm in single- and two-storey housing.)

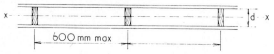

21 *Loadbearing stud frame wall construction. 50 mm 75 mm or 44 mm × 100 mm are generally found to be adequate for usual spans in single- and two-storey domestic buildings*

Beams

3.13 Timber beams behave generally in conformity with classical elastic theory but with modifications of permissible stresses resulting from the nature of the material, which are detailed in CP 112. These apply in particular to: duration of loading, as timber can support much greater loads for a short time than over a period of years; bearing stresses, owing to the anisotropic structure of timber and its relatively low resistance to compression perpendicular to the grain; shear in flexural members notched at the ends. Reduced stresses are also stipulated for deep beams.

3.14 For given conditions of loading, spans and end fixity, beam sizes are determined from the usual formulae in standard text books on strength of materials generally, or on the design of timber structures (see bibliography at end of this section). Check that the beam sizes withstand bending stress, shear and deflection (generally assumed at 0·003 of span).

Solid rectangular sections

3.15 Solid section beams of appropriate species and grade will generally prove to be the most economical **22**. Where deflection governs design, a low stress grade of a particular species will be more economical than a higher grade, as the code of practice specifies the same modulus of elasticity for all stress grades within a given species. Where bending strength or shear govern design and particularly where depth of beam is to be minimised the use of a higher grade and/or different species may well be justified. Information sheet TIMBER 7 contains three load/span charts for sawn and surfaced beams of common softwoods of a low grade (50) in widths of 38 mm, 50 mm and 75 mm (basic) and depths up to 225 mm (basic).

22 *Solid rectangular beam*

3.16 The charts in Information sheet TIMBER 7 show clearly that, except for very short spans and very high loads, deflection generally controls the size of beams. In these conditions there is no justification in using high grade materials, for the code allows the same value of E for all

grades. Differential E values for different grades have, however, been established by PRL/BRE, for European and Canadian softwoods and where these are used in design high stress grades may occasionally be justified.

3.17 As a rule, sawn timber beams with a maximum permissible depth-to-breadth ratio (which depends on the degree of lateral support and can be any figure between 2, where there is no lateral support, and 7, where both ends are held in position and both edges held firmly in line), will be the most economical within the range of sizes and lengths available for the various species (see tables III to V of Information sheet TIMBER 1).

Circular section beams

3.18 Round beams may be used in pole construction (eg barns) **23**. The code allows for an increase in stress value as a 'form factor'. An even greater increase is allowed for square section beams loaded diagonally **24**. With the increased stress value, it can be shown that in both cases the beams will have a bending strength equal to that of a square beam of the same cross-sectional area but normally loaded.

23 *Circular section beam (form factor = 1·18)*
24 *Square section beam loaded diagonally (form factor = 1·41)*

Trussed beams

3.19 Trussed beams are economical when used for long spans and large loads, but are limited in application because of the great depth of beam required and reduced headroom **25a, b**. They are usually built of solid rectangular timber beams and struts with steel rods. Attention should be paid to the bearing stress of struts on beams. Where headroom is limited but space above the beam is adequate, an inverted form of trussed beam may be used **25c, d**. In this case all members except hangers are timber. An approximate method of calculating trussed beams for estimating purposes makes use of the following formulae:

Uniform loading (W)		*Single strut*	*Double strut*
Tension in rod	$T_r =$	$\dfrac{0·312Wh}{3r}$	$\dfrac{Wh}{3r}$
Compression in strut	$C_s =$	$0·625W$	$\dfrac{W}{3}$
Compression in beam	$C_b =$	$\dfrac{0·312Wl}{2r}$	$\dfrac{Wl}{9r}$

Concentrated load (P) over strut			
Tension in rod	$T_r =$	$\dfrac{Ph}{2r}$	$\dfrac{Ph}{9r}$
Compression in strut	$C_s =$	P	P
Compression in beam	$C_b =$	$\dfrac{Pl}{4r}$	$\dfrac{Pl}{3r}$

Built-up solid beams

3.20 Beams are sometimes built up of timbers of relatively small dimensions to reduce effects of seasoning, to utilise available smaller sizes or to reduce effect of knots and other strength reducing features. Various types have been made and tested. Table II shows some of the main types used and indicates their relative efficiency in strength and stiffness, compared with solid beams of the same dimensions.

Table 11 Data for built-up solid beams

	Efficiency (compared with 100 per cent for solid single beams)	
	Strength (per cent)	**Stiffness (per cent)**
Vertically laminated bolted (small beams, say up to 10in deep, may be nailed except at ends)	100	100
Horizontally, bolted and keyed with ring connectors	95	80
Horizontally, bolted and keyed with oak or cast iron keys	70 (oak) 80 (cast iron)	50
Diagonally sheathed compound beam	75	50

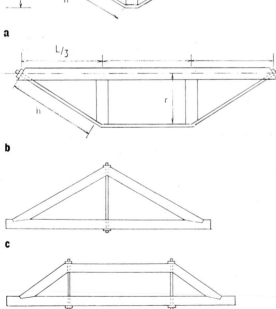

25a *Single strut trussed beam;* **b** *double strut beam;* **c** *single strut inverted beam;* **d** *double strut inverted beam*

Glued laminated rectangular sections

3.21 Principles of design and construction of glued laminated beams **26** and tables permitting choice of suitable sections of straight square section beams, are given in Information sheet TIMBER 12. The shapes and names of laminated beam types (according to the nomenclature of the American Institute of Timber Construction) are shown in **27**.

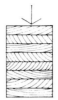

26 *Glued laminated beam (cross-section)*

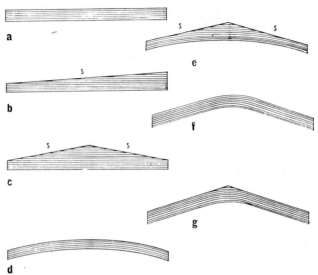

27 *Types of laminated beam (s denotes sawn surface, this should be avoided on tension side of beam. Pitched or curved beams should be manufactured with laminations parallel with tension):* **a** *straight;* **b** *single tapered straight;* **c** *double tapered straight;* **d** *curved;* **e** *double tapered curved (ridge built up on curve);* **f** *pitched;* **g** *double tapered pitched (ridge built up on pitch).*

3.22 Straight beams are the most economical. Pitched beams are designed to provide roof slopes and/or minimum depth at supports. Provision to counteract dead load should be made by cambering (before gluing) and not by cutting laminations on tension side. For the curved portions of laminated members stress reduction factors in accordance with CP 112 should be applied.

Laminated I-beams

3.23 I-section beams are somewhat more efficient than rectangular beams, as material is concentrated in the compression and tension zones of the beams. They may be manufactured with horizontal laminations only **28a** or by gluing (or gluing and nailing) battens to horizontally laminated beams **28b**, or—less usually—by vertical lamination **28c**. Horizontal shear may be the controlling factor in such beams. To prevent buckling web stiffeners become essential in deep beams with thin webs. Provision of camber is more difficult in **28b** and in **28c** than in **28a**.

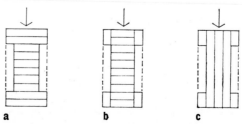

28 *Laminated I-beams:* **a** *horizontally laminated;* **b** *horizontally laminated and with applied battens to form the flanges;* **c** *vertically laminated with applied flange battens.*

29 *Laminated curved frames (school in Copenhagen, architect V. Jacobsen)*

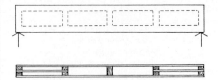

31 *Plan and elevation of plywood and timber beams*

I- and box-section beams are available (one type with corrugated web). Plywood beams are usually much cheaper than glued laminated beams, but their fire resistance is much lower than that of either solid or laminated beams. For details of manufacturers and design see Information sheet TIMBER 14.

Boarded web beams

3.28 Built-up beams and girders generally of I-section employ a web of two or more layers of boards laid diagonally to the edges **32** and at right angles to each other, with solid or laminated timber flanges and stiffeners, nailed, glued and nailed, or screwed.

3.29 Members produced in this way are light-weight and more economical than laminated timber beams but are subject to excessive deflection if built of green material. Spans of up to 30 m long have been produced satisfactorily.

Economical spans

3.24 The *Timber construction manual*[8] gives the limits for primary beams of sawn (solid) timber and glued laminated timber shown in table III. The upper limits for solid beams are unusual lengths, not commonly available in commercial timbers, while the upper limits for laminated straight beams for floors are somewhat low.

Table III Limits of spans for primary beams of sawn (solid) timber and glued laminated timber

	For roofs (m)	For floors (m)
Single span		
solid	1·8 to 12	1·8 to 6
laminated (straight)	3 to 30	1·8 to 12
tapered or curved	7·5 to 30	
Cantilever beams		
solid	7·2	
laminated	3 to 27	
	(usually more economical than single span over 12 m)	
Continuous span		
glued laminated	3 to 15	7·5 to 12
solid	3 to 15	7·5 to 12

Plywood and timber beams

3.25 Beams can be built up with single **30a** or multiple plywood webs and solid or laminated timber flanges or chords. They are usually nailed, glued, or glued and nailed, and are characterised by great stiffness-to-weight ratios. The thin webs make stiffeners essential but in box beams **30b** and **c** they are concealed.

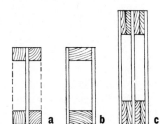

30 *Built-up plywood and timber beams:* **a** *single plywood web;* **b** *and* **c** *plywood box beams*

3.26 The number or thickness of webs in box beams may be increased near supports so as to increase resistance to shear without affecting appearance **31**. Design skills are low to moderate, depending on whether simplified or more accurate calculations are made*.

3.27 Plywood box girders of up to 30 m long have been manufactured for lightly loaded structures. More usual spans are 6 m to 18 m. Proprietary types of plywood and timber

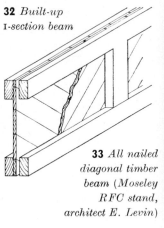

32 *Built-up I-section beam*

33 *All nailed diagonal timber beam (Moseley RFC stand, architect E. Levin)*

Steel reinforced beams

3.30 The most usual forms of reinforced timber beams are flitched beams made up with either steel plate or rolled steel sections such as channels or angles, coach screwed or bolted to the timber members **34**. These have been in common use in building for many years. Good timber to steel connections are particularly important to ensure that there is no buckling of the steel plates or members.

3.31 Flitch beams are not used often because they are uneconomical. They are generally considered only for conditions where reinforcement of a timber beam is required to increase its loadbearing capacity without a corresponding increase in depth (eg a lintel over a wide opening, carrying a wall and restricted in depth). Design is generally based on the assumption that the timber and the steel sections will each carry loads proportionate to their stiffness, which depends in each case on the moment of inertia of the section relative to the axis of bending and the modulus of elasticity of the material.

a **b** **c**

34 *Steel reinforced beams with* **a** *solid steel plates,* **b** *angles or* **c** *channels*

*Simplified design assumptions are to be found in various text books, eg *Wood handbook*[3].

Cambered beams

3.32 Beams and girders may be precambered to counteract deflection. In the absence of any special considerations dictating greater or less deflection, this is limited by CP 112 to 0·003 of the span. This deflection is permitted under live or intermittent loads in circumstances where the camber counteracts deflection caused by the dead loads. Canadian Institute of Timber Construction standards recommend a camber to counteract twice the dead load for general purposes.

Framed floors and flat roofs

3.33 The most common framing for flat roofs and lightly loaded floors takes the form of closely spaced joists spanning between walls, framed panels, beams or girders. Such joists are sheathed with boarding spanning across them. This may be butted plain edge or tongued and grooved softwood boarding or hardwood strip (flooring) or plywood or particle board. All of them will generally be secured by nailing (concealed, in the case of strip flooring) **35**.

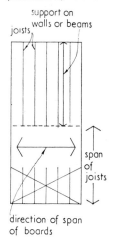

35 *Conventional joist and board construction for either (lightly loaded) floors or for flat roof panels*

3.34 Joist spacing is generally governed by the maximum permissible span for the sheathing material and the limit of 610 mm for load sharing systems. Information sheets TIMBER 6 and 7 provide respectively details of design and modular metric joist span tables for domestic floors and flat roofs, including limiting spans for planed timber boards, plywood and particle board of various thicknesses. (Tables for other loading conditions are also available from TRADA.)

End bearing

3.35 Ends of joists should be well secured in position and the bearing should be checked for adequacy in bearing stress. It must be borne in mind that tolerances and movement of components and supporting members may significantly reduce the length of end bearing of beams and joists so that adequate additional bearing over the minimum bearing length calculated as structurally necessary should be provided.

Generally notches made in joists not exceeding 250 mm in depth need not be calculated, but the effect of deep square end notches should be checked in shear: a notch of half joist depth reduces shear strength by three-quarters. Splayed notches reduce strength less and shear strength may be computed for the net cross-section above the notch, **36a**. Square notches should be limited to between one-quarter and three-eighths of the depth **36b**.

3.36 Joists are lapped or fish plated over supports to maintain alignment but are generally designed as if they were simply supported. At junctions with beams, ledgers or hangers are commonly used for support **37**. Joist and board floors may be prefabricated in sections but no structural advantage is obtained unless they are designed as stressed skin panels.

Deflection

3.37 Deflection usually governs the design of joisted floors and roofs. The following limitations are suggested:

Generally (CP 112 recommendation)	0·003
Domestic floors:	
over 4·8 m span	0·003
1·8 m to 4·8 m span	0·0035
under 1·8 m	0·004
Roofs:	
with dry lined ceilings	0·005
with wet applied ceilings	0·004
with skim coat or flushed joints	0·004

36 a *splayed notch;* **b** *square notch*

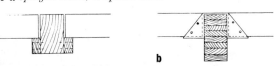

37 *Method of supporting joists at junctions with beams:* **a** *timber ledgers ;* **b** *metal hanger*

Tension members

3.38 Calculating members subjected to axial tensile stress (eg hangers or ties) is the simplest of all; the total tension force divided by the effective cross section (ie after deducting any notches, bolt holes or sinkings at or near the calculated section) indicates the tensile stress which must not exceed that permitted for the grade and species. Where there is combined tension and bending (eg a truss tie supporting a ceiling) the sum of ratios of actual-to-permitted stresses in bending and tension parallel to grain must not exceed these limits.

4 Trusses and girders

4.01 There is a great variety of roof trusses suitable for construction in timber. They are generally classified by their shape and arrangement of members. Statically determinate trusses are characterised by the fact that individual members are connected to form triangular spaces or panels of various sizes and proportions. Some of the common truss types used for medium and large spans are shown in **38**.

4.02 Scissors trusses are often used for raised ceiling height or open roofs but they are more limited in span than other forms. There are many combinations and variations of these basic types.

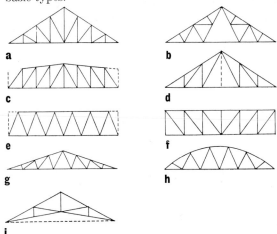

38 *Common trusses:* **a** *Howe;* **b** *Fink;* **c** *quadrangular Howe;* **d** *Pratt;* **e** *flat top Warren;* **f** *flat top Pratt;* **g** *Belgian;* **h** *bowstring;* **j** *scissors*

39 *Small flat top Pratt trusses*

40 *Huge bowstring trusses (hangar, Idaho, USA)*

41 *Large Belgian truss under test loading* (TRADA)

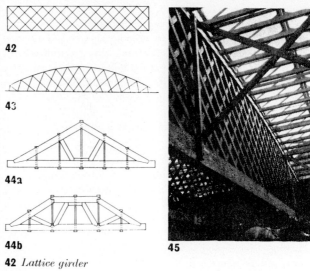

42 *Lattice girder*
43 *and* 45 *Belfast truss*
44a *Howe truss with rods;* **b** *queen rod truss*

Obsolete types

4.03 Lattice girders **42** and Belfast trusses **43** are rarely used nowadays, as they are not readily amenable to stress analysis and, although they use small section material, are expensive in labour. Similarly, trusses made up with solid large section single members housed at the connections, such as king post trusses, are rare except for churches and similar structures. Some types of solid-member trusses with steel rods incorporated to form all the tension members are used occasionally **44**. In these trusses rafters are usually birdsmouthed into the bottom member. Fabrication is generally more costly than that of modern types of truss.

Design considerations

Joint connections

4.04 Connections for modern truss joints may be nailed, glued and nailed or bolted, or connected by toothed plates, split rings or other types of connector. For details of methods of jointing see Technical study TIMBER 5. Nailed or glued joints generally require the use of gussets, usually of plywood, and are more frequently employed for small or medium span trusses than for large ones.

Members

4.05 Except in the case of small trusses, multiple piece members are commonly used so as to obtain more efficient connections at joints and enable smaller sections of timber to be used. Both practices save money. Top chords of bowstring trusses may be laminated to obtain the curvature or made up from solid segmental pieces. The shape is usually a segmental arc with a radius approximately the same as the span.

Comparative efficiency

4.06 Choice of the most suitable and economical type of truss, apart from other architectural considerations, depends on various factors such as availability of timbers in different sizes, their relative cost, cost of fabrication, cost of installation and cost of roofing.

4.07 Under normal loading conditions, it is generally considered that bowstring trusses are the most economical form for long spans, say of 25 m or over, and pitched trusses for medium spans. Flat top triangulated girders are usually more costly than trusses because stresses in web members are considerably higher and their connection more complicated and expensive.

Span ranges

4.08 The practical range of spans of trusses and girders spaced at between 4·6 m and 6·1 m will usually be between the following limits:

Pitched trusses	9 m and 27 m
Flat top girders	15 m and 45 m
Bowstring (with continuous chord)	15 m and 75 m

Exceptionally larger trusses are made in each category.

Truss proportions

4.09 In the design of trusses, depth-to-span ratios can vary considerably but in order to obtain a reasonably balanced stress distribution in members and to avoid awkward connections, certain proportions are recommended. These are:

Depth to span ratios:

Pitched trusses	$\frac{1}{5}$ or steeper
Flat or quadrangular girders	$\frac{1}{8}$ to $\frac{1}{10}$
Bowstring trusses	$\frac{1}{6}$ to $\frac{1}{8}$

Deflection and camber

4.10 Deflection in trusses can be minimised by using stouter members of low or medium grade, large panels (ie few connection points), and relatively stiff connections (ie less joint slip). Calculation of the necessary rise of the bottom chord to prevent an appearance of sag is complex but there are simple empirical formulae for the various types of truss (See *Timber construction manual*[3]).

Light trusses

4.11 Light trusses for domestic and other small roof spans are now frequently designed for close spacing of 600 mm or less. This avoids a need for purlins, intermediate rafters or ceiling joists. Even for wider spacing of about 1·8 m to 2·4 m trusses spanning up to 9 m are comparatively light and easy to handle. The w shape is the most common **46**. Individual design of such trusses is not often necessary as a wide range of standard designs for a variety of pitches and roof coverings is available from TRADA as well as from manufacturers of roof trusses. See Information sheet TIMBER 10.

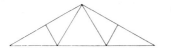

46 *Light truss*

Design procedure

4.12 The design procedure is the same for large or small statically determinate trusses, but rather more laborious and requiring greater skill and experience for the more complex multi-member trusses:

1 Select type and geometry of truss, allowing camber for deflection

2 Determine dead and live forces, with separate analysis of stresses in members and nodes by algebraic or graphical methods (using summation of moments or forces)

3 Select timber species and grade, type and size of connector

4 Design in detail joint carrying largest load, arranging fasteners in accordance with code requirements for spacing, end and edge distances etc and hence member sizes, giving general idea of other members

5 Check each member size for stress

6 Design each joint in detail

7 Check entire design for erection loads.

5 Portal frames and arches

5.01 All types of portal frames and arches—two-hinged, three-hinged and rigid or semi-rigid frames—have been built of timber and plywood. The range of constructional methods employed covers every type of solid, laminated and built-up assembly encountered in simpler forms, including diagonally boarded and plywood webs as well as box sections.

Trussed construction

5.02 Trussed portals and arches are probably the most economical in cost. In detail of design and construction they follow the general pattern of triangulated roof trusses. From the simplest form of knee-braced dutch barn **47** to the more sophisticated frames for warehouses or exhibition halls **48**, **49**, triangulated framing can fill a wide variety of purposes, sizes and shapes.

Rigid portal frames

5.03 Rigid frames for farm buildings of spans of up to 12 m have been designed by TRADA, using solid timber section and metal plate reinforcement at knees **51**. (See Information sheet TIMBER 11.)

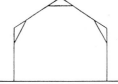
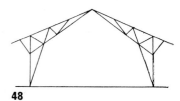

47 *Dutch barn of round or sawn timber (semi-rigid frame)*
48 *Frame structure for warehouse building (three-pin portal)*

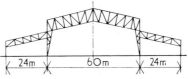

24m ⟵ 60m ⟶ 24m

49 *Frame structure for exhibition building with side aisles*

50 *Danish example of plywood web portal*

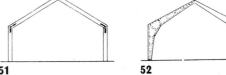

51　　　　**52**
51 *Rigid portal frame with metal plate inserts at knees*
52 *Plywood box portal frame (three-pin)*

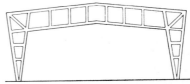

53 *Boarded web portal frame (diagonal boarding not shown)*

54 *Diagonal boarded web portal frame (Antwerp)*

Plywood and cross-boarded portals

5.04 Both plywood and cross-boarded web portals are generally competitive in costs with laminated construction. Both two- and three-pin portals have been built up by this method, but curved shapes are generally avoided. When using plywood, box structures are the more popular. Their sheathing conceals a framework which plays an important part in the design of the portals, especially at the knees, where bending moments are greatest **52**. Boarded web portals of proprietary design are available in spans of up to 46 m **53**.

Laminated timber

5.05 The advantages of laminated timber as compared with solid timber for the design of wide span arches and portals are that curved shapes of the most suitable forms can be produced with a minimum depth of members and give an unencumbered and pleasing interior. Two-pinned arches and tied arches are used where relatively low roofs are required. Three-pinned arches and portals are more useful for high rise roofs. A comparative range of economical spans is given below **55** (based on *Timber construction manual*[8]).

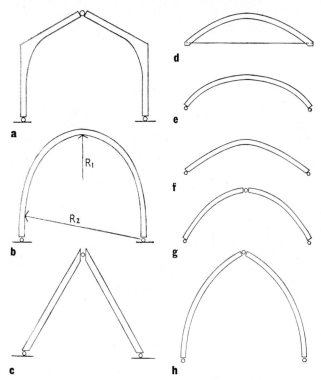

55a *Tudor arch (three-pin, 6 m to 36 m span);* **b** *three-centred arch (two-pin, 12 m to 75 m span);* **c** *A-frame (three-pin, 6 m to 30 m span);* **d** *tied arch (15 m to 60 m span);* **e** *radial (two-pin, 15 m to 60 m span; three-pin, 12 m to 75 m span);* **f** *shallow parabolic arch (two-pin, 15 m to 60 m span);* **g** *parabolic arch (three-pin, 12 m to 25 m span);* **h** *Gothic arch (three-pin, 12 m to 27 m span).* (*Load-span tables for laminated three-pin portals, span 10 m to 20 m, pitch 15° to 60°, are in Information sheet* TIMBER 13.)

56 *World's largest glue laminated arches (Florida, USA)*

57

58

57 *Laminated Gothic arches (Ontario, Canada)*
58 *Laminated shallow parabolic arch*
59 *Lamella roof (Seattle, USA)*
60 *Laminated ribbed dome (Canada)*

59

60

Ribbed domes and lamella roofs

5.06 Ribbed domes of laminated timber have been built with diameters over 100 m, **60** and **61**. Lamella roofs **59** are built up with short timber members of uniform length, bevelled and bored at the ends and bolted together so as to form a series of diagonally intersecting arches which can be covered by boarding or sheet material. Lamella frameworks can be used for various forms of vault—segmental, parabolic or Gothic. Shallow vaults must be tied or buttressed.

61 *Ribbed dome*

Design skills
5.07 These can be rated as medium for laminated arches and portals of rectangular section. The code provides modification factors for the grade bending stress, depending on the extent of induced curvature. Furthermore, the radial stress in the member resulting from the applied bending moment is limited to a fraction of the permissible shear parallel to grain or compression perpendicular to grain, depending on whether the moment tends to increase or reduce the curvature. Different geometrics of section and plywood or cross boarded portals would require more complex analyses for detailing the structure.

6 Spatial structures

Post and beam construction
6.01 The simplest three-dimensional structures to design are post and beam buildings. If spans, spacings and loading requirements allow the use of sawn sections, these provide the most economical framing systems. Built up, laminated or plybox sections are only used when wider spacings of supports are essential. Sometimes the lower storey is built in post and beam construction with the beams supporting joists and boarding or laminated slabs and the upper storey spanned by trusses.

Heavy timber construction
6.02 Post and beam construction is generally used in North American heavy timber construction ('slow burning mill construction') which has a high fire rating if the columns are not less than 200 mm × 200 mm in cross section, the beams not less than 150 mm × 200 mm and the planking 75 mm thick. Junctions of columns and beams usually have heavy metal caps, spreaders and pintels **62a** and **b**.

Framed roof systems
6.03 There are many systems of framing for pitched roofs and various possible methods of classification. Classifying by mode of support for the roofing material is the most useful in the present context:
1 rafter system **63, 64**
2 purlin system **65**
3 trusses only (trussed rafters) **66, 67**
4 trusses with purlins **68**
5 trusses with purlins and rafters **69**
6 rafters with purlins **70**.

Rafter systems
6.04 Unsupported rafter systems are suitable only for short spans because of limitation in lengths of available sawn timbers and the uneconomical depths of members needed above spans of around 4·8 m for monopitched roofs and 6·3 m for the couple close type, at 400 mm to 600 mm centres. Similar considerations apply to collar tie roofs, although the ties help to reduce thrust and, applied to every pair of rafters, can with hangers support a ceiling.
6.05 Simple roof systems of these types, like the collar and hanger type **68**, are popular on the Continent where they are used with proprietary thin webbed or latticed beams.

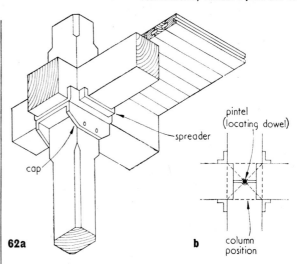

62a   **b**

62a *Junction of columns, beams and planking in heavy framed multi-storey building and* **b** *plan through connection*

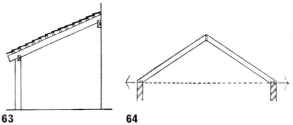

63   **64**

63 *Rafter system of roof construction: monopitch roof supporting tiles on battens*
64 *Rafter system: close couple roof: walls must resist thrust or a tie is required.*

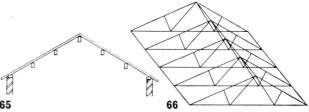

65   **66**

65 *Purlin system: purlins span from wall to wall carrying planking or roof sheeting.*
66 *Trussed rafter system (trusses only)*

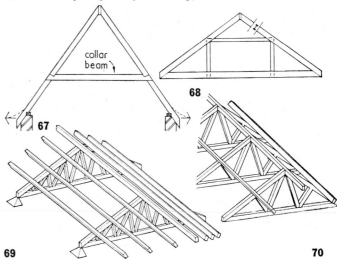

67   **68**

69   **70**

67 *Truss system using collar beams, a rudimentary form of truss but inefficient as it does not eliminate thrust.*
68 *Collar and hanger roof system using proprietary I-section beams.*
69 *Truss system: trusses, supported on walls or columns, carry purlins which in turn support the roofing.*
70 *Truss system: trusses support purlins which in turn carry common rafters (rafters not shown).*

These permit wider spans of up to about 12 m and leave an unencumbered roof space.

Trussed rafter systems

6.06 Trussed rafter roofs, without purlins, are used for light roof construction—generally for housing. With spacing limited to a maximum of 600 mm, components can be designed as parts of a load-sharing system with increase of stress allowance and high stiffness value according to CP 112. Trussed rafters are now being mass produced to suit various spans and pitches, usually 4·8 m to 9 m spans and 15°-35° pitches. They are rapidly replacing composite truss-purlin-rafter roofs.

6.07 The latter, still a popular conventional roof system for domestic buildings, usually has light trusses at about 1·8 m to 2·4 m apart supporting purlins carrying common rafters at 400 mm to 600 mm centres. Ceiling joists are supported from the trusses by means of hangers and benders. For purlin sizes and TRADA trussed rafters see Information sheet TIMBER 10.

6.08 Trusses and purlins are very commonly used for medium and large buildings such as halls, warehouses and industrial buildings. Purlin spacing is dictated by the roofing material and purlin size by the spacing adopted. In general, as spans increase, the more economical it becomes to space main trusses far apart: 4·8 m to 6 m apart is the limit usually set by available sawn purlin sizes, but the use of proprietary built-up purlins, or purlin strutting **71** frequently justifies wider spacing.

Purlin systems

6.09 Rafter systems, with or without trusses, generally require support on front and rear walls, or eaves girders. In the case of cross-wall construction, purlins which transmit roof loads to the cross-walls are more desirable. Simple purlin roofs **65** which use sawn timber purlins spanning between the walls are uneconomical for spans exceeding 4·8 m. Systems of built-up purlin girders with light rafters are more commonly used. A cross-section of such a roof is shown in **72**. (Alternative designs of roofs for cross-wall construction are obtainable from TRADA.)

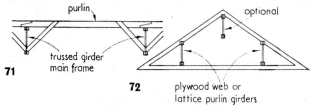

71 *Main roof framing comprising trussed girders. Note that purlins are strutted to permit increased spans.*
72 *Plywood web or lattice type purlin girders*

Design skill

6.10 The design skill required for a light roofing system which is safe and economical is considerable and the best results in this field are obtained from prototype tests—especially for small and medium span components.

Lateral stability

6.11 Buildings composed of planar elements, such as post and beam construction, or an array of trusses on columns, or of portal frames requires bracing in the lateral and often transverse direction to resist wind loading and impart stability to the building. Diagonal bracing is generally provided in the vertical and horizontal planes and sometimes, in the case of trussed roofs, by linking top and bottom chords to provide a space grid. The extent of the bracing required depends on exposure rating and on the

reliance that can be placed on sheathing, cladding and roofing materials to provide some or all of the necessary resistance to the racking loads which tend to deform the rectangular parallel planes of the structure. **73** is a typical example of bracing for walls and a trussed roof system of a large rectangular building.

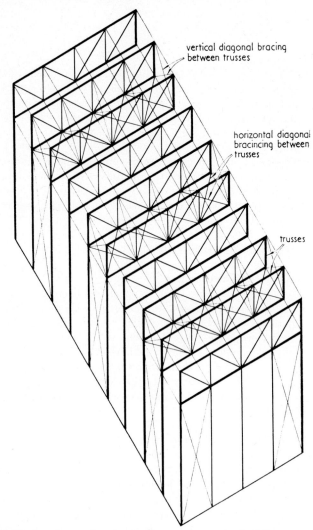

73 *Wall and truss bracing for large rectangular building*

Structural diaphragms

6.12 Structural diaphragms are thin walled elements, usually rectangular, designed to resist shear and transfer the load of lateral forces such as wind to the ground, either directly or through other resisting elements. The sheathings of walls and partitions may act as vertical diaphragms and those of floors and roofs as horizontal, inclined or curved diaphragms. As the shear forces are transferred through the edges of the diaphragms, efficiency of connections to perimeter framing, and also to intermediate framing members, is all important. In the case of in situ assembled structural elements, connections will be generally by nailing. The spacing and pattern of nailing should be determined by calculations or recommendations based on tests. Nailing or stapling or glue-nailing is often used for factory made elements.

6.13 The construction of diaphragm skins is listed below in order of increasing efficiency:

1 Transverse boarding, ie boarding set at right angles to studs, rafters, etc. Shear resistance is low. It depends entirely on the moment couple of the nails—hence wide boarding is more efficient **74**. Boards 19 mm to 50 mm thick are generally used

2 Diagonal boarding. Although some of the bending moment is taken by the boards, the main resistance to loads is in

direct tension or compression, axially. These stresses are highest at the perimeter and efficient connections there are vital **75**

3 Two layers of diagonal boarding set at right angles to one another. This is considerably stiffer and acts almost as a continuous sheet. One layer is designed to take compression axially and the other tension **76**

4 Plywood sheathing. As with double diagonal boarding, no bending moments are introduced into perimeter members which need be designed only for axial loading. Depending upon loading, it is practicable to reduce the thickness of sheeting to 6 mm

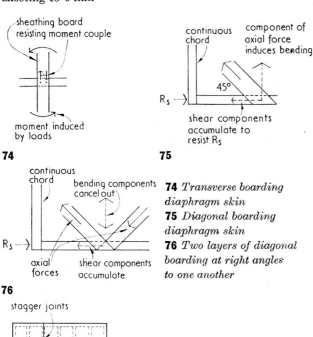

74

75

76

74 Transverse boarding diaphragm skin
75 Diagonal boarding diaphragm skin
76 Two layers of diagonal boarding at right angles to one another

77 Staggered joints in a plywood sheathed diaphragm

6.14 In the arrangement of sheets in a plywood sheathed diaphragm it is advisable to stagger joints on intermediate members **77**. In structures where a continuous membrane is required as a back support or final surface material, its design as a structural diaphragm usually provides an economical solution and permits the use of simpler connections for framing members. Continuous diaphragm construction is tending to replace diagonal bracings in light structures, and is now used extensively for stud frame houses and similar structures (See *Timber frame housing design guide*[9] and *Timber construction manual*[8]).

7 Surface structures

Stressed skin panels

Floors and flat roofs

7.01 Stressed skin panels consist of plywood sheets glued to top and bottom of longitudinal framing members, or sometimes top only, so that the whole structure acts as a series of joined T-beams or I-beams **78**. Framing members (webs) must not be spaced at more than twice the width of effective plywood flanges. Width depends on the kind of plywood, its thickness, number of plies and ratio of total thickness to thickness of plies in a longitudinal direction. Shear in framing members must be checked as well as the 'rolling shear' of plywood glued to them.

7.02 Continuity of longitudinal framing members and of the plywood fixed to them is essential, so is adequate bond.

Plywood is best scarf jointed in a factory, otherwise butt joints should be supported by glued plywood strips and full blocking. End members are essential. Intermediates are usually placed at one-third or one-quarter points. Panels may be press glued or glue-nailed. Carefully calculated pattern nailing has also been used successfully for relative small and lightly loaded panels.

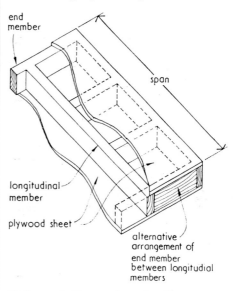

78 Stressed skin panel construction

7.03 Stressed skin panels are extremely light in terms of their loadbearing capacity. The double skin type in particular has a much reduced depth compared with conventional joist and board construction. This is an advantage where floor zones are limited in depth. The size of stressed skin panels is limited only by available lengths of framing members, consideration of economy in use of the more expensive plywood, and transport and handling problems: 900 mm and 1200 mm widths are common and, for house construction, lengths of up to 5000 mm are usual.

7.04 Particle boards may also be used in stressed panel construction, but their structural properties must be known quantitatively to permit design by calculation or determined by prototype testing. Creep under long term loading should be considered with particular care.

Folded plate roof construction

7.05 Folded structures composed of flat panels acting as diaphragms or stressed skin units can be so arranged as to derive strength and stiffness from their geometrical shape, individual panels being arranged to give each other mutual support. Such panels can be rectangular, triangular or trapezoid. In timber construction they consist generally of single or double plywood membranes with solid or laminated timber used for perimeter framing and intermediary members.

7.06 The most common type of folded structure is the continuous folded roof, consisting of inclined rectangular plates resting against each other and having common top and bottom chords **79a**. Roofs of this type may be designed with members acting as plates spanning between end supports or as slabs spanning from ridge to eaves. Alternatively, they may be designed as a series of v-beams bending about a neutral axis about midway between ridges and valleys **79b**. Connections to perimeter members, and in particular bearing rafters at ends, are of critical importance. For unbalanced loading conditions and, in general, for edge plates, ties are required to counteract thrust. Roofs of this type have been designed economically for spans from 9 m to more than 30 m.

7.07 Double pitch roofs using the folded plate technique have been designed by TRADA for buildings of cross-wall construction where front and rear walls are non-loadbearing **80**. Another common form of plywood folded roof is that in which inclined planes are separated by horizontal panels, connecting alternatively the top and bottom chords of inclined plates **81**. Timber and plywood pyramidal forms occur with a variety of polygonal bases—square, hexagonal and octagonal **82**, **86**. Junctions at the perimeter base have to be designed to be strong enough to counteract roof thrusts, or else ties must be introduced.

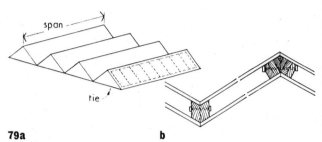

79a   **b**
79a *Continuous folded roof structure and* **b** *section through valley and ridge*

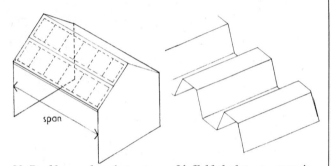

80 *Double panel roof structure*   **81** *Folded plate structure in which inclined planes are separated by horizontal panels*

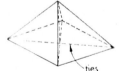

82 *Square-based pyramidal folded roof structure*

Timber shell roofs

7.08 Timber shells are curved diaphragms or stressed skin structures which derive their stiffness from the curvature of the particular geometrical shape used. Because radii of curvature are generally great in relation to the thickness of timber members used for these membranes, there is no difficulty in conforming to required shapes by slightly bending or twisting individual boards when fixing them in position.

7.09 It has also been possible to use strips of plywood of up to 600 mm wide to form shells. Shell roofs are generally built up with two or more layers of tongued or grooved boarding 20 mm to 25 mm thick, laid diagonally to the perimeter members and at right angles to one another. Perimeter beams may be solid for small shells but for larger ones are generally laminated, sandwiching the membrane between their top and bottom parts **87**. To ensure full membrane action layers must be properly fixed together. Glue-nailing throughout results in stiffer shells than partial gluing. Efficient connections between membranes and edge beams is also essential.

83 *First British hyperbolic paraboloid, Wilton carpet factory (architect: Robert Townsend)*

84 *Hyperbolic paraboloids at Sprites Lane School, Ipswich (architects: Johns, Slater & Haward)*

85

86

85 *Laboratory for testing shell roofs, constructed of three conoid shells on plywood box (architect: E. Levin TRADA)*
86 *Octagonal-based folded roof structure*

7.10 Many forms of shell are possible and various types have been built in timber. As a rule, double curvature shells such as hyperbolic paraboloids or conoids are stiffer than single curvature shells such as barrel vaults and, given adequate curvature, do not require stiffening ribs. Shells of greater curvature (or rise) are also stiffer than shells of similar type and size but lesser curvature. Some widely used types of timber shells are shown in **83-6** and **88-90**; common range of spans and recommended limits of rise are shown in table IV.

Table IV *Timber shell roofs*

Shell form	Span range (m)	Rise	Remarks
Barrel vault	9 to 36	1 : 3 to 1 : 5	Ratio of rise to cross-dimension of vault
Hyperbolic-paraboloid	6 to 24	1 : 4 to 1 : 5	Ratio of rise to side. Span indicated is for single plate. Double for arrangement in groups where four shells support one another
Conoid	6 to 15	1 : 3 to 1 : 4	Ratio of rise to span (ie between arches or bowstring trusses)

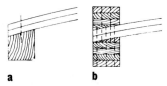

87 *Perimeter beams in shell roof construction:* **a** *solid edge beam with top shaped to curvature of shell;* **b** *laminated edge beam*

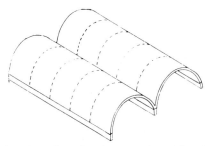

88 *Barrel vaults, these require stiff end diaphragms for support (not illustrated)*

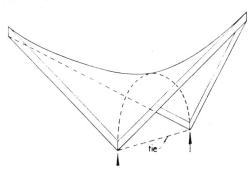

89 *Hyperbolic paraboloid*

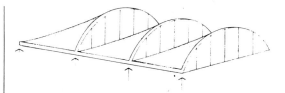

90 *Conoid shells spanning between bottom chord of bowstring truss and top chord of adjacent truss*

References

1 BS CP 3: Chapter V: Loading: Part 2: 1972 Wind loads [(K4)]

2 BS CP 112: Part 2: 1971 The structural use of timber [(2–) Yi (K)]

3 US FOREST PRODUCTS LABORATORY Wood handbook. Washington DC, 1955 [Yi]

4 HANSEN, H. J. Timber engineering handbook. New York, 1948 [Yi] O/P

5 TIMBER ENGINEERING COMPANY (TECO) Timber design and construction handbook, New York, 1959 (2nd edition) [Yi]

6 KARLSEN, G. G. (editor) Wooden structures. Moscow, 1967, MIR [(2–) Yi]

7 BS 1990: Part 2: 1971 Wood poles for overhead lines [Hi]

8 AMERICAN INSTITUTE OF TIMBER CONSTRUCTION Timber construction manual. New York, 1966 [Yi]

9 TRADA Timber frame housing design guide. [81 (28) Yi]

90 *Blimp hangar, Lakehurst, New Jersey, USA. When constructed 25 years ago, this was the world's largest timber structure. It remains the largest timber framed arch*

Technical study
Timber 4

Section 7 **Structural material: Timber**

Timber and the Building
Regulations 1976

This study by EZRA LEVIN *and* P. R. HERMAN *provides a brief general summary of the position concerning timber in the various elements of a building under the Building Regulations, and is supplemented by an Information sheet* (TIMBER 2) *which gives a quick, tabular guide to where timber may be used. Both are primarily concerned with the English Building Regulations 1976 but also include brief notes on Regulations in London and Scotland*

1 The Building Regulations 1976

1.01 The Building Regulations 1965 for England and Wales (revised in 1972 and 1976) introduced many changes affecting the use of timber. There is greater reliance on space separation and compartmentation and new constraints on internal spread of flame. However, greater freedom is allowed for structural applications of timber, particularly in external walls and separating elements. The frequent amendments since the inception of the regulations have not substantially modified this position.

External walls
1.02 The Regulations permit timber framed external walls for all buildings up to 15 m high if the wall is 1 m or more from a boundary and does not support a compartment floor. (For institutional buildings*, purpose group 2, all floors above ground must be compartment floors.)

1.03 A timber framed external wall can be built within 1 m of a boundary, and right up to it if the building is a house (or flats or maisonettes) up to three storeys high, or is not more than 15 m high in the case of all other non-institutional buildings (subject to upper limits of floor area)†. Such a wall must be completely faced with cladding of class 0 and give ½h resistance to internal *and external* fire. (The Regulations give schedules of various timber framed constructions which do this; see p71-73, The Guide to the Building Regulations 1976[2]).

1.04 Claddings of softwood, hardwood or plywood are permitted on any external walls up to a height of 15 m provided that they are 1 m or more from the nearest boundary. Tables and methods of calculation are given in the Regulations to determine the minimum distances to the boundary (which vary chiefly with height and length of wall but also with the class of building) for differing percentages of openings and claddings. To give one example, a windowless end wall of an office building, clad entirely in timber, 15 m high × 15 m long would have to be 6·5 m from the side boundary. The maximum distance from a boundary within which any form of control applies as to combustability is 55·5 m; and this only occurs in buildings over 120 m long and 27 m high.

Separating walls
1.05 Provided that adequate fire resistance (generally 1h) can be achieved, timber construction is permitted for walls separating:

*An institutional building is defined as one in which disabled persons or children under five years are accommodated.

†See table p82, The Guide to the Building Regulations 1976[2].

1 one-, two- or three-storey houses, of any size
2 one-storey buildings of all other classes, of floor areas not exceeding 500 m² to 3000 m² (depending on the purpose group of the building)
3 buildings of all classes except institutional, not exceeding 7·5 m high, but of floor areas limited to between 150 m² and 500 m², again depending on the purpose group.
The moderate fire resistance requirements of the Regulations governing separating walls in the above cases will ensure that such walls can be built at reasonable cost and using a variety of board or infill materials.

Compartment walls and floors
1.06 Regulations introduce the notion of *compartment floors* and *compartment walls*. These by definition isolate flats or maisonettes from each other in multi-storey buildings and isolate parts of buildings falling within different purpose groups. Generally they also subdivide into smaller *compartments* institutional buildings, shops, factories, storage-buildings and residential (other than small) buildings of more than one storey and exceeding certain limits of height, floor area or volume.

1.07 Compartment walls and floors are generally required to be non-combustible but timber construction is permitted for compartment walls in blocks of flats or maisonettes up to three storeys high and for floors in blocks up to four storeys (except walls in and floors over basements) and for compartment walls of single-storey buildings of all purpose groups within certain limits of floor area.

Timber floors
1.08 Timber construction of floors is permitted for:
1 one-, two-, or three-storey houses, including floors over basement not exceeding 100 m²
2 floors of all buildings up to 7·5 m high, except institutional
3 all floors of buildings up to 28 m high (except between flats and maisonettes and in institutions and except for floors which separate parts used for different purposes)
4 floors up to 9 m above ground in buildings over 28 m high (except between flats and maisonettes and in institutions)
5 all floors within maisonettes (ie not separating one unit from another), regardless of height of building.

Structural frameworks (columns and beams)
1.09 The fire resistance requirements for loadbearing frameworks of buildings are at least those of the elements they support. The same general principle is followed for non-combustibility requirements of these elements (see IS TIMBER 2).

Partitions

1.10 The treatment of loadbearing walls and partitions is similar. There are no special requirements for fire precautions for non-loadbearing partitions, apart from the general Regulation about surface spread of flame on internal linings, dealt with in para **1.14** below.

Small garages

1.11 The by-law restrictions on small garages were a constant source of complaint. The Building Regulations liberalise the requirements in several respects:

1 Permitted size (for exemption from the main Fire Regulations) is increased from 28 m² to 40 m²

2 Distance to house or boundary below which restrictions on combustible walls apply is reduced from 3 m to 2 m.

3 If the walls of the building are externally non-combustible, of a minimum fire resistance of ½h and contain no large openings the non-combustibility requirements do not apply

4 Exposed timber framing is permitted internally.

Similarly, there is a relaxation in respect of garages built within the houses; separating walls and floors need have only ½h fire resistance and there is no requirement for non-combustibility as in the by-laws.

Roofs

1.12 The Regulations (E17) take into account the 16 different designations of fire resistance of roofs in accordance with BS 476: Part 3[3]. This BS has been revised but to date Regulations still refer to earlier version. The roof designations depend not only on covering materials but also on underlay, support and structure. Tables in the Regulations list a wide variety of specifications deemed to comply with many of the designations. Designations of low resistance are not allowed in the Regulations for factories, storage buildings or terraces of houses, nor for any other buildings over 1500 m³. Distances to boundaries are not stipulated for roofs of high designation, but for other types they vary between 6 m, 12 m and 22 m irrespective of building height, and the 12 m and 22 m distances may only be used in small separate areas.

1.13 Many timber constructions are described in schedule D (Notional Designations of roof coverings) for the various designations from highest (AA) to the lowest (CC) including stressed skin plywood decks and chipboard on timber rafters. Other combinations than those listed will also satisfy the requirements of the various designations. Wood shingles are forbidden for factories, warehouses, terrace houses and for all buildings over 1500 m³, also for any building closer than 12 m to a boundary except in small separate areas. (See the *Guide to the Building Regulations* 1976).

Internal linings

1.14 The Regulations (E15) permit the use of untreated timber boarding, plywood or particle board (with flame spread no lower than class 3 as defined by BS 476) on a proportion of the walls of any room in a house, block of flats or institutional building, provided it does not exceed half the floor area of that room or 20 m², whichever is the less. In the case of a room in any other type of building (ie groups 4 to 8) the limit is again half the floor area or 60 m², whichever is the less.

1.15 Apart from this general provision, untreated timber products (of class 3) may be used for all ceilings of houses up to two storeys, whatever their size, and for the walls and ceilings of what are defined as 'small rooms' in the Regulations. These are very small indeed in all residential and institutional buildings for the upper limit of floor area is only 4 m². In all other types of building, including offices, shops and assembly buildings, the limit is 30 m². The walls of all larger rooms and circulation spaces in one- or two-storey houses can be lined with timber or wood products provided these have been impregnated or surface treated to reduce their risk of flame spread and improve their classification to class 1. The circulation spaces of other types of building (ie all except group 1), as well as the walls of institutional buildings, must have a still higher resistance to flame spread. This can be achieved if the wood based boards are faced with 3 mm non-combustible material or if 0·8 mm wood veneers are stuck to a non-combustible base material to give class O. (BS 476: Parts 6 and 7).

Timber used externally

1.16 The Regulations for externally exposed timber are more specific and more exacting. Softwood boarding (except western red cedar and sequoia) must, when used as the weather resistant part of a wall, be impregnated with preservative using one of the six systems specified in table 5 to schedule 5 of the Regulations, some of which are based on BS 913 or BS 4072[4]. Paint protection is not regarded as adequate. The minimum thickness of boarding is 16 mm, for external plywood 8 mm or if feather edged from 6 mm to 16 mm. Many hardwoods can be used without any treatment if heart wood is used.

Treatment of roof timbers

1.17 To guard against longhorn beetle attack, the Regulations follow previous by-law requirements that softwood timber used in roofs, including ceiling joists within the roof space shall be adequately treated. Refer to Reg. B3 in TIMBER 16.

Resistance to ground moisture

1.18 The Regulations follow closely the requirements of the previous by-laws but are stricter and more specific. Suspended timber ground floors, described in the 'deemed to satisfy' clauses, are raised a little higher above the concrete oversite slab. There are specifications of concrete floors incorporating timber finishes laid either as wood blocks or as strip flooring. There is no suggestion that on suitable sites (dry) a site covering may be omitted and no alternatives to concrete are listed as 'deemed to satisfy' (suitable alternatives have been investigated by TRADA and published in AJ). However, the 'deemed to satisfy' specification does not exclude any other solution which may meet functional requirements.

Thermal Insulation

1.19 Part FF[9], which came into operation on 1.6.75 concerns conservation of fuel and power in buildings other than dwellings and requires higher insulation values for them. However, these requirements can be easily met in timber frame construction by introducing insulating materials in the cavities formed between members.

2 Other building regulations

2.01 The Regulations discussed above are applicable to the whole of England and Wales, with the exception of those areas of the Greater London Council formerly administered by the LCC, where the London Building Acts 1930-39 and the constructional by-laws passed by the LCC and revised from time to time are still in force[5]. Northern Ireland also retained its own building by-laws and Scotland has its own Building Regulations, the latest version of which came into operation in 1970[6]. The following brief notes are intended to indicate the main differences between them.

London Constructional By-laws

2.02 The provisions of these by-laws are similar in essence to those of the Building Regulations as regards the structural uses of timber (ie joists, beams, and so on). They contain rules of calculation based BS CP 112: Part 2: 1971 *Structural use of timber*[7]. The district surveyors retain considerable discretionary powers particularly over the use of species and grades not stipulated in the by-laws.

2.03 Provisions as to flame spread are similar to those of the Building Regulations but by no means identical. They are less stringent on the use of timber in entrance halls, foyers and circulation spaces. The by-laws are generally more rigorous about external walls and separating walls, as regards both non-combustibility requirements and the periods of fire resistance laid down.

2.04 The GLC has also relaxed the non-combustibility requirements for separating walls. Between houses of one and two storeys, timber frame construction may be permitted by the waiver procedure. This brings the by-laws closer to the Building Regulations. Explanatory memorandum 2 *Guide to the use of timber in small residential buildings in the former LCC area*[8], explains in greater detail the LCC requirements for house construction.

The Building Standards (Scotland) (Consolidation) Regulations 1971

2.05 In the principles behind them and their formal structure these Regulations[6] are similar to those of England and Wales particularly in their stipulation of functional requirements and performance standards and their treatment of 'deemed to satisfy' clauses and schedules as being non-mandatory. The Scottish Regulations contain a more detailed classification of buildings by occupancy, and the enabling Act of Parliament has also made it possible to include provisions for means of escape and certain other provisions absent from the England and Wales Regulations (which have a bearing on the overall requirements for fire resistance). Consequently they are more sensitive in their fire resistance which for certain classes of buildings are less onerous. Proposals for including these in the Building Regulations for England and Wales have now been published.

2.06 Separating floors in timber are permitted in blocks of flats up to four storeys high. However, the Scottish Regulations do not generally permit timber framed party walls without non-combustible cores. However this is amended by a 'class relaxation' allowing hollow timber framed separating walls similar to those of England and Wales, between single occupancy houses up to two storeys. Recent relaxations have allowed three storey flats of timber frame construction to be built without the requirement of non-combustible compartment walls. With regard to preservative treatment of timber claddings, the Scottish Regulations make a sensible distinction between accessible and non-accessible parts.

Northern Ireland

2.07 The Building Regulations (Northern Ireland) Order 1972 gave powers to the Northern Ireland Ministry of Finance to make Building Regulations applying throughout the province, in replacement of the hitherto prevailing local by-laws. The new Northern Irish Regulations[9] are substantially those of England and Wales, with minor modifications. They will be enforced by the district councils.

References

1 The Building Regulations 1976, and the Building (First Amendment) Regulations 1978 (A3j)

2 ELDER, A. J. The guide to the Building Regulations 1972. The Architectural Press 1977 [(A3j) (F7)]

3 BS 476 Fire tests on building materials and structures. Part 3: 1958 External fire exposure roof tests [Yy (R4) (Aq)]; Part 6: 1975 Fire propagation test for materials; Part 7: 1971 Sunfree spread of flame tests for materials

4 BS 913: 1973 Wood preservation by means of pressure creosoting, BS 4072: 1972 Wood preservation by water-borne copper/chrome/arsenic compositions

5 GLC London Building (Constructional) Bylaws 1972

6 Building Standards (Scotland) (Consolidation) Regulations 1971 [(A3j)]

7 BS CP 112: 1952 Structural use of Timber in Buildings. See also BS CP 112: Part 1: 1967 and BS CP 112: Part 2: 1971 [(2–) Yi (K)]

8 Building Regulations (Northern Ireland) 1973 [(A3j)]

Technical study
Timber 5

Joints in structural timber

This study by C. J. METTEM *is concerned essentially with joints in loadbearing timber structures, although many of the fasteners have other applications in the assembly and fixing of timber components. The* AJ *Handbook of fixings and fastenings[1] describes the fixings themselves in great detail and covers considerations such as anchorage to masonry or concrete. This study, giving useful data and references on structural timber joints, is supplemented by two Information sheets (*TIMBER *3 and 4) which classify solid timber joints and summarise the principal characteristics of mechanical fasteners and fixing devices commonly used in timber frame construction. The adhesives considered here are cold setting assembly glues which are suitable for structural use. A number of other adhesives, generally hot setting, are associated with use in timber, but they are outside the present scope. The section on jointing of timber with plywood attached by nails or staples contains previously unpublished information.*

1 Introduction

1.01 Joints in structural timber are made either with mechanical fixings—nails, staples, bolts etc—or by adhesives. Factors affecting the strength of timber reflect in the strength of joints; eg the structural strength varies according to whether the force in the members meeting at the joint is primarily tensile, compressive or shear. It will also vary according to the manner in which the force is transmitted relative to the grain of the wood. Of course other factors, ie timber species, density and moisture content, strongly influence the strength of timber and hence the structural timber joints.

1.02 Permissible loads for structural timber joints are based upon ultimate values carried in strength tests. Unless failure occurs in the fastener itself, these test results will reflect the natural variability associated with timber strength data. A statistical estimate of a lower-limit strength value is obtained to take this into account. Furthermore a safety factor is applied that allows in part for duration of load effects, always significant in timber engineering, and also for variations in workmanship and other parameters usually associated with a safety factor. In certain instances, if excessive joint deformation must not occur, and thus the amount of slip must be restricted at design load stage, it may be necessary to set a further limit below the value derived from ultimate strength.

1.03 BS CP 112[2] contains much essential information on permissible working loads and spacings of various types of structural timber fastener. Any timber structure must be designed with reference to this code, on which Booth and Reece[3] provide an excellent commentary as well as explaining many of the background reasons for allocating strength values and factors, and giving much information on joints. For the purposes of joint design, softwood and hardwood species with similar strength properties are grouped together in the Code. (This grouping is not that used for grade stresses of the timbers themselves, but the lower three groups, J2, J3 and J4, correspond with the S1, S2 and S3 strength groups respectively, while a higher joint strength group, J1, has been added to include certain structural hardwoods such as greenheart, keruing, opepe and jarrah. It will be found

that the ranking of the joint strength groups corresponds closely with the density of the various species.)

1.04 Permissible load figures and modification factors are given in the Code for nails, screws, bolts, toothed plate, split ring and shear plate connectors. Also a limited amount of information is included on glued joints. Spacing rules are laid down. These should be strictly followed, unless strong test evidence or prototype proving indicates otherwise.

1.05 Certain types of connector can not easily have their design values laid down in a code of practice, and new devices appear regularly on the market. In all instances reliable test information should be obtained before starting design; a reputable manufacturer with a new device will let it be tested by an independent authority, such as TRADA or the Princes Risborough Laboratory of BRE. Alternatively an Agrément Board Certificate may be preferred; this offers a guarantee of high standing for a novel device which has not reached the stage of acceptance within a standard, and the Certificate verifies the quality of both materials and production.

Choice of joint

1.06 Consideration should be given to methods of assembly, connection details, function and quality of fastenings, permissible stresses and spacings for a given type of loading. There are usually several methods of connection for a given joint that could be equally efficient from a purely structural point of view. Choice will generally be dictated by considerations of economy (depending on ease of manufacture or assembly by the various methods), or cost of devices and accessories and on availability of necessary skills and suitable premises for manufacture. In general, mechanical joints require less rigorous conditions of application and skilled supervision than glued joints. Certain types of proprietary connector, such as punched metal plate fasteners, require special equipment and are therefore suitable only for factory production. On the other hand a small batch production alternative can nearly always be found. In this instance, perforated nailable steel plates might be used, or plywood gussets.

2 Structural jointing by adhesive

2.01 Gluing provides a method which, if used correctly, offers the advantages of strong, rigid and durable joints that may be formed over large or small areas and from which it is possible to omit visible fastenings, thus providing a neat appearance. An adhesive must maintain an adequate bond under the conditions and period of service experienced. The durability requirement, which in terms of structural members may possibly mean resistance to full weather exposure for more than 50 years, obviously gives rise to an important way of classifying adhesives. It is not possible for the performance over a long period to have been obtained for the modern synthetic resin adhesives, consequently recommendations are made on the basis of performance over the past 20 to 25 years, with a knowledge of the actual formulations involved. Table I, based on information published recently by Princes Risborough Laboratory, shows the types of cold setting structural assembly glues recommended for different exposure conditions.

Table 1 Adhesives suitable for different exposure categories

Exposure category	Exposure conditions	Examples	Adhesive	BS reference[4]
Exterior				
High hazard[3]	Full exposure to weather	Marine structures. Exterior components or assemblies where the glue line is exposed to the elements	Resorcinol-formaldehyde (RF) Phenol-formaldehyde (cold setting) (PF)	WBP WBP
Low hazard[2]	Exposed to weather but protected from sun and rain	Inside the roofs of open sheds and porches	Resorcinol-formaldehyde (RF) Phenol-formaldehyde (cold setting) (PF) Melamine/urea-formaldehyde[1] (MF/UF)	WBP WBP BR
Interior				
High hazard[2]	In closed buildings with warm and damp conditions and/or cyclical variations of temperature and humidity	Laundries, enclosed roof spaces	Resorcinol-formaldehyde (RF) Phenol-formaldehyde (cold setting) (PF) Melamine/urea-formaldehyde[1] (MF/UF)	WBP WBP BR
Low hazard	In inhabited buildings provided with ventilation and with heat either whole or part-time	Inside dwelling houses, halls, churches, heated farm buildings	Resorcinol-formaldehyde (RF) Phenol-formaldehyde (cold setting) (PF) Melamine/urea-formaldehyde[1] (MF/UF) Urea-formaldehyde (UF) Casein	WBP WBP BR MR Type A
Special	Chemically polluted atmospheres	Structures in the neighbourhood of chemical plants or associated with the manufacture of electrical batteries, dye-works, swimming baths	Resorcinol-formaldehyde (PF) Phenol-formaldehyde (cold setting) (PF)	WBP WBP

[1] Little data are available on long-term performance of cold-setting MF/UF glues, but assurance is given by glue manufacturers that they are considered suitable for categories indicated
[2] It is understood from the glue manufacturers that the improvements in formulation of currently available UF glues should permit their use in exterior low hazard and interior high hazard categories. In such cases, the assurance of glue manufacturer should be obtained of their suitability for particular use contemplated
[3] In structures where expected life is short, eg not more than 10 years, then MF/UF types are considered adequate
[4] All refer to BS 1204 : 1964 : Part 1 except casein which refers to BS 1444

2.02 Durability classification should be by no means the only consideration when selecting a suitable structural adhesive. Other factors such as the possibility of adverse effects caused by ageing, changing roof temperatures or fluctuating humidity conditions must be taken into account. Ease of use in manufacture is another important matter. Furthermore, the characteristic of certain types of adhesive to exhibit plastic flow under sustained loading renders such glues unsuitable for normal structural jointing.

2.03 Prior to a supply shortage a few years ago, a gap-filling resorcinol-formaldehyde adhesive would have been the automatic choice to overcome all the possible objections mentioned, and to provide a joint with the highest durability rating, WBP. Although difficulties of availability now appear to have decreased, this problem led to the recommended substitution in a number of instances of cold setting phenol-formaldehydes which give equal performance but call for more care in handling and use.

2.04 Among the adhesives not covered by Table 1, special mention should be made of epoxy resins which may be used to fasten timber, plywood and other wood-based boards to prepared metal surfaces and to other materials such as glass and plastic. Another class of adhesives attracting attention by timber researchers are the elastomerics. These are essentially non-structural according to normal definition and display distinct plastic-flow behaviour under load, indeed this characteristic gives rise to their name. Structural uses may be found for these, however, in the resistance of short-term loads such as those caused by wind loading or impact, and also to gain increased stiffness when used in conjunction with mechanical fasteners in composite forms of construction such as stressed skin roof and floor panels.

3 Jointing of timber with plywood attached by nails or staples

3.01 An easy and effective means of jointing timber is by the use of plywood attached with nails or staples. Structural components, such as trussed rafters, plywood web beams, ply gussetted portals and stud walls sheathed with plywood to provide wind-resisting shear walls, are some examples that can make use of nail or staple attachment. Plywood plates and gussets in frameworks provide rigidities at joints and are capable of being designed to carry moments and transfer shear and direct forces.

3.02 Higher load transfers can often be achieved by nailing rather than by using glue, as the strength of glued joints is limited by a low permissible shear stress in the ply laminations (rolling shear stress). Other important features of nailed or stapled joints are that they can be made with less specialised labour than the glued type and they can be seen to be effectively made, whereas for glued joints reliance must be placed in inspection during manufacture.

Dense nailing

3.03 TRADA has carried out a series of dense nailing tests. Nails of different sizes were driven through plywood into various species of timber and it was found that little or no splitting had occurred, even though the nail spacings were closer than those recommended in CP 112. The clauses on nailing in this Code deal with the jointing of solid timber and do not give recommended nail spacings for fastening plywood to timber. The conclusion derived from the test work is that the recommended edge, side and along-the-grain spacings for nails driven without pre-drilling given in table 24 of CP 112 are conservative for nailing through plywood.

3.04 The standard nailing pattern which has been adopted for trials is shown in **1**, and table II indicates the largest diameter 50 mm long nail which has been successfully hand driven without pre-drilling through plywood into various species using this pattern. Proneness to splitting appears to be a timber property susceptible to many variables, such as moisture content of the timber at the time of fastening, the presence of defects and nature of the grain, and the size and degree of support of the piece being nailed.

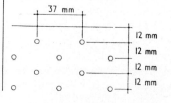

1 Nailing pattern adopted for TRADA tests

Table II *Largest diameter 50 mm nail driven without significant splitting to pattern shown in* **1** *through 12 mm hardwood plywood into various timbers*

Timbers	Diameter (gauge)					
	13	12	11	10	9	SWG
	2·3	2·6	2·9	3·3	3·7	mm
Softwoods						
European redwood					*	
European whitewood					*	
Western hemlock				*		
Douglas fir		*				
Hardwoods						
Afara				*		
Antiaris				*		
Ayan		*				
Berlinia		*				
Brown Sterculia		*				
Ceiba				*		
Celtis				*		
Dahoma				*		
Greenheart		*				
Guibourtia		*				
Kapur		*				
Keruing		*				
Missanda		*				
Okan		*				
Okwen	*					
Opepe				*		

3.05 Summary recommendations suitable for all softwoods and most hardwoods are given in table III. These spacings apply whether the nails are driven right through and clenched, in double shear, or driven from each side. In the latter case the spacings may be used on each side of the joint and an attempt should be made to ensure that the nail pattern is staggered on opposite faces. This is not difficult to achieve if some form of template is used to set out the nailing. For timbers not covered by table II or in cases of doubt nailing tests should be made on short pieces of timber.

Table III *Minimum spacing of nails without pre-boring when attaching plywood to solid timber*

Spacing	Times nail diameter
End distance	14 d
Edge distance	5 d
Side spacing between lines of nails*	10 d
Spacing along grain between adjacent nails	7 d

*or laterally adjacent pairs if staggered pattern is used

Ply thickness

3.06 To obtain the maximum nail force in the attachment of plywood to solid timber, the nail diameter should be the greatest that will not cause splitting. A further factor to be considered is the appropriate thickness of plywood to develop the full capacity of the nail strength. The thicknesses shown in table IV are the minimum required to develop the full capacity of the nail sizes shown against them in Douglas fir plywood fixed to softwood timbers. Clearly this information does not hold for plywoods made from stronger species in which case thinner material will develop the nail strength, or for more dense timbers, where joints will fail by nail heads pulling through the sheet material.

Table IV *Minimum thickness of Douglas fir plywood to develop full load capacity of nail sizes shown*

Ply thickness	Nail diameter	
	SWG	mm
8·0 mm ($\frac{5}{16}$ in)	12	2·6
9·5 mm ($\frac{3}{8}$ in)	10	3·3
12·5 mm ($\frac{1}{2}$ in)	9	3·7
16·0 mm ($\frac{5}{8}$ in)	7	4·5

Lateral load values

3.07 Dry basic lateral load values for the attachment of plywood to solid timber are not given in CP 112. However, experience has shown that dry basic lateral loads per nail may normally be extracted conservatively from the Code table dealing with nailed solid timber joints. The appropriate value for the load duration factor K_{21}, as explained in the Code (para 3.19.3 *et seq*), should be used with ply to timber joints taking values as for solid timber. In rare cases, for instance when attaching a softwood plywood to a dense, hardwood, this procedure might lead to too high a load being selected, thus causing failure through nail-head pulling; normally when using such timbers strong plywood would also be selected in order to obtain balanced design.

Staples

3.08 The permissible lateral load per fastener may be calculated by the following method:
A staple may be assumed to have a value equal to that of a nail $1\frac{1}{2}$ times its diameter. For example, assuming 50 mm × 1·6 mm (16 SWG) two-leg wire staples in J3 group timber, $1\frac{1}{2}$ times the diameter = $1·6 \times 3/2 = 2·4$ mm. This is still less than the diameter of the smallest gauge nail given in CP 112, so the permissible load for a 2·6 mm (12 SWG) nail is reduced by a ratio of the diameter squared:

$$\frac{2·4^2}{2·6^2} \times 178 = 152 \text{ N.}$$

3.09 In using this method, resistance to corrosion should be considered in comparison with that of the larger diameter of the equivalent nail. Also for staples driven into dense timbers the tensile strength of the staple could be the limiting factor, and the method given above is not appropriate. If in doubt consult staple manufacturers.

Nailing dense hardwoods

3.10 It is often thought that structural joints cannot be made in dense timbers unless the holes are pre-drilled. In fact pre-drilling can usually be eliminated as extremely strong joints can be made by using plywood gussets nailed to the solid timber by small-size nails. Eg, 12 mm plywood may be fastened to, say, Celtis using a close pattern of 50 mm long nails up to 3·3 mm (10 SWG) in diameter, and the joint design may be arranged so that advantage is taken of this possibility to eliminate all pre-drilling.

4 Classification of joints

4.01 It is possible to classify joints in a number of ways. The method by which the joint is fabricated, whether it be cut, mechanically fastened or glued, for example, may be used as a key. Alternatively joints might be classified according to the type of fastener or fixing employed. All of these methods present difficulties when the desire is primarily to view the joint functionally. Cut, fastened or glued joints may all be used in the heel joint of trusses for example, neither can this joint be easily classified by force-transmission, as it carries both tensile and compressive members and transmits both shear and moment. The classification in IS TIMBER 3 is according to functional form.

References

1 LAUNCHBURY, B. AJ handbook of fixings and fastenings. London, 1971, The Architectural Press [Xt7]
2 BS CP 112: Part 1: 1967 and Part 2: 1971 The structural use of timber [(2–) Yi (K)]
3 BOOTH, L. G. and REECE, P. O. The structural use of timber: A commentary on CP 112: 1967. London, 1967, Spon [(2–) Yi (K)]

Technical study
Timber 6

Section 7 **Structural material: timber**

Contractual, manufacturing and assembly procedures

This study by EZRA LEVIN *and* W. L. FOSTER *summarises design and contract procedures, and advises on problems concerning the fabrication, erection and maintenance of timber structures. Photographs do not relate specifically to text but illustrate assembly techniques*

1 Method of design

1.01 Factors which may determine the most suitable method of design to be adopted include: extent and complexity of the structural design involved, availability of adequate standard designs or design aids in the form of tables or nomograms (eg from TRADA), and the possibility of obtaining ready made timber components or entire buildings. Choice will depend to some extent on the architect's own knowledge and expertise and what other specialist advice he seeks. The possibilities and limitations of the various methods available are given below.

Design by architect
1.02 Generally limited to the design of simple columns and beams, joisted floors and roofs, stud walls and partitions, conventional pitched roofs and the selection of standard trusses or portal frames etc from available literature.

Design by structural engineering consultant
1.03 Non-standard structures of a complexity beyond the architect's ability or experience must be designed by a structural engineering consultant. In either case there is a choice of obtaining competitive tenders or nominating a specialist supplier or subcontractor for the timber component or structure. Information about the services of structural engineers may be obtained from the professional institutions, or from TRADA's register of consultants specialising in timber structures. TRADA also provides an independent consultancy service combined with testing facilities.

Design by specialist firms
1.04 Some timber engineering firms have their own design staff and others employ consultants; in both cases designs and quotations can be obtained. Some firms limit their design work to a standard range of buildings or components, but will modify or adapt them to suit special requirements. Lists of manufacturers may be obtained from TRADA[1] or the BWMA[2].

2 Contractual procedures

2.01 Choice of contracting method for timber structures or components will depend on: size of contract, nature and complexity of the structure, extent of other trades involved and whether proprietary components or systems are to be used. Specialised timber components such as glue laminated beams or portal frames are generally best dealt with as prime cost items or by nominated suppliers or sub-contractors.

1

2a

2b

1 *Erecting prefabricated timber school building without mechanical aids. All elements, including plywood box beams, are light enough to be handled by small gang.*
2a *Timber shell roof at Wilton carpet factory (architect: Robert Townsend, engineering design: TRADA); b roof during construction. Each of four hyperbolic paraboloid sections of roof built to formwork of steel scaffolding.*

2.02 Many timber engineering firms will supply *and* erect, whether providing their own systems or producing to architect's or engineer's designs. Where they are to be employed as subcontractors, it is important to specify the extent of the services to be provided by the main contractor (such as access, scaffolding and plant) as well as any return visits required by the specialist subcontractor during the maintenance period—eg for final tightening of bolts and making good any defects which may have developed.

2.03 Tender documents should include, in addition to the normal conditions of contract, specification, and information mentioned in para 2.02, sufficient information about the intended programme and procedure to make an allowance in the price for any special requirements. In particular responsibilities for dimensional accuracy should be defined.

2.04 When a tender is accepted a programme should be agreed with the appointed specialist contractor, the main contractor and the engineering consultant. Consideration should be given to the erection programme, date of ordering and delivery, tests and samples which will be required, quality control procedures and any independent checking of calculations. Arrange for site storage, protection of components, limits of responsibility and warranty.

3 Methods of production and workmanship

3.01 The extent to which the architect is concerned with methods of production depends on the procedure adopted for structural design. Where proprietary components or systems are specified, his interest is limited to their performance, appearance and cost. Where however, the manufacturer is to produce to designs prepared by the architect or his consultant, an understanding of the basic manufacturing procedure is necessary—in particular of the factors which can affect costs. Where complex structures are involved or where there is a substantial repetition of certain components it is advisable to consult manufacturers and to check their suggestions for modification against performance and appearance requirements at an early design stage.

Manufacturing principles

3.02 Maximum economy in fabrication is obtained by minimum handling and the smallest number of individual operations. The manufacturer will achieve minimum handling by organising assembly to suit his fixed and mobile equipment, but the number of operations required depends almost entirely on the design.

3.03 Timber components, unless standard such as doors and windows, are almost invariably produced by batch production methods. Even when assembly for very large continuous orders is by flow-line methods, machining (eg cutting, planing, drilling) is in batches. In general, each different section size or profile etc constitutes a new operation (although some modern woodworking machines are able to produce two or more profiles in one pass and occasionally specials can be produced from standard stock profiles at little extra cost).

3.04 Each new batch costs a fixed amount for setting up and preparation and this adds significantly to overall costs if the runs are small. This increased cost can often be greater than any saving in material resulting from a slight reduction in section. Graph **3** shows how unit costs increase when machining is done in two batches instead of one. For example the cost of sawing 300 m in two batches instead of one increases from index 300 to 360, ie 20 per cent. If the material is also planed and moulded to a standard profile the cost of this part of the process is increased by 60 per cent.

Particularly for sections of low material content these increases far outweigh any possible saving from using two different sections.

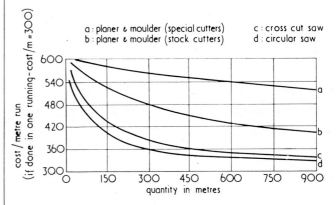

3 *Machining costs related to quantity production.*

4 *Laminated bow-string arches under test at* TRADA.

5 *Laminated three-hinged parabolic arches, 24 m span and 27 m in height, being crane erected during construction of Radar Test Laboratory, Dayton, Ohio, US.*

4

5

6a

6b

7

6 *Conoidal timber shell roof under construction at Oxford Street Station, Manchester:* **a** *end wall in laminated timber arch was pre-assembled on the ground and lifted by crane into position to support end bay of roof;* **b** *shell elements were built in situ on steel scaffolding formwork.*

7 *Lamella roof of standard precast ribs, manually assembled in situ, using light temporary staging (workshop in Seattle, Washington, US).*

3.05 The minimum economic size for a batch varies for different machines and kinds of operations, and also between fabricators. The designer must always strive for maximum standardisation of section, profiles, lengths, drilling, notching, etc. The smaller the job the more important standardisation is, for the sake of economy. Any component design which requires more than, say, six or seven different sections or profiles should be carefully reconsidered for possible simplification.

Assembly principles

3.06 Much the same kind of consideration affects the assembly of members into components and the designer should aim to reduce the number of separate operations involved. Operations which may need to be considered when fabricating a structural wall panel are shown in **8**, which illustrates fabrication on a horizontal plane. In this case if access is required to both faces of the panel it requires to be turned over (shown at G in the diagram). This is an operation which can disrupt production and may be expensive. The diagram shows that if insulation and windows can be designed so that both are installed from the sheathing side, turning the panel over can be eliminated from the sequence of operations. If access to both sides is required in any case (eg to fix services and inner linings) insulation and windows can of course be designed to be fitted from either side. Some fabricators carry out part of the assembly of this kind of panel in a vertical plane and this demonstrates the advisability of consulting fabricators, at least when substantial quantities are to be produced.

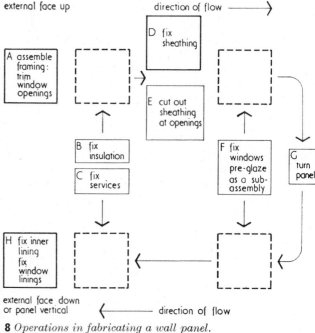

8 *Operations in fabricating a wall panel.*

4 Erection factors

4.01 Although contractors are entirely responsible for the methods they employ when erecting components the architect, or his engineering consultant, must have a clear idea at the design stage of how the component can be erected. The cost implications of heavy plant or equipment, or of complicated site processes are too considerable to be left to chance.

4.02 Because timber has a high strength/weight ratio, many timber components are designed to be erected without mechanical assistance. There is little authoritative guidance available for designers as to practicable weights and sizes which can be man-handled by a team. This is partly because much depends on the type of hand-hold provided by the component, the underfoot conditions in which men will be working, height above ground, period for which the weight must be supported, simplicity and position of fixings etc.

4.03 Some guidance on the weights and sizes of wall panels, curtain walling etc which can be handled by men working on the ground is given in table I. It may be used as a guide for similar erection problems. The possible effect of wind and poor underfoot conditions have been allowed for but more conservative assumptions would be prudent for work above ground level.

Table I Manual handling of timber components

Type of handling	Panel weight	Panel size	Other limitations and considerations
Lift and lay: floor and roof panels	Per man 40 kg, maximum 160 kg	Maximum within limits of weight	
Lift and erect: wall and partition panels	Per man 36 kg, maximum 290 kg	Height 2·6 m, length within limits of weight	Smooth level working platform
Carry or slide in vertical position			
Lift and erect on prepared base: walls and partition panels on upstand	Per man 34 kg, maximum 290 kg	Height 3·6 m less height of base. Length within limits of weight	Height of base B ≯ 600 mm Thickness of panel ≯ 75 mm. Panel framing exposed at least one side. Smooth level working platform
Tilt and erect: wall and partition panels	Per man 45 kg, maximum 270 kg	Height 6·0 m, length within limits of weight	Length of prop ≮ 0·80 height. Maximum pivot reaction: $RH = \dfrac{W}{4}$

Taken from TRADA Timber frame housing design guide

Cranes and tackles

4.04 Sometimes it is possible to lift components approximately into position by site crane, but to finally position and fix without its assistance. This should be considered at the design stage because considerable savings over full machine erection are possible. A normal prerequisite is that there should be storage space available at the level at which the component is to be fixed, otherwise continuity of work cannot be assured for the fixing team. The designer should consider also the work stage at which the component is to be fixed, and the situation and condition of scaffolds and working platforms at that time, the safety of the men handling and securing components, and any special conditions associated with final or closing components.

9a *Glued laminated arches and diagonal boarded nailed portal frames (HB system), awaiting transport outside*

9a

9b

10

11

fabrication plant of Engineered Timber Products, AG, Burlington, Ontario; **b** *interior of large hall built by Engineered Timber Products AG, of Burlington, Ontario, using diagonal boarded nailed portal frames (HB system).*
10 *Crane erecting 70 m span bow-string truss roof over hangar at Minneapolis, Minnesota, US.*
11 *Crane erection of laminated radial arches and laminated purlins. Swimming pool at Dearborn, Michigan, US (by Timber Structures Inc, NY).*

4.05 Particularly for portal frames, long span trusses and beams, and folded plate and shell roofs, erection may be by special tackle or appliances designed to suit a particular design or situation. In this case the design of the components, the erection sequences and the method of erection are all interdependent and must be a joint exercise between architect, structural consultant, and the contractors.

4.06 Where the usual construction lifting equipment is to be used the designer should consider the size and shape of components and the possible effect of wind on them while they are being hoisted and fixed. This is generally a greater problem than their dead weight. It is also necessary, in most cases, to specify the points at which lifting gear may be attached and in some cases attachments must be provided during assembly of the components. Structural consultants should be asked to specify points of attachment to ensure that erection stresses do not exceed the permissible stresses in any member.

5 Final inspection and maintenance

5.01 Final inspection should consist of checking that the correct materials have been used, that any accidental damage has been made good, that finishes are in good condition and that all bolts and other connections have been tightened. The alignment of frames and members should be checked and structure examined for excessive deflections.

5.02 Deflections can give some indication of the stresses in individual members. This is sometimes important in the case of components such as roof trusses which have been designed on the basis of prototype tests to the requirements of BS CP 112. Tests are generally limited to the behaviour of the components as a whole and do not necessarily validate the performance of individual members, eg deflection of rafters between joints. In cases where actual deflection appears to be excessive the structural consultant and specialist subcontractor should be informed. Where remedial measures are undertaken, short of replacement of faulty members or components (eg by additional strutting), they should be carefully designed and executed and their effect on the redistribution of stresses in the structure checked.

5.03 Fissures and distortions in the members require detailed examination. Some will have been present in the material before erection and within the limits permitted for the grade specified, or may have progressed due to drying shrinkage without however impairing structural strength. Others may have developed due to excessive shrinkages or stresses (eg faulty jointing details) and may require replacement of members and/or modifications of details. Fissures and distortions may however worry a client and the acceptable limits in terms of both magnitude and position should be laid down in a maintenance manual handed to the client after final inspection to help him evaluate the significance of any increased splits or distortions during the lifetime of the building.

5.04 The maintenance manual should also contain the names and addresses of suppliers of proprietary finishes used, with recommendations for inspection and periodic renewal of the finish. It should draw attention to any fixings which may require tightening when the mc of the timber has reached equilibrium with the environment and notes on any provision made in the design for ventilation of parts of the structure.

Sources of information

1 Timber Research and Development Association (TRADA), Hughenden Valley, High Wycombe, Bucks (Naphill 3091)
2 British Woodworking Federation (BWF), 82 Cavendish Street, London W1 (01–6369075)

12

13a

13b

12 *Simultaneous use of two cranes for erecting pre-assembled sections of bow-string trussed roof. US Naval Air Station, Lakehurst, New Jersey, US.*
13a *Wide span glass house with laminated timber arches, designed by* TRADA *for an agricultural college at Bayfordbury;*
b *all prefabricated members, including arches, were designed for manual assembly on site.*

Information sheet
Timber 1

Section 7 **Structural material: timber**

Summary characteristics and structural properties of timber and wood products

This sheet gives seven tables of properties of imported softwood, homegrown timbers, imported hardwood, plywood, blockboard, chipboard and fibreboard.

Table I Characteristics and strength properties of standard Finnish birch blockboard and laminboard

Type of board	Nominal thickness mm	Weight (at average mc of 9 per cent) kg/m²	Permissible bending stress N/mm² Parallel to face grain	Perpendicular to face grain	Permissible moduli of elasticity kN/mm² Parallel to face grain	Perpendicular to face grain
Standard blockboard 5-ply	12 16 18	7·6 9·5 10·5	11·80	6·87	10·80	4·41
	22 25	12·5 14·2	7·85	6·87	7·85	5·39
Standard blockboard 3-ply	12 16 18	7·5 8·8 9·5	10·8	8·93	9·81	4·90
	22 24 25	11·7 12·4 12·9	7·85	6·87	6·87	6·38
Laminboard 5-ply	12 16 18	8·1 10·2 11·5	13·70	6·87	10·80	4·90
	22 24 25	13·4 14·4 14·9	7·85	6·87	9·81	6·38
Laminboard 3-ply	12 16 18	8·1 10·2 11·5	9·80	21·6	8·84	7·36

Above stress values, based on tests carried out at State Institute for Technical Research, Helsinki, and published by Finnish Plywood Development Association (Technical Publication No 13, 1970), apply to short duration loading, at mc of 9% ± 3%. If design criteria given in CP 112 for plywood are adopted, stresses would require modification by factor of 0·66 if they are to apply to long-term loads (eg dead + permanent imposed, as for shelving, flooring).

Table II Types and structural properties of wood chipboards (excluding extruded boards) complying with BS 5669: 1979

Type	Thickness (mm)	Mean bending strength (N/mm²)	Mean modulus of elasticity (N/mm²)	Tensile strength perpendicular to plane (N/mm²)	Edge screw holding (N)	Impact resistance (mm)*
Type I— Standard	6 mm or less over 6 to 19 mm over 19 to 25 mm (inclusive)	18·0 13·8 12·5	2000 2000 1850	0·8 0·34 0·25	360 250	
Type II— Flooring	under 18 mm 18 mm 19 mm to 25 mm (inclusive)	——not specified—— 17·0 17·0	2500 2500	0·5 0·5		525 525
Type III— Improved moisture resistance	6 mm or less over 6 to 19 mm over 19 to 25 mm (inclusive)	——not specified—— 19·0 17·0	2750 2750	0·5 0·5		

* The value denotes the height of free fall of a 4·5 kg block with spherical end on the board, which produces impact rupture.
Type II/III board combines the strength properties of II and the moisture resistance of III.
Densities are not specified in the standard as the values of the properties are not necessarily related to them, although in general the heavier boards tend also to be stronger.
Board sizes depend on manufacturers' presses and a wide range is available. 2440 × 1220 is a common size but boards of much larger dimensions up to 2500 × 12 500 in various thicknesses can also be obtained.

Table III *Characteristics and structural properties of imported softwoods*

Species, common and botanical names	Colour	Texture	Average density at 18 per cent moisture content kg/m³	Dura-bility	Mois-ture move-ment	Working qualities	Range of sizes mm	Range of lengths m	Supply availa-bility	Price level	Permissible basic stresses for timber (N/mm²)						Mean modulus of elasticity	
											Bending and tension parallel to grain		Compression parallel to grain		Compression perpendicular to grain			
											Green*	Dry†	Green*	Dry†	Green*	Dry†	Green*	Dry†
Douglas fir or British Columbian pine (*Pseudotsuga menziesii*)	Pinkish brown	Fine	540	Medium	Small	Good	100-300 wide 25-100 thick	1.8-9.0 Longer lengths to order	Limited	Medium/high	15·2	18·6	11·0	14·5	1·74	2·62	10 300	11 700
Western hemlock (unmixed) (*Tsuga heterophylla*)	Light brown	Fine	540	Medium	Small	Good	100-300 wide 50-100 thick	1.8 and up Longer lengths available	Good	Medium	13·1	15·9	10·3	12·4	1·38	2·07	9000	10 000
Western hemlock (commercial) (*Tsuga heterophylla*) (mixed with *abies*)	Light brown	Fine	530	Medium	Small	Good	100-300 wide 50-100 thick	1.8 and up Longer lengths available	Good	Medium	11·7	14·5	9·0	11·0	1·38	2·07	8600	9300
Parana pine (*Araucaria angustifolia*)	Light brown Some-times varie-gated with red streaks	Fine, irregular	560	Low	Large	Good	150-350 wide 19-75 thick ‡ se	3-6 and up	Good	Medium/high	11·7	14·5	10·3	12·4	1·52	2·21	8300	9000
Western red cedar (*Thuja plicata*)	Reddish brown	Medium/coarse	380	High	Small	Good	100 and wider 19-150 thick	2.4-7.3	Good	Medium	9·0	11·0	6·2	9·0	1·03	1·52	6200	6900
Canadian spruce (*Picea species*)	Whitish	Fine	450	Low	Small	Good	76-304 wide 25-76 thick	2.4-5.2	Good	Low	11·0	13·8	8·3	11·0	1·38	2·9	8300	9000
European redwood (*Pinus sylvestris*)	Pinkish brown	Fine/medium	540	Low	Medium	Good	75-275 wide 16-100 thick	1.8-6.3	Good	Low	11·7	14·5	8·3	11·0	1·52	2·21	7600	8300
European whitewood (*Picea abies*)	Whitish	Fine	510	Low	Small	Good	75-280 wide 19-100 thick	1.8-6.3	Good	Low	11·7	14·5	8·3	11·0	1·38	2·07	6900	8300
Pitch pine (*Pinus palustris* *Pinus ellicttii* *Pinus caribaea*)	Pinkish brown	Fine	720	Medium	Medium	Good	350 × 350 450 × 450 Flitches any size cut	5.4-10.8	Good	Medium	15·2	18·6	11·0	14·5	1·79	2·62	10 300	11 700

* Green: moisture content exceeding 18 per cent
† Dry: moisture content not exceeding 18 per cent
‡ se: square edge

Table IV Characteristics and structural properties of home-grown softwoods and hardwoods

Species, common and botanical names	Colour	Texture	Average density at 18 per cent moisture content kg/m³	Durability	Moisture movement	Working qualities	Range of sizes mm	Range of lengths m	Supply availability	Price level	Permissible basic stresses for timber N/mm²						Mean modulus of elasticity	
											Bending and tension parallel to grain Green*	Dry†	Compression parallel to grain Green*	Dry†	Compression perpendicular to grain Green*	Dry†	Green*	Dry†
Softwoods																		
Douglas fir (*Pseudotsuga menziesii*)	Pinkish brown	Fine to moderately coarse	560	Medium	Small	Good	100-300 wide 25-100 thick	1·8-9·0 Longer to order	Limited	Medium	14·5	17·9	10·3	13·8	1·72	2·48	9000	10 000
Larch (*Larix decidua*) (*Larix leptlepsis*)	Reddish brown	Fine to moderately coarse	560	Medium	Large	Good	100-250 wide 16-75 thick	1·4-4·2	Limited	Medium	13·8	17·2	9·7	13·1	1·79	2·62	9000	9700
Scots pine (*Pinus sylvestris*)	Pinkish brown	Fine	540	Low	Medium	Good	75-275 wide 16-100 thick	1·8-3·6	Good	Low	11·0	13·2	8·3	11·7	1·72	2·45	8300	9700
European spruce (*Picea abies*)	Whitish	Fine	380	Low	Small	Good	75-200 wide 16-75 thick	1·8-3·6	Good	Low	8·3	11·0	6·2	9·0	1·10	1·65	5900	6900
Sitka spruce (*Picea sitchensis*)	Pinkish white	Coarse	400	Low	Small	Good	100-200 wide 16-75 thick	1·8-3·6	Good	Low	7·6	10·3	5·5	8·3	1·10	1·65	6600	7200
Hardwoods																		
Ash (*Fraxinus excelsior*)	Whitish brown	Medium	720	Low	Medium	Good	150 and up ‡ se and tt ** 16-75 thick	1·8 and up	Fair	Low	17·2	22·8	11·0	15·2	3·10	4·48	10 000	11 400
Beech (*Fagus sylvatica*)	Pale brown	Fine	720	Low	Large	Good	150 and up ‡ se and tt ** 16-75 thick	1·8 and up	Good	Medium	17·2	22·8	11·0	15·2	3·10	4·48	10 000	11 400
Oak (*Quercus robur*) (*Quercus petraea*)	Light brown	Medium	720	High	Medium	Medium to difficult	150 and up ‡ se and tt ** 16-75 thick	1·8 and up	Good	Medium to high	17·2	20·7	11·0	15·2	3·10	4·48	8600	9700

* Green: moisture content exceeding 18 per cent
† Dry: moisture content not exceeding 18 per cent

‡ se: square edge
** tt: through and through

Table v Characteristics and structural properties of imported hardwoods

Species, common and botanical names	Colour	Texture	Average density at 18 per cent moisture content kg/m³	Durability	Moisture movement	Working qualities	Range of sizes mm	Range of lengths m	Supply availability	Price level	Permissible basic stresses for timber N/mm²						Mean modulus of elasticity	
											Bending and tension parallel to grain		Compression parallel to grain		Compression perpendicular to grain			
											Green*	Dry†	Green*	Dry†	Green*	Dry†	Green*	Dry†
Abura (*Mitragyna ciliata*)	Pinkish brown	Fine to medium	590	Low	Small	Good	Logs, se‡ 150 and up 25-100 thick	1·8-6·0 Longer to order	Good	Low	13·8	16·5	10·3	13·8	2·34	3·45	8300	9300
African mahogany (*Khaya species*)	Reddish brown	Medium	590	Medium	Small	Medium	Logs, se‡, tt** 75 and up ‡50 and up 19-50 thick	1·5 and up	Good	Medium	12·4	15·2	9·7	13·1	2·07	3·10	7900	9600
Afrormosia (*Afrormosia elata*)	Light brown	Medium to fine	720	High	Small	Medium	Logs, se‡ 150 and up wide 25-50 thick	1·8 and up	Good	High	22·1	26·2	15·9	22·1	4·14	6·21	10300	12100
Greenheart (*Ocotea rodiaei*)	Olive green	Fine	1060	High	Medium	Difficult	600 × 600 × 300 × 450 Flitches any size cut	Up to 18·0	Good	Medium	37·9	41·4	27·6	30·3	6·20	9·31	17200	18600
Gurjun/keruing (*Dipterocarpus species*)	Reddish brown	Medium to coarse	720	Medium	Large	Medium	150-375 wide 25-150 thick	1·8-7·2	Good	Low	17·2	22·8	13·8	19·3	2·34	4·48	12400	13800
Iroko (*Chlorophora excelsa*)	Yellowish to dark brown	Medium	690	High	Small	Medium to difficult	150-300 and up wide 19-100 thick se‡, logs	1·8 and up	Good	Medium	20·7	23·4	15·2	19·3	4·14	6·21	9000	10300
Jarrah (*Eucalyptus marginata*)	Dark red	Medium	910	High	Medium	Difficult	100 and up wide 25-100 thick cs††	1·8 and up	Limited	Medium	19·3	23·4	15·9	20·7	4·14	6·21	10300	12100
Karri (*Eucalyptus diversicolor*)	Reddish brown	Medium	930	High	Large	Medium to difficult	20-100 thick 150 and up wide, se‡, cs††	1·8 and up	Limited specified import	Medium	22·1	26·2	16·5	22·1	4·83	7·24	13800	15500
Opepe (*Nauclea diderrichii*)	Orange yellow	Medium	780	High	Small	Medium	150-450 wide 25-150 and thicker se‡, cs††	1·8 and up	Good	Medium	25·5	29·0	22·1	24·8	5·52	8·27	12400	13800
Red meranti/ Red seraya (*Shorea species*)	Light red	Medium	540	High	Small	Good	150 and up wide 25-100 thick ‡se	1·8-7·3	Good	Low to medium	12·4	15·2	9·7	13·1	1·79	2·62	7600	8300
Sapele (*Entandrophragma cylindricum*)	Reddish brown	Fine/medium	690	Medium	Medium	Medium	150 and up wide 19 and up thick	1·8 and up	Good	Medium	19·3	23·4	15·9	20·7	4·14	6·21	9700	11000
Teak (*Tectona grandia*)	Light brown	Medium	720	High	Small	Medium	150-350 wide 25-100 thick Wide variety of dimension stock available	1·8 and up	Good	High	22·1	26·2	16·5	22·1	4·14	6·21	11000	12400

* Green: moisture content exceeding 18 per cent
† Dry: moisture content not exceeding 18 per cent
‡se: square edge
**tt: through and through
††cs: construction sizes

Table VI Characteristics and structural properties of imported and British made plywoods

Species or type	Colour	Texture	Durability	Range of thicknesses generally available (mm)	Number of plies	Average weight (kg/m³)	Range of board sizes mm	Design stresses and moduli of elasticity with face grain parallel to span in N/mm²				
								Extreme fibres in bending	Tension	Compression	Rolling shear in plane of plies	Modulus of elasticity in bending
Canadian Douglas fir plywood (*unsanded*)	Reddish brown	Fine to medium	Medium to low	7.5	3	4.48	Common size:	9.99	6.86	8.98	0.403	10800
				9.5	3	5.49	2440 ×	9.45	5.41	7.99	0.403	10300
				12.5	3		1220	10.30	6.86	8.98	0.298	11300
				12.5	4		(96 × 48 in)	8.34	4.11	5.39	0.403	9190
				12.5	5			8.17	5.34	7.00	0.403	9290
				15.5	5	8.91	also in	7.27	4.36	5.72	0.403	8370
				18.5	5		modular	6.55	4.21	5.52	0.298	7560
				18.5	6		metric	6.64	4.81	6.30	0.298	7700
				18.5	7		2400 ×	7.23	4.81	6.30	0.403	8380
				20.5	5		1200	6.70	4.70	6.16	0.298	7750
				20.5	6			6.21	4.70	6.16	0.298	7210
				20.5	7			6.70	4.34	5.69	0.403	7750
Finnish birch plywood (*sanded*)	Whitish	Fine	Low	6.5	5	4.5	1220 ×	16.40	7.92	6.06	0.69	11700
				9.3	7	6.4	1220 to	12.60	4.37	4.92	0.69	9650
				12.0	9	8.2	3660	12.00	4.16	4.58	1.31	9650
				14.4	9	9.6	and 1525	11.00	3.96	4.40	1.27	8950
				14.8	11	10.0	× 1525 to	11.00	3.96	4.40	1.27	8950
				17.6	13	11.7	3600 grain	9.92	3.86	4.27	1.24	8620
				18.0	11	11.5	direction	9.92	3.86	4.27	1.24	8620
				20.4	15	13.5	correspond	8.82	3.75	4.16	1.24	8270
				21.6	13	13.8	to the	8.82	3.75	4.16	1.24	8270
				23.2	17	15.3	first sizes	8.05	3.68	4.00	1.24	7930
				25.2	15	15.7	above	8.05	3.68	4.00	1.24	7930
				26.0	19	17.1	scarfed	7.58	3.65	3.86	1.20	7930
				28.8	17	17.9	sizes also	7.58	3.65	3.86	1.20	7930
British made plywood from tropical hardwoods (*sanded*)	Red to yellowish	Fine to medium	High: Agba, utile, makore. Low: gaboon, obeche.	6.5	3	3.91	widths to	13.1	11.4	7.58	0.552	6210
				12.5	5	7.81	1830	13.1	11.4	9.65	0.552	6210
				19.0	7	11.7	lengths	13.1	11.4	9.65	0.552	6210
				25.5	9	16.6	to 3050	13.1	8.65	9.65	0.552	6210
				7.0	3		and flush					
				12.0	5		door size					
				15.0	9							
				25.0	11							

Strength values are given for a range of available sizes and for the grades most commonly used. Information on Finnish plywoods relates to birch faced but inner plies of fir. That for Canadian plywood applies to Douglas fir faced plywood with inner plies of several possible species, including Western hemlock, True fir, Sitka spruce, Western white spruce, Western larch, Western white pine, Ponderosa pine, Lodgepole pine and Douglas fir; the latter is now rarely available. This structural plywood is marketed as COFI EXTERIOR.

Table VII Characteristics and structural properties of fibre building board

Type	Density range kg/m³	Specified manufacturing thickness mm	Minimum mean bending strength N/mm²	Effective of relative humidity change from 33 to 90 per cent max mean increase, per cent		Colour	Texture	Range of thicknesses generally available mm	Average weight kg/m²	Range of board sizes mm
				Length and width	thickness					
TE and TN Tempered hardboard	Normally exceeding 960	2 to 3·2 >3·2 to >10	TE 59 / TN 52 45 / 40	TE 0·30 / TN 0·35	TE 10 / TN 15	Dark brown	Fine	3·2 4·8 6·4 12·0	30·7 46·0 61·4 115·2	1830 2440 1220 × 2745 3050 3660
S Standard hardboard	Normally exceeding 800	2 to 3·2 >3·2 to >10	38 30	S 0·35	S 10	Light to medium brown	Fine	3·2 4·8 6·4	25·2 37·8 50·4	1220 × 1600 up to 3660 also 1525 × 2440
HME and HMN High-density medium board	560-800	6·4 to 10 >10 to <13 >13 to <16 >16	HME / HMN 20 / 15 17 / 12 15 / 10 13 / 8	HME 0·25 / HMN 0·30	HME 7 / HMN 10	Medium brown	Fine	6·4 9·0 12·0 16·0	43·5 61·2 81·6 108·8	1220 × 1830 to 3660 also 915 × 2440 and 3660 and 1830 × 3660
LME and LMN Low-density medium board	350-560	6·4 >6·4—10 >10	LME / LMN 14 / 12 11 / 10 9 / 8	LME 0·30 / LMN 0·40	LME 5 / LMN 8	Medium brown	Fine	6·4 9·0 12·0 16·0	28·8 40·5 -54·0 72·0	1220 × 1830 to 3660 also 915 × 2440 and 3600 and 1830 × 3660
Insulating board (softboard)	Not exceeding 350	up to 10 >10 to 13 >13 to 16 >16 to 19 >19 to 25	2·0 1·8 1·7 1·5 1·0	0·4	7	Grey buff	Medium	10 13 16 19 25·4	35 45·6 56·2 66·7 89·0	610, 915, 1220 widths × standard lengths 1830 to 3660
Bitumen impregnated insulating board	Not exceeding 400	up to 10 >10 to 13 >13 to 16 >16 to 19 >19 to 25	2·0 1·8 1·7 1·5 1·0	0·4	7	Grey	Medium	13 16 19 25·4	51·0 62·8 74·7 99·8	610, 915 and 1220 widths × standard lengths 1830 to 3660

TE, TN, S, HME, HMN, LME and LMN are standard grades of fibreboard

Information sheet
Timber 2

Permitted use of timber in eight occupancy groups

The tables in this sheet show at a glance permitted and restricted uses of timber in each of the eight purpose groups defined in the Regulations.

1 Introduction

1.01 The eight tables of this information sheet are:
I Small residential buildings
II Institutional buildings
III Other residential buildings
IV Offices
V Shops
VI Factories
VII Other places of assembly
VIII Storage and general
These relate to the eight purpose groups defined in The Building Regulations 1976.
1.02 The controlling dimensions at the head of the table relate to maximum sizes of *compartments*. By reducing the size of compartments in large buildings, limitations on use of wood can be minimised. The indications are for compartment elements within each purpose group. Where two purpose groups are concerned, the requirement of the most onerous prevails.

Symbols
1.03 The following symbols are used in the tables:
W: Timber may be used
Wa : Timber may be used only up to 15 m high; above 15 m, class 0 (surface spread of flame) is required

Wb : Timber may be used but not above a basement of more than 50 m²
Wc : Where a wall is far enough from a boundary to be 100 per cent unprotected and where it is non-loadbearing, it could consist entirely of windows (with wood frames)
Wd : Timber may be used for those parts of a structure that do not carry compartment floors or walls or loadbearing external walls
We : Timber may be used for those parts of a structure that do not carry compartment floors or external walls
Wf : Wood may be used for the framework of non-loadbearing walls of buildings up to 15 m high
Wg : Timber may be used for floors that are not more than 9 m above ground
Wp : Timber is permitted for stair construction but not within a protected shaft
R : Timber may be permitted for two-storey buildings under relaxation precedure
Ra : Timber may be permitted under relaxation procedure
Rb : Timber may be permitted under relaxation procedure for stairs at a change of level or to a gallery
NA : Not applicable

Table I Small residential buildings (purpose group 1)

	Height (storeys above ground)				
	Single storey	**Two storey**	**Three storey**	**Four storey**	**Any number of storeys**
Floor area (m²)	**No limit**	**No limit**	**No limit**	**250**	**No limit**
Capacity (m³)	**No limit**	**No limit**	**No limit**	**No limit**	**No limit**
External walls:					
Timber or plywood cladding 1 m or more from boundary	W	W	W	W	Wa
Framework to walls less than 1 m from boundary	W	W	W	—	—
Framework to walls 1 m or more from boundary	W	W	W	W	Wa
Separating walls	W	W	W	—	—
Floors	W	W	W*	W*	W*
Structural members	W	W	W	W	W
Roof construction	W	W	W	W	W
Stair construction	W	W	W	W	—
Linings to walls:					
Timber treated to class 1	W†	W†	W†	W†	W†
Linings to ceilings: ††					
Timber untreated (class 3)	W	W	W†	W†	W†
Timber treated to class 1			W	W**	W**

*Timber may not be used in the construction of a floor immediately over a basement which exceeds 100 m² in area.
†Untreated wood to class 3 in small rooms not exceeding 4 m² only.
**Circulation spaces and protected shafts may contain timber in composite boards of class 0 performance.
††Walls of other rooms may incorporate timber in class 3 form (or better) up to 20 m² or 50 per cent of the floor area, whichever is lesser.

Table II *Institutional buildings (purpose group 2)*

	Single storey	Height of buildings above ground	
		Up to 28 m	Over 28 m
Floor area (m²)	3000	2000	2000
Capacity (m³)	No limit	No limit	No limit
External walls:			
Timber or plywood cladding 1 m or more from boundary	W	Wa	Wa
Class 0 cladding (incorporating wood)	W	W	W
Framework to walls less than 1 m from boundary	W	—	—
Framework to walls 1 m or more from boundary	W	Wf	
Compartment walls	W	—	—
Floors (compartment)	NA	R	—
Floors (other)	W*	W*	W*
Structural members	W	R	—
Roof construction	W	W	W
Stair construction	W	—	—
Linings: ††			
Timber treated to class 1	W†	W†	W†
Timber in class 0 boards	W**	W**	W**

*Ground floor only. †Ceilings generally, except in circulation spaces and protected shafts which require class 0. **Walls generally, except in small rooms (up to 4 m²) which may be class 1.
††An area of timber of class 3 (or better) is permitted on walls of other rooms—up to 20 m² or 50 per cent of the floor area, whichever is lesser.

Table III *Other residential buildings (purpose group 3)*

	Single storey	Two storey	Three storey	28 m above ground	Any number of storeys
Floor area (m²)	3000	500	250	3000	2000
Capacity (m³)	No limit	No limit	No limit	8500	5500
External walls:					
Timber or plywood cladding 1 m or more from boundary	W	W	W	Wa	Wa
Class 0 cladding (incorporating timber)	W	W	W	W	W
Framework to walls less than 1 m from boundary	W	W	W	—	—
Framework to walls 1 m or more from boundary	W	W	W	Wc	Wc
Compartment walls	W	W	W	—	Ra*
Floors (compartment)	NA	W†	W†	—	W*
Floors (other)	W	W	W	W	W
Structural members	W	W	W	—	Ra*
Roof construction	W	W	W	W	W
Stair construction	W	W	W	—	—
Linings: ††					
Timber treated to class 1 (walls and ceilings)	W**	W**	W**	W**	W**

*Up to four storeys. †Except floors immediately over basements. **Except in circulation spaces and protected shafts requiring class 0 linings which may contain timber, and in small rooms (4 m²) which may be constructed to class 3. ††Walls of other rooms may incorporate timber in class 3 flamespread untreated, up to 20 m² or 50 per cent of the floor area, whichever is the lesser.

Table IV *Offices (purpose group 4)*

	Single storey		Height of building above ground (m)				
			7·5	7·5	15	28	No limit
Floor area (m²)	3000	No limit	250	500	No limit	5000	No limit
Capacity (m³)	No limit	No limit	No limit	No limit	3500	14 000	No limit
External walls:							
Timber or plywood cladding 1 m or more from boundary	W	W	W	W	W	—	—
Class 0 cladding (incorporating timber)	W	W	W	W	W	W	W
Framework to walls less than 1 m from boundary	W	—	W	—	—	—	—
Framework to walls 1 m or more from boundary	W	W	W	W	W	Wc	Wc
Compartment walls	W	—	W	W	—	—	—
Floors (compartment)	NA	NA	Wb	W*	—	—	—
Floors (other)	W	W	W	W	W	W	Wg
Structural members	W	W	W	W	Wd	Wd	—
Roof construction	W	W	W	W	W	W	W
Stair construction	W	Rb	W	W	—	—	—
Linings: ††							
Timber treated to class 1 (walls and ceilings)	W†	W†	W†	W†	W†	W†	W†

*Except over basement storey. †Except in circulation spaces and protected shafts which require class 0 (which may, however, contain timber), and in small rooms (up to 30 m²) which may be untreated timber to class 3.††Walls of other rooms may incorporate timber in class 3 form (or better) up to 60 m² or 50 per cent of the floor area, whichever is lesser.

Table v *Shops* (*purpose group* 5)

	Single storey			Height of building above ground (m)				
				7·5	7·5	15	28	No limit
Floor area (m²)	2000	3000	No limit	150	500	No limit	1000	2000
Capacity (m³)	No limit	No limit	No limit	No limit	No limit	3500	7000	7000
External walls:								
Timber or plywood cladding 1 m or more from boundary	W	W	W	W	W	W	—	—
Class 0 cladding (incorporating timber)	W	W	W	W	W	W	W	W
Framework to walls less than 1 m from boundary	W	—	—	W	—	—	—	—
Framework to walls 1 m or more from boundary	W	W	W	W	W	W	Wc	Wc
Compartment walls	W	—	—	W	W	—	—	—
Floors (compartment)	NA	NA	NA	Wb	Wb†	—	—	—
Floors (other)	W	W	W	W	W	W	W	Wg
Structural members	W	W	W	W	W	Wd	Wd	—
Roof construction	W	W	W	W	W	W	W	W
Stair construction	W	Wp	Wp	W	W	Wp	Wp	Wp
Linings:** Timber treated to class 1 (walls and ceilings)	W*	W*	W*	W*	W*	W*	W*	W*

*Class 0 required in circulation spaces and protected shafts, but may contain timber; untreated timber to class 3 permitted in small rooms (up to 30 m²).
†Not immediately over a basement. **Walls of other rooms may incorporate timber in class 3 form (or better) up to 60 m² or 50 per cent of the floor area, whichever is lesser.

Table vi *Factories* (*purpose group* 6)

	Single storey			Height of building above ground (m)					
				7·5	7·5	15	28	28	Over 28
Floor area (m²)	2000	3000	No limit	250	No limit	No limit	No limit	No limit	2000
Capacity (m³)	No limit	No limit	No limit	No limit	1700	4250	8500	28 000	5500
External walls:									
Timber or plywood cladding 1 m or more from boundary	W	W	W	W	W	W	—	—	—
Class 0 cladding (incorporating timber)	W	W	W	W	W	W	W	W	W
Framework to walls less than 1 m from boundary	W	—	—	W	—	—	—	—	—
Framework to walls 1 m or more from boundary	W	W	W	W	W	W	Wc	Wc	Wc
Compartment walls	W	—	—	W	W	—	—	—	—
Floors (compartment)	NA	NA	NA	Wb	W†	—	—	—	—
Floors (other)	W	W	W	W	W	W	W	W	Wg
Structural members	W	W	W	W	W	W	We	We	—
Roof construction	W	W	W	W	W	W	W	W	W
Stair construction	W	Rb	Rb	W	W	—	—	—	—
Linings:** Timber treated to class 1 (walls and ceilings)	W*	W*	W*	W*	W*	W*	W*	W*	W*

*Except in circulation spaces and protected shafts which require class 0 (but this may contain timber); and small rooms (up to 30 m²) which may be untreated timber to class 3.
†Not immediately over a basement. **Walls of other rooms may incorporate timber in class 3 form (or better) up to 60 m² or 50 per cent of the floor area, whichever is lesser.

Table vii *Other places of assembly* (*purpose group* 7)

	Single storey		Height of building above ground (m)				
			7·5	7·5	15	28	No limit
Floor area (m²)	3000	No limit	250	500	No limit	5000	No limit
Capacity (m³)	No limit	No limit	No limit	No limit	3500	14 000	No limit
External walls:							
Timber or plywood cladding 1 m or more from boundary	W	W	—	—	W§	—	—
Class 0 cladding (incorporating timber)	W	W	W	W	W	W	W
Framework to walls less than 1 m from boundary	W	—	W	—	—	—	—
Framework to walls 1 m or more from boundary	W	W	W	W	W	Wc	Wc
Compartment walls	W	—	W	—	—	—	—
Floors (compartment)	NA	NA	W	W	—	—	—
Floors (other)	W	W	W	W†	W	W	Wg
Structural members	W	W	W	W	W	Wd	—
Roof construction	W	W	W	W	W	W	W
Stair construction	W	Rb	W	W	—	—	—
Linings** Timber treated to class 1 (walls and ceilings)	W*	W*	W*	W*	W*	W*	W*

*Except in circulation spaces and protected shafts which require class 0, but this may contain timber, and in small rooms (up to 30 m²) which may be untreated to class 3.
†Not immediately over a basement. **Walls of other rooms may incorporate timber in class 3 form (or better) up to 60 m² or 50 per cent of the floor area, whichever is lesser.
§Above 7·5 m from the ground.

Table VIII *Storage and general (purpose group* 8)

	Single storey				Height of building above ground (m)						
					7·5	7·5	15	15	28	28	Over 28
Floor area (m²)	500	1000	3000	No limit	150	300	No limit	No limit	No limit	No limit	1000
Capacity (m³)	No limit	No limit	No limit	No limit	No limit	No limit	1700	3500	7000	21 000	No limit
External walls:											
Timber or plywood cladding 1 m or more from boundary	W	W	W	W	W	W	W	W	—	—	—
Class 0 cladding (incorporating timber)	W	W	W	W	W	W	W	W	W	W	W
Framework to walls less than 1 m from boundary	W	—	—	—	W	—	—	—	—	—	—
Framework to walls 1 m or more from boundary	W	W	W	W	W	W	W	W	Wc	Wc	Wc
Compartment walls	W	—	—	—	W	W	—	—	—	—	—
Floors (compartment)	NA	NA	NA	NA	Wb	W	—	—	—	—	—
Floors (other)	W	W	W	W	W	W	W	W	W	W	Wg
Structural members	W	W	W	W	W	W	Wd	Wd	Wd	Wd	—
Roof construction	W	W	W	W	W	W	W	W	W	W	W
Stair construction	W	Rb	Rb	Rb	W	W	—	—	—	—	—
Linings: **											
Timber treated to class 1 (walls and ceilings)	W*	W*	W*	W*	W*	W*	W*	W*	W*	W*	W*

*Except in circulation spaces and protected shafts which require class 0, but this may contain timber, and in small rooms (up to 30 m²) which may be class 3 (untreated timber). †Not immediately over a basement. ** An area of timber of class 3 (or better) is permitted on other walls, up to 60 m² or 50 per cent of the floor area whichever is lesser.

Information sheets
Timber 3 and 4

Information sheet 3:
Classification of solid timber joints

Information sheet 4:
Mechanical fasteners

These two sheets relate directly to Technical study TIMBER 5. *Information sheet* TIMBER 3 *consists of a table which classifies joints accordingly to functional form. Thus, irrespective of other attributes, structural end joints are considered first. Node joints are dealt with second and in this category are joints in structural frameworks such as trusses, portals and trussed rafters, where two or more members come together at a node point. These joints have been subdivided into those in which the members are laterally displaced and overlap at the connections, and the simple single-plane joints which tend to be currently preferred. Finally, joints in framing members are categorised and these cover instances where two or more members meet at right angles and with axis in more than one plane.*

Information sheet TIMBER 4 *summarises in tabular form the principal characteristics of the types of mechanical fastener most commonly used for structural joints in timber frame construction.*

Embedding timber connectors in joints of timber lattice girder

References for Information sheet 3
1 BRITISH WOODWORK MANUFACTURERS' ASSOCIATION Finger joints in timber. London, November 1968, The Association [Xt6]
2 FOREST PRODUCTS RESEARCH LABORATORY Technical note 19 End joints in timber. Revised April 1971, Princes Risborough, Aylesbury [(9–) Yi]
3 FOREST PRODUCTS RESEARCH LABORATORY Timberlab paper 29 1970 Recommendations for trussed rafters for domestic roofs. Revised 1972, Princes Risborough, Aylesbury [(27) Hi]
4 TRADA E/IB/17 Span tables for ridged frame portals in solid timber [26 (28) Yi (F4j)]

References for Information sheet 4
1 LAUNCHBURY, W. AJ Handbook of fixings and fastenings. London, 1971, The Architectural Press [Xt7] £1
2 BS CP 112: 1971 The structural use of timber [(2–) Yi (K)] £3.00
3 BOOTH, L. G. and REECE, P. O. The structural use of timber: A commentary on CP 112: 1967. London, 1967, Spon [(2–) Yi (K)]
4 BS 1202 Nails: Part 1: 1966 Steel nails; Part 2: 1966 Copper nails; Part 3: 1962 Aluminium nails [Xt6]
5 BROCK, G. R. The strength of nailed joints. Forest Products Research Bulletin 41, London, 1957, HMSO [(9–) Xt6]
6 US DEPARTMENT OF AGRICULTURE Wood handbook. Forest Products Laboratory. Washington, 1955, US Government Printing Office [Yi]
7 TRADA Research Report E/19 The resistance to withdrawal of five types of 2½in long, 9 swg nails in European redwood. High Wycombe 1964, The Association [Xt6]
8 TRADA Research Report E/32. The rotational rigidity of nailed joints with plywood gussets subjected to short-term loading. High Wycombe 1969, The Association [(9–) Xt6]
9 TRADA Information Bulletin E/17 Span tables for ridged frame portals in solid timber. High Wycombe 1970, The Association [26 (28) Yi (F4j)]
10 TRADA Test Memorandum E/64 The resistance to withdrawal and lateral loading of 2in ringed shank nails and 2in square twisted nails in European redwood. High Wycombe 1962, The Association [Xt6]
11 SHAW, M. Annular fasteners: An assessment of their properties and potential. Rylands-Whitecross & Co (Xt6)
12 BS 1210: 1963 Wood screws [Xt6]
13 WILEY, J. Timber construction manual. New York, 1966, American Institute of Timber Construction [(2–) Yi (K)]
14 Timber design and construction handbook. New York, 1959, Timber Engineering Co, F. W. Dodge Corporation [(2–) Yi (K)]
15 BS 1579: 1960 Connectors for timber [Xt6]
16 Pneumatic and mechanically driven building construction fasteners. Pasadena, California, 1970, January, International Conference of Building Officials, Report 2403 [Xt6]
17 STERN, E. G. Holding power of 2½in 15 gauge SENCO staples. Virginia Polytechnic Institute Research Division, Bulletin no 97, 1970, November, Blacksburg, Virginia [Xt6]
18 STERN, E. G. Lateral load transmission by 2½in 15 gauge SENCO staples. Virginia Polytechnic Institute Research Division, Bulletin no 92, 1970, June, Blacksburg, Virginia [Xt6]

Information sheet 3: Classification of solid timber joints

Joint	Typical strength characteristics	Type of fixing used	Comments and references
1 End joint			
Butt joint	Only about one-twentieth of the strength of unjointed wood in tension	Adhesive. BS 4169 : 1970 *Glued laminated timber structural members* gives extensive information on structural adhesives, and the manufacturing requirements for glued structural timber components	Practical application limited to non-structural applications or in laminated work where the joints are staggered and the laminate strength at the butt joint is discounted
Plain scarf joint	Plain scarf joints with slope of 1 in 12 or flatter may be permitted up to 85 per cent of the strength of unjointed timber, dependent on grade	As above	The plain scarf may provide strongest of all end joints, but it has two disadvantages, first amount of material wasted in forming joint, and second the difficulty of locating joint during fabrication
Hooked and stepped scarf joint	Canadian practice is to permit the same reduction factors as for plain scarf joints butt to reduce the nett cross-section by the depth of the step. The step creates a significant stress raiser. Typically such joints have about 50 per cent efficiency	As above	In scarf joint with notches and a step, both halves of joint are completely self-locating in longitudinal direction. Because of the reduction in strength caused by end steps, a modified form is sometimes used with a hook only
Finger joint	Structural finger joints must have strength ratio expressed as a proportion of the basic strength of the clear timber equal to the grade stress percentage, eg a finger joint in 40 grade must have 40 per cent efficiency	As above	Major advantages of finger joints are that they are completely self-locating and minimise timber wastage. The BWMA paper[1] defines four classes of use : structural ; structural laminated, both with a minimum efficiency equal to the grade in which the joint is used ; semi-structural and loadbearing joinery with a minimum efficiency of 40 per cent ; and non-structural. Factors affecting these joints and methods of meeting the required performance are covered. A BS specification for finger joints in structural softwoods is in preparation. Numerous joint profiles have been developed, one unusual form has fingers in an angled plane relative to both faces of timber. A typical profile is illustrated, with a length of 50 mm, pitch 12 mm, and tip width of 2 mm. So-called mini-finger joints have smaller dimensions and more fingers in a given width than that shown, achieving even greater economy of material. These joints should not be confused with die-formed finger joint dealt with in next section. Reference 2 covers all four of the end joints so far mentioned

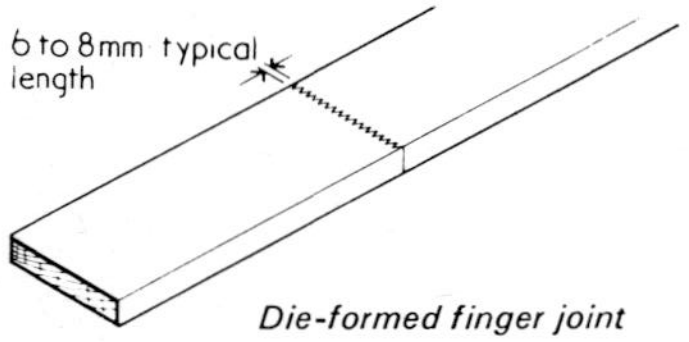

Die-formed end-joints pressed on cold dies and cured with radio frequency heating offer rapid and economical method of producing joints of moderate strength for joinery and semi-structural applications

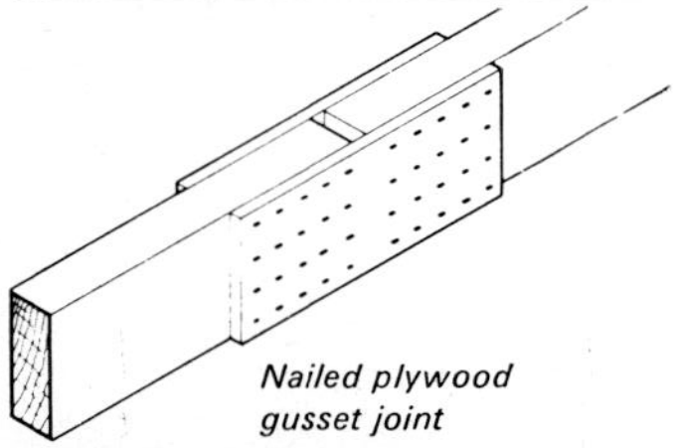

Tensile strengths of up to 40 per cent of the strength of clear timber have been achieved in tests with European redwood

6 to 8 mm typical length

Die-formed finger joint

As above

A common means of splicing ceiling ties in trussed rafters. Nails may be driven from both sides, or more commonly driven through all three layers and clinched or riveted. Such joints should be designed to carry direct tension or compression forces and a moment equal to half the permissible moment of resistance of cross-section of solid timber. They should be positioned close to a point of contraflexure in the structure

Typical average ultimate load 40 kN in a joint made with twenty 3·3 mm diam (10 SWG) nails in double shear in each half. Members were 38 mm × 100 mm Baltic whitewood and gussets of 9 mm Finnish birch plywood

Nailed plywood gusset joint

Plain wire, galvanised or pneumatically-driven nails are found in this type of joint, 2·6 mm to 3·7 mm diam (12 to 9 SWG) being typical. Similar joints are made with punched metal plate fasteners and nailed metal plates. Further information will be found below

2 Node joints—Lap jointed members

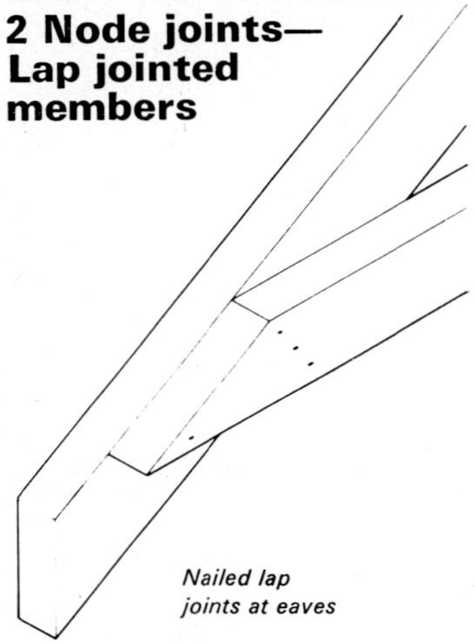

In practice the poor load-carrying capacity of this type of joint limits its usefulness. It is mainly found in fixing of common rafters and ties in a spaced truss roof system or in traditional framing construction where it is rarely calculated

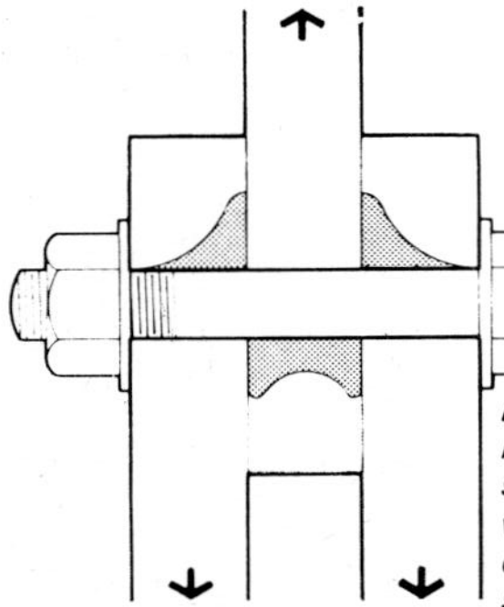

Thrust in rafter of 22·5° pitch would be restricted to about 3 kN when joining 50 mm × 100 mm members because of limited area available for nail spacing, whereas with design governed by member sizes only, three times as much force could be transmitted

Common wire nails: 3·7 mm (9 SWG) × 90 mm typical

Nailed lap joints at eaves

Essentially the improvement in performance achieved by using a bolt in place of a nail is only due to the increased diameter of the fastener. Marked concentrations of stress and a tendency of fastener to bend are equally present in either system. Clamping effects which may at first be achieved with a bolt are not beneficial as they must be discounted because of likely timber shrinkage. For these reasons connectors are nearly always used with bolts except in lightly loaded structures or in cases where bolts are used only for assembly purposes

Typical dry basic load 2·43 kN on one bolt in a two-member joint, parallel with grain. This will drop by about one-third at right angles to the grain. In the three-member joint illustrated, bolt is in double shear and basic load is twice that of two member joint, although there are certain restrictions relating to relative thickness of members

Values quoted as typical are for a 12·7 mm bolt in 50 mm softwood members

Diagram of bolted double shear lap joint with stress distribution shaded

Joint	Typical strength characteristics	Type of fixing used	Comments and references

Typical strength characteristics (column header area)

Design forces in members of the illustrated joints are as follows:

187 kN
53°
136 kN
100×100 mm
240 kN
127 kN
2/50×125 mm

Truss joint with split ring connections

Type of fixing used

Example is for 64 mm split ring connector with 12·7 mm diam bolt. Design principles are similar for split ring, shear plate and toothed plate connectors. Basic loads, factors and spacings are given in CP 112

Comments and references

The principal advantage of timber connectors is high joint efficiency and rigidity. In addition skilled site labour can often be eliminated, the joints being shop fabricated and simply assembled on site. Compared with fasteners such as plain bolts and nails, connectors frequently permit a reduction in member sizes; nevertheless sizes may still be governed by connector spacing rather than stress in the members. During the last 20 years a wide variety of trusses, mainly in the 6 m to 9 m range, have been designed by TRADA for domestic roofs using toothed plate and split ring connectors. Such trusses are generally spaced at 1·8 m centres and support purlins carrying intermediate rafters. Designs are also available for industrial trusses of up to 21 m span and very large spans are possible using bow string trusses with the largest sizes of connector. For domestic use the connectored trusses have in many instances been superseded by light trussed rafters with members in a single plane and joints formed from ply gussets or punched metal plate fasteners. A small but steady demand for the spaced truss continues however for small developments of an individual nature, or in instances where an unrestricted space is required

---

3 Node joints—members in a single plane

Typical strength characteristics

Lateral resistance depends on the effective number of nails in the joint, species of timber, moisture content, duration of load and direction of bearing of the pressed-out nails in the metal plate relative to the timber grain.

Typical parallel with the grain ultimate load value for a softwood joint made with two 75 mm × 150 mm (20 gauge) plates is in the order of 22 kN for a 1 mm (20 gauge) plate. The strength of the fastener material itself, both in shear and tension, must be considered in design

Apex joint of trussed rafter with punched metal plate fastener (teeth in plate illustrated are exaggerated)

Type of fixing used

Proprietary types of punched metal plate fastener. Profiles vary among manufacturers. Thicknesses range from 1 mm (20 gauge) to 2 mm (14 gauge). Manufactured from galvanised mild steel strip which is cut and punched to form a large number of integral nails or teeth projecting at right angles to one face of the plate

Comments and references

These fasteners have probably been the most significant recent development in the assembly of light timber structures. They combine the functions of gusset plate and fixing device, the plate overlapping two or more members at a joint and being embedded under pressure. Assembly requires the use of jigs and presses, in some systems one large press under which the component is rolled closes all the joints, while in other cases a separate press is used at each station.

The plates are used mainly in the manufacture of trussed rafters with spans up to 11 m and spacings of not more than 600 mm. Spans for ranges of trussed rafter using this type of joint are given in reference 3. A BS Code of Practice relating to trussed rafters for roofs of dwellings is in preparation. It is forecast that punched metal plate fasteners will come into use for other light framing applications such as the assembly of wall frame units and end jointing of joists. Trials have shown that these fasteners can be satisfactorily embedded in structural hardwoods

---

Typical strength characteristics

Average maximum load for end-jointed softwood specimens measuring 35 mm × 95 mm with 150 mm × 175 mm plates angled to minimise nail spacing was 32 kN with one proprietary type. Design values obtainable from manufacturers are quoted per nail in parallel and perpendicular grain directions. Advice regarding edge distances and strength of the plate material itself can also be obtained

Strut and tie joint in trussed rafter with proprietary nailed steel gusset plate

Type of fixing used

Plates are usually 1 mm thick (20 gauge) in galvanised mild steel. Square twisted nails are favoured for the attachment of these plates, 3·7 mm × 32 mm being typical. These nails, which are a driving fit in the plate holes, develop a higher lateral load than plain nails and are claimed to have less tendency to split the timber

Comments and references

A successful alternative to punched metal plate fasteners in trussed rafters and other forms of light framing have the advantage of requiring only hand tools, although pneumatically driven nails are often employed for labour economy

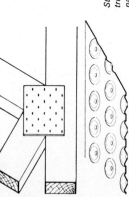

Nailed gussetted construction is particularly appealing to builders who wish to make their own trusses or trussed rafters, as special pneumatic or hydraulic equipment is not essential. Trusses may be built either on a shop-sited jig table or on the building site, in some instances the house floor itself has proved a satisfactory setting-out surface. Precision cutting of members is not always called for and provided that gusset stresses are calculated accordingly, some members may be left square-ended, although it is important to mitre the rafter joints.

Nail density should be calculated in design procedure and the stipulated spacings followed during fabrication. Glued plywood gussets are an alternative proposition, although currently less favoured in practice. In the glued form of construction, fixity from fastenings used to apply bonding pressure should be discounted during design, as disparity in stiffness between the two methods of fixing prevents the mechanical device from taking effect

A new era in the use of timber for low-cost farm buildings has been opened by the completion of an ambitious portal design programme at TRADA. Subsequent work has shown that the same designs can be used for industrial buildings. Using solid timber members joined at the eaves and ridge by nailed plywood gusset plates, the standard form of design is applicable to any span in the range 6-15 m, at any bay spacing. Different types of timber may be used, and the new system comprises literally thousands of designs yet all are basically the same.

A most important economic aspect of timber portals is the ease with which they can be positioned and erected. The manufacture of nailed plywood gusset portals requires some direction but is basically simple and does not make a demand on highly skilled labour. Although the nailed plywood haunch and ridge gusset joints have a high structural duty, the design and manufacture of the portal frame is based on the humble nail. The function of the gussets is to provide rigid joint connections, and an easy and effective means of attaching plywood to solid timber is by dense pattern nailing.

Higher load transfers can often be achieved by nailing rather than by using glue, as the strength of glued joints is limited by a low permissible shear stress in the ply laminations (rolling shear stress). Nailed plywood joints can be made by less specialised labour than the glued type as they do not require the same skill.

Reference 4 gives details of the design and construction of nailed plywood gusset portals, including joint construction and extensive tables of member sizes

When a joint of this nature is limited in load-carrying capacity by compression perpendicular to the grain strength, then it is worthwhile considering the possibility of using a hardwood species for the sole plate. Design tables using softwood studs with hardwood plates are available from TRADA

Typical design values as follows:

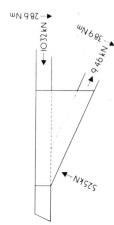

Example uses a gusset 600 mm long measured on the rafter in 10 mm Douglas fir plywood with 2·6 mm × 45 mm nails driven from one side and clinched are also available and commonly used

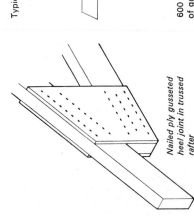

Nailed ply gusseted heel joint in trussed rafter

---

600 mm centres, 8·4 m span and 30° pitch (both types of gusset shown are commonly used)

Typical design values shown

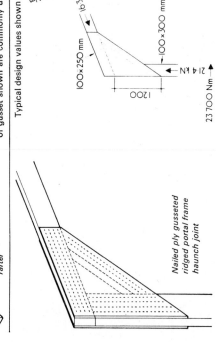

Example is with 19 mm Douglas fir plywood gussets and a total of 130 round wire nails in each side of the joint, 68 in the rafter and 62 in the column. The size of these nails is 50 mm × 3·3 mm (10 SWG). In practice it is found that a large number of rails such as this can rapidly be driven by hand even in dense timbers, showing a considerable saving in labour compared with the fabrication time spent on other methods of forming such a heavy-duty joint. Pre-drilling is not necessary when the relatively small diameter nails are started in plywood gussets

Nailed ply gusseted ridged portal frame haunch joint

---

4 Joints in Framing members

In the case of a stud bearing on a sole plate the design of this type of joint is likely to be governed by compression perpendicular to the grain strength in the area of the sole plate covered by the cross-section. Typically this might restrict the stud capacity to a value between one-half and two-thirds of that which could be carried otherwise. Where design must include precautions against racking effects, and the sole plate is anchored against uplift, then it may be necessary to provide a fixing in this type of joint which will resist vertical tensile forces

Joint may be made by skew nailing, end nailing, with a nail plate, plywood gusset or punched metal fastener across the joint. Alternatively framing anchors may be employed to form a positive fixing

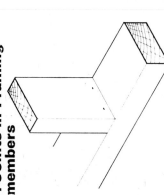

Simple butt joint occurring in sole plate, stud to header, etc

Joint	Typical strength characteristics	Type of fixing used	Comments and references
	Short-term design load in the direction shown is 1·8 kN. Values for other directions and conditions of loading may be obtained from manufacturers of framing anchors *Framing anchor used in attachment of purlin support*	Framing anchors are manufactured from galvanised or sherardised mild steel strip and attached with small galvanised clout nails, 32 mm × 2·9 mm (11 SWG) being typical	These versatile connectors may be used for a number of applications such as the anchorage of truss heel joints to a wall plate, the fixing of purlins, as shown, in joist trimming or the anchorage of stud framing
	Typical end reaction of a floor joist to be carried in this manner would be about 4 kN. Strength tests on various types of joist hanger have shown average values ranging from 8·9 to 17·8 kN according to the shape of the hanger straps, the number of nails used and steel thickness *Trimming joint as used for floor and roof joists. A metal hanger*	Galvanised mild steel strip ranging in thickness from 1 to 2 mm is typical material for these fixings which are attached with similar nails to those used for framing anchors	Joist hangers are commonly used today in place of the traditional tusk tenon joint cut by a skilled carpenter. In certain situations such as over a plastered ceiling, a large deflection of the supported joist would be highly undesirable and possible deflection of the joint should be guarded against both in design and construction
	In designing trimming joint, effect of notch upon shear strength should be checked using recommendations and factors from the Code of Practice *Square housed*	Tenon cut by tenon machine with cut off saw, mortise by chisel mortiser. Joint is fixed by hand nailing	
	Comments above apply equally to this form of joint. Strength of fixings attaching ledger strip to trimmer must also be checked in lateral strength *On ledger*	Ledger may be hand or power nailed, screwed, bolted or glued to trimmer. Joist is skew nailed to trimmer	Metal angle ledgers may also be used on occasion. Where joist trims on to deep beam such as a plywood box, ledger will be above lower edge of the trimmer beam

Information sheet 4: Mechanical fasteners

Type		Composition and surface finish	Comparative strength	Application	References
1 *Nails* **a** *round plain-head wire;* **b** *round lost-head wire;* **c** *clout or slate nails;* **d** *pneumatically driven nails—clipped head (illustrated), tee etc* a b c d		Mild steel (bright, galvanised or sherardised finish) Copper and aluminium also used, especially where staining or chemical action is likely to take place. **Strength values for** nails of these materials not tabulated in the structural design code	Design should ensure that these nails are loaded laterally whenever possible as withdrawal strength is rather low. As nails can be rapidly applied in dense patterns, a strong and stiff joint may be made	Securing constructional timbers. One of most common forms of fastening. Pneumatic tools have revolutionised the economics of their application	1 (For description of types of nail and tabular information on available sizes) ; 2 (for basic loads, factors and spacings of plain wire nails) ; 3 (for background to derivation of Code values) ; 4 (the relevant British Standard) ; 5 (for details of the research upon which nail loadings are based) ; 6 (discusses factors affecting nail strength) ; 7 (resistance to withdrawal of five types of nail) ; 8 (rotational rigidity of nailed joints) ; 9 (design of nailed gusset joints in portal frames).
2 *Twisted shank nails*		Mild steel (bright or galvanised finish). Yield strength not less than 375 N/mm² if advantage to be taken of increase factors in CP 112	CP 112 permits 1·25 times lateral load and 1·5 times withdrawal of ordinary nails if yield strength as stipulated	Originally for roofing, especially corrugated sheet; have proved very successful for attaching proprietary pierced metal plate fasteners.	1, 2, 3, 10
3 *Annular ring-shanked nails*		Mild steel (bright or galvanised or sherardised)	Have shown up to 250 per cent better withdrawal resistance than plain wire nails in test	May be used for similar purposes to twisted shank nails; also useful for plywood floor panels in partially eliminating 'nail popping'	10, 11 (Manufacturer's publication giving lateral and withdrawal strength)
4 *Staples*		Steel (zinc coated, resin coated)	Not covered by present code. As an approximate rule a staple may be assumed to have a lateral load value equal to that of a nail 1½ times the diameter of one leg of the staple	Alternative to pneumatically driven nails for fixing plywood, fibreboard, plasterboard, insulation etc. Most frequently fixed with powered tools although hand tools are available. Tools driven from compressed air units carried as back-packs are available. In many US systems timber frame housing is now assembled almost exclusively with on site pneumatic stapling and nailing equipment. Manufacturers' advice should be sought on suitable type of staple as these vary greatly, eg gauge, width of crown, length of leg, straight or divergent, type of point, coating etc	16 (US Uniform Building Code report recommending lateral and withdrawal load values for staples) 17, 18 (A list of more than 90 different VPI Bulletins relating to fixings and fastenings will be found at the end of these two references which deal with 2½in 15 gauge staples).
5 *Screws* **a** *countersunk head;* **b** *roundhead;* **c** *raised countersunk head;* **d** *recessed head:*		Steel, bright, zinc, plated, sherardised, various enamelled or plated finishes. Also in brass, stainless steel and other alloys	Compared with round wire nails, screws have very superior withdrawal resistance while lateral loads are slightly less	Greater cost and slower rate of application limit their use in purely structural functions	1 (For description of types of screw and sizes) : 2 (for basic loads, factors and spacings) ; 3, 6, 12 (the relevant British Standard)

Type	Composition and surface finish	Comparative strength	Application	References
6 *Bolts* **a** *hexagonal head:* **b** *square head:* **c** *coach bolts with cup, square or round head*	Steel, black finish usual	Stresses in a bolted joint are confined to a comparatively small area in the contact region of the bolt shank, and there is a tendency for the bolt to bend. Use of bolts without connectors is restricted to cases where forces in joints do not exceed the relatively low loadbearing capacity of the bolt or the timber	For assembly work or in lightly loaded structures. Sometimes a pair of bolts are added to 'stitch' a connected joint	2 (For basic loads, factors and spacings) ; 3 (for background to above)
7 *Coach screws*	Steel, black finish usual	Lateral load capacity similar to that of bolts, provided that correct size of lead-in hole and proper penetration achieved	Sometimes a useful means of fixing heavy 'planted on' sections. Tend to be overlooked as a rather old-fashioned form of fixing. Have also been used to fix splice plates in butt joints of heavy timber sections	Called lag screws in US timber design handbooks where strength values may be found ; 13
8 *Connectors* **a** *double-sided round toothed plate;* **b** *single-sided round toothed plate;* **c** *double-sided square toothed plate;* **d** *single-sided square toothed plate;* **e** *split ring;* **f** *shear plate* (see illustration p511)	Steel, usually zinc coated. Malleable cast-iron (shear plates)	Lateral load capacity is in general at least doubled by use of some form of connector compared with a plain bolt. A number of factors that affect strength of connectored joint such as form of joint and connector, diameter of bolt, size and species of timber, angles of loading, end distances and spacing and so on. These are comprehensively dealt with in the Code	Before introduction of punched metal plate fasteners and development of ply gussetted designs, the connectored joint was most common form in timber roof structures. Toothed plate connectors can be easily bedded into softwoods and trials have shown that using a ratchet spanner and ball-bearing washer, they can be bedded into more dense hardwoods than was previously believed, and also into plywood. Split ring connectors consist of circular bands of steel placed in pre-cut grooves in contact faces of joint members. They form stronger joints than toothed plates and can be used in most dense timbers but need a special cutting tool. Shear plates are most frequently used for timber to structural steel joints	2 (For basic loads, factors and spacings) ; 3 (for very comprehensive background and descriptions of types dealt with in the Code) ; 6, 13, 14 (for US views on factors and design) ; 15 (the relevant British Standard)

Information sheet
Timber 5

Section 7 **Structural material: Timber**

An introduction to structural calculations for timber

The examples by R. W. WANDS *calculated here are of a general kind, stopping short of those larger and more complex design problems normally requiring the advice of an engineer. The methods of calculation expanded here are applicable to any timber grades and plywoods of known strength characteristics.*

1 Introduction

1.01 BS CP 112[1], *The structural use of timber*, advises on the design of timber members, components and structures using solid timber, laminated timber, plywood and composite members of timber and plywood.

1.02 Even granted a fair knowledge of structural theory and design, the architect could make full use of the code's various timber stress modification factors (there are 26 of them and 56 tables) only after an exhaustive study. The scope here is therefore confined to an outline of broad principles.

2 Timber characteristics affecting design

Availability of timber and plywood

2.01 The code lists permissible stresses for 14 softwood species and 15 hardwood species but few of these are readily available in a sufficient number of sizes to allow their use to be specified with confidence. For example, Canadian hemlock and Douglas fir, softwoods widely used in the past in the UK, are now in short supply. European redwood or whitewood will be preferable. BS 4471[2] specifies softwood dimensions and table I therein lists the combinations of thickness and width normally cut—hence the reason for using the 'European origin' range when designing and specifying. The standard also deals with sawing and planing tolerances.

2.02 The two main sources of plywood are Douglas fir plywood from Canada and birch plywood from Finland. Permissible stresses and strengths are listed in section 4 of CP 112.

Timber strength—stress grading

2.03 During its lifetime, a tree develops a variety of growth characteristics affecting shape, colour and strength so that wood from the saw mills, although cut to the same cross section, may widely differ in its load carrying capacity. Various characteristics such as the number, size and position of knots and splits, and the slope of the grain, must be taken into account in estimating the strength of the piece. Such evaluation is called 'stress grading' and the rules listed for this in the code produce four strength grades designated 75, 65, 50 and 40. (The figures represent roughly the percentage of timber cross section free from defect.) A piece of clear, defect-free timber forms the basis for strength assessment and the stress applicable to such a piece is termed the *basic stress*. The *permissible stress* for the various grades is proportionately lower, 40 grade being the lowest admissible grade and consequently the one which is apportioned the lowest permissible stresses.

2.04 A typical parcel of 100 pieces of vt commercial quality softwood would comprise on average 10 pieces at 75 grade, 15 at 65 grade, 50 at 50 grade, 20 at 40 grade and 5 rejects. Therefore it will usually be best to specify 50 grade softwood and to use the stresses appropriate to this grade in design. However grading by machine is now becoming more widespread and the appropriate M-grade and higher permissible stresses may be used where a supply of machine graded timber is available.

Moisture content

2.05 Timber is weaker and more elastic when wet and consequently the code prints two permissible stress tables, one listing green stresses and the other reproduced in this study as table I, listing stresses applicable to wood at or below 18 per cent moisture content. Almost all softwood brought into the UK is shipped at a moisture content of about 20 per cent which will reduce somewhat during storage, so unless the timber is likely to be used in a wet situation, the dry permissible stresses may be used. (See also Technical study TIMBER 2 para 2.08.)

3 Design of housing components

3.01 A fairly typical two storey structure is illustrated in **1**. In calculating the loading to be carried, the roof load on slope is normally taken to equal the plan load with pitches less than say 25°; otherwise plan load = $\dfrac{\text{slope load}}{\text{cosine pitch}}$

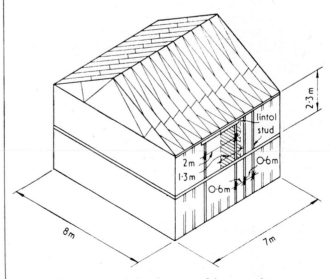

1 *Typical two-storey timber frame used in example calculation, in text.*

Table I *Dry stresses and moduli of elasticity (extracted from* BS CP 112, *table 4)*

Standard name	Bending and tension parallel to grain (N/mm²)					Compression parallel to grain (N/mm²)					Compression perpendicular to grain (N/mm²)			Shear parallel to grain (N/mm²)					Modulus of elasticity for all grades	
	Basic	75 grade	65 grade	50 grade	40 grade	Basic	75 grade	65 grade	50 grade	40 grade	Basic	75/65 grades	50/40 grades	Basic	75 grade	65 grade	50 grade	40 grade	Mean	Minimum
Softwoods																				
Imported																				
Douglas fir	18·6	13·1	11·0	8·6	6·6	14·5	10·3	8·6	6·6	5·2	2·62	2·34	1·93	1·93	1·34	1·21	0·90	0·76	11 700	6 600
Western hemlock (unmixed)	15·9	11·4	9·3	7·6	5·9	12·4	9·3	7·9	6·2	4·8	2·07	1·72	1·52	1·65	1·21	1·07	0·86	0·66	10 000	5 900
Western hemlock (commercial)	14·5	10·0	8·6	6·6	5·2	11·0	8·3	6·9	5·2	4·1	2·07	1·72	1·52	1·52	1·14	0·97	0·76	0·62	9 300	5 500
Parana pine	14·5	10·0	8·6	6·6	5·2	12·4	9·3	7·9	6·2	4·8	2·21	1·93	1·65	1·65	1·21	1·07	0·83	0·60	9 000	4 800
Pitch pine	18·6	13·1	11·0	8·6	6·6	14·5	10·3	8·6	6·6	5·2	2·62	2·34	1·93	1·93	1·34	1·21	0·90	0·76	11 700	6 600
Redwood	14·5	10·0	8·6	6·6	5·2	11·0	7·9	6·6	4·8	3·8	2·21	1·93	1·65	1·52	1·14	0·97	0·76	0·62	8 300	4 500
Whitewood	14·5	10·0	8·6	6·6	5·2	11·0	7·9	6·6	4·8	3·8	2·07	1·72	1·52	1·52	1·14	0·97	0·76	0·62	8 300	4 500
Canadian spruce	13·8	9·7	7·9	6·2	5·2	11·0	7·9	6·6	4·8	3·8	2·07	1·72	1·52	1·52	1·14	0·97	0·76	0·62	9 000	5 500
Western red cedar	11·0	7·6	6·6	5·2	3·8	9·0	5·9	4·8	3·4	2·8	1·52	1·31	1·10	1·38	0·97	0·83	0·69	0·55	6 900	4 100
Home-grown																				
Douglas fir	17·9	12·4	10·7	8·3	6·6	13·8	10·0	8·3	6·2	4·8	2·48	2·21	1·93	1·52	1·14	0·97	0·76	0·62	10 000	4 800
Larch	17·2	12·1	10·3	7·9	6·2	13·1	9·3	7·6	5·5	4·5	2·62	2·34	1·93	1·72	1·21	1·07	0·83	0·66	9 700	4 800
Scots pine	15·2	9·7	7·9	6·2	5·2	11·7	7·9	6·6	4·8	3·8	2·48	2·21	1·93	1·52	1·14	0·97	0·76	0·62	9 700	5 500
European spruce	11·0	7·2	5·9	4·8	3·4	9·0	5·9	4·8	3·4	2·8	1·65	1·38	1·24	1·34	0·90	0·76	0·62	0·45	6 900	3 800
Sitka spruce	10·3	6·6	5·5	4·5	3·4	8·3	5·2	4·1	3·1	2·4	1·65	1·38	1·24	1·24	0·90	0·76	0·62	0·45	7 200	3 800
Hardwoods																				
Imported																				
Abura	16·5	12·1	10·3	7·9	6·2	13·8	10·0	8·3	6·2	4·8	3·45	3·10	2·48	2·41	1·65	1·45	1·14	0·90	9 300	4 800
African mahogany	15·2	10·7	9·0	7·2	5·5	13·1	9·3	7·6	5·5	4·5	3·10	2·62	2·21	1·93	1·34	1·21	0·90	0·76	8 600	4 500
Afrormosia	26·2	19·3	15·9	12·4	9·7	22·1	15·2	12·4	9·3	7·6	6·21	5·17	4·48	2·76	2·07	1·79	1·38	1·10	12 100	7 900
Greenheart	41·4	31·0	26·9	20·7	16·5	30·3	22·8	19·7	15·2	12·1	9·31	7·93	6·90	5·52	3·93	3·38	2·62	2·14	18 600	13 400
Gurjun/keruing	22·8	14·8	12·4	9·7	7·9	19·3	13·1	11·0	8·3	6·6	4·48	3·79	3·45	2·62	1·86	1·65	1·28	0·97	13 800	9 300
Iroko	23·4	17·6	15·2	11·7	9·3	19·3	14·5	12·1	9·0	7·2	6·21	5·17	4·48	2·62	1·86	1·65	1·28	0·97	10 300	6 900
Jarrah	23·4	16·9	14·1	11·0	8·6	20·7	15·2	12·4	9·3	7·6	6·21	5·17	4·48	2·62	1·86	1·65	1·28	0·97	12 100	7 900
Karri	26·2	19·3	15·9	12·4	9·7	22·1	15·9	13·1	9·7	7·9	7·24	6·21	5·17	2·76	2·07	1·72	1·34	1·10	15 500	9 700
Opepe	29·0	22·4	18·6	14·5	11·7	24·8	18·6	15·9	12·4	9·7	8·27	7·24	6·21	3·72	2·48	2·21	1·65	1·34	13 800	9 300
Red meranti/red seraya	15·2	10·7	9·0	7·2	5·5	13·1	9·3	7·6	5·5	4·5	2·62	2·34	1·93	1·72	1·21	1·07	0·83	0·66	8 300	4 500
Sapele	23·4	16·9	14·1	11·0	8·6	20·7	15·2	12·4	9·3	7·6	6·21	5·17	4·48	2·76	1·86	1·65	1·28	0·97	11 000	6 900
Teak	26·2	19·3	15·9	12·4	9·7	22·1	15·9	13·1	9·7	7·9	6·21	5·17	4·48	2·62	1·86	1·65	1·28	0·97	12 400	7 900
Home-grown																				
European ash	22·8	14·8	12·4	9·7	7·9	15·2	10·3	8·6	6·6	5·2	4·48	3·79	3·45	3·10	2·28	2·00	1·52	1·24	11 400	7 200
European beech	22·8	14·8	12·4	9·7	7·9	15·2	10·3	8·6	6·6	5·2	4·48	3·79	4·35	3·10	2·28	2·00	1·52	1·24	11 400	7 200
European oak	20·7	13·8	11·7	9·0	7·2	15·2	10·3	8·6	6·6	5·2	4·48	3·79	3·45	3·10	2·07	1·72	1·34	1·10	9 700	5 200

Note 1. These stresses apply to timber having a moisture content not exceeding 18 per cent.
Note 2. When calculating deflection, the mean value of the modulus of elasticity should be used for rafters, floor joists and other systems where it can be shown that transverse distribution of load is achieved. The minimum value should be used for principles, binders and other components acting alone.

Hence: to find roof load

Dead load (from BS 648[3] unless otherwise stated)

Tiles (from maker's catalogue) = 12 lb/sq ft
Battens and felt = 1
Truss weight (typical value) = 2
Plaster and skim ceiling = 5

———
20

Superimposed loads (from BS CP 3: Chapter V; Part 1[4])

Rafter (including snow) = 15
Ceiling = 5

———
20

Total ≒ 40 lb/sq ft
= 1920 N/m²

Select the truss arrangement

3.02 This load can be supported either by trussed rafters at 600 mm centres spanning between back and front walls or by a purlin spanning from cross-wall to cross-wall carrying trussed rafters or common rafters and ceiling joists. Trussed rafter sizes are given in Information sheet TIMBER 10 tables I to VIII or BRE Digest No 147[5] (supplementing Timberlab paper 29 issued by Princes Risborough Laboratory.) The size of connecting plates can be left to the truss manufacturer.

Loading at wall head

3.03 Assuming trussed rafters supported at eaves level are to be used, the loading per metre run along the top of first floor walls will be:
4 m (half span) × 1920 N/m² (from para 3.01) = 7680 N/m. This load will normally be transmitted through the wall plate to the brickwork in the front and rear walls, and although the trusses are at 600 mm centres, it is generally accepted as uniformly distributed.

Lintel dimensions

3.04 Loading = 2 m (span) × 7680 N/m = 15360 N
The constraint here is deflection, which should be limited to 6 mm over a window

$$\text{deflection} = \frac{5\,\text{WL}^3}{384\,\text{EI}} \text{ where}$$

W = total load = 15360 N
L = span = 2000 mm
E = minimum* modulus of elasticity
= 4500 N/mm² (table I assume European whitewood)
I = moment of inertia (or second moment of area) required

$$\text{Hence: I required} = \frac{5 \times 15360\,\text{N} \times (2000\,\text{mm})^3}{384 \times 4500\,\text{N/mm}^2 \times 6\,\text{mm}}$$

= 59 × 10⁶ mm⁴

The member size with the required Ixx value can then be selected from table II. For example a nominal 75 mm × 225 mm planed to a finished size of 72 mm × 219 mm would suffice.

3.05 Adequate support must be provided at the ends of the lintel, the minimum bearing area being found from the formula

$$\text{Minimum area} = \frac{\text{load on end of lintel}}{\text{permissible bearing stress}}$$
(compression perpendicular to grain)
$$= \frac{7680}{1\cdot52} \text{ (from table I)} = 5052\,\text{mm}^2$$

As the lintel width is 72 mm, the minimum bearing length will be $\frac{5052}{72}$ = 70 mm

*See note to table 1

If the end of the lintel is bearing on brickwork, it is recommended that a minimum bearing of 150 mm is used to avoid concentration of load at any point.

Shear stress check

3.06 Timber is weak in its resistance to horizontal shear, ie shear acting parallel to the direction of the grain and it is therefore advisable to check that the permissible stress is not exceeded. The load producing maximum shear is the end vertical reaction of 7680 N. Average shear stress

$$= \frac{\text{load}}{\text{area}} = \frac{7680\,\text{N}}{72 \times 219} = 0\cdot48\,\text{N/mm}^2$$

but the maximum horizontal shear stress across the rectangular section is 0·48 × 1·5 (see **2**) = 0·72 N/mm²
The permissible shear stress from table I = 0·76 N/mm²
Therefore the lintel is satisfactory.

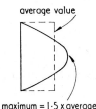

average value

maximum = 1·5 × average

2 *Distribution of shear stress across rectangular timber section.*

Stud dimensions

Actual bending stress

3.07 If timber stud walling is to be used, the load from the lintel will be carried by the studs next to the window opening. These studs will also transmit the wind load on the walls to the points of lateral restraint, namely the roof and first floor levels. Wind loading is covered in BS CP 3: Chapter V; Part 2[6], but calculation methods are fairly involved and it is sufficient here to assume a typical wind load of 488 N/m² (approximately 10 lb/sq ft). Then the total wind load on stud
= force × area (see **1**)
= 488 N/m² × 1·3 m × 2·3 m = 1459 N

$$\text{Bending moment} = \frac{\text{WL}}{8} = \frac{1459 \times 2\cdot3\,\text{m}}{8} = 419\cdot5\,\text{Nm}$$

$$\text{Actual bending stress} = \frac{\text{bending moment}}{\text{section modulus (from table II)}}$$
converting to Nmm and with 50 × 100 mm stud

$$= \frac{419\cdot5 \times 10^3}{73\cdot7 \times 10^3\,\text{mm}^3} = 4\cdot64\,\text{N/mm}^2$$

Actual compressive stress—axial loading

3.08 Load from lintel = 7680 N. Load from walling, external tile cladding. internal plasterboard and window, will average about 10lb/sq ft (480 N/m²)
load × area = 480 × 1·3 × 2·3 = 1435 N
While this will be distributed down the length of the stud, it is convenient to accept that it is applied at the top and with the lintel reaction forms an axial load of 7680 + 1435 = 9115 N

$$\text{Actual compressive stress} = \frac{\text{axial load}}{\text{area}}$$

$$\text{Assuming planed sizes} = \frac{9115\,\text{N}}{47\,\text{mm} \times 97\,\text{mm}} = 2\,\text{N/mm}^2$$

Slenderness ratio

3.09 The vertical loading will tend to buckle the stud outwards or inwards, as the restraint of the plywood facing or battens will prevent buckling on the weaker yy axis. The simplest method of designing a member for compound loading is to select a size and then calculate the stresses induced by the loading system. If the member proves to be overstressed or obviously understressed then a new section can be selected and the calculation reworked. Factors other than structural affect the choice. Here, for example, the

Table II *Geometrical properties of processed timber (reproduced from table 56 CP 112)*

Basic size mm	Minimum size mm	Area 10^3 mm²	Section modulus About x–x 10^3 mm³	About y–y 10^3 mm³	Second moment of area (I) About x–x 10^6 mm⁴	About y–y 10^6 mm⁴	Radius of gyration About x–x mm	About y–y mm
40 × 75	37 × 72	2·66	32·0	16·4	1·15	0·304	20·8	10·7
40 × 100	37 × 97	3·59	58·0	22·1	2·81	0·409	28·0	10·7
40 × 125	37 × 120	4·44	88·8	27·4	5·33	0·507	34·6	10·7
40 × 150	37 × 145	5·36	130	33·1	9·40	0·612	41·9	10·7
40 × 175	37 × 169	6·25	176	38·5	14·9	0·713	48·8	10·7
40 × 200	37 × 194	7·18	232	44·3	22·5	0·819	56·0	10·7
40 × 225	37 × 219	8·10	296	50·0	32·4	0·924	63·2	10·7
44 × 75	41 × 72	2·95	35·4	20·2	1·28	0·414	20·8	11·8
44 × 100	41 × 97	3·98	64·3	27·2	3·12	0·557	28·0	11·8
44 × 125	41 × 120	4·92	98·4	33·6	5·90	0·689	34·6	11·8
44 × 150	41 × 145	5·94	144	40·6	10·4	0·833	41·9	11·8
44 × 175	41 × 169	6·93	195	47·3	16·5	0·971	48·8	11·8
44 × 200	41 × 194	7·95	257	54·4	24·9	1·11	56·0	11·8
44 × 255	41 × 219	8·98	328	61·4	35·9	1·26	63·2	11·8
44 × 250	41 × 244	10·0	407	68·4	49·6	1·40	70·4	11·8
44 × 300	41 × 294	12·1	591	82·4	86·8	1·69	84·9	11·8
50 × 75	47 × 72	3·38	40·6	26·5	1·46	0·623	20·8	13·6
50 × 100	47 × 97	4·56	73·7	35·7	3·57	0·839	28·0	13·6
50 × 125	47 × 120	5·64	113	44·2	6·77	1·04	34·6	13·6
50 × 150	47 × 145	6·82	165	53·4	11·9	1·25	41·9	13·6
50 × 175	47 × 169	7·94	224	62·2	18·9	1·46	48·8	13·6
50 × 200	47 × 194	9·12	295	71·4	28·6	1·68	56·0	13·6
50 × 225	47 × 219	10·3	376	80·7	41·1	1·89	63·2	13·6
50 × 250	47 × 244	11·5	466	89·9	56·9	2·11	70·4	13·6
50 × 300	47 × 294	13·8	677	108	99·5	2·54	84·9	13·6
63 × 100	60 × 97	5·82	94·1	58·2	4·56	1·75	28·0	16·3
63 × 125	60 × 120	7·20	144	72·0	8·64	2·16	34·6	17·3
63 × 150	60 × 145	8·70	210	87·0	15·2	2·61	41·9	17·3
63 × 175	60 × 169	10·1	286	101	24·1	3·04	48·8	17·3
63 × 200	60 × 194	11·6	376	116	36·5	3·49	56·0	17·3
63 × 225	60 × 219	13·1	480	131	5·25	3·94	63·2	17·3
75 × 100	72 × 97	6·98	113	83·8	5·48	3·02	28·0	20·8
75 × 125	72 × 120	8·64	173	104	10·4	3·73	34·6	20·6
75 × 150	72 × 145	10·4	252	125	18·3	4·51	41·9	20·8
75 × 175	72 × 169	12·2	343	146	29·0	5·26	48·8	20·8
75 × 200	72 × 194	14·0	452	168	43·8	6·03	56·0	20·8
75 × 225	72 × 219	15·8	576	189	63·0	6·81	63·2	20·8
75 × 250	72 × 244	17·6	714	211	87·2	7·59	70·4	20·8
75 × 300	72 × 294	21·2	1040	254	152	9·14	84·9	20·8
100 × 100	97 × 97	9·41	152	152	7·38	7·38	28·0	28·0
100 × 150	97 × 145	14·1	340	227	24·6	11·0	41·9	28·0
100 × 200	97 × 194	18·8	608	304	59·0	14·8	56·0	28·0
100 × 250	97 × 244	23·7	962	383	117	18·6	70·4	28·0
100 × 300	97 × 294	28·5	1400	461	205	22·4	84·9	28·0
150 × 150	145 × 145	21·0	508	508	36·8	36·8	41·9	41·9
150 × 200	145 × 194	28·1	910	680	88·2	49·3	56·0	41·9
150 × 300	145 × 294	42·6	2090	1030	307	74·7	84·9	41·9
200 × 200	194 × 194	37·6	1220	1220	118	118	56·0	56·0
250 × 250	244 × 244	59·5	2420	2420	295	295	70·4	70·4
300 × 300	294 × 294	86·4	4240	4240	623	623	84·9	84·9

stud must be wide enough to allow plasterboard to be butt-jointed and fixed to its narrower face. The stud depth should be either 75 mm or 100 mm nominal, as these two sizes are commonly available. A nominal 50 mm × 100 mm stud will be reduced by planing to 47 mm × 97 mm and should satisfy most practical requirements.

Here slenderness ratio $= \dfrac{\text{length}}{\text{depth}} = \dfrac{2300 \text{ mm}}{97 \text{ mm}} = 24$

Permissible bending stress

3.10 Bending stress is due entirely to wind load which is

Table III *Modification factor K_{12} for duration of loading on flexural members and members in tension*

Duration of loading	Value of K_{12}
Long term (eg dead + permanent imposed)	1·00
Medium term (eg dead + snow, dead + temporary loads)	1·25
Short term (eg dead + imposed + wind, dead + imposed + snow + wind)	1·5

essentially short-term. Hence, from table III, the permissible bending stress of 6·6 N/mm², found from table I, may be increased: $6·6 \times 1·5 = 9·9$ N/mm².

Permissible compressive stress (Cp)

3.11 Table IV lists modification factors for values $\dfrac{\text{length}}{\text{breadth}}$ from 1·4 to 72·2. By interpolation on the table an l/b value of 24 gives a medium term load modification factor of 0·75. The medium term column is used because the load producing compression in the stud is partly from snow (see table III). $0·75 \times 480$ N/m² (from para 3.08) = 360 N/m².

Compound effect

3.12 The actual bending and compressive stresses cannot be simply added and compared with the permissible stresses. Instead the ratios must be added and should be equal to or less than 0·9.

Table IV *Modification factor* K_{18} *for slenderness ratio and duration of loading on compression members of 40 grade and 50 grade softwood*

Slenderness ratio		Values of K_{18}		
Length/radius of gyration	Length/ breadth	Long-term loads	Medium-term loads	Short-term loads
Less than 5	1·4	1·00	1·25	1·50
5	1·4	0·99	1·24	1·49
10	2·9	0·98	1·23	1·47
20	5·8	0·96	1·20	1·44
30	8·7	0·94	1·17	1·40
40	11·5	0·91	1·13	1·34
50	14·4	0·87	1·08	1·27
60	17·3	0·83	1·00	1·16
70	20·2	0·77	0·90	1·01
80	23·0	0·70	0·79	0·86
90	26·0	0·61	0·68	0·72
100	28·8	0·53	0·58	0·60
120	34·6	0·40	0·42	0·44
140	40·4	0·31	0·32	0·33
160	46·2	0·24	0·25	0·25
180	52·0	0·20	0·20	0·20
200	57·7	0·16	0·16	0·17
220	63·5	0·13	0·14	0·14
240	69·2	0·11	0·12	0·12
250	72·2	0·10	0·11	0·11

$$\frac{\text{Actual compressive stress (ca)}}{\text{Permissible compressive stress (cp)}} +$$

$$\frac{\text{Actual bending stress (fa)}}{\text{Permissible bending stress (fp)}} = 0 \cdot 9$$

$$\frac{\text{ca}}{\text{cp}} + \frac{\text{fa}}{\text{fp}} = \frac{2}{3 \cdot 6} + \frac{4 \cdot 64}{9 \cdot 9} = 0 \cdot 56 + 0 \cdot 468 = 1 \cdot 028$$

3.13 The ratio of actual to permissible stresses is therefore too high and a larger member is required. As the particular stud chosen carries more vertical and horizontal loading than the ordinary intermediate studs, the 50 × 100 section selected will obviously be adequate for all but this special case. Therefore it is more economical to leave the ordinary studs as 50 × 100 and to double the member under consideration. Two 50 × 100 studs nailed together will suffice. This will also provide the required bearing for the window lintel at the top of the stud.

Floor joists

3.14 The superimposed loading once found (from BS CP 3[3]) joists may be selected either from the Building Regulations schedule 6, table I, or in the case of floors spanning between end walls, from Information sheet TIMBER 8 of this handbook. Otherwise design may be from first principles (as for the lintel calculation above) and is as follows:

3.15 Superimposed load   = 1500 N/m²
Dead load of floor (typical value) = 480 N/m²
Total   = 1980 N/m²
Joist and stud spacing can be coincided to ensure direct transfer of load, hence load on joist (assuming mid support)
= centering × span × total load
= 0·45 m × 3·5 m × 1980 N/m² = 3119 N
As with the lintel (para 3.04) deflection will be limiting constraint:

$$\text{deflection} = \frac{5 \times \text{total load} \times \text{span}^3}{384 \times E \times I}$$

Here deflection would normally be limited to 0·003 × span, and the equation can be rewritten

$$\text{I required} = \frac{5 \times \text{total load} \times \text{span}^3}{384 \times E \times 0 \cdot 003 \times \text{span}}$$

assuming whitewood or redwood joists, then from table I the value of E is 8300 N/mm² (the mean value is taken

because the floor joist is part of a load sharing system). Hence I required (mm⁴)

$$= \frac{5 \times 3119 \text{ N} \times (3500 \text{ mm})^3}{384 \times 8300 \times 0 \cdot 003 \times 3500 \text{ mm}} = 19 \cdot 98 \times 10^6 \text{ mm}^4$$

The floor joists will be in the sawn state and therefore appropriate sizes may be selected from table V. A 50 mm × 175 mm section will suffice, having a second moment of area (Ixx value) of $22 \cdot 3 \times 10^6$ mm⁴. Spacing 450 mm.

3.16 Bending stress check

$$\text{Bending moment} = \frac{\text{WL}}{8} = \frac{3119 \times 3500}{8} = 1\ 364\ 563 \text{ Nmm}$$

$$\text{Actual bending stress} = \frac{\text{bending moment}}{\text{section modulus}} \text{ (from table V)}$$

$$= \frac{1364563}{255\ 000} \text{ N/mm}^2$$

$$= 5 \cdot 35 \text{ N/mm}^2$$

Permissible bending stress = 6·6 N/mm² ∴ satisfactory

Bearing check
3.17 Required bearing length

$$= \frac{\text{end reaction}}{\text{width} \times \text{permissible bearing stress (compression perpendicular to grain)}}$$

$$= \frac{1560}{50 \text{ mm} \times 1 \cdot 52} \text{ (from table I)}$$

$$= 20 \cdot 5 \text{ mm}$$

a minimum of 50 mm bearing is recommended

Shear check
3.18 Loading producing shear = 1560 N
Horizontal shear = 1·5 × vertical shear load (see para 3.06 this study)

$$= 1 \cdot 5 \times 1560 = 2340 \text{ N}$$

$$\text{Actual shear} = \frac{\text{horizontal shear load}}{\text{area resisting shear}} = \frac{2340}{50 \times 175}$$

$$= 0 \cdot 27 \text{ N/mm}^2$$

Permissible shear stress parallel to grain (from table I) = 0·76 N/mm²
Therefore section is satisfactory.

Ground floor studs
3.19 Design here is as for first floor studs but allowing for the additional loading from the first floor and ground floor walls. In designing for wind loading, studs can be assumed to act as simply supported beams between ground slab and

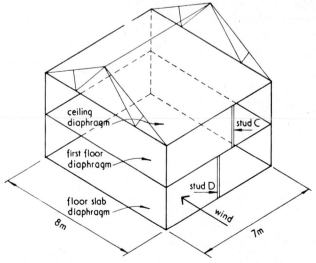

3 *Roof, first and ground floor diaphragms provide stability against wind action.*

Table v *Geometrical properties of sawn timber (extracted from table 55, CP 112, part 2)*

| Basic size | Area | Section modulus | | Second moment of area | | Radius of gyration | |
| | | About x–x | About y–y | About x–x | About y–y | About x–x | About y–y |
mm	10^3 mm^2	10^3 mm^3	10^3 mm^3	10^6 mm^4	10^6 mm^4	mm	mm
44 × 75	3·30	41·2	24·2	1·55	0·532	21·7	12·7
44 × 100	4·40	72·3	32·3	3·67	0·710	28·9	12·7
44 × 125	5·50	114	40·3	7·16	0·887	36·1	12·7
44 × 150	6·60	165	48·4	12·4	1·06	43·3	12·7
44 × 175	7·70	225	56·5	19·7	1·24	50·5	12·7
44 × 200	8·80	293	64·5	29·3	1·42	57·7	12·7
44 × 225	9·90	371	72·6	41·8	1·60	65·0	12·7
44 × 250	11·0	458	80·7	57·3	1·77	72·2	12·7
44 × 300	13·2	660	96·8	99·0	2·13	86·6	12·7
50 × 75	3·75	46·9	31·2	1·76	0·781	21·7	14·4
50 × 100	5·00	83·3	41·7	4·17	1·04	28·9	14·4
50 × 125	6·25	130	52·1	8·14	1·30	36·1	14·4
50 × 150	7·50	188	62·5	14·1	1·56	43·3	14·4
50 × 175	8·75	255	72·9	22·3	1·82	50·5	14·4
50 × 200	10·0	333	83·3	33·3	2·08	57·7	14·4
50 × 225	11·2	422	93·8	47·5	2·34	65·0	14·4
50 × 250	12·5	521	104	65·1	2·60	72·2	14·4
50 × 300	15·0	750	125	112	3·12	86·6	14·4
63 × 100	6·30	105	66·2	5·25	2·08	28·9	18·2
63 × 125	7·88	164	82·7	10·3	2·60	36·1	18·2
63 × 150	9·45	236	99·2	17·7	3·13	43·3	18·2
63 × 175	11·0	322	116	28·1	3·65	50·5	18·2
63 × 200	12·6	420	132	42·0	4·17	57·7	18·2
63 × 225	14·2	532	149	59·8	4·69	65·0	18·2
75 × 100	7·50	125	93·8	6·25	3·52	28·9	21·7
75 × 125	9·38	195	117	12·2	4·39	36·1	21·7
75 × 150	11·2	281	141	21·1	5·27	43·3	21·7
75 × 175	13·1	383	164	33·5	6·15	50·5	21·7
75 × 200	15·0	500	188	50·0	7·03	57·7	21·7
75 × 225	16·9	633	211	71·2	7·91	65·0	21·7
75 × 250	18·8	781	234	97·7	8·79	72·2	21·7
75 × 300	22·5	1120	281	169	10·5	86·6	21·7
100 × 100	10·0	167	167	8·33	8·33	28·9	28·9
100 × 150	15·0	375	250	28·1	12·5	43·3	28·9
100 × 200	20·0	667	333	66·7	16·7	57·7	28·9
100 × 250	25·0	1040	417	130	20·8	72·2	28·9
100 × 300	30·0	1500	500	225	25·0	86·6	28·9

first floor level. As far as possible the spacing of ground and first floor joists should be made to coincide. This should be possible with a 600 mm module except where large window openings occur.

4 General stability

4.01 The rear and front walls and party or gable walls form a 'box', stiffened by the truss, ceiling, first floor and ground slab diaphragms **3**. The whole should be firmly anchored to the ground slab against wind action. Overturning in the longitudinal direction will be adequately resisted by the front and rear stud walls, but care should be taken with the detailing of cross walls and gables. Introducing diagonal bracing will prevent the panel distorting along its plane, even when subjected to forces from the strongest gale. For example a panel 2·3 m deep with 50 mm × 100 mm studs and bracing will resist a horizontal force of 25 000 N applied

to the top corner, provided the sill member is anchored against overturning **4**. This allows a high safety factor considering that the force along the plane of the much larger gable in the two-storey building considered earlier, would be of the order of 8000 N only, assuming a wind force of 488 N/m^2 on the front or rear walls.

4.02 Research on the racking resistance (resistance to distortion along the plane, as opposed to buckling) of plasterboard-sheathed, timber stud walls has been undertaken by the DOE's Princes Risborough Laboratory and will form the basis of a code of practice on the racking resistance of various types of timber panel.

4.03 A large part of the fabrication and jointing of components in timber frame construction will be by nailing. Pre-boring should be avoided as it is costly and often impractical. The gauge and length of nail should be selected from table 22 of BS CP 112. As far as possible, nails should be staggered rather than in line along the grain to avoid splitting.

References

1 BS CP 112: Part 2: 1971 The structural use of timber. Metric units. [(2–) Yi (K)]

2 BS 4471: Part 1: 1978 Sizes of sawn and planed timber [Yi2 (F4j)]

3 BS 648: 1964 Schedule of weights of building materials [Yy (F4)]

4 BS CP 3: Chapter v: Part 1: 1967 Dead and imposed loads [(K4)]

5 BRE Digest 147: 1972 Permissible spans for trussed rafters. HMSO [(27) Xi (F4j)]

6 BS CP 3: Chapter v: Part 2: 1970 Wind loads [(K4f)]

diagonal braces accurately notched to bear on studs

wind

cill member anchored to floor slab using rag bolts or similar

4 *Elevation of party wall with diagonal bracing.*

Information sheet
Timber 6

Section 7 **Structural material: Timber**

Selection of columns

This information sheet by H. J. BURGESS *contains one table for use in designing solid rectangular columns with no lateral load, a table for use where there is no lateral load as with an internal column not subject to wind load, and five tables for designing columns with a uniformly distributed lateral wind load. The chart and tables apply to softwoods grade* S2-50 *in accordance with* BS CP 112: *Part 2: 1971* The structural use of timber.

1 Design considerations for a compression member

Effective length
1.01 The effective length of a column, as distinct from actual length, is the length assumed for design purposes according to the degree of restraint at the supports. The factors which determine what is to be taken as the effective length are set out in Technical study TIMBER 3, para 3.03.

Stress grades
1.02 BS CP 112: Part 2: 1971 contains rules for determining the stress grade of timber. Working stresses are reckoned as percentages of basic stresses (ie those stresses which would apply were the timber free from defects). These percentages are 75, 65, 50 and 40. In addition, in order to simplify their specification, all softwoods are classified in three groups, referred to as S1, S2 and S3. See IS TIMBER 5.
1.03 For design purposes, all softwoods in a group are assumed to have the same working stress and modulus of elasticity; consequently a softwood can be identified by quoting its species and the grade required. For example, a grade S2—50 means a softwood in the S2 group whose working stress is 50 per cent of the basic stress for that group. If, as anticipated, the new GS grade in the revised BS 1860 (soon to be published) has stress values similar to 50 grade in the code, these tables will remain valid.

Duration of load
1.04 BS CP 112: Part 2: 1971 specifies loads as being long term, medium term and short term. Long term loads include all loads which always act on the structure, eg the self weight of the timber; medium term loads include temporary loads, eg snow; short term loads include wind loads.

2 Columns subject to lateral load

Use of the tables
2.01 Tables I to VI show permissible axial loads per 10 mm of column thickness for columns with a section depth from 38 to 300 mm. It is assumed that the columns are restrained by cladding or sheeting rails against buckling in a direction perpendicular to this section dimension. Thus the figures may be considered as applying to a column only 10 mm thick, restrained in such a way that buckling can only take place in the stronger direction—effectively 'the column-in-the-wall'.

Data
2.02 The tables refer to softwoods group S2—50 having a moisture content not exceeding 18 per cent. Permissible axial loads are long term but the lateral loads have been increased by a factor of 1·5 as permitted by BS CP 112: Part 2: 1971 for short term loading of flexural members. The axial load may be limited by the permissible bearing stress perpendicular to the grain of a horizontal timber member upon which the column bears at its ends. This should be checked from tables III, IV and V in Information sheet TIMBER 1. In the tables I to VI the loads to the right of the zig-zag line would exceed the bearing capacity of a supporting member of the same timber species and grade as the column; in such cases, a harder timber or other material must be used for transmitting load to the column if its full buckling capacity must be attained.

Table I Sizing of columns not subject to a lateral load

| Effective length (m) | Permissible axial load N per 10 mm breadth on column of size stated (mm) | | | | | | | | | | | | |
	38	50	63	75	100	125	150	175	200	225	250	275	300
1·2	688	1410	2220	2900	4190	5320	6540	7700	8910	10 100	11 300	12 500	13 700
1·5	467	1040	1840	2600	3990	5170	6410	7580	8790	10 000	11 200	12 400	13 600
1·8	334	766	1450	2210	3730	4980	6250	7440	8670	9880	11 100	12 300	13 500
2·1	250	581	1140	1820	3400	4740	6070	7290	8530	9750	11 000	12 200	13 400
2·4	193	454	906	1480	3010	4440	5850	7100	8370	9610	10 800	12 100	13 300
2·7	154	364	733	1220	2610	4080	5570	6890	8190	9450	10 700	11 900	13 100
3·0	126	298	603	1010	2250	3680	5250	6640	7980	9280	10 500	11 800	13 000
3·3	104	248	504	850	1930	3280	4880	6340	7750	9080	10 400	11 600	12 900
3·6	88	210	428	724	1670	2910	4480	6000	7470	8850	10 200	11 400	12 700
3·9	75	179	367	623	1450	2570	4080	5620	7150	8590	9950	11 300	12 500
4·2	65	155	318	542	1270	2280	3700	5220	6800	8290	9700	11 000	12 300
4·5	57	136	279	475	1120	2030	3340	4820	6420	7970	9420	10 800	12 100
4·8		120	246	420	994	1810	3020	4430	6020	7610	9110	10 500	11 900
5·1			219	373	888	1630	2730	4060	5620	7220	8770	10 200	11 700
5·4				334	797	1470	2480	3720	5220	6820	8410	9930	11 400
5·7					719	1330	2260	3410	4850	6420	8020	9580	11 100
6·0						1210	2060	3140	4490	6020	7620	9210	10 700

Table II *Sizing of columns subject to a lateral load* 100 N/m *per* 10 mm *of thickness*

Effective length (m)	Thickness 38	50	63	75	100	125	150	175	200	225	250	275	300
1·2		527	1280	1960	3250	4360	5520	6620	7740	9860	11 100	12 300	13 500
1·5		81	729	1430	2820	4000	5210	6340	7490	8620	9740	12 100	13 300
1·8			253	875	2320	3580	4850	6020	7200	8350	9490	10 600	11 700
2·1				389	1780	3100	4440	5650	6870	8050	9220	10 400	11 500
2·4				6	1220	2560	3970	5240	6500	7720	8910	10 100	11 200
2·7					719	2010	3460	4780	6090	7350	8570	9760	10 900
3·0					289	1460	2910	4290	5650	6950	8200	9420	10 600
3·3						952	2350	3760	5160	6510	7800	9050	10 300
3·6						507	1810	3200	4650	6040	7370	8650	9900
3·9						124	1290	2640	4110	5530	6900	8220	9500
4·2							826	2100	3550	5000	6410	7770	9080
4·5							412	1580	2990	4460	5900	7290	8630
4·8							49	1100	2450	3900	5360	6780	8160
5·1								672	1930	3340	4810	6250	7660
5·4								285	1440	2790	4250	5710	7140
5·7									988	2270	3690	5150	6610
6·0									578	1770	3140	4590	6060

Table III *Sizing of columns subject to a lateral load* 200 N/m *per* 10 mm *of thickness*

Effective length (m)	Thickness 38	50	63	75	100	125	150	175	200	225	250	275	300
1·2			566	1310	2730	3930	5160	6300	7470	9620	10 900	12 100	13 300
1·5				519	2050	3350	4660	5860	7060	8240	9400	11 800	13 000
1·8					1290	2670	4070	5340	6600	7820	9010	10 200	11 300
2·1					490	1920	3410	4740	6060	7330	8560	9770	10 900
2·4						1130	2680	4080	5470	6790	8060	9300	10 500
2·7						340	1900	3360	4810	6190	7510	8790	10 000
3·0							1100	2600	4110	5540	6910	8240	9520
3·3							315	1810	3360	4850	6270	7640	8960
3·6								1010	2570	4110	5580	7000	8360
3·9								231	1780	3340	4850	6310	7720
4·2									980	2540	4090	5600	7050
4·5									207	1740	3310	4850	6340
4·8										952	2510	4070	5600
5·1										183	1710	3280	4830
5·4											925	2480	4050
5·7											159	1680	3250
6·0												899	2450

Table IV *Sizing of columns subject to a lateral load* 300 N/m *per* 10 mm *of thickness*

Effective length (m)	Thickness 38	50	63	75	100	125	150	175	200	225	250	275	300
1·2				660	2220	3500	4800	5990	7190	9370	10 600	11 900	13 200
1·5					1280	2700	4110	5380	6640	7860	9060	11 500	12 800
1·8					250	1770	3300	4660	5990	7280	8520	9740	10 900
2·1						754	2380	3840	5260	6610	7910	9170	10 400
2·4							1390	2930	4430	5860	7220	8530	9800
2·7							341	1950	3530	5030	6460	7830	9150
3·0								913	2570	4140	5630	7060	8430
3·3									1550	3180	4740	6230	7660
3·6									500	2180	3800	5340	6830
3·9										1140	2800	4410	5940
4·2										84	1780	3430	5010
4·5											726	2410	4040
4·8												1370	3040
5·1												311	2000
5·4													950

Table V *Sizing of columns subject to a lateral load* 400 N/m *per* 10 mm *of thickness*

Effective length (m)	Thickness 38	50	63	75	100	125	150	175	200	225	250	275	300
1·2				11	1700	3070	4440	5680	6920	9130	10 400	11 700	13 000
1·5					510	2050	3550	4890	6220	7490	8720	11 200	12 500
1·8						867	2520	3970	5390	6740	8030	9290	10 500
2·1							1360	2930	4450	5880	7250	8570	9850
2·4							95	1770	3400	4930	6370	7760	9090
2·7								528	2250	3870	5400	6860	8250
3·0									1030	2730	4350	5880	7340
3·3										1520	3210	4820	6350
3·6										251	2010	3690	5290
3·9											755	2500	4170
4·2												1260	2980
4·5													1750
4·8													478

Table VI Sizing of columns subject to a lateral load 500 N/m per 10 mm of thickness

Effective length (m)	Thickness 38	50	63	75	100	125	150	175	200	225	250	275	300
1·2					1180	2640	4080	5360	6640	8880	10 200	11 500	12 800
1·5						1400	3000	4410	5790	7100	8330	10 900	12 200
1·8							1740	3290	4790	6200	7550	8850	10 100
2·1							329	2020	3650	5160	6600	7970	9300
2·4								617	2370	4000	5530	6980	8380
2·7									974	2710	4350	5890	7360
3·0										1330	3060	4690	6250
3·3											1680	3410	5040
3·6											226	2040	3750
3·9												590	2390
4·2													950

Example

2.03 Find the permissible axial load on a column of 100 mm × 200 mm section with an effective length of 3 m if the wind loading applied by sheeting to the smaller face of the column is 500 N/m² and the column spacing is 3 m.

Load per m run on column = 3 × 500 = 1500 N

Lateral load per 10 mm of thickness

$$= \frac{1500}{10} = 150 \text{ N/m per 10 mm thick}$$

Tables II and III give permissible axial loads of 5650 N and 4110 N for lateral loads of 100 and 200 N/m respectively; the average value 4880 may be taken for a lateral load of 150 N/m and the total for the 100 mm thick column will be 4880 × 10 = 48 800 N

3 Columns not subject to lateral load

Use of the chart

3.01 The chart 1 shows permissible axial loads on columns of various cross sections with effective lengths ranging from 1·2m to 6m. Alternatively, where the load is known, a suitable section can be chosen for a given effective length. The chart is applicable only where there is no lateral load on the column, as in the c se of an internal column not subjected to wind loading.

3.02 Each curve shows how resistance to buckling of a particular section varies with the effective length. This resistance depends on the length of the column and the shape of its cross section. If the end of a column bears upon a horizontal member of the same timber as the column, its load-carrying capacity will often be limited by the permissible compressive stress perpendicular to the grain of the horizontal member. This is the case wherever a curve is shown dotted.

3.03 If advantage is to be taken of the whole buckling strength of the column in such cases, a harder material must be used to transmit the load to its ends; this may be a denser timber than that forming the column, or another material such as concrete. Where the curves are solid lines, the bearing stress does not exceed that which may be taken perpendicular to the grain on timber the same as that of which the column is made.

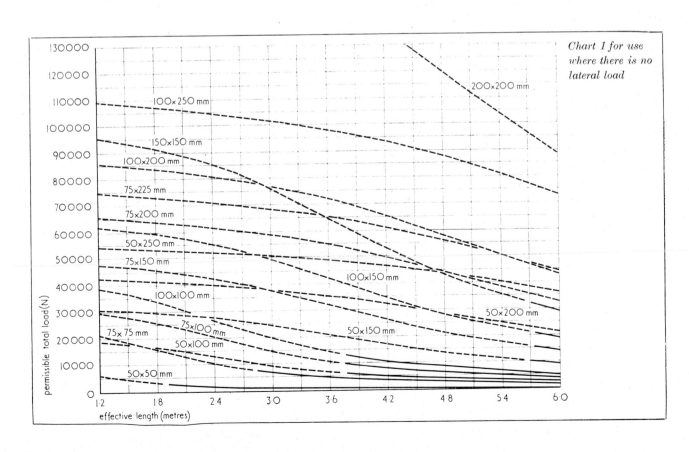

Chart 1 for use where there is no lateral load

Information sheet
Timber 7

Section 7 **Structural material: Timber**

Beam selection charts for softwoods grade S2-50

This information sheet by H. J. BURGESS *contains three charts for use in designing beams of softwood grade* s2-50 *in accordance with* BS CP 112: *Part 2: 1971. The charts cover the following range of timber sections: 38 mm wide ×* 100 *mm to 225 mm (chart 1); 50 mm × 100 mm to 225 mm (chart 2); 75 mm × 100 mm to 225 mm (chart 3).*

Stress grades

1.01 For information on working stresses see BS CP 112 Part 2: 1971 and Information sheets TIMBER 5 and 6.

Use of charts

1.02 With these charts the designer can choose the correct beam section when both the span and the total load are known. If the section is known and the span decided, the permissible load can be determined. Linear interpolation is permissible. The radial broken lines show loads per m run, but here the interpolation is non-linear:
From chart 1 a beam 38 mm × 200 mm which spans 3 m carries 2410 N or about 803 N/m including the weight of the beam.
From chart 2 a beam 50 mm × 200 mm which spans 3·6 m carries 2280 N or about 633 N/m, including the weight of the beam.
From chart 3 a beam 75 mm × 200 mm which spans 4·2 m carries 2580 N or about 615 N/m, including the weight of the beam.

Data

1.03 The charts refer to softwoods groups S2 as listed in BS CP 112[2] of 50 grade and a moisture content not exceeding 18 per cent. Softwoods in the group include western hem-lock, parana pine, redwood, whitewood, Canadian spruce and home-grown Scots pine.

1.04 For most of the chart area, the design is controlled by the permissible deflection of 0·003 of the span. Two zones at the left are marked in which the criteria are bending stress and shear stress, with the values 6·2 N/mm² and 0·76 N/mm² respectively.

1.05 The graphs are for flexural members acting individually, and the 'minimum' value of the modulus of elasticity is used: ie 4500 N/mm². For load-sharing systems with the members spaced at 0·6 m centres or closer, that would be conservative and it is best to refer to other design aids—see the joist tables in Information sheets TIMBER 8 and 9.

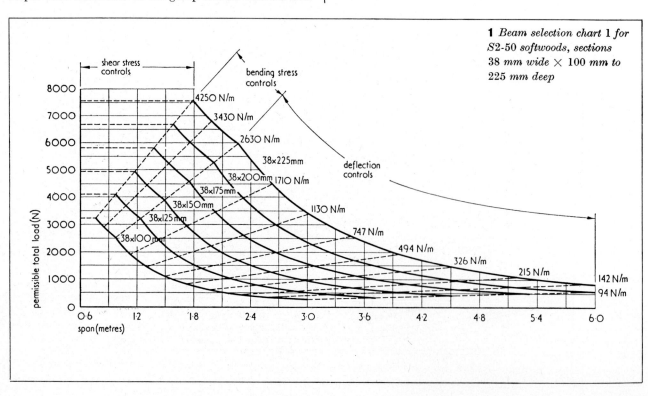

1 *Beam selection chart 1 for S2-50 softwoods, sections 38 mm wide × 100 mm to 225 mm deep*

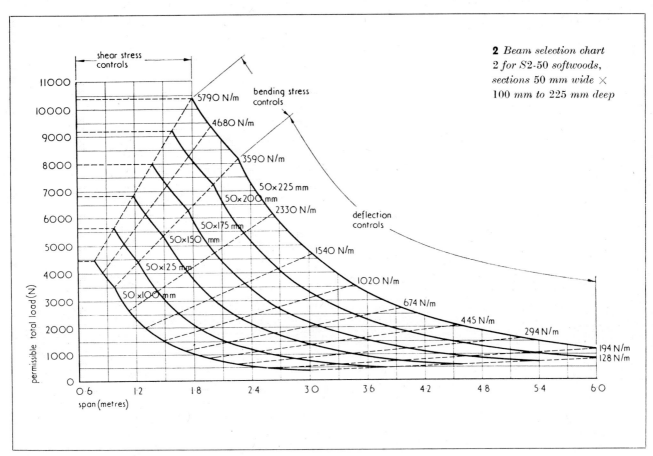

2 *Beam selection chart 2 for S2-50 softwoods, sections 50 mm wide × 100 mm to 225 mm deep*

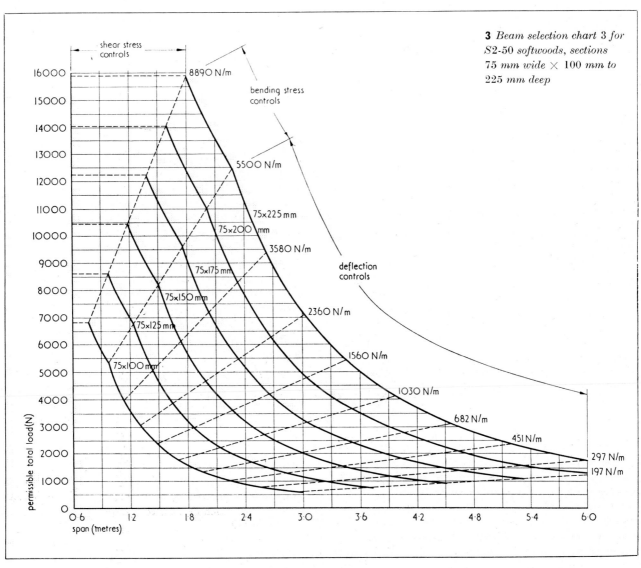

3 *Beam selection chart 3 for S2-50 softwoods, sections 75 mm wide × 100 mm to 225 mm deep*

Information sheet
Timber 8

Section 7 **Structural material: timber**

Design of domestic suspended floors

This information sheet by H. J. BURGESS *explains the requirements of the Building Regulations* 1972, BS CP 3: *Chapter* V: *Part* 1: 1967 *and* BS CP 112: *Part* 2: 1971. *It contains two charts, two span/size tables on the selection of joist sizes and two supplementary tables on the maximum spans for softwood boards and for plywood/chipboard floorings. The information applies to softwoods grade* S2–50 *in accordance with* BS CP 112: *Part* 2: 1971 The structural use of timber.*

1 Construction of timber joist and board floors

Floor loadings

1.01 The Building Regulations 1976 require imposed loading for floors of single occupation houses of not more than three storeys to be one of these two types:

1 In accordance with CP 3: Chapter V: Part 1: 1967

Intensity of distributed load:

$$\frac{kN/m^2}{1\cdot5} \text{ or } \frac{kgf/m^2}{153}$$

or

Concentrated load on any 300 mm square:

$$\frac{kN}{1\cdot4} \text{ or } \frac{kgf}{14\cdot3}$$

whichever is the more critical when the floor is designed as a slab not capable of effective lateral distribution of the concentrated load.

2 In accordance with:

Intensity of distributed load	Alternative minimum distributed load	
	Slabs	*Beams*
kN/m²	kN/m width	kN
1·44	3·5	8·5

The alternative minimum distributed loads in the two right-hand columns are to be used for short spans where members would be very small if calculated at only 1·44 kN/m², and not strong enough for the point loads of furniture or the weight of a person.

1.02 For a joisted floor, the minimum slab loading should be applied to each metre width of floor; this is the only imposed load to be applied on spans of less than 2·44 m. At a span of 2·44 m either the slab loading or the alternative 1·44 kN/m² will give the same result. Above this span, only the 1·44 kN/m² in the left-hand column is applied. Similarly, the minimum load for beams can only be applied on areas less than 5·9 m² since 1·44 × 5·9 = 8·5 kN; beams or joists spaced at not more than 1 m centres may be calculated for slab loadings.

Dead load

1.03 The Building Regulations require floors to support not only the imposed load but their own weight plus the flooring and ceiling materials they carry and the weight of any other parts of the permanent construction supported by the floor. Where partitions are shown on drawings, their weights should be included in the dead load; if partitions are to be provided for but their positions are not shown on the plan, an increased dead load must be adopted in the calculations.

Typical loading

1.04 This example (it does not allow for partitions or services) is a typical calculation of the total loading on a 50 mm × 200 mm first floor with joists at 0·6 m spacing:

Own weight of joists

$$\frac{50 \times 200}{0\cdot6\times10^6} \times 1\text{ m} \times 540\text{ kg/m}^3 = 9\cdot0\text{ kg/m}^2$$

21 mm t & g flooring

$$\frac{21}{1000} \times 1\text{ m} \times 1\text{ m} \times 540 = 11\cdot4\text{ kg/m}^2$$

13 mm plasterboard ceiling $= 11\cdot2\text{ kg/m}^2$

Total dead load $31\cdot6\text{ kg/m}^2 \times 9\cdot81 = 310\text{ N/m}^2$

1.05 1920 N/m² is often taken as total floor load in calculations for two-storey housing.

2 Types of suspended floor

2.01 Two types of suspended timber floor are here considered:

1 Ground floors supported on sleeper walls or other intermediate supports

2 Upper floors or ground floors spanning between end walls

The main differences between these types are:

a Longer lengths and deeper sections are generally required for 2

b There will be a saving in timber in 1 but it may be offset by more building work in foundations and intermediate supports.

In both cases it is assumed that joists are simply supported in single spans in the calculations and tables given.

3 Structural design

3.01 The following methods are possible for arriving at suitable sizes:

1 Design to CP 112: Part 2: 1971, the code of practice for the structural use of timber following basic principles or use data such as joist span tables which have themselves been worked out in accordance with the code. The work below refers throughout to CP 112: Part 2: 1971.

2 Adopt the deemed-to-satisfy provision of Regulation D12 of the Building Regulations 1976, by using the softwood member sizes given in schedule 6 for single-occupancy houses of not more than three storeys. In all other respects follow CP 112: Part 2: 1971.

3 Make use of proprietary floor systems that have been shown by calculation or tests to conform with the requirements of CP 112.

291

Information sheet **Timber 8 para 3.01 to 3.07**

4 Test in accordance with the procedure set out in cp 112.
3.02 It is useful to note that cp 112[2] has only deemed-to-satisfy status, and that calculations not entirely in accordance with the code may be accepted if it can be shown that they satisfy the Regulations on general stability, strength and stiffness.
3.03 Where maximum economy is sought or where design loadings exceed the statutory minimum, it is best to design from first principles. **1** should be used first to indicate the approximate conclusions of all such calculations, and to show on which occasions the design is controlled by permissible deflection, by bending stress or by shear stress.
3.04 The curved lines in **1** are similar to those shown in Information sheet TIMBER 7 but this chart shows 'load-sharing systems' such as joisted floors where the flexural members are spaced at centres not exceeding 600 mm and the flooring (boards or ply) allows the load to be distributed laterally over several joists: the code in these cases (boards or ply) allows that:
1 The higher 'mean' value of the modulus of elasticity (E) may be used instead of the much lower 'minimum' value (see Information sheet TIMBER 1) applied to isolated beams acting singly
2 For the particular grade of timber suggested for floor joists below, the working stress in bending may be increased by 20 per cent.

Grade of timber
3.05 In **1** the curved lines give permissible loads per 25 mm of thickness on joists of various depths. For each depth there are three distinct zones: for the longer spans, the design is controlled by deflection; over an intermediate region bending stress controls; and for very small spans (usually too small to be of any importance in design) shear stress controls the design. These criteria will be examined more closely below in connection with domestic joist design. For the present, **1** is used as a guide to the type of timber that is suitable for floor joists.
3.06 The code ascribes the same value of E to all the four structural grades of timber; these are the grades having 75, 65, 50 and 40 per cent of the strength of timber free of all defects as explained in Technical study TIMBER 2 and Information sheet TIMBER 5. As for most conditions

the design of floor joists is governed by deflection, which depends on the value of E, there is little or nothing to be gained by using expensive high-grade timber. The 40 grade is therefore generally indicated, but when bending stress controls and at least three-quarters of the members comply with the 50 grade and the remaining 25 per cent with 40 grade, the working stress in bending may be increased by 20 per cent instead of the 10 per cent otherwise allowed by the code for load-sharing systems.

Choice of section shape for joists and beams
3.07 For maximum economy, the deepest permissible rectangular section should be chosen, consistent with the following factors limiting the depth:
1 Availability of timber in the thickness selected (see also Technical study TIMBER 2). At present the commonest basic breadths of sawn timbers are 38, 44, 50, 63 and 75 mm. Depths (widths) are available in 25 mm steps but with some exceptions as shown in **2**.
2 Nailing surface. Opinions vary on what is a suitable minimum face width for fastenings of various kinds. The case for accepting 38 mm is stronger for flooring than in studs, for example, and still stronger than in industrial purlins where 63 mm is commonly used.
3 Permissible depth-to-breadth ratios which are laid down in cp 112: 1971 for various degrees of lateral support as described later.
4 Available depth of floor zone reduced by the combined thickness of flooring and ceiling.
5 Assembly with other members. For example, where joists and trimmers need to be the same depth or to have some other relation between their depths.

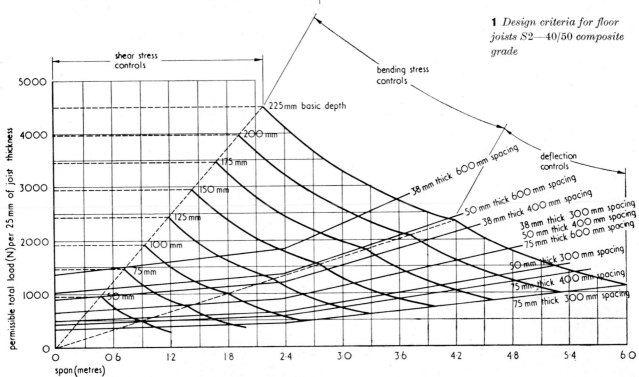

1 *Design criteria for floor joists S2—40/50 composite grade*

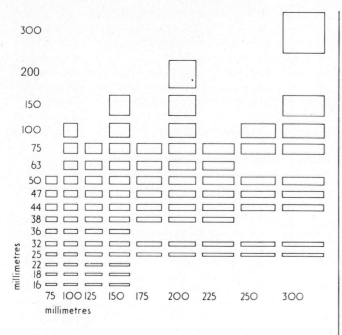

2 *BS range of sizes—basic breadths 38, 44, 50, 63 and 75 mm*

4 Hardwood joists and beams

4.01 High strength hardwoods may be used for structural reasons where they can compete on cost grounds with softwoods because of the reduced volume of timber required; the imported hardwood keruing has been used for long-span floor and roof joists where large quantities of uniform length were required. Its air-dry modulus of elasticity has the value 13 800 N/mm², compared with 8300 N/mm² for S2 softwood which allows the joist thickness to be reduced by 40 per cent under conditions where deflection controls the design.

4.02 For beams, the minimum values of E are 9300 for keruing and 4500 for S2 softwoods, and the reduction in breadth is 52 per cent. These reductions can only be applied where the softwood joist or beam would have a squat section; otherwise the resulting hardwood beam would have too large a depth-to-breadth ratio. The ratio is limited by CP 112[2] as explained below.

5 Lateral support

5.01 CP 112[2] requires lateral support of joists and beams in domestic floors in this way:

	Maximum depth-to-breadth ratio
1 Joists with top edge held in line by flooring	5
2 Beams with top edge held in line by adequately nailed joists	5
3 Beams as in 2 with adequate bridging or blocking at intervals not exceeding six times the depth.	6 (eg 38 × 225)

The code allows a ratio of 7 if the ends of the beams are held in position and both top and bottom are held firmly in line—in all cases the ends of the joists or beams must be held vertically in position.

6 Design to schedule 6 Building Regulations

Joists

6.01 Schedule 6, table I of the Building Regulations, gives timber floor joist sizes and maximum spans for a range of dead loads and joist spacings. The table is based on Baltic redwood stress values.

6.02 For ease of reference the size/span tables of the Building Regulations have been recalculated as span/size tables, giving suitable timber sections for a range of spans from 1·2 m upwards in 100 mm increments. These are shown in tables I to IV calculated for a range of dead loads (excluding the weight of the floor joists) of 240 and 480 N/m² and for GS grade (tables I and II) and SS grade (tables III and IV).

Table I Floor loading 1440 N/m² imposed + 240 N/m² dead

Span	Spacing (mm) 300	350	400	450	500	550	600
1200	38×75	44×75	50×75	50×75	38×100	38×100	38×100
1300	38×75	44×75	50×75	38×100	38×100	38×100	44×100
1400	44×75	50×75	38×100	38×100	38×100	44×100	50×100
1500	44×75	50×75	38×100	38×100	44×100	44×100	50×100
1600	50×75	38×100	38×100	44×100	44×100	50×100	38×125
1700	38×100	38×100	38×100	44×100	50×100	38×125	38×125
1800	38×100	38×100	44×100	44×100	50×100	38×125	38×125
1900	38×100	38×100	44×100	50×100	38×125	38×125	44×125
2000	38×100	44×100	50×100	50×100	38×125	44×125	44×125
2100	38×100	44×100	50×100	38×125	38×125	44×125	50×125
2200	38×100	44×100	50×100	38×125	44×125	44×125	50×125
2300	44×100	50×100	38×125	38×125	44×125	50×125	50×125
2400	50×100	38×125	38×125	44×125	44×125	50×125	38×150
2500	38×125	38×125	38×125	44×125	50×125	38×150	44×150
2600	38×125	38×125	44×125	50×125	38×150	44×150	44×150
2700	38×125	44×125	44×125	50×125	44×150	44×150	50×150
2800	38×125	44×125	50×125	38×150	44×150	50×150	50×150
2900	44×125	50×125	38×150	44×150	50×150	50×150	44×175
3000	50×125	38×150	38×150	44×150	50×150	44×175	44×175
3100	50×125	38×150	44×150	50×150	38×175	44×175	50×175
3200	38×150	38×150	44×150	50×150	44×175	44×175	50×175
3300	38×150	44×150	50×150	38×175	44×175	50×175	44×200
3400	38×150	50×150	38×175	44×175	50×175	50×175	44×200
3500	44×150	50×150	44×175	44×175	50×175	44×200	44×200
3600	50×150	38×175	44×175	50×175	44×200	44×200	50×200
3700	50×150	38×175	44×175	50×175	44×200	50×200	50×200
3800	38×175	44×175	50×175	44×200	44×200	50×200	44×225
3900	38×175	44×175	50×175	44×200	50×200	44×225	44×225
4000	44×175	50×175	44×200	44×200	50×200	44×225	50×225
4100	44×175	50×175	44×200	50×200	44×225	44×225	50×225
4200	50×175	44×200	44×200	50×200	44×225	50×225	75×200
4300	50×175	44×200	50×200	44×225	44×225	50×225	75×200
4400	44×200	44×200	50×200	44×225	50×225	75×200	75×200
4500	44×200	50×200	44×225	44×225	50×225	75×200	50×250
4600	44×200	50×200	44×225	50×225	75×200	75×200	50×250
4700	44×200	44×225	44×225	50×225	75×200	50×250	75×225
4800	50×200	44×225	50×225	50×225	75×200	50×250	75×225
4900	50×200	44×225	50×225	75×200	50×250	75×225	75×225
5000	44×225	44×225	50×225	50×250	50×250	75×225	75×225
5100	44×225	50×225	50×250	50×250	75×225	75×225	75×250
5200	44×225	50×225	50×250	50×250	75×225	75×250	75×250
5300	50×225	50×250	50×250	50×250	75×250	75×250	75×250
5400	50×225	50×250	50×250	75×225	75×250	75×250	75×250
5500	75×200	50×250	50×250	75×250	75×250	75×250	75×250
5600	50×250	50×250	75×225	75×250	75×250	75×250	75×300
5700	50×250	50×250	75×250	75×250	75×250	75×250	75×300
5800	50×250	50×250	75×250	75×250	75×250	75×300	75×300
5900	50×250	75×225	75×250	75×250	75×250	75×300	75×300
6000	50×250	75×250	75×250	75×250	75×250	75×300	75×300
6100	75×225	75×250	75×250	75×250	75×300	75×300	75×300
6200	75×250	75×250	75×250	75×300	75×300	75×300	75×300
6300	75×250	75×250	75×300	75×300	75×300	75×300	75×300
6400	75×250	75×250	75×300	75×300	75×300	75×300	75×300
6500	75×250	75×250	75×300	75×300	75×300	75×300	75×300
6600	75×250	75×300	75×300	75×300	75×300	75×300	75×300
6700	75×250	75×300	75×300	75×300	75×300	75×300	
6800	75×250	75×300	75×300	75×300	75×300	75×300	
6900	75×300	75×300	75×300	75×300	75×300		
7000	75×300	75×300	75×300	75×300	75×300		
7100	75×300	75×300	75×300	75×300	75×300		
7200	75×300	75×300	75×300	75×300	75×300		
7300	75×300	75×300	75×300				
7400	75×300	75×300	75×300				
7500	75×300	75×300					
7600	75×300	75×300					
7700	75×300	75×300					
7800	75×300						
7900	75×300						
8000	75×300						
8100	75×300						

Note
Dead load = 240 N/m² (5·0 lb per sq ft)
Imposed load = 1440 N/m² (30 lb per sq ft)

Table II Floor loading 1440 N/m² imposed + 480 N/m² dead

Span	300	350	400	450	500	550	600
1200	38×75	44×75	50×75	38×100	38×100	38×100	44×100
1300	44×75	50×75	38×100	38×100	38×100	44×100	50×100
1400	44×75	50×75	38×100	38×100	44×100	50×100	50×100
1500	50×75	38×100	38×100	44×100	44×100	50×100	38×125
1600	38×100	38×100	38×100	44×100	50×100	38×125	38×125
1700	38×100	38×100	44×100	50×100	38×125	38×125	44×125
1800	38×100	38×100	44×100	50×100	38×125	44×125	44×125
1900	38×100	44×100	50×100	38×125	38×125	44×125	50×125
2000	38×100	44×100	50×100	38×125	44×125	44×125	50×125
2100	44×100	50×100	38×125	38×125	44×125	50×125	38×150
2200	44×100	50×100	38×125	44×125	50×125	50×125	38×150
2300	50×100	38×125	38×125	44×125	50×125	38×150	44×150
2400	38×125	38×125	44×125	50×125	50×125	38×150	44×150
2500	38×125	38×125	44×125	50×125	38×150	44×150	50×150
2600	38×125	44×125	44×125	50×125	38×150	44×150	50×150
2700	38×125	44×125	38×150	44×150	44×150	50×150	44×175
2800	44×125	50×125	38×150	44×150	50×150	38×175	44×175
2900	50×125	38×150	44×150	50×150	38×175	44×175	50×175
3000	38×150	38×150	44×150	50×150	44×175	44×175	50×175
3100	38×150	44×150	50×150	44×175	44×175	50×175	44×200
3200	38×150	44×150	50×150	44×175	50×175	50×175	44×200
3300	44×150	50×150	44×175	44×175	50×175	44×200	50×200
3400	44×150	38×175	44×175	50×175	44×200	44×200	50×200
3500	50×150	44×175	44×175	50×175	44×200	50×200	50×200
3600	38×175	44×175	50×175	44×200	44×200	50×200	44×225
3700	38×175	44×175	50×175	44×200	50×200	44×225	50×225
3800	44×175	50×175	44×200	50×200	50×200	44×225	50×225
3900	44×175	50×175	44×200	50×200	44×225	50×225	50×225
4000	50×175	44×200	44×200	50×200	44×225	50×225	75×200
4100	50×175	44×200	50×200	44×225	50×225	50×225	75×200
4200	44×200	44×200	50×200	44×225	50×225	75×200	75×200
4300	44×200	50×200	44×225	50×225	75×200	75×200	50×250
4400	44×200	50×200	44×225	50×225	75×200	75×200	75×225
4500	44×200	44×225	50×225	50×225	75×200	50×250	75×225
4600	25×200	44×225	50×225	75×200	50×250	75×225	75×225
4700	50×200	50×225	50×225	75×200	50×250	75×225	75×225
4800	44×225	50×225	75×200	50×250	75×225	75×225	75×250
4900	44×225	50×225	50×250	50×250	75×225	75×225	75×250
5000	44×225	50×225	50×250	75×225	75×225	75×250	75×250
5100	50×225	50×250	50×250	75×225	75×225	75×250	75×250
5200	50×225	50×250	50×250	75×225	75×250	75×250	75×250
5300	75×200	50×250	75×225	75×250	75×250	75×250	75×300
5400	50×250	50×250	75×225	75×250	75×250	75×300	75×300
5500	50×250	50×250	75×250	75×250	75×250	75×300	75×300
5600	50×250	75×225	75×250	75×250	75×250	75×300	75×300
5700	50×250	75×250	75×250	75×250	75×300	75×300	75×300
5800	50×250	75×250	75×250	75×250	75×300	75×300	75×300
5900	75×250	75×250	75×250	75×300	75×300	75×300	75×300
6000	75×250	75×250	75×250	75×300	75×300	75×300	75×300
6100	75×250	75×250	75×300	75×300	75×300	75×300	75×300
6200	75×250	75×250	75×300	75×300	75×300	75×300	75×300
6300	75×250	75×300	75×300	75×300	75×300	75×300	
6400	75×250	75×300	75×300	75×300	75×300	75×300	
6500	75×250	75×300	75×300	75×300	75×300	75×300	
6600	75×300	75×300	75×300	75×300	75×300		
6700	75×300	75×300	75×300	75×300	75×300		
6800	75×300	75×300	75×300	75×300			
6900	75×300	75×300	75×300	75×300			
7000	75×300	75×300	75×300				
7100	75×300	75×300	75×300				
7200	75×300	75×300	75×300				
7300	75×300	75×300					
7400	75×300	75×300					
7500	75×300						
7600	75×300						
7700	75×300						
7800	75×300						

Note

Dead load = 480 N/m² (10·0 lb per sq ft)
Imposed load = 1440 N/m² (30 lb per sq ft)

Selection of dead load

6.03 In using tables I to IV the dead load should first be worked out as described earlier but the self weight of the joists should be left out as this is already allowed for in the tables; for domestic floors, only the lower values of the dead load will normally be found applicable.

Floor boarding

6.04 Table V gives maximum spans for softwood floorboards complying in all respects with BS 1297: 1970 for tongued and grooved boards. The use of plain edged boards is now uncommon, but these have been retained as they appear in the 1965 edition of the Building Regulations and in CP 112². The thicknesses are adequate for domestic floor loading with allowance for wear, the effect of point loads, and, in the case of plain edged boards, the lack of 'load sharing' between adjacent boards.

6.05 Table VI gives the required thicknesses of plywood and chipboard for selected floor joist spacings. In this case the physical properties of the materials (determined from tests on samples) have been used to calculate the spans and thicknesses for economical use. A further advantage that is also taken into account is that the sheets will always tend to spread the loading.

Table V Minimum planed floor board thicknesses to meet requirements of the Building Regulations 1971

Minimum finished thickness (mm)	Maximum modular span (mm)	
	T&g, minimum face width 56 mm	Plain edged, minimum face width 144 mm
16	450	350
19	600	400
21	see note below	450
28	800	550

Note: for 600 mm joist spacing a stiffer floor may be obtained if required by using 21 mm t&g boarding.

Table VI Recommended thicknesses for flooring

Material	Thickness of material (mm) for various centres of support			Notes
Plywood British made plywood, ABPVM specification BP 101/65	16	16	12·5	Special constructions available to order. Face grain of plywood at right angles to supports. Headers and blocking required to support edges.
Canadian Fir plywood, t&g profile	16	16	12·5	Face grain of plywood at right angles to supports. Headers and blocking not required.
Canadian Fir plywood select sheathing	16	16	12·5	Face grain of plywood at right angles to supports. Headers and blocking required to support edges.
Finnish birch t&g plywood flooring	15	12	12	T&g on all edges. Headers and blocking not required. Face qualities for clear finishing or overlaying with other materials.
Finnish birch exterior plywood	15	12	12	Face grain of plywood at right angles to supports. Headers and blocking required to support edges.
Chipboard Flooring grade to BS 2604: 1963, amendment 3, July 1966	22	19	19	Thicknesses given are typical and manufacturers' recommendations should be sought. Headers and blocking required to support edges.

Information sheet
Timber 9

Section 7 **Structural material: Timber**

Domestic flat roofs

This sheet by H. J. BURGESS *notes some general points in flat roof construction including tables on the variation of joist size with span and loading. It should be read in conjunction with Information sheet* TIMBER 8 *'Domestic suspended floors'.*

1 Loading

Imposed loads

1.01 Flat roofs (pitch up to and including 10°) for single occupation houses not more than three storeys high and with limited access for maintenance and repair only, are required by the Building Regulations 1976 to be designed for either of these two types of imposed loading:

1 In accordance with BS CP 3: Chapter V: Part 1: 1967

Uniformly distributed load *or*		Concentrated load on any 300 mm square	
kN/m²	kgf/m²	kN	kgf
0·75	76·5	0·9	91·8

2 Uniformly distributed load
kN/m²
0·72

1.02 Type 2 gives no minimum concentrated load. Tables I to III, however, allow for a uniformly distributed minimum load of 1·75 kN per metre width when considering short span joists less than 2·44 m as the calculated member sizes would be very small if only 0·72 kN/m² were allowed for, and would be inadequate for point loads caused by the weight of a person during cleaning or repair.

Dead load

1.03 In addition to imposed load, the Regulations require the roof supports to be designed for their own weight plus the weight of roofing and ceiling materials and any other parts of the permanent construction they carry.

Typical loading

1.04 The following is a typical calculation of the total loading, including self-weight, on 50 mm × 200 mm roof joists at 600 mm (0·6 m) centres. There is no provision for the weight of services. Unit weights of materials are as from BS 648 'Schedule of weights of building materials'.

Self weight joists

$$= \frac{\text{volume (m}^3) \times \text{weight per unit volume (kg)}}{\text{centering (m)}}$$

$$= \frac{(0·05 \times 0·2 \times 1) \times 540\,\text{kg}}{0·6} = 9·0\,\text{kg/m}^2$$

Roof covering and insulation	= 25·8 kg/m²
13 mm plasterboard ceiling	= 11·2 kg/m²
Total dead load	= 46·0 kg/m²
Converting to N/m²	= 46·0 × 9·81
	= 451 N/m²
Imposed load (from para 1·01)	= 720 N/m²
Total load	= 720 + 451
	= 1171 N/m²

2 Types of domestic roof

2.01 The simplest roof uses joists spanning from cross wall to cross wall with the roof covering directly attached. Where cross wall spans are too great for readily available timber sections, including hardwood timbers, intermediate support may be provided by loadbearing partitions or beams. But it should be remembered that this will complicate the structure, with corresponding loss of economy. The load borne by intermediate partitions must be carried down to ground level, requiring heavy first floor joists if the ground and first floor partitions are not aligned—and even if they are aligned, beams above openings must be capable of transmitting part of the roof load. Use of intermediate beams in an all-timber construction requires columns built into the external walls to carry the loads to the foundations. Of the two solutions, partitions should normally be used in stud frame construction and intermediate beams in post and beam construction.

3 Structural design

3.01 The methods already described for the design of domestic floors (see Information sheet TIMBER 8 para 3.01) may also be applied in flat roof design. One general point here is that deflection rather than bending stress is likely to be the critical factor in roof joists.

Lateral support

3.02 The requirements of BS CP 112 clause 3.13.3 for lateral stability of joists and beams may be interpreted as follows for domestic roofs:

Table I Permissible depth to breadth ratios for varying degrees of joist lateral restraint

	Maximum depth-to-breadth ratio
Joists with well-fitted noggings for edge joints of woodwool slabs	4
Joists with top edge held in line by plywood or board roofing	
Beams with top edge held in line by adequately nailed joists	5
As above, with adequate bridging or blocking at intervals not exceeding six times the depth	6
Stringers of stressed skin panels covered with plywood on both sides	7

In all cases the ends of the joists or beams must be held vertically in position eg by blocking pieces between; by building in; or by placing in galvanised shoes provided the shoes are themselves restrained to the brickwork or other framing.

Roof joist span tables

3.03 These are included in the TRADA *Timber frame housing design guide*. They are available in metric as a separate TRADA publication (TBL 41), where they differ from the imperial tables in allowing for the self-weight of the joists, which need not therefore be included in the dead weight calculation. Also the exact minimum length of bearing required has been deducted from the span, thus avoiding the use of an arbitrary figure.

3.04 As with floor joists, the tables in schedule 6 of the Building Regulations may be used as an alternative, and a recalculated version of these is given in tables II to IX. The tables, which allow for the self-weight of the joists, are for the various imposed and dead loads as shown.

Table II Limited access only. Dead load 480 N/m², imposed load 720 N/m²

FELT ON WOODWOOL SLABS WITHOUT CHIPPINGS

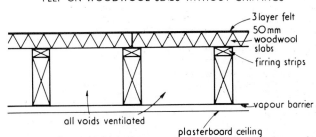

Span	Spacing (mm)						
	300	350	400	450	500	550	600
1200	38×75	38×75	38×75	38×75	38×75	38×75	38×75
1300	38×75	38×75	38×75	38×75	38×75	38×75	38×75
1400	38×75	38×75	38×75	38×75	38×75	44×75	44×75
1500	38×75	38×75	38×75	38×75	44×75	50×75	50×75
1600	38×75	38×75	44×75	44×75	50×75	38×100	38×100
1700	38×75	44×75	50×75	50×75	38×100	38×100	38×100
1800	44×75	50×75	38×100	38×100	38×100	38×100	38×100
1900	44×75	38×100	38×100	38×100	38×100	38×100	38×100
2000	50×75	38×100	38×100	38×100	38×100	38×100	44×100
2100	38×100	38×100	38×100	38×100	44×100	44×100	50×100
2200	38×100	38×100	38×100	44×100	44×100	50×100	38×125
2300	38×100	38×100	44×100	44×100	50×100	38×125	38×125
2400	38×100	38×100	44×100	50×100	38×125	38×125	38×125
2500	38×100	44×100	50×100	38×125	38×125	38×125	38×125
2600	44×100	50×100	38×125	38×125	38×125	38×125	44×125
2700	50×100	38×125	38×125	38×125	44×125	44×125	50×125
2800	38×125	38×125	38×125	44×125	44×125	50×125	38×150
2900	38×125	38×125	44×125	44×125	50×125	38×150	38×150
3000	38×125	38×125	44×125	50×125	38×150	38×150	38×150
3100	38×125	44×125	50×125	38×150	38×150	38×150	44×150
3200	44×125	50×125	38×150	38×150	38×150	44×150	50×150
3300	44×125	38×150	38×150	38×150	44×150	50×150	50×150
3400	50×125	38×150	38×150	44×150	50×150	50×150	38×175
3500	38×150	38×150	44×150	50×150	50×150	38×175	38×175
3600	38×150	38×150	44×150	50×150	38×175	38×175	44×175
3700	38×150	44×150	50×150	38×175	38×175	44×175	50×175
3800	44×150	50×150	38×175	38×175	44×175	44×175	50×175
3900	44×150	50×150	38×175	44×175	44×175	50×175	44×200
4000	50×150	38×175	38×175	44×175	50×175	44×200	44×200
4100	50×150	38×175	44×175	50×175	44×200	44×200	44×200
4200	38×175	44×175	44×175	50×175	44×200	44×200	44×200
4300	38×175	44×175	50×175	44×200	44×200	44×200	50×200
4400	44×175	50×175	44×200	44×200	44×200	50×200	44×225
4500	44×175	50×175	44×200	44×200	50×200	50×200	44×225
4600	50×175	44×200	44×200	44×200	50×200	44×225	44×225
4700	50×175	44×200	44×200	50×200	50×200	44×225	44×225
4800	44×200	44×200	50×200	44×225	44×225	44×225	50×225
4900	44×200	44×200	50×200	44×225	44×225	50×225	50×225
5000	44×200	50×200	44×225	44×225	44×225	50×225	50×250
5100	44×200	50×200	44×225	44×225	50×225	75×200	50×250
5200	44×200	44×225	44×225	50×225	50×225	50×250	50×250
5300	50×200	44×225	44×225	50×225	50×250	50×250	50×250
5400	50×200	44×225	50×225	75×200	50×250	50×250	50×250
5500	44×225	44×225	50×225	50×250	50×250	50×250	75×225
5600	44×225	50×225	75×225	50×250	50×250	50×250	75×250
5700	44×225	50×225	50×250	50×250	50×250	75×225	75×250
5800	44×225	75×200	50×250	50×250	75×225	75×250	75×250
5900	50×225	50×250	50×250	50×250	75×250	75×250	75×250
6000	50×225	50×250	50×250	75×225	75×250	75×250	75×250
6100	50×250	50×250	50×250	75×250	75×250	75×250	75×250
6200	50×250	50×250	75×225	75×250	75×250	75×250	75×300
6300	50×250	50×250	75×250	75×250	75×250	75×250	75×300
6400	50×250	50×250	75×250	75×250	75×250	75×300	75×300
6500	50×250	75×225	75×250	75×250	75×250	75×300	75×300
6600	50×250	75×250	75×250	75×250	75×300	75×300	75×300
6700	50×250	75×250	75×250	75×250	75×300	75×300	75×300
6800	75×250	75×250	75×250	75×300	75×300	75×300	75×300
6900	75×250	75×250	75×250	75×300	75×300	75×300	75×300
7000	75×250	75×250	75×300	75×300	75×300	75×300	75×300
7100	75×250	75×250	75×300	75×300	75×300	75×300	75×300
7200	75×250	75×300	75×300	75×300	75×300	75×300	75×300
7300	75×250	75×300	75×300	75×300	75×300	75×300	75×300
7400	75×250	75×300	75×300	75×300	75×300	75×300	
7500	75×300	75×300	75×300	75×300	75×300	75×300	
7600	75×300	75×300	75×300	75×300	75×300		
7700	75×300	75×300	75×300	75×300	75×300		
7800	75×300	75×300	75×300	75×300			
7900	75×300	75×300	75×300	75×300			
8000	75×300	75×300	75×300				
8100	75×300	75×300	75×300				
8200	75×300	75×300	75×300				
8300	75×300	75×300					
8400	75×300	75×300					
8500	75×300	75×300					
8600	75×300						
8700	75×300						
8800	75×300						

Table III *Limited access only. Dead load 480 N/m²; imposed load 720 N/m². Timber grade* SS

FELT ON WOODWOOL SLABS WITHOUT CHIPPINGS:

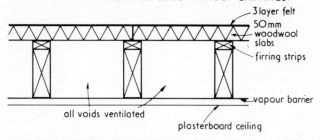

Table IV *Limited access only. Dead load 600 N/m²; imposed load 720 N/m². Timber grade* GS

FELT ON COMPRESSED STRAWBOARD WITH CHIPPINGS

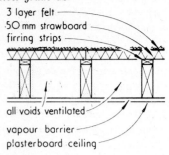

Table III

Span	300	350	400	450	500	550	600
1200	38×75	38×75	38×75	38×75	38×75	38×75	38×75
1300	38×75	38×75	38×75	38×75	38×75	38×75	38×75
1400	38×75	38×75	38×75	38×75	38×75	38×75	38×75
1500	38×75	38×75	38×75	38×75	38×75	44×75	44×75
1600	38×75	38×75	38×75	38×75	38×75	38×75	38×75
1700	38×75	38×75	38×75	44×75	50×75	38×100	38×100
1800	38×75	38×75	44×75	50×75	38×100	38×100	38×100
1900	38×75	44×75	50×75	38×100	38×100	38×100	38×100
2000	44×75	50×75	38×100	38×100	38×100	38×100	38×100
2100	50×75	38×100	38×100	38×100	38×100	38×100	44×100
2200	38×100	38×100	38×100	38×100	38×100	44×100	50×100
2300	38×100	38×100	38×100	38×100	44×100	50×100	50×100
2400	38×100	38×100	38×100	44×100	50×100	38×125	38×125
2500	38×100	38×100	44×100	50×100	38×125	38×125	38×125
2600	38×100	44×100	50×100	38×125	38×125	38×125	38×125
2700	44×100	50×100	38×125	38×125	38×125	38×125	44×125
2800	50×100	38×125	38×125	38×125	38×125	44×125	50×125
2900	50×100	38×125	38×125	38×125	44×125	50×125	50×125
3000	38×125	38×125	38×125	44×125	50×125	38×150	38×150
3100	38×125	38×125	44×125	50×125	38×150	38×150	38×150
3200	38×125	44×125	50×125	38×150	38×150	38×150	44×150
3300	38×125	44×125	50×125	38×150	38×150	44×150	44×150
3400	44×125	50×125	38×150	38×150	44×150	44×150	50×150
3500	50×125	38×150	38×150	44×150	44×150	50×150	38×175
3600	50×125	38×150	38×150	44×150	50×150	38×175	38×175
3700	38×150	38×150	44×150	50×150	38×175	38×175	38×175
3800	38×150	44×150	44×150	50×150	38×175	38×175	44×175
3900	38×150	44×150	50×150	38×175	38×175	44×175	50×175
4000	44×150	50×150	38×175	38×175	44×175	50×175	50×175
4100	44×150	50×150	38×175	44×175	44×175	50×175	44×200
4200	50×150	38×175	38×175	44×175	50×175	44×200	44×200
4300	50×150	38×175	44×175	50×175	44×175	44×200	44×200
4400	38×175	38×175	44×175	50×175	44×200	44×200	44×200
4500	38×175	44×175	50×175	44×200	44×200	44×200	50×200
4600	38×175	44×175	44×200	44×200	44×200	50×200	50×200
4700	44×175	50×175	44×200	44×200	50×200	50×200	44×225
4800	44×175	44×200	44×200	44×200	50×200	44×225	44×225
4900	50×175	44×200	44×200	50×200	44×225	44×225	44×225
5000	50×175	44×200	44×200	50×200	44×225	44×225	50×225
5100	44×200	44×200	50×200	44×225	44×225	44×225	50×225
5200	44×200	44×200	50×200	44×225	44×225	50×225	75×200
5300	44×200	50×200	44×225	44×225	50×225	50×225	50×250
5400	44×200	50×200	44×225	44×225	50×225	50×250	50×250
5500	50×200	44×225	44×225	50×225	75×200	50×250	50×250
5600	50×200	44×225	44×225	50×225	50×250	50×250	50×250
5700	44×225	44×225	50×225	50×250	50×250	50×250	50×250
5800	44×225	44×225	50×225	50×250	50×250	50×250	75×225
5900	44×225	50×225	50×250	50×250	50×250	75×225	75×250
6000	44×225	50×225	50×250	50×250	50×250	75×225	75×250
6100	44×225	75×200	50×250	50×250	75×225	75×250	75×250
6200	50×225	50×250	50×250	50×250	75×250	75×250	75×250
6300	50×225	50×250	50×250	75×225	75×250	75×250	75×250
6400	50×250	50×250	50×250	75×250	75×250	75×250	75×250
6500	50×250	50×250	75×225	75×250	75×250	75×250	75×300
6600	50×250	50×250	75×250	75×250	75×250	75×250	75×300
6700	50×250	50×250	75×250	75×250	75×250	75×300	75×300
6800	50×250	75×225	75×250	75×250	75×250	75×300	75×300
6900	50×250	75×250	75×250	75×250	75×300	75×300	75×300
7000	50×250	75×250	75×250	75×250	75×300	75×300	75×300
7100	75×225	75×250	75×250	75×300	75×300	75×300	75×300
7200	75×250	75×250	75×250	75×300	75×300	75×300	75×300
7500	75×250	75×250	75×300	75×300	75×300	75×300	75×300
7700	75×250	75×300	75×300	75×300	75×300	75×300	75×300
7800	75×250	75×300	75×300	75×300	75×300	75×300	
8000	75×300	75×300	75×300	75×300	75×300		
8200	75×300	75×300	75×300	75×300			
8400	75×300	75×300	75×300				
8500	75×300	75×300	75×300				
8600	75×300	75×300	75×300				
8800	75×300	75×300					
9000	75×300						
9200	75×300						
9300	75×300						

Table IV

Span	300	350	400	450	500	550	600
1200	38×75	38×75	38×75	38×75	38×75	38×75	38×75
1300	38×75	38×75	38×75	38×75	38×75	38×75	44×75
1400	38×75	38×75	38×75	38×75	38×75	44×75	50×75
1500	38×75	38×75	38×75	44×75	50×75	50×75	38×100
1600	38×75	38×75	44×75	50×75	38×100	38×100	38×100
1700	38×75	44×75	50×75	38×100	38×100	38×100	38×100
1800	44×75	50×75	38×100	38×100	38×100	38×100	38×100
1900	50×75	38×100	38×100	38×100	38×100	38×100	44×100
2000	38×100	38×100	38×100	38×100	38×100	44×100	50×100
2100	38×100	38×100	38×100	38×100	44×100	50×100	38×125
2200	38×100	38×100	38×100	44×100	50×100	38×125	38×125
2300	38×100	38×100	44×100	50×100	38×125	38×125	38×125
2400	38×100	44×100	50×100	38×125	38×125	38×125	38×125
2500	44×100	50×100	38×125	38×125	38×125	38×125	44×125
2600	50×100	38×125	38×125	38×125	38×125	44×125	50×125
2700	38×125	38×125	38×125	44×125	44×125	50×125	38×150
2800	38×125	38×125	38×125	44×125	50×125	38×150	38×150
2900	38×125	38×125	44×125	50×125	38×150	38×150	38×150
3000	38×125	44×125	50×125	38×150	38×150	38×150	44×150
3100	44×125	50×125	38×150	38×150	38×150	44×150	50×150
3200	44×125	38×150	38×150	38×150	44×150	50×150	50×150
3300	50×125	38×150	38×150	44×150	50×150	50×150	38×175
3400	38×150	38×150	44×150	50×150	50×150	18×175	38×175
3500	38×150	44×150	44×150	50×150	38×175	38×175	44×175
3600	38×150	44×150	50×150	38×175	38×175	44×175	50×175
3700	44×150	50×150	38×175	38×175	44×175	50×175	50×175
3800	44×150	50×150	38×175	44×175	44×175	50×175	44×200
3900	50×150	38×175	44×175	44×175	50×175	44×200	44×200
4000	38×175	38×175	44×175	50×175	44×200	44×200	44×200
4100	38×175	44×175	50×175	44×200	44×200	44×200	50×200
4200	38×175	44×175	50×175	44×200	44×200	44×200	50×200
4300	44×175	50×175	44×200	44×200	44×200	50×200	44×225
4400	44×175	50×175	44×200	44×200	50×200	44×225	44×225
4500	50×175	44×200	44×200	50×200	50×200	44×225	44×225
4600	50×175	44×200	44×200	50×200	50×200	44×225	50×225
4700	44×200	44×200	50×200	44×225	44×225	44×225	50×225
4800	44×200	44×200	50×200	44×225	44×225	50×225	75×200
4900	44×200	50×200	44×225	44×225	50×225	50×225	50×250
5000	44×200	50×200	44×225	44×225	50×225	50×250	50×250
5100	50×200	44×225	44×225	50×225	75×200	50×250	50×250
5200	50×200	44×225	44×225	50×225	50×250	50×250	50×250
5300	44×225	44×225	50×225	50×250	50×250	50×250	75×225
5400	44×225	44×225	50×225	50×250	50×250	50×250	75×225
5500	44×225	50×225	50×225	50×250	50×250	75×225	75×250
5600	44×225	50×225	50×250	50×250	50×250	75×225	75×250
5700	50×225	50×250	50×250	50×250	75×225	75×250	75×250
5800	50×225	50×250	50×250	75×225	75×250	75×250	75×250
5900	75×200	50×250	50×250	75×250	75×250	75×250	75×250
6000	50×250	50×250	75×225	75×250	75×250	75×250	75×300
6100	50×250	50×250	75×225	75×250	75×250	75×250	75×250
6200	50×250	50×250	75×250	75×250	75×250	75×300	75×300
6300	50×250	75×225	75×250	75×250	75×250	75×300	75×300
6400	50×250	75×250	75×250	75×250	75×250	75×300	75×300
6500	50×250	75×250	75×250	75×250	75×250	75×300	75×300
6600	75×225	75×250	75×250	75×300	75×300	75×300	75×300
6700	75×250	75×250	75×250	75×300	75×300	75×300	75×300
6800	75×250	75×250	75×250	75×300	75×300	75×300	75×300
6900	75×250	75×250	75×300	75×300	75×300	75×300	75×300
7000	75×250	75×250	75×300	75×300	75×300	75×300	75×300
7100	75×250	75×250	75×300	75×300	75×300	75×300	75×300
7200	75×250	75×250	75×300	75×300	75×300	75×300	
7300	75×250	75×300	75×300	75×300	75×300	75×300	75×300
7400	75×300	75×300	75×300	75×300	75×300		
7500	75×300	75×300	75×300	75×300	75×300		
7600	75×300	75×300	75×300	75×300	75×300		
7700	75×300	75×300	75×300	75×300	75×300		
7800	75×300	75×300	75×300				
8000	75×300	75×300	75×300				
8100	75×300	75×300					
8300	75×300	75×300					
8400	75×300						
8600	75×300						

Table v *Limited access only. Dead load 600 N/m²;*
imposed load 720 N/m². Timber grade ss

FELT ON COMPRESSED
STRAWBOARD
WITH CHIPPINGS

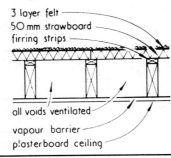

Table vi *Pedestrian access. Dead load 240 N/m²;*
imposed load 1440 N/m². Timber grade ss

FELT ON COMPRESSED
STRAWBOARD WITHOUT
CHIPPINGS

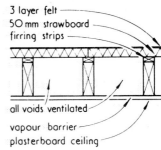

Table V

Span	300	350	400	450	500	550	600
1200	38×75	38×75	38×75	38×75	38×75	38×75	38×75
1300	38×75	38×75	38×75	38×75	38×75	38×75	38×75
1400	38×75	38×75	38×75	38×75	38×75	38×75	44×75
1500	38×75	38×75	38×75	38×75	38×75	44×75	50×75
1600	38×75	38×75	38×75	44×75	50×75	50×75	38×100
1700	38×75	38×75	44×75	50×75	38×100	38×100	38×100
1800	38×75	44×75	50×75	38×100	38×100	38×100	38×100
1900	44×75	50×75	38×100	38×100	38×100	38×100	38×100
2000	50×75	38×100	38×100	38×100	38×100	38×100	44×100
2100	38×100	38×100	38×100	38×100	38×100	44×100	44×100
2200	38×100	38×100	38×100	38×100	38×100	50×100	50×100
2300	38×100	38×100	38×100	44×100	50×100	50×100	38×125
2400	38×100	38×100	44×100	50×100	38×125	38×125	38×125
2500	38×100	44×100	50×100	38×125	38×125	38×125	38×125
2600	44×100	50×100	38×125	38×125	38×125	38×125	44×125
2700	44×100	38×125	38×125	38×125	38×125	44×125	44×125
2800	50×100	38×125	38×125	38×125	44×125	50×125	50×125
2900	38×125	38×125	38×125	44×125	50×125	50×125	38×150
3000	38×125	38×125	44×125	50×125	38×150	38×150	38×150
3100	38×125	44×125	50×125	38×150	38×150	38×150	44×150
3200	38×125	44×125	50×125	38×150	38×150	44×150	44×150
3300	44×125	50×125	38×150	38×150	44×150	44×150	50×150
3400	50×125	38×150	38×150	44×150	44×150	50×150	38×175
3500	50×125	38×150	38×150	44×150	50×150	38×175	38×175
3600	38×150	38×150	44×150	50×150	38×175	38×175	44×175
3700	38×150	44×150	50×150	50×150	38×175	38×175	44×175
3800	38×150	44×150	50×150	38×175	38×175	44×175	50×175
3900	44×150	50×150	38×175	38×175	44×175	50×175	50×175
4000	44×150	50×150	38×175	44×175	50×175	50×175	44×200
4100	50×150	38×175	44×175	44×175	50×175	44×200	44×200
4200	38×175	38×175	44×175	50×175	44×200	44×200	44×200
4300	38×175	44×175	50×175	50×175	44×200	44×200	44×200
4400	38×175	44×175	50×175	44×200	44×200	44×200	50×200
4500	44×175	50×175	44×200	44×200	44×200	50×200	44×225
4600	44×175	50×175	44×200	44×200	50×200	50×200	44×225
4700	50×175	44×200	44×200	44×200	50×200	44×225	44×225
4800	50×175	44×200	44×200	50×200	44×225	44×225	44×225
4900	44×200	44×200	50×200	50×200	44×225	44×225	50×225
5000	44×200	44×200	50×200	44×225	44×225	50×225	50×225
5100	44×200	50×200	44×225	44×225	44×225	50×225	50×250
5200	44×200	50×200	44×225	44×225	50×225	75×200	50×250
5300	44×200	44×225	44×225	50×225	50×225	50×250	50×250
5400	50×200	44×225	44×225	50×225	50×250	50×250	50×250
5500	50×200	44×225	50×225	75×200	50×250	50×250	50×250
5600	44×225	44×225	50×225	50×250	50×250	50×250	75×225
5700	44×225	44×225	75×200	50×250	50×250	50×250	75×250
5800	44×225	50×225	50×250	50×250	50×250	75×225	75×250
5900	44×225	50×225	50×250	50×250	75×225	75×250	75×250
6000	50×225	50×250	50×250	50×250	75×225	75×250	75×250
6100	50×225	50×250	50×250	75×225	75×250	75×250	75×250
6200	50×225	50×250	50×250	75×225	75×250	75×250	75×250
6300	50×250	50×250	50×250	75×250	75×250	75×250	75×250
6400	50×250	50×250	75×225	75×250	75×250	75×250	75×300
6500	50×250	50×250	75×250	75×250	75×250	75×300	75×300
6600	50×250	75×225	75×250	75×250	75×250	75×300	75×300
6800	50×250	75×250	75×250	75×250	75×300	75×300	75×300
6900	75×250	75×250	75×250	75×300	75×300	75×300	75×300
7000	75×250	75×250	75×250	75×300	75×300	75×300	75×300
7100	75×250	75×250	75×300	75×300	75×300	75×300	75×300
7200	75×250	75×250	75×300	75×300	75×300	75×300	75×300
7400	75×250	75×300	75×300	75×300	75×300	75×300	75×300
7500	75×250	75×300	75×300	75×300	75×300	75×300	75×300
7600	75×250	75×300	75×300	75×300	75×300	75×300	
7700	75×300	75×300	75×300	75×300	75×300	75×300	
7800	75×300	75×300	75×300	75×300	75×300		
8000	75×300	75×300	75×300	75×300			
8100	75×300	75×300	75×300	75×300			
8200	75×300	75×300	75×300				
8400	75×300	75×300	75×300				
8500	75×300	75×300					
8700	75×300	75×300					
8800	75×300						
9000	75×300						

Table VI

Span	300	350	400	450	500	550	600
1200	38×75	38×75	38×75	38×75	44×75	44×75	50×75
1300	38×75	38×75	38×75	38×75	44×75	50×75	38×100
1400	38×75	38×75	44×75	44×75	50×75	38×100	38×100
1500	38×75	44×75	50×75	50×75	38×100	38×100	38×100
1600	38×75	50×75	38×100	38×100	38×100	38×100	38×100
1700	44×75	38×100	38×100	38×100	38×100	38×100	44×100
1800	50×75	38×100	38×100	38×100	38×100	38×100	44×100
1900	38×100	38×100	38×100	38×100	44×100	44×100	50×100
2000	38×100	38×100	38×100	44×100	44×100	50×100	38×125
2100	38×100	38×100	44×100	44×100	50×100	38×125	38×125
2200	38×100	38×100	44×100	50×100	38×125	38×125	38×125
2300	38×100	44×100	50×100	38×125	38×125	38×125	38×125
2400	44×100	50×100	38×125	38×125	38×125	38×125	44×125
2500	44×100	38×125	38×125	38×125	38×125	44×125	44×125
2600	50×100	38×125	38×125	38×125	44×125	50×125	50×125
2700	38×125	38×125	38×125	44×125	50×125	38×150	38×150
2800	38×125	38×125	44×125	50×125	38×150	38×150	38×150
2900	38×125	44×125	50×125	38×150	38×150	38×150	44×150
3000	44×125	50×125	38×150	38×150	38×150	44×150	50×150
3100	44×125	50×125	38×150	38×150	44×150	50×150	50×150
3200	50×125	38×150	38×150	44×150	50×150	50×150	38×175
3300	38×150	38×150	44×150	50×150	50×150	38×175	38×175
3400	38×150	38×150	44×150	50×150	38×175	38×175	44×175
3500	38×150	44×150	50×150	38×175	38×175	44×175	50×175
3600	44×150	50×150	38×175	38×175	44×175	50×175	50×175
3700	44×150	50×150	38×175	44×175	44×175	50×175	44×200
3800	50×150	38×175	44×175	44×175	50×175	44×200	44×200
3900	38×175	38×175	44×175	50×175	44×200	44×200	44×200
4000	38×175	44×175	50×175	44×200	44×200	44×200	50×200
4100	38×175	44×175	50×175	44×200	44×200	50×200	44×225
4200	44×175	50×175	44×200	44×200	50×200	50×200	44×225
4300	44×175	50×175	44×200	44×200	50×200	44×225	44×225
4400	50×175	44×200	44×200	50×200	44×225	44×225	44×225
4500	50×175	44×200	44×200	50×200	44×225	44×225	50×225
4600	44×200	44×200	50×200	44×225	44×225	50×225	50×225
4700	44×200	44×200	50×200	44×225	44×225	50×225	75×200
4800	44×200	50×200	44×225	44×225	50×225	75×200	50×250
4900	44×200	50×200	44×225	44×225	50×225	50×250	50×250
5000	50×200	44×225	44×225	50×225	50×250	50×250	50×250
5100	50×200	44×225	44×225	50×225	50×250	50×250	50×250
5200	44×225	44×225	50×225	50×250	50×250	50×250	75×225
5300	44×225	44×225	75×200	50×250	50×250	50×250	75×250
5400	44×225	50×225	50×250	50×250	50×250	75×225	75×250
5500	44×225	50×225	50×250	50×250	50×250	75×225	75×250
5600	50×225	50×250	50×250	50×250	75×225	75×250	75×250
5700	50×225	50×250	50×250	75×225	75×250	75×250	75×250
5800	75×200	50×250	50×250	75×250	75×250	75×250	75×250
5900	50×250	50×250	75×225	75×250	75×250	75×250	75×300
6000	50×250	50×250	75×250	75×250	75×250	75×250	75×300
6100	50×250	50×250	75×250	75×250	75×250	75×300	75×300
6200	50×250	75×225	75×250	75×250	75×250	75×300	75×300
6300	50×250	75×250	75×250	75×250	75×300	75×300	75×300
6400	75×225	75×250	75×250	75×300	75×300	75×300	75×300
6600	75×250	75×250	75×250	75×300	75×300	75×300	75×300
6800	75×250	75×250	75×300	75×300	75×300	75×300	75×300
6900	75×250	75×250	75×300	75×300	75×300	75×300	75×300
7000	75×250	75×300	75×300	75×300	75×300	75×300	
7100	75×250	75×300	75×300	75×300	75×300		
7200	75×300	75×300	75×300	75×300	75×300		
7300	75×300	75×300	75×300	75×300	75×300		
7400	75×300	75×300	75×300	75×300			
7600	75×300	75×300	75×300	75×300			
7700	75×300	75×300	75×300				
7800	75×300	75×300	75×300				
7900	75×300	75×300					
8100	75×300	75×300					
8200	75×300						
8500	75×300						

Table VII *Pedestrian access. Dead load 480 N/m²; imposed load 1440 N/m². Timber grade* SS

Table VIII *Pedestrian access. Dead load 720 N/m²; imposed load 1440 N/m². Timber grade* GS

FELT ON COMPRESSED
STRAWBOARD WITHOUT
CHIPPINGS

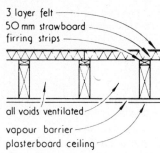

MASTIC ASPHALT
TOPPING ON BOARDS

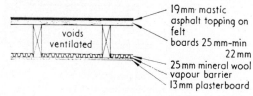

	Spacing (mm)						
Span	300	350	400	450	500	550	600
1200	38×75	38×75	38×75	38×75	44×75	50×75	50×75
1300	38×75	38×75	38×75	44×75	50×75	38×100	38×100
1400	38×75	38×75	44×75	50×75	38×100	38×100	38×100
1500	38×75	44×75	50×75	38×100	38×100	38×100	38×100
1600	44×75	50×75	38×100	38×100	38×100	38×100	44×100
1700	50×75	38×100	38×100	38×100	38×100	44×100	44×100
1800	38×100	38×100	38×100	38×100	44×100	44×100	50×100
1900	38×100	38×100	38×100	44×100	44×100	50×100	38×125
2000	38×100	38×100	44×100	44×100	50×100	38×125	38×125
2100	38×100	38×100	44×100	50×100	38×125	38×125	38×125
2200	38×100	44×100	50×100	38×125	38×125	38×125	38×125
2300	44×100	50×100	38×125	38×125	38×125	38×125	44×125
2400	44×100	38×125	38×125	38×125	38×125	44×125	50×125
2500	50×100	38×125	38×125	38×125	44×125	50×125	38×150
2600	38×125	38×125	38×125	44×125	50×125	38×150	38×150
2700	38×125	38×125	44×125	50×125	38×150	38×150	38×150
2800	38×125	44×125	50×125	38×150	38×150	38×150	44×150
2900	44×125	50×125	38×150	38×150	38×150	44×150	50×150
3000	44×125	38×150	38×150	38×150	44×150	50×150	38×175
3100	05×125	38×150	38×150	44×150	50×150	38×175	38×175
3200	38×150	38×150	44×150	50×150	38×175	38×175	44×175
3300	38×150	44×150	50×150	38×175	38×175	44×175	44×175
3400	38×150	44×150	50×150	38×175	44×175	44×175	50×175
3500	44×150	50×150	38×175	44×175	44×175	50×175	44×200
3600	50×150	38×175	38×175	44×175	50×175	44×200	44×200
3700	50×150	38×175	44×175	50×175	44×200	44×200	44×200
3800	38×175	44×175	44×175	50×175	44×200	44×200	44×200
3900	38×175	44×175	50×175	44×200	44×200	44×200	50×200
4000	44×175	50×175	44×200	44×200	44×200	50×200	44×225
4100	44×175	50×175	44×200	44×200	50×200	44×225	44×225
4200	50×175	44×200	44×200	50×200	50×200	44×225	44×225
4300	50×175	44×200	44×200	50×200	44×225	44×225	50×225
4400	44×200	44×200	50×200	44×225	44×225	44×225	50×225
4500	44×200	44×200	50×200	44×225	44×225	50×225	75×200
4600	44×200	50×200	44×225	44×225	50×225	75×200	50×250
4700	44×200	50×200	44×225	50×225	50×225	50×250	50×250
4800	50×200	44×225	44×225	50×250	50×250	50×250	50×250
4900	50×200	44×225	50×225	75×200	50×250	50×250	50×250
5000	44×225	44×225	50×225	50×250	50×250	50×250	75×225
5100	44×225	50×225	75×200	50×250	50×250	75×225	75×250
5200	44×225	50×225	50×250	50×250	50×250	75×225	75×250
5300	44×225	75×200	50×250	50×250	75×225	75×250	75×250
5400	50×225	50×250	50×250	50×250	75×250	75×250	75×250
5500	50×225	50×250	50×250	75×225	75×250	75×250	75×250
5600	50×250	50×250	50×250	75×250	75×250	75×250	75×250
5700	50×250	50×250	75×225	75×250	75×250	75×250	75×300
5800	50×250	50×250	75×250	75×250	75×250	75×300	75×300
5900	50×250	75×225	75×250	75×250	75×250	75×300	75×300
6000	50×250	75×250	75×250	75×250	75×300	75×300	75×300
6100	50×250	75×250	75×250	75×250	75×300	75×300	75×300
6200	75×225	75×250	75×250	75×300	75×300	75×300	75×300
6300	75×250	75×250	75×250	75×300	75×300	75×300	75×300
6400	75×250	75×250	75×300	75×300	75×300	75×300	75×300
6500	75×250	75×250	75×300	75×300	75×300	75×300	75×300
6600	75×250	75×250	75×300	75×300	75×300	75×300	75×300
6700	75×250	75×300	75×300	75×300	75×300	75×300	75×300
6800	75×250	75×300	75×300	75×300	75×300	75×300	
6900	75×250	75×300	75×300	75×300	75×300		
7000	75×300	75×300	75×300	75×300	75×300		
7100	75×300	75×300	75×300	75×300			
7300	75×300	75×300	75×300	75×300			
7400	75×300	75×300	75×300				
7500	75×300	75×300	75×300				
7600	75×300	75×300					
7800	75×300	75×300					
7900	75×300						
8000	75×300						
8200	75×300						

	Spacing (mm)						
Span	300	350	400	450	500	550	600
1200	44×75	50×75	38×100	38×100	38×100	44×100	44×100
1300	44×75	50×75	38×100	38×100	44×100	44×100	50×100
1400	50×75	38×100	38×100	44×100	44×100	50×100	38×125
1500	38×100	38×100	38×100	44×100	50×100	38×125	38×125
1600	38×100	38×100	44×100	50×100	38×125	38×125	44×125
1700	38×100	44×100	50×100	50×100	38×125	44×125	44×125
1800	38×100	44×100	50×100	38×125	44×125	44×125	50×125
1900	44×100	50×100	38×125	38×125	44×125	50×125	50×125
2000	44×100	50×100	38×125	44×125	44×125	50×125	38×150
2100	44×100	38×125	38×125	44×125	50×125	38×150	44×150
2200	50×100	38×125	44×125	50×125	50×125	38×150	44×150
2300	38×125	38×125	44×125	50×125	38×150	44×150	44×150
2400	38×125	44×125	50×125	38×150	44×150	44×150	50×150
2500	38×125	44×125	50×125	38×150	44×150	50×150	38×175
2600	44×125	50×125	38×150	44×150	50×150	50×150	44×175
2700	44×125	50×125	44×150	50×150	50×150	44×175	44×175
2800	50×125	38×150	44×150	50×150	44×175	44×175	50×175
2900	38×150	44×150	50×150	38×175	44×175	50×175	50×175
3000	38×150	44×150	50×150	44×175	50×175	50×175	44×200
3100	44×150	50×150	44×175	44×175	50×175	44×200	44×200
3200	44×150	50×150	44×175	50×175	44×200	44×200	50×200
3300	50×150	38×175	44×175	50×175	44×200	50×200	50×200
3400	50×150	44×175	50×175	44×200	44×200	50×200	44×225
3500	38×175	44×175	50×175	44×200	50×200	44×225	50×225
3600	44×175	50×175	44×200	50×200	50×200	44×225	50×225
3700	44×175	50×175	44×200	50×200	44×225	50×225	50×225
3800	44×175	44×200	50×200	50×200	44×225	50×225	75×200
3900	50×175	44×200	50×200	44×225	50×225	75×200	75×200
4000	44×200	44×200	50×200	44×225	50×225	75×200	75×200
4100	44×200	50×200	44×225	50×225	75×200	75×200	50×250
4200	44×200	50×200	44×225	50×225	75×200	50×250	75×225
4300	44×200	44×225	50×225	75×200	75×200	75×225	75×225
4400	50×200	44×225	50×225	75×200	50×250	75×225	75×225
4500	50×200	44×225	50×225	75×200	75×225	75×225	75×225
4600	44×225	50×225	75×200	50×250	75×225	75×225	75×250
4700	44×225	50×225	75×200	50×250	75×225	75×225	75×250
4800	44×225	50×225	75×250	75×225	75×225	75×250	75×250
4900	50×225	75×250	50×250	75×225	75×225	75×250	75×250
5000	50×225	75×250	75×225	75×250	75×250	75×250	75×300
5100	75×200	50×250	75×225	75×250	75×250	75×250	75×300
5200	50×250	50×250	75×225	75×250	75×250	75×250	75×300
5300	50×250	50×250	75×250	75×250	75×250	75×300	74×300
5400	50×250	75×225	75×250	75×250	75×300	75×300	75×300
5500	50×250	75×250	75×250	75×250	75×300	75×300	75×300
5600	50×250	75×250	75×250	75×250	75×300	75×300	75×300
5700	75×225	75×250	75×250	75×300	75×300	75×300	75×300
5800	75×250	75×250	75×250	75×300	75×300	75×300	75×300
5900	75×250	75×250	75×300	75×300	75×300	75×300	75×300
6000	75×250	75×250	75×300	75×300	75×300	75×300	
6100	75×250	75×300	75×300	75×300	75×300	75×300	
6200	75×250	75×300	75×300	75×300	75×300		
6300	75×250	75×300	75×300	75×300	75×300		
6400	75×300	75×300	75×300	75×300	75×300		
6500	75×300	75×300	75×300	75×300			
6600	75×300	75×300	75×300	75×300			
6700	75×300	75×300	75×300	75×300			
6800	75×300	75×300	75×300				
6900	75×300	75×300	75×300				
7000	75×300	75×300					
7100	75×300	75×300					
7200	75×300	75×300					
7300	75×300						
7400	75×300						
7500	75×300						

Table ix *Pedestrian access. Dead load 720 N/m²; imposed load 1440 N/m². Timber grade* ss

MASTIC ASPHALT
TOPPING ON BOARDS

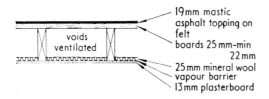

19mm mastic asphalt topping on felt
boards 25mm-min 22mm
25mm mineral wool
vapour barrier
13mm plasterboard

| Span | Spacing (mm) | | | | | | |
	300	350	400	450	500	550	600
1200	38×75	38×75	38×75	44×75	50×75	50×75	38×100
1300	38×75	38×75	44×75	50×75	50×75	38×100	38×100
1400	38×75	44×75	50×75	38×100	38×100	38×100	38×100
1500	44×75	50×75	38×100	38×100	38×100	38×100	44×100
1600	50×75	38×100	38×100	38×100	38×100	44×100	44×100
1700	38×100	38×100	38×100	38×100	44×100	44×100	50×100
1800	38×100	38×100	38×100	38×100	44×100	50×100	38×125
1900	38×100	38×100	38×100	44×100	50×100	38×125	38×125
2000	38×100	38×100	44×100	50×100	38×125	38×125	38×125
2100	38×100	44×100	50×100	38×125	38×125	38×125	44×125
2200	44×100	50×100	38×125	38×125	38×125	38×125	44×125
2300	50×100	38×125	38×125	38×125	38×125	44×125	50×125
2400	50×100	38×125	38×125	38×125	44×125	50×125	38×150
2500	38×125	38×125	38×125	44×125	50×125	38×150	38×150
2600	38×125	38×125	44×125	50×125	38×150	38×140	38×150
2700	38×125	44×125	50×125	38×150	38×150	38×150	44×150
2800	44×125	50×125	38×150	38×150	44×150	44×150	50×150
2900	50×125	38×150	38×150	44×150	44×150	50×150	38×175
3000	50×125	38×150	38×150	44×150	50×150	38×175	38×175
3100	38×150	38×150	44×150	50×150	38×175	38×175	44×175
3200	38×150	44×150	50×150	38×175	38×175	44×175	44×175
3300	44×150	50×150	38×175	38×175	44×175	44×175	50×175
3400	44×150	50×150	38×175	44×175	44×175	50×175	44×200
3500	50×150	38×175	44×175	44×175	50×175	44×200	44×200
3600	50×150	38×175	44×175	50×175	44×200	44×200	44×200
3700	38×175	44×175	50×175	44×200	44×200	44×200	50×200
3800	38×175	44×175	50×175	44×200	44×200	50×200	50×200
3900	44×175	50×175	44×200	44×200	50×200	50×200	44×225
4000	44×175	44×200	44×200	44×200	50×200	44×225	44×225
4100	50×175	44×200	44×200	50×200	44×225	44×225	44×225
4200	44×200	44×200	50×200	44×225	44×225	44×225	50×225
4300	44×200	44×200	50×200	44×225	44×225	50×225	75×200
4400	44×200	50×200	44×225	44×225	50×225	50×225	50×250
4500	44×200	50×200	44×225	44×225	50×225	50×250	50×250
4600	50×200	44×225	44×225	50×225	75×200	50×250	50×250
4700	50×200	44×225	50×225	50×225	50×250	50×250	50×250
4800	44×225	44×225	50×225	50×250	50×250	50×250	75×225
4900	44×225	50×225	75×200	50×250	50×250	75×225	75×250
5000	44×225	50×225	50×250	50×250	50×250	75×225	75×250
5100	44×225	50×225	50×250	50×250	75×225	75×250	75×250
5200	50×225	50×250	50×250	50×250	75×250	75×250	75×250
5300	50×225	50×250	50×250	75×225	75×250	75×250	75×250
5400	50×250	50×250	50×250	75×250	75×250	75×250	75×250
5500	50×250	50×250	75×225	75×250	75×250	75×250	75×300
5600	50×250	50×250	75×250	75×250	75×250	75×300	75×300
5700	50×250	75×225	75×250	75×250	75×250	75×300	75×300
5800	50×250	75×250	75×250	75×250	75×300	75×300	75×300
5900	50×250	75×250	75×250	75×250	75×300	75×300	75×300
6000	75×225	75×250	75×250	75×300	75×300	75×300	75×300
6100	75×250	75×250	75×250	75×300	75×300	75×300	75×300
6200	75×250	75×250	75×300	75×300	75×300	75×300	75×300
6300	75×250	75×250	75×300	75×300	75×300	75×300	75×300
6400	75×250	75×300	75×300	75×300	75×300	75×300	75×300
6500	75×250	75×300	75×300	75×300	75×300	75×300	
6600	75×250	75×300	75×300	75×300	75×300	75×300	
6700	75×300	75×300	75×300	75×300	75×300		
6800	75×300	75×300	75×300	75×300	75×300		
6900	75×300	75×300	75×300	75×300			
7000	75×300	75×300	75×300	75×300			
7100	75×300	75×300	75×300				
7300	75×300	75×300	75×300				
7400	75×300	75×300					
7600	75×300	75×300					
7700	75×300						
7900	75×300						

Type	Span	Drawing ref no

Type F

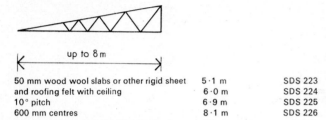

up to 12 m

	Span	Drawing ref no
Tiles not exceeding 102 kg/m²	6·0 m	SDS 217
(21lb/sq ft)	7·2 m	SDS 218
15° to 20° pitch	8·4 m	SDS 219
1·8 m centres	9·6 m	SDS 220
	10·8 m	SDS 221
	12·0 m	SDS 222

Table III TRADA *trussed rafters*

Type W

up to 8 m

	Span	Drawing ref no
50 mm wood wool slabs or other rigid sheet	5·1 m	SDS 223
and roofing felt with ceiling	6·0 m	SDS 224
10° pitch	6·9 m	SDS 225
600 mm centres	8·1 m	SDS 226

Type W

up to 8 m

	Span	Drawing ref no
50 mm wood wool slabs or other rigid sheet	5·1 m *	SDS 227
and roofing felt with ceiling	6·0 m	SDS 228
15° pitch	6·9 m	SDS 229
600 mm centres *	8·1 m	SDS 230
20° pitch	5·1 m	SDS 231
600 mm centres	6·0 m	SDS 232
	6·9 m	SDS 233
**	8·1 m	SDS 234
25° pitch	5·1 m	SDS 235
600 mm centres	6·0 m	SDS 236
	6·9 m	SDS 237
**	8·1 m	SDS 238

*May be used for tiles or slates up to 53·6 kg/m² on slope and pitch up to 17½°
**May be used for tiles or slates up to 50·3 kg/m² on slope
Designs for farm buildings also available

Table IV TRADA *industrial roof trusses*

Type	Span	Spacing (centre to centre)	Drawing ref no

Type A

up to 20 m

	Span	Spacing	Drawing ref no
Sheet material not exceeding	5 m	3·8 m	SDS 41
15·6 kg/m² (3·2lb/sq ft)	7·5 m	3·8 m	SDS 42
with purlins at 1370 mm	10·2 m	3·8 m	SDS 43
centres	12·6 m	3·8 m	SDS 44
22½° pitch	15·3 m	3·8 m	SDS 45
	17·4 m	4·25 m	SDS 46
	20·1 m	4·25 m	SDS 47

Table V TRADA *industrial building*

Type	Span	Drawing no	Notes

Type B

up to 20 m

	Span	Drawing no	Notes
Sheet materials not exceeding	5 m	3·8 m	SDS 70
15·6 kg/m² /3·2lb/sq ft)	7·5 m	3·8 m	SDS 71
with purlins at 1370 mm	10·2 m	3·8 m	SDS 72
centres with ceiling	12·6 m	3·8 m	SDS 73
22½° pitch	15·3 m	3·8 m	SDS 74
	17·4 m	4·25 m	SDS 75
	20·1 m	4·25 m	SDS 76

Type C

up to 15 m

	Span	Drawing no	Notes
Sheet materials not exceeding	5·1 m	SDS 87	Sides open
15·6 kg/m² (3·2lb/sq ft) with	5·1 m	SDS 88	Sides clad
purlins at 1370 mm centres	5·1 m	SDS 89	Gable frame
22½° pitch	7·6 m	SDS 90	Sides open
3·35 m centres	7·6 m	SDS 91	Sides clad
	7·6 m	SDS 92	Gable frame
	9·9 m	SDS 93	Sides open
	9·9 m	SDS 94	Sides clad
	9·9 m	SDS 95	Gable frame
4·25 m centres	12·6 m	SDS 102	Sides clad
	12·6 m	SDS 103	Gable frame
	12·6 m and 15 m	SDS 104	details for SDS 103 and SDS 106
	15 m	SDS 105	Sides clad
	15 m	SDS 106	Gable frame

Table VI BRE *Maximum permissible spans (m) for rafters for fink trussed rafters (timber to BS 4471 and of composite grade to CP 112)*

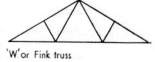

'W' or Fink truss

Basic size (mm)	Actual size (mm)	Pitch (degrees)								
		15	17½	20	22½	25	27½	30	32½	35
		Maximum spans (m)								
38× 75	35× 72	6·03	6·16	6·29	6·41	6·51	6·60	6·70	6·80	6·90
100	97	7·48	7·67	7·83	7·97	8·10	8·22	8·34	8·47	8·61
125	120	8·80	9·00	9·20	9·37	9·54	9·68	9·82	9·98	10·16
44× 75	41× 72	6·45	6·59	6·71	6·83	6·93	7·03	7·14	7·24	7·35
100	97	8·05	8·23	8·40	8·55	8·68	8·81	8·93	9·09	9·22
125	120	9·38	9·60	9·81	9·99	10·15	10·31	10·45	10·64	10·81
50× 75	47× 72	6·87	7·01	7·13	7·25	7·35	7·45	7·53	7·67	7·78
100	97	8·62	8·80	8·97	9·12	9·25	9·38	9·50	9·66	9·80
125	120	10·01	10·24	10·44	10·62	10·77	10·94	11·00	11·00	11·00

Information sheet
Timber 10

Section 7 **Structural material: Timber**

Structural design of domestic and industrial pitched roofs

This sheet by H. J. BURGESS *covers some of the commoner designs for domestic and industrial pitched roof trusses.*

1 Roof loadings

Dead load

1.01 Dead loading applied to domestic roof structure varies with the type of tile or other roof covering, and to a lesser extent with the type of framing used. Calculations commonly assume 493 N/m² on slope for low pitch interlocking tiles. Thus the corresponding load on plan where

the pitch $= 22\frac{1}{2}° = \dfrac{493}{\cos 22\frac{1}{2}°} = \dfrac{493}{0 \cdot 924} = 534$ N/m²

For 30° pitch, the plan load would be 570 N/m².

Imposed load

1.02 BS CP 3: Chapter V, gives a snow loading of 750 N/m² on plan for roofs of up to 30° pitch with limited access. This reduces by 50 N/m² for every 3° increase in pitch over 30°, reaching zero at 75° **1**. Also any roof with pitch less than 45° must be capable of carrying a 900 N concentrated load. The Building Regulations allow houses up to three storeys for single occupation a design loading of 720 N/m² instead of 750, reducing by 48 N/m² for each 3° of pitch. In this case there is no alternative concentrated load.

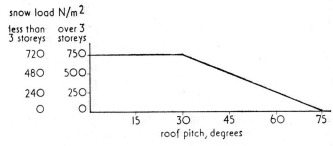

1 *Imposed load on slope varies with roof pitch*

Imposed ceiling load

1.03 The 1976 Building Regulations require a 720 N/m² imposed ceiling load, or (by reference to CP 3: Chapter V) a concentrated load of 900 N if the ceiling supports may have to support the weight of a man. CP 112: Part 3: 1973 trussed rafters for roofs of dwellings require a 250 N/m² distributed load in addition to the concentrated load. Water tank loading should be distributed over a number of joists or carried by a partition wall under.

Wind load

1.04 Wind loading effects are not usually calculated in detail on timber framed domestic roofs, as it has been shown that they make negligible difference to the design. With lightweight roofs care should be taken that holding-down details are adequate to resist the uplift caused by wind suction **2**. The roof must be stabilised by diagonal timber bracing nailed up to underside of rafters **3**.

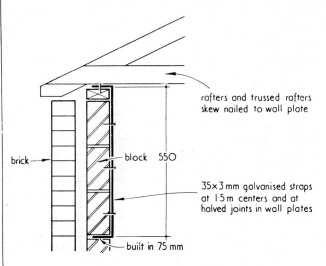

2 *Trussed roof perimeter requires holding down against wind suction uplift*

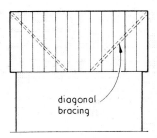

3 *Timber gable ends diagonally braced against wind action*

Total load calculation

1.05 This will depend on whether or not the frame carries the ceiling as well as the roof covering. If it does, the following is a common calculation for the total distributed load applied to trusses up to 30° pitch.

Tile and batten (12lb/sq ft)*	576 N/m²
Frame self-weight (3lb/sq ft)*	144
Total dead	720
Imposed (snow) (see fig 1)	720
Total roof	1440
Ceiling dead (5lb/sq ft)*	240
Ceiling imposed (see para 1.03)	240
Total ceiling	480
Total	1920 N/m² on plan

2 Types of pitched roof

Simple purlin roofs

2.01 The term 'traditional' is often used to describe a pitched roof where rafters are trussed only by the ceiling joists, deflection of the rafters being prevented by solid purlins (or lattice beams) running at right angles along the length of the roof and propped off internal walls or the gable ends. Similarly the ceiling joists can be assisted by dropped hangers. An often overlooked advantage is that this type of construction leaves a clear area of usable attic space **5**.

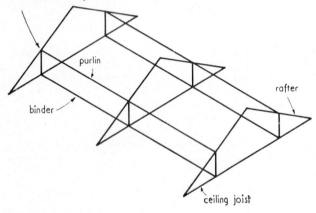

5 *Simple purlin roof. This type leaves a clear attic space*

2.02 In the simplest case, the load per metre run of purlin in this type of roof may be obtained by multiplying the total plan load for roof and ceiling per square metre by the purlin spacing. The typical plan load of 1920 N/m² given earlier comprises 960 N/m² dead load and 960 N/m² imposed load. Table I is an extract from the more extensive data on purlin spans for this loading published by TRADA (*Span tables for domestic purlins* 1970, E/IB/14). Suitable sizes for short-span softwood purlins and rafters are given in schedule 6 of the Building Regulations.

Truss and purlin roofs

2.03 Here trusses are generally spaced at 1·8 m centres, with rafters between them spanning in the same direction and at 450 mm centres. The rafters are supported at their mid points by a purlin, which is itself supported by the trusses **6**. The common ceiling joists are similarly supported from suspended binders.

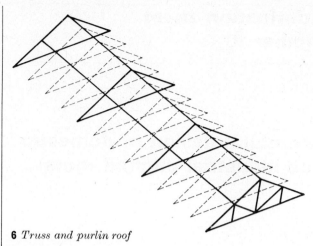

6 *Truss and purlin roof*

Trussed rafter roofs

2.04 The trussed rafter is a light truss used at 600 mm centres, eliminating the structural need for purlins and binders **7**. [75 mm × 50 mm binders are often incorporated, however, to assist in construction and location, diagonal bracing to underside of rafters to prevent buckling is also recommended. The trusses generally lie in the same plane, fastened by glued or nailed plywood gussets or by proprietary toothed or pierced metal plates.

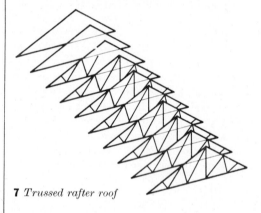

7 *Trussed rafter roof*

2.05 TRADA Standard design sheets cover trusses for both domestic and industrial use, in the latter case including notes on purlin sizes and spacing, knee-braced trusses on timber columns, and trusses spanning up to 9·8 m with or without ceiling joists. There is also an extensive range of designs for farm buildings. Extracts from these design sheets are shown in tables II to V and working drawings may be obtained from the Association for a small charge by quoting reference numbers.

2.06 The TRADA information is the result of calculation but proprietary designs depend generally on prototype testing. The BRE Princes Risborough Laboratory has published tables showing the maximum permissible span of trussed rafters using various member sizes—the joints being assumed adequate for the required loading (see tables VI to IX).

*Commonly assumed values translated into metric equivalent.

Table 1 Typical purlin data. Span (in effect truss centering) varies with purlin spacing and size. The plan load in this case is 1920 N/m²

Spacing (mm)	S.50 grade softwood (dry)					Keruing hardwood (dry)				
	1200	1500	1800	2100	2400	1200	1500	1800	2100	2400
	span (m)									
50 × 100	1·232	1·144	1·077	1·023	0·978	1·572	1·460	1·374	1·305	1·248
125	1·541	1·430	1·346	1·278	1·223	1·966	1·825	1·717	1·631	1·560
150	1·849	1·716	1·615	1·534	1·467	2·359	2·190	2·060	1·957	1·872
175	2·157	2·002	1·884	1·790	1·712	2·752	2·555	2·404	2·283	2·184
200	2·465	2·288	2·153	2·045	1·956	3·145	2·919	2·747	2·610	2·496
225	2·773	2·574	2·422	2·301	2·201	3·538	3·284	3·091	2·936	2·808
250	3·081	2·860	2·692	2·557	2·445	3·931	3·649	3·434	3·262	3·120
275	3·389	3·146	2·961	2·812	2·690	4·324	4·014	3·778	3·588	3·432
300	3·697	3·432	3·230	3·068	2·935	4·717	4·379	4·121	3·915	3·744
75 × 125	1·764	1·637	1·541	1·463	1·400	2·250	2·089	1·966	1·867	1·786
150	2·116	1·965	1·849	1·756	1·680	2·700	2·506	2·359	2·241	2·143
175	2·469	2·292	2·157	2·049	1·960	3·150	2·924	2·752	2·614	2·500
200	2·822	2·619	2·465	2·341	2·240	3·600	3·342	3·145	2·987	2·857
225	3·174	2·947	2·773	2·634	2·519	4·050	3·760	3·538	3·361	3·214
250	3·527	3·274	3·081	2·927	2·799	4·500	4·177	3·931	3·734	3·572
275	3·880	3·602	3·389	3·219	3·079	4·950	4·595	4·324	4·108	3·929
300	4·232	3·929	3·697	3·512	3·359	5·400	5·013	4·717	4·481	4·286
325	4·585	4·256	4·005	3·805	3·639	5·850	5·431	5·110	4·854	4·643
100 × 150	2·329	2·162	2·035	1·933	1·849	2·972	2·759	2·596	2·466	2·359
175	2·717	2·523	2·374	2·255	2·157	3·467	3·218	3·029	2·877	2·752
200	3·106	2·883	2·713	2·577	2·465	3·962	3·678	3·461	3·288	3·145
225	3·494	3·243	3·052	2·899	2·773	4·458	4·138	3·894	3·699	3·538
250	3·882	3·604	3·391	3·221	3·081	4·953	4·598	4·327	4·110	3·931
275	4·270	3·964	3·730	3·543	3·389	5·448	5·058	4·759	4·521	4·324
300	4·658	4·324	4·069	3·866	3·697	5·943	5·517	5·192	4·932	4·717
325	5·047	4·685	4·409	4·188	4·005	6·439	5·977	5·625	5·343	5·110
350	5·435	5·045	4·748	4·510	4·314	6·934	6·437	6·057	5·754	5·504

Table II TRADA *domestic trusses with ceiling joists*

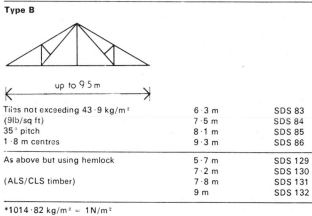

Type	Span	Drawing ref no
Type A		
up to 9.5 m		
Tiles not exceeding 68·4 kg/m²*	5·1 m	SDS 77
(14lb/sq ft)	6·3 m	SDS 78
40° pitch	7·2 m	SDS 79
1·8 m centres	8·1 m	SDS 80
	9·3 m	SDS 81
As above but using hemlock	6 m	SDS 125
	7·2 m	SDS 126
(ALS/CLS timber)	8·1 m	SDS 127
	9 m	SDS 128
Type B		
up to 9.5 m		
Tiles not exceeding 43·9 kg/m²	6·3 m	SDS 83
(9lb/sq ft)	7·5 m	SDS 84
35° pitch	8·1 m	SDS 85
1·8 m centres	9·3 m	SDS 86
As above but using hemlock	5·7 m	SDS 129
	7·2 m	SDS 130
(ALS/CLS timber)	7·8 m	SDS 131
	9 m	SDS 132

*1014·82 kg/m² = 1N/m²

Type	Span	Drawing ref no
Type C		
up to 11 m		
Tiles not exceeding 51·3 kg/m²	6·0 m	SDS 200
(10·5/sq ft)	7·2 m	SDS 201
22° to 30° pitch	8·4 m	SDS 202
1·8 m centres	9·6 m	SDS 203
	10·8 m	SDS 204
Type D		
up to 12 m		
Tiles not exceeding 78 kg/m²	6·0 m	SDS 205
(16lb/sq ft)	6·9 m	SDS 206
20° to 27° pitch	8·1 m	SDS 207
1·8 m centres	9·0 m	SDS 208
	10·2 m	SDS 209
	11·1 m	SDS 210
Type E		
up to 12 m		
Tiles not exceeding 73 kg/m²	6·0 m	SDS 211
(15lb/sq ft)	6·9 m	SDS 212
28° to 35° pitch	8·1 m	SDS 213
1·8 m centres	9·0 m	SDS 214
	10·2 m	SDS 215
	11·1 m	SDS 216

Type	Span	Drawing ref no
Type F		

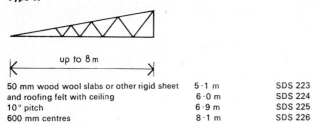

up to 12 m		
Tiles not exceeding 102 kg/m²	6·0 m	SDS 217
(21 lb/sq ft)	7·2 m	SDS 218
15° to 20° pitch	8·4 m	SDS 219
1·8 m centres	9·6 m	SDS 220
	10·8 m	SDS 221
	12·0 m	SDS 222

Table III TRADA *trussed rafters*

Type W

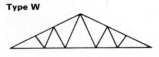

up to 8 m

50 mm wood wool slabs or other rigid sheet	5·1 m	SDS 223
and roofing felt with ceiling	6·0 m	SDS 224
10° pitch	6·9 m	SDS 225
600 mm centres	8·1 m	SDS 226

Type W

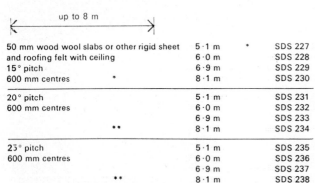

up to 8 m

50 mm wood wool slabs or other rigid sheet	5·1 m	* SDS 227
and roofing felt with ceiling	6·0 m	SDS 228
15° pitch	6·9 m	SDS 229
600 mm centres	* 8·1 m	SDS 230
20° pitch	5·1 m	SDS 231
600 mm centres	6·0 m	SDS 232
	6·9 m	SDS 233
**	8·1 m	SDS 234
25° pitch	5·1 m	SDS 235
600 mm centres	6·0 m	SDS 236
	6·9 m	SDS 237
**	8·1 m	SDS 238

*May be used for tiles or slates up to 53·6 kg/m² on slope and pitch up to 17½°
**May be used for tiles or slates up to 50·3 kg/m² on slope
Designs for farm buildings also available

Table IV TRADA *industrial roof trusses*

Type	Span	Spacing (centre to centre)	Drawing ref no
Type A			

up to 20 m

Sheet material not exceeding	5 m	3·8 m	SDS 41
15·6 kg/m² (3·2 lb/sq ft)	7·5 m	3·8 m	SDS 42
with purlins at 1370 mm	10·2 m	3·8 m	SDS 43
centres	12·6 m	3·8 m	SDS 44
22½° pitch	15·3 m	3·8 m	SDS 45
	17·4 m	4·25 m	SDS 46
	20·1 m	4·25 m	SDS 47

Table V TRADA *industrial building*

Type	Span	Drawing no	Notes
Type B			

up to 20 m

Sheet materials not exceeding	5 m	3·8 m	SDS 70
15·6 kg/m² /3·2 lb/sq ft)	7·5 m	3·8 m	SDS 71
with purlins at 1370 mm	10·2 m	3·8 m	SDS 72
centres with ceiling	12·6 m	3·8 m	SDS 73
22½° pitch	15·3 m	3·8 m	SDS 74
	17·4 m	3·8 m	SDS 75
	20·1 m	4·25 m	SDS 76

Type C

up to 15 m

Sheet materials not exceeding	5·1 m	SDS 87	Sides open
15·6 kg/m² (3·2 lb/sq ft) with	5·1 m	SDS 88	Sides clad
purlins at 1370 mm centres	5·1 m	SDS 89	Gable frame
22½° pitch	7·6 m	SDS 90	Sides open
3·35 m centres	7·6 m	SDS 91	Sides clad
	7·6 m	SDS 92	Gable frame
	9·9 m	SDS 93	Sides open
	9·9 m	SDS 94	Sides clad
	9·9 m	SDS 95	Gable frame
4·25 m centres	12·6 m	SDS 102	Sides clad
	12·6 m	SDS 103	Gable frame
	12·6 m and 15 m	SDS 104	details for SDS 103 and SDS 106
	15 m	SDS 105	Sides clad
	15 m	SDS 106	Gable frame

Table VI BRE *Maximum permissible spans (m) for rafters for fink trussed rafters (timber to* BS *4471 and of composite grade to* CP *112)*

'W' or Fink truss

Basic size (mm)	Actual size (mm)	Pitch (degrees)								
		15	17½	20	22½	25	27½	30	32½	35
					Maximum spans (m)					
38 × 75	35 × 72	6·03	6·16	6·29	6·41	6·51	6·60	6·70	6·80	6·90
100	97	7·48	7·67	7·83	7·97	8·10	8·22	8·34	8·47	8·61
125	120	8·80	9·00	9·20	9·37	9·54	9·68	9·82	9·98	10·16
44 × 75	41 × 72	6·45	6·59	6·71	6·83	6·93	7·03	7·14	7·24	7·35
100	97	8·05	8·23	8·40	8·55	8·68	8·81	8·93	9·09	9·22
125	120	9·38	9·60	9·81	9·99	10·15	10·31	10·45	10·64	10·81
50 × 75	47 × 72	6·87	7·01	7·13	7·25	7·35	7·45	7·53	7·67	7·78
100	97	8·62	8·80	8·97	9·12	9·25	9·38	9·50	9·66	9·80
125	120	10·01	10·24	10·44	10·62	10·77	10·94	11·00	11·00	11·00

Table VII BRE *Maximum permissible spans* (m) *for rafters for fan trussed rafters* (*timber to* BS 4471 *and of composite grade to* CP 112)

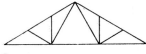

Fan truss

Basic size (mm)	Actual size (mm)	Pitch (degrees)								
		15	17½	20	22½	25	27½	30	32½	35
		Maximum spans (m)								
38× 75	35× 72	8·03	8·38	8·64	8·97	9·08	9·27	9·46	9·65	9·85
100	97	9·89	10·37	10·67	10·96	11·00	11·00	11·00	11·00	11·00
125	120	11·00	11·00	11·00	11·00	11·00				
44× 75	41× 72	8·65	9·00	9·25	9·48	9·73	9·89	10·08	10·30	10·48
100	97	10·71	11·00	11·00	11·00	11·00	11·00	11·00	11·00	11·00
125	120	11·00								
50× 75	47× 72	9·26	9·62	9·86	10·10	10·36	10·53	10·70	10·93	11·00
100	97	11·00	11·00	11·00	11·00	11·00	11·00	11·00	11·00	

Table VIII BRE *Maximum permissible spans* (m) *for ceiling ties for fink and fan trussed rafters* (*timber to* BS 4471 *and of composite grade to* CP 112)

Basic size (mm)	Actual size (mm)	Pitch (degrees)								
		15	17½	20	22½	25	27½	30	32½	35
		Maximum spans (m)								
38× 75	35× 72	5·07	5·31	5·53	5·74	5·94	6·12	6·31	6·50	6·67
100	97	7·03	7·36	7·68	7·99	8·27	8·54	8·81	9·06	9·33
125	120	8·66	9·10	9·49	9·88	10·24	10·59	10·93	11·00	11·00
150	145	10·17	10·71	11·00	11·00	11·00	11·00	11·00		
44× 75	41× 72	5·53	5·78	6·03	6·26	6·48	6·69	6·89	7·08	7·28
100	97	7·53	7·90	8·25	8·59	8·90	9·19	9·48	9·75	10·04
125	120	9·13	9·60	10·04	10·46	10·86	11·00	11·00	11·00	11·00
150	145	10·52	11·00	11·00	11·00	11·00				
50× 75	47× 72	5·92	6·20	6·46	6·72	6·94	7·17	7·39	7·60	7·81
100	97	7·93	8·33	8·71	9·06	9·38	9·70	10·02	10·32	10·62
125	120	9·42	9·94	10·40	10·86	11·00	11·00	11·00	11·00	11·00
150	145	10·59	11·00	11·00	11·00					

Table IX *Maximum permissible spans* (m) *for rafters of fink trusses* (*timbers to* BS 4978 *grades*)

Species and grade of timber	Actual size (mm)	Pitch (degrees)								
		15	17½	20	22½	25	27½	30	32½	35
M50 grade (Table 1 species)	35 × 72	5·60	5·76	5·92	6·09	6·25	6·41	6·57	6·74	6·90
	97	7·03	7·23	7·42	7·62	7·82	8·02	8·22	8·41	8·61
Canadian spruce-pine-fir (No 1)	120	8·35	8·58	8·80	9·03	9·26	9·48	9·71	9·93	10·16
	145	9·76	10·04	10·29	10·56	10·80	11·06	11·34	11·58	11·85
Home grown Sitka spruce (M75)	47 × 72	6·46	6·63	6·79	6·95	7·12	7·29	7·45	7·62	7·78
	97	8·10	8·31	8·53	8·74	8·95	9·16	9·38	9·59	9·80
	120	9·41	9·67	9·93	10·19	10·45	10·70	10·96	11·22	11·47
SS/MSS grades (Table 1 species)*	35 × 72	5·96	6·12	6·30	6·46	6·63	6·80	6·96	7·13	7·25
	97	7·50	7·71	7·92	8·12	8·33	8·54	8·74	8·94	9·00
Canadian spruce-pine-fir (No 1)	120	8·71	8·95	9·20	9·42	9·66	9·89	10·12	10·36	10·60
	145	10·25	10·54	10·80	11·07	11·35	11·63	11·90	12·00	12·00
	47 × 72	6·87	7·01	7·14	7·28	7·42	7·55	7·69	7·82	7·96
	97	8·64	8·81	8·99	9·17	9·35	9·52	9·70	9·87	10·05
	120	9·81	10·07	10·32	10·58	10·83	11·10	11·34	11·60	11·85
	145	11·10	11·41	11·73	12·00	12·00	12·00	12·00	12·00	12·00
M75 grade (Table 1 species)*	35 × 72	6·90	6·98	7·07	7·16	7·25	7·25	7·25	7·25	7·25
	97	8·44	8·55	8·71	8·85	8·98	9·00	9·00	9·00	9·00
	120	9·55	9·74	9·93	10·11	10·30	10·49	10·68	10·86	11·05
	145	10·92	11·17	11·41	11·66	1·90	12·00	12·00	12·00	12·00
	47 × 72	7·57	7·65	7·73	7·80	7·88	7·96	8·04	8·12	8·20
	97	9·20	9·35	9·49	9·64	9·78	9·92	10·07	10·21	10·35
	120	10·75	10·94	11·13	11·31	11·50	11·69	11·88	12·00	12·00

* Table 1 species are: Western hemlock (commercial); European redwood; European whitewood; Canadian spruce; and Douglas fir (imported).

Information sheet Timber 11

Section 7 **Structural material: Timber**

Solid wood portals

V. C. JOHNSON *briefly outlines the use of solid timber portals in lightweight low-cost structures*

1.01 A simple, low-cost construction for ridged portal frames has been developed by TRADA, using solid timber members joined at the eaves and ridge by nailed plywood gusset plates **1**. The portals are particularly suited to buildings where economy is of prime importance, but only for spans of 15 m and less.

1.02 Standard design sheets are available from TRADA covering 9 and 12 m span portals for agricultural buildings with lightweight roof claddings. Other publications relate to the selection of member and gusset sizes from load-span tables to cope with heavier loadings.

1.03 The following load-span table is given for preliminary cost and planning purposes. Member sizes are based on a typical lightweight loading of 1·4 kN/m² and the use of the common structural hardwood, keruing.

Table 1 Load-span table for solid, 20° to 25° pitch keruing portals with 3·6 m eaves height and 1·4 kN/m² design load

Frame spacing (m)	3·6 m		4·8 m	
Span (m)	Column	Rafter	Column	Rafter
	Member sizes (mm)			
6·0	75 × 275	75 × 250	100 × 275	100 × 250
9·0	100 × 300	100 × 250	125 × 300	125 × 250
12·0	125 × 325	125 × 275	150 × 350	150 × 300
15·0	125 × 400	125 × 350	150 × 425	150 × 350

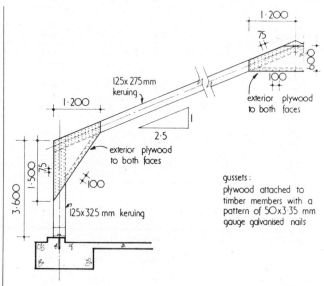

1 *Typical low cost solid timber portal with plywood gussets*

Information sheet
Timber 12

Section 7 **Structural material: timber**

Glued laminated beams

This sheet by v. c. johnson *notes basic properties and provides a table for calculation of laminated member sizes, based on the laminating grades to* cp 112: *Part 2: 1971.*

1 Materials

Species and grades

1.01 Most laminated structural timber is either in softwood, Baltic and European whitewood, Baltic redwood, Douglas fir or Western hemlock. However, hardwoods are occasionally used because of their high strength and stiffness, natural durability and appearance.

1.02 bs cp 112 (para 3.7) defines the use of three grades of horizontally laminated timber: la, lb and lc and in table 5 lists for each grade, and for the total number of laminations used, the various modification factors by which the basic stresses of the component timbers must be modified to grade stress in order to evaluate the strength of the laminated product. Table 6 gives modification factors where the laminate is composed of two different basic grades combined, eg la/lb/la or lb/lc/lb. Such a combination allows the higher grade to be placed at the top and bottom of the beam where bending stresses are greatest, with the cheaper lower grade between. The grade stresses for horizontally laminated members are based on the product of the basic stress for a particular type of timber, and the modification factor given in tables v and vi of cp 112. The modification factor depends on the grade or grades used and the number of laminates making up the structural member. Hence one grade of material may be used or two grades in combination.

Size

1.03 Laminated beams are generally produced from sawn boards of widths 100, 125, 150, 175, 200 and 225 mm; thicknesses are generally 33, 44, 50 and 63 mm (but reduced 4 to 5 mm by planing).

Glues

1.04 Adhesives for glued-laminated members relate to the intended use.

Casein
Suitable for interior structures where moisture content of timber in use does not exceed 18 per cent for prolonged periods.

Urea-formaldehyde
Suitable for interior conditions, although moisture content of the timber may rise above 18 per cent provided that the temperature of the glued line does not exceed 50°c.

Resorcinol and phenolic
Suitable for both interior and exterior structures without restriction on temperature and humidity.

Preservative treatment

1.05 Laminated timber is normally treated after manufacture as the chemicals used are often incompatible with the glue curing compounds. Timber can be satisfactorily glued after preservative treatment but the manufacturers of both glue and preservative should first be consulted. For example treatment with water-borne preservatives will permit use of resorcinol or phenolic adhesives only (see also Information sheet timber 16).

2 Structural design

Lateral restraint

2.01 Table i (reprinted from bs cp 112) lists the maximum permissible depth-to-breadth ratios for solid and laminated beams of rectangular cross section for various degrees of lateral support. The fifth type of construction is particularly common.

Table i: Maximum depth-to-breadth ratios (solid and laminated members)

Degree of lateral support	Maximum depth-to-breadth ratio
No lateral support	2
Ends held in position	3
Ends held in position and member held in line, as by purlins or tie rods	4
Ends held in position and compression edge held in line, as by direct connection of sheathing, deck or joists	5
Ends held in position and compression edge held in line, as by direct connection of sheathing, deck or joists, together with adequate bridging or blocking spaced at intervals not exceeding 6 times the depth	6
Ends held in position and both edges firmly held in line	7

Deflection and built-in camber

2.02 bs cp 112 gives 0·003 × span as a common limit to deflection. Members may be precambered to absorb the deflection under dead load, in which case the deflection under live load should not exceed 0·003 × span.

Load-span table

2.03 Table ii provides a means of selecting simply supported laminated beam sizes for various spans and uniformly distributed loads. The figures were calculated for 33·3 mm net thickness laminations of lb grade Baltic redwood or Baltic/European whitewood. Dry basic stresses are for timber with a moisture content of not more than 18 per cent—thus the beam sizes derived from the table will only be suitable for interior conditions. The permissible loads are in N/m run on a hypothetical breadth of beam of 1 cm

(10 mm). Hence the permissible load on an actual beam can be found by multiplying the load given in the table by the beam breadth in cm. The 'load conditions' L and M indicate respectively the long and medium term duration loads (defined in BS CP 112 para 3.12.1.1 and Information sheet TIMBER 5) at maximum permissible stress, and D the load which the beam will support with the deflection limited to $0 \cdot 003 \times$ span.

Example 1
2.04 A building of 12 m span has laminated beams spaced at 3 m centres supporting a roof made up of three layer felt on 75 mm timber decking. The beam is to be precambered for the deflection owing to dead load. Select a size of beam to limit the imposed load deflection to $0 \cdot 003$ of the span and provide sufficient strength for the total design load.

Determine loading/m²
Assume (or possibly choose for other design reasons) a beam breadth of 135 mm,
Dead loading (long-term)*

Felt	$150 \cdot 0$
Decking	$340 \cdot 0$
Self-weight beam* assumed as	$100 \cdot 0$
	$590 \cdot 0$ N/m²

Imposed loading including snow
medium-term load (external) † 750·0

$1340 \cdot 0$ N/m²

To use the table determine the load/m run/cm
Imposed load only

Load per beam	750 N/m² $\times$ 3 m
Breadth of beam	135 mm

Therefore load on 1 cm breadth of joist $750 \times 3 \times \dfrac{10}{135}$

$= 167$ N/m run

Similarly total design load $= 1340 \times 3 \times \dfrac{10}{135}$

$= 298$ N/m run

Enter the table for 12 m span and select a depth which will satisfy the imposed load (167 N/m run) for deflection limit 0.003 (lower figure D ie 183.8). Check the total design load (298 N/m run) does not exceed the middle figure (medium term load) 430·8. A beam of 135 $\times$ 600 mm will be required. The depth-to-breadth ratio $\dfrac{600}{135} < 5$, therefore special calculations are not required for proof of sufficient lateral restraint. If the ratio was greater, the beam would be recalculated for a greater breadth or some lateral stability introduced in the design eg by insertion of blocking pieces. As a further check the actual self-weight of the 135 $\times$ 600 mm beam may be found using BS 648* and the calculation reworked.

Example 2
2.05 Laminated beams spanning 9 m and spaced at 2·4 m centres support a timber joisted floor. The construction dead load with an allowance for self-weight of beam is 500 N/m² and the imposed floor load is 2000 N/m². Select a size of beam to limit the deflection to $0 \cdot 003$ of the span and provide sufficient strength for the total design load. Consider a beam breadth of 150 mm.
Total design load in N/m run on beam 10 mm (1 cm) breadth

$= \text{load/m}^2 \times \text{centering (m)} \times \dfrac{\text{width considered (mm)}}{\text{total beam width (mm)}}$

$= 2500 \times 2 \cdot 4 \times \dfrac{10}{150} = 400$ N/m run

Hence a beam of 150 $\times$ 600 will be required. This will carry a load of 435·8 N/m/cm limiting deflection (lower figure). For strength purposes alone the beam will support 612·4 N/m/cm (top figure for long-term load).

*BS 648 'Schedule of Weights of Building Materials'.
†CP 3: Chapter v: Part 1.

Table II *Permissible loads in N/m run 1 cm breadth 'Gluelam' beam**

Key to table
L long term load
M medium term load
D deflection limit

Span (m)	Loading* Condition	200	300	400	500	600	700	800	900	1000	1100	1200	1300
		Permissible loads (N/m run)											
3	L	626·6	1467·4										
	M	783·1	1834·2										
	D	420·6	1439·0										
4	L	352·3	825·2	1423·0									
	M	440·2	1031·5	1779·0									
	D	177·2	607·3	1452·1									
5	L	225·2	528·1	910·9	1395·4								
	M	281·8	660·3	1138·3	1744·8								
	D	90·8	310·9	743·1	1461·5								
6	L	156·9	336·8	632·7	969·0								
	M	195·4	458·4	790·3	1211·7								
	D	52·3	180·1	430·0	845·5								
7	L	114·8	269·5	464·9	711·9								
	M	143·8	337·0	581·1	890·0								
	D	33·4	113·3	270·9	532·5								
8	L		206·3	355·9	545·5								
	M		257·9	444·6	681·4								
	D		75·4	181·6	356·7								
9	L		162·7	281·1	430·7	612·4							
	M		204·1	351·6	538·3	765·6							
	D		53·0	127·1	250·6	435·8							
10	L		132·2	227·4	348·7	496·1	670·5						
	M		164·9	284·7	435·8	620·4	838·3						
	D		38·5	93·0	182·3	317·4	507·0						
11	L		109·0	188·1	288·4	409·7	554·2						
	M		136·6	235·3	360·3	512·1	693·0						
	D		29·0	69·7	137·3	239·0	381·4						
12	L		91·5	158·3	242·6	344·3	465·6	605·1	763·5	942·9	1136·8	1349·7	1581·4
	M		114·8	197·6	303·0	430·8	581·9	765·2	954·5	1178·2	1420·9	1686·7	1976·6
	D		22·5	53·7	106·0	183·8	293·5	439·5	627·6	864·4	1149·9	1493·5	1898·8
13	L			134·4	206·3	293·5	396·6	515·0	650·9	803·4	968·3	1149·9	1347·5
	M			168·5	257·9	366·8	496·1	644·3	813·6	1003·9	1210·9	1437·6	1684·6
	D			42·1	82·8	144·5	231·0	345·8	494·0	679·9	904·4	1174·6	1493·5
14	L			116·2	178·0	252·8	342·1	444·6	560·8	693·0	835·4	991·6	1161·5
	M			145·3	222·3	316·0	427·9	555·7	701·0	865·9	1043·9	1239·3	1452·1
	D			34·1	66·8	115·5	184·5	276·8	395·1	544·1	724·2	940·7	1195·7
15	L				154·7	220·1	297·8	387·2	488·9	603·6	727·1	863·7	1011·9
	M				193·9	275·3	372·6	483·8	611·0	754·0	909·5	1079·4	1264·7
	D				54·5	94·4	150·4	225·2	321·8	442·4	589·1	764·2	971·9
16	L					194·0	262·2	340·0	429·3	530·3	639·2	759·1	889·1
	M					241·9	327·6	425·0	536·8	663·2	799·1	948·7	1112·1
	D					77·7	123·5	185·2	265·1	364·7	485·2	629·8	801·2
17	L					171·4	231·7	301·5	380·6	470·0	566·6	672·7	788·2
	M					214·3	289·8	377·0	475·8	587·0	708·2	840·5	985·0
	D					64·6	103·1	154·7	220·8	303·6	404·6	525·2	667·6
18	L						207·0	268·8	339·2	419·1	504·8	600·0	702·4
	M						258·6	336·3	424·2	523·7	631·2	749·7	878·2
	D						87·2	130·0	186·0	255·7	340·7	442·4	562·2
19	L						186·0	241·2	304·4	376·3	453·3	538·3	630·5
	M						232·4	301·5	380·6	470·0	566·6	672·7	788·2
	D						71·4	110·4	158·3	218·0	289·8	376·3	478·0
20	L							217·9	274·6	339·2	409·0	486·0	569·5
	M							272·4	343·6	424·2	511·4	607·3	711·9
	D							95·2	135·9	186·7	248·4	322·5	410·4

*Beams assumed to have 33·3 mm net thickness laminations of LB grade Baltic redwood or European whitewood with moisture content less than 18 per cent.

Information sheet
Timber 13

Section 7 **Structural material: Timber**

The selection of glue laminated portals

In this sheet V. C. JOHNSON *covers methods for determining minimum cross-sectional sizes of three hinged portal frames. The frames are available from proprietary companies with a normal delivery period of 4-5 weeks.*

1 The three hinged portal

Member sizes

1.01 This is one of the most commonly used types of portal frame, where the column and rafter taper from the curved haunch (or knee) to the base and apex respectively.

Load span tables

1.02 Although the amount of taper is partly dependent on visual appearance, tables II to VI are provided as a general guide to shape design for preliminary cost and planning purposes. They give for varying loads, spans, and degrees of pitch the minimum section depth at rafter mid span (on plan) **1** and the minimum cross-sectional size at the tangent point to the curved haunch (ie where the inside curve of the haunch meets the straight underside of the rafter or column). The data is based on the use of LB grade Baltic redwood or Baltic and European whitewood.

Table I Recommended minimum radius of haunch curvature in metres (see also 1)

Thickness of laminate (mm)		Haunch radius (m) varies with roof pitch				
Nominal	Actual	15°	22½°	30°	45°	60°
19	15	2·9	2·8	2·6	2·3	2·0
25	20	3·8	3·6	3·4	3·1	2·7
32	27	6·1	5·8	5·4	4·8	4·1
38	33·3	9·0	8·5	7·9	6·8	5·7

2 Example use of tables II–V

2.01 A 20 m span building with 15° pitched, double-ridged roof and 5 m eaves height is to be constructed with glued laminated timber portal frames at 3·6 m centres **2**. The timber purlin roof is lined externally with troughed aluminium sheet and internally with 16 mm tongued and grooved boards. Lightweight insulating material fills the resulting cavity. For preliminary cost and planning select a member size from the tables.

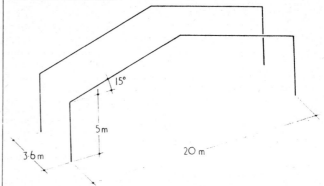

2 *Glue-laminated portal frame construction (see 2.01)*

First calculate dead load on slope of roof*

Aluminium cladding	30
Timber purlins	50
Insulation	33
22 mm tongued and grooved boarding	110
Allow for self-weight of frame	120
	343 N/m²

343 N/m² load on 15° slope will give a load on plan of

$$\frac{343}{\cos 15°} = \frac{343}{0·9659} = 355$$

$$Imposed\ load\ (snow) = \frac{750}{1105}\ N/m²$$

Load/m run on plan of rafter

$$= 1105 \times 3·6 = 3978\ N/m\ run$$

Thus from table II the cross-sectional size of the rafter (or column) at the tangent point to the haunch is 630 × 160 mm and the section depth of rafter at mid span should be not less than 390 mm. If the laminate thickness is 15 mm, the minimum recommended radii of haunch curvature is 2·9 m.

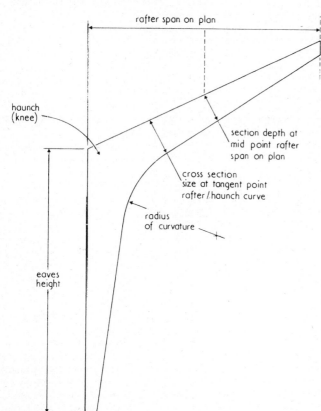

1 *Three hinged laminated portal*

Haunch curvature

1.03 The radius of curvature of the knee-haunch varies with the thickness of the laminate and the angle between the column and rafter. Table I recommends minimum radii for a range of laminate thicknesses and roof pitches.

*Loadings from BS 648 'Schedule of Weights of Building Materials'

Table II† *15 degrees pitch*

Load N/m		3600	4400	5100	5900	7400	
Span (m)	Eaves height (m)	Cross-sectional size at haunch (mm)					Minimum rafter depth midway between eaves and apex
10	2·5	300×105	330×105	360×105	390×105	435×105	
	3·5	345×105	390×105	420×105	450×105	495×105	240
	5·0	405×105	450×105	480×105	510×105	570×105	
15	2·5	360×115	405×115	435×115	465×115	525×115	
	3·5	465×115	510×115	540×115	580×115	660×115	330
	5·0	540×115	585×115	630×115	690×115	765×115	
20	2·5	375×160	405×160	435×160	480×160	540×160	
	3·5	480×160	525×160	570×160	615×160	690×160	390
	5·0	570×160	630×160	660×160	720×160	810×160	

Table III† *22½ degrees pitch*

Load N/m		3600	4400	5100	5900	7400	
Span (m)	Eaves height (m)	Cross-sectional size at haunch (mm)					Minimum rafter depth midway between eaves and apex
10	2·5	300×105	330×105	360×105	390×105	435×105	
	3·5	360×105	390×105	420×105	450×105	510×105	240
	5·0	405×105	435×105	480×105	510×105	570×105	
15	2·5	375×115	420×115	450×115	480×115	540×115	
	3·5	450×115	495×115	540×115	585×115	645×115	345
	5·0	510×115	570×115	615×115	660×115	750×115	
20	2·5	405×160	420×160	450×160	495×160	555×160	
	3·5	480×160	525×160	570×160	615×160	690×160	405
	5·0	570×160	630×160	675×160	720×160	810×160	

Table IV† *30 degrees pitch*

Load N/m		3600	4400	5100	5900	7400	
Span (m)	Eaves height (m)	Cross-sectional size at haunch (mm)					Minimum rafter depth midway between eaves and apex
10	2·5	315×105	345×105	375×105	405×105	480×105	
	3·5	360×105	405×105	435×105	465×105	525×105	240
	5·0	405×105	450×105	480×105	510×105	570×105	
15	2·5	390×115	435×115	465×115	510×115	570×115	
	3·5	450×115	510×115	540×115	585×115	660×115	345
	5·0	510×115	570×115	615×115	660×115	735×115	
20	2·5	405×160	435×160	480×160	510×160	570×160	
	3·5	465×160	510×160	555×160	600×160	660×160	405
	5·0	525×160	585×160	630×160	675×160	750×160	

Table V† *45 degrees pitch*

Load N/m		3600	4400	5100	5900	7400	
Span (m)	Eaves height (m)	Cross-sectional size at haunch (mm)					Minimum rafter depth midway between eaves and apex
10	2·5	285×105	315×105	330×105	360×105	435×105	
	3·5	330×105	360×105	375×105	420×105	465×105	225
	5·0	375×105	405×105	435×105	465×105	525×105	
15	2·5	345×115	375×115	405×115	450×115	495×115	
	3·5	405×115	435×115	480×115	510×115	570×115	360
	5·0	465×115	510×115	540×115	585×115	660×115	
20	2·5	435×160	435×160	435×160	450×160	495×160	
	3·5	435×160	450×160	480×160	525×160	585×160	435
	5·0	465×160	525×160	555×160	600×160	675×160	

Table VI† *60 degrees pitch*

Load N/m		3600	4400	5100	5900	7400	
Span (m)	Eaves height (m)	Cross-sectional size at haunch (mm)					Minimum rafter depth midway between eaves and apex
10	2·5	285×105	285×105	300×105	330×105	360×105	
	3·5	285×105	315×105	345×105	360×105	405×105	285
	5·0	315×105	360×105	375×105	405×105	465×105	
15	2·5	420×115	420×115	420×115	420×115	450×115	
	3·5	420×115	420×115	420×115	450×115	510×115	420
	5·0	420×115	435×115	480×115	510×115	570×115	
20	2·4	495×160	495×160	495×160	495×160	495×160	
	3·5	495×160	495×160	495×160	495×160	510×160	495
	5·0	495×160	495×160	495×160	495×160	585×160	

†Tables II to VI for three-hinged laminated portal giving minimum cross-sectional size at haunch (knee) and minimum rafter depth midway between eaves and apex (see also **1**).

Information sheet Timber 14

Plywood web beams and selection tables

In this sheet C. J. METTEM *outlines the manufacture and uses of ply-web beams and tabulates their likely sectional dimensions for various spans and loads. The beams are available from proprietary suppliers but can easily be made by builder or contractor.*

1 Geometrical advantages

1.01 In ply-web beams the solid top or bottom members or flanges are spaced well apart by use of highly shear-resistant plywood connecting webs. The resulting I or box arrangement provides great geometrical efficiency in bending when compared with a rectangular cross section. In a section under load the material near the centre is necessary to transmit shear forces but contributes relatively little to bending strength and stiffness. The ply-web arrangement eliminates most of the material close to the centroid by separating the shear-carrying and compression/tension functions. Consequently the strength and stiffness-to-weight ratios are vastly improved.

Construction

1.02 Typical box and I-section construction is shown in **1**. The upper and lower flange members, either of sawn timber or more rarely of plywood or laminated material, are joined by plywood webs fastened with glue or other mechanical means. Where glue and nails or staples are used together, the latter only provide gluing pressure as the difference in stiffness of the two fixing methods prevents summation of their attachment values. Short vertical stiffeners are inserted at intervals of twice the clear distance between flanges and/or at positions of heavy point loads. Both flanges and webs may be jointed, the former with scarf or finger joints and the latter with scarf or butt joints backed by plywood straps. Multiple flange and web arrangements are possible **2**.

2 *Multiple flange and web arrangements*

Uses

Roof construction

1.03 The beams' most common use is in roof construction spanning to 15 m or more, ie outside the normal range of solid timber. The beams can take a varying profile, perhaps to cope with differing shear forces and bending moments along their length, or simply to provide a fall for roof drainage.

Floor construction

1.04 Here the beams can either occur at eg 600 mm centres, acting as long span joists, or at some wider centering (probably dependent on internal planning arrangements) with secondary solid joists spanning between. The latter can be carried on ledgers or hangers to reduce overall floor depth.

Portals

1.05 Ply-web construction has also been successfully used in large-span portal frames spanning up to 36 m. Recent TRADA designed portals spanning up to 21 m use a proprietary form of corrugated web between the flanges to increase lateral stability (see Technical study TIMBER 1).

2 Load tables for glued plywood box beams

2.01 Tables I and II give permissible loads for box beams with various top and bottom flange sizes and spans. The selected beam depth of 395 mm on both tables is the most commonly used, fitting conveniently into standard plywood sheet sizes with an allowance for cutting and trimming after assembly. Two figures are given against each flange/span combination, the upper being the permissible load as governed by the stresses owing to bending, panel shear and rolling shear, and the lower giving the calculated deflection under that load in millimetres. The latter may be compared with the 0·003L, 0·004L and 0·005L values at the right of the table, according to the deflection criterion required.

2.02 The tables are not applicable to all-nailed construction, but assume the use of structural adhesives (see Technical study TIMBER 5). Also the permissible loads are for long-term loading—for medium and short term loading the figures should be multiplied by 1·25 or 1·5 respectively.

To illustrate use of tables

2.03 Beams at 1·8 m centres are to span 6·0 m and take a dead and imposed load of 125 kg/m².

Load per metre = 125 × 1·8   = 225 kg

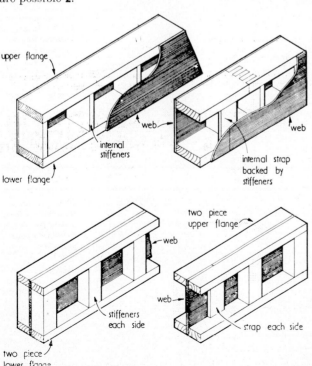

1 *Construction of ply box and I-section beams*

This is a 'medium term' load (as partly comprised of intermittent imposed loads). Converting to the long term equivalent:

$$\frac{225}{1\cdot25} = 180 \text{ kg/m run}$$

Hence, from tables I and II suitable 395 mm depth beams include:

120 × 47 S2 softwood flanges with 9·0 mm Finnish birch faced ply webs *or* 72 × 47 keruing flanges with 12·5 mm Canadian Douglas fir ply webs.

2.04 Where anticipated loadings fall outside the range of the tables, ie beyond the capacity of the 395 mm beam sizes discussed, doubling up the beams or reducing their centering may be preferable to increasing the section depth.

Table I *Permissible loads (kg/m run) and actual deflections (mm) for glued plywood box beams*

Flanges	Group S2 softwoods, SS grade, air-dry timber
Webs	12·5 mm Canadian Douglas fir plywood (Good one side grade) or 9·0 mm Finnish birch faced plywood (face grain vertical)
Beam depth	395 mm

Flange size mm	Canadian Douglas fir plywood			Finnish birch faced plywood			Deflection limits (mm)		
	72×47	97×47	120×47	72×47	97×47	120×47	0·003L	0·004L	0·005L
Span m	**Long term load (kg) and actual deflection (mm)**								
3·0	545	501	475	514	670	756			
	3·2	2·5	2·1	5·0	5·8	6·0	9·0	12·0	15·0
3·6	375	415	394	354	462	562			
	4·1	3·9	3·2	6·0	6·8	7·5	10·8	14·4	18·0
4·2	273	352	336	258	337	410			
	5·2	5·6	4·7	7·2	8·0	8·7	12·6	16·8	21·0
4·8	207	267	292	195	255	311			
	6·5	7·0	6·7	8·5	9·4	10·1	14·4	19·2	24·0
5·4	161	209	252	152	200	243			
	8·0	8·5	8·9	10·1	11·0	11·7	16·2	21·6	27·0
6·0	129	167	202	122	160	195			
	9·7	10·2	10·6	11·8	12·7	13·5	18·0	24·0	30·0
6·6	105	136	165	99	130	159			
	11·5	12·1	12·5	13·7	14·7	15·5	19·8	26·4	33·0
7·2	86	112	136	82	108	132			
	13·5	14·2	14·7	15·8	16·8	17·7	21·6	28·8	36·0
7·8	72	94	114	68	90	110			
	15·6	16·4	16·9	18·0	19·2	20·1	23·4	31·2	39·0
8·4	61	79	97	57	76	94			
	18·0	18·8	19·4	20·5	21·7	22·6	25·2	33·6	42·0
9·0	52	68	83	49	65	80			
	20·5	21·4	22·1	23·1	24·4	25·4	27·0	36·0	45·0
9·6	44	58	71	42	56	69			
	23·2	24·2	24·9	25·9	27·2	28·3	28·8	38·4	48·0

Table II *Permissible loads (kg/m run) and actual deflections (mm) for glued plywood box beams*

Flanges	Keruing 65 grade, air-dry timber
Webs	12·5 mm Canadian Douglas fir plywood (Good one side grade) or 9·0 mm Finnish birch faced plywood (face grain vertical)
Beam depth	395 mm

Flange size mm	Canadian Douglas fir plywood			Finnish birch faced plywood			Deflection limits (mm)		
	72×47	97×47	120×47	72×47	97×47	120×47	0·003L	0·004L	0·005L
Span m	**Long term load (kg) and actual deflection (mm)**								
3·0	484	454	437	766	731	712			
	2·4	1·9	1·7	6·8	5·8	5·2	9·0	12·0	15·0
3·6	402	376	361	623	607	591			
	3·6	2·9	2·4	9·2	7·9	7·0	10·8	14·4	18·0
4·2	343	321	308	455	519	504			
	5·2	4·1	3·5	10·6	10·4	9·2	12·6	16·8	21·0
4·8	298	279	267	346	453	440			
	7·3	5·7	4·8	12·3	13·6	11·9	14·4	19·2	24·0
5·4	264	246	236	271	358	389			
	10·0	7·7	6·4	14·2	15·6	15·1	16·2	21·6	27·0
6·0	214	221	211	218	288	349			
	12·0	10·1	8·4	16·3	17·8	18·9	18·0	24·0	30·0
6·6	175	199	191	178	236	289			
	14·2	13·0	10·8	18·6	20·2	21·5	19·8	26·4	33·0
7·2	145	182	173	148	196	241			
	16·5	16·5	13·6	21·2	22·8	24·1	21·6	28·8	36·0
7·8	122	161	159	125	165	203			
	19·1	19·9	16·9	23·9	25·6	27·0	23·4	31·2	39·0
8·4	104	137	147	106	141	173			
	21·8	22·8	20·8	26·9	28·6	30·1	25·2	33·6	42·0
9·0	89	118	136	91	121	149			
	24·8	25·8	25·2	30·1	31·9	33·4	27·0	36·0	45·0
9·6	77	102	125	79	105	129			
	27·9	29·0	29·8	33·6	35·4	37·0	28·8	38·4	48·0

Information sheet
Timber 15

Section 7 **Structural material: Timber**

Stressed skin plywood floor and roof panels

C. J. METTEM *outlines the particular applications of these panels and comments on their design and manufacture.*

1 Introduction

Description

1.01 Plywood stressed skin roof and floor panels comprise plywood sheets glued or mechanically fastened to the top, or top and bottom, of longitudinal timber stringers **1**. The whole assembly acts integrally, utilising the load carrying capacity of both skin and stringers to resist bending, provided that the joint between the two is strong enough to transmit the resulting longitudinal shear forces.

Uses

1.02 The panels can be used as roof, floor or wall units and so form the basic component in many building systems. A common width is 1·2 m, while practical spans range from 1·2 m to 12 m with particular economies between 3·5 m and 7 m. Other advantages are: strength-to-weight ratio is high; construction depth, particularly of the double skin panel, is less than with traditional joist and boarding, suggesting use of the panels where the depth of floor zones is limited; the ply surfaces offer a readymade finish; and the cavity between skins can be useful in containing services or insulating/soundproofing material.

Design points in construction

1.03 *Grain* The face grain of the plywood should preferably lie parallel to the stringers

1 *Plywood stressed skin panel. Longitudinal stringers run in direction of span*

Structural continuity Skins and stringers must be continuous for the full length of the panel with any joints in the plywood butted or scarfed and backed with splice plates

Panel width This should conform to commercially available widths of plywood sheets, usually 1200 mm

Stability End headings and intermediate bridging are required to provide lateral support to the stringers and to assist in transverse distribution of loads

Top skin This must be designed to span between stringers (ie to carry imposed load) in addition to performing its function as a stressed skin

Gluing Types of adhesive and gluing method are critical in ensuring that the joint has a greater horizontal shear strength than the timber itself. Gluing pressure may be achieved by nailing or better still by using a glue press. Only processed timber should be used with moisture content of less than 16 per cent. Suitable adhesives are specified in Information sheet TIMBER 12 and in all cases their curing times and temperatures should be obtained from the manufacturers

Splices In practice the overall panel length is influenced by economic considerations in cutting from standard sheets of plywood. For longer panels the plywood must be spliced and it is advisable to make the splices outside the middle half of the span length. 1:12 scarf joints may be used (Information sheet TIMBER 3) but butt jointing with plywood splice pieces glued to the inner surface of the panel is more common. The splice piece must have the same thickness as the material being spliced and have its face grain parallel to the length of the panel. Splices of this type will have a strength equal to about 90 per cent of that of the plywood. Large-size, factory jointed boards are available from plywood manufacturers and stringers can either be spliced by factory-made finger joints or scarf to 1:12 slope.

Design method

1.04 For the analysis and design of stressed skin panels see TRADA research reports 1/79 and 2/79.

2 Load span tables

2.01 Tables I and II give spans for 1·2 m wide stressed skin panels with and without top skin used in domestic floors and roofs. The spans are measured from the centre of the bearings. The design load for floors is taken as 1·91 kN/m² (1·44 kN/m² imposed load plus 0·47 kN/m² dead load) and for roofs as 1·44 kN/m² (0·72 kN/m² imposed load plus 0·72 kN/m² dead load). Stringer centerings and dimensions vary as shown but stringers of other widths may be substituted, provided that the total width of stringer per panel is maintained.

Table I Permissible spans (m) of stressed skin floor panels

Thickness of plywood top skin (mm)			12·5	12·5	16·0	16·0	16·0	16·0	19·0	19·0	19·0	19·0
Thickness of plywood bottom skin (mm)			none	6·5	none	none	6·5	6·5	none	none	6·5	6·5
Stringer dimensions (mm)	Stringer centres	(mm)	400	400	400	600	400	600	400	600	400	600
38 × 75			1·950	3·125	2·050	1·425	3·300	3·225	2·175	1·500	3·375	3·300
38 × 100			2·650	3·750	2·750	2·250	3·900	3·750	2·850	2·325	3·975	3·825
38 × 125			3·300	4·375	3·375	2·775	4·575	4·425	3·450	2·850	4·650	4·500
38 × 150			3·825	4·975	3·975	3·300	5·250	5·100	4·050	3·375	5·325	5·175
38 × 175			4·425	5·625	4·575	3·825	5·850	5·700	4·650	3·900	5·950	5·775
38 × 200			4·975	6·225	5·175	4·350	6·450	6·300	5·250	4·425	6·550	6·375
38 × 225			5·550	6·825	5·850	4·875	7·050	6·900	5·850	4·950	7·200	6·975

Design load 1·91 kN/m² (dead and imposed load)
Span Table gives effective span from centre to centre of bearings
Plywood skin Douglas fir unsanded sheathing grade
Stringers S2-50 grade processed softwood, stresses and geometrical properties in accordance with CP 112: Part 2: 1971
Assembly Glue with synthetic resin adhesive of resorcinol/phenol formaldehyde type

Table II Permissible spans (m) of stressed skin roof structures

Thickness of plywood top skin (mm)			12·5	12·5	16·0	16·0	16·0	16·0	19·0	19·0	19·0	19·0
Thickness of plywood bottom skin (mm)			none	6·5	none	none	6·5	6·5	none	none	6·5	6·5
Stringer dimensions (mm)	Stringer centres	(mm)	400	400	400	600	400	600	400	600	400	600
38 × 75			2·400	3·450	2·475	2·125	3·600	3·525	2·550	2·225	3·675	3·600
38 × 100			3·000	4·075	3·075	2·775	4·275	4·200	3·150	2·850	4·350	4·275
38 × 125			3·600	4·800	3·750	3·375	5·025	4·950	3·825	3·450	5·250	5·025
38 × 150			4·250	5·475	4·425	3·975	5·775	5·600	4·500	4·050	5·850	5·700
38 × 175			4·875	6·150	5·050	4·575	6·450	6·275	5·125	4·650	6·525	6·375
38 × 200			5·475	6·825	5·700	5·175	7·125	6·900	5·775	5·275	7·200	7·050
38 × 225			6·075	7·500	6·300	5·700	7·800	7·500	6·425	5·800	7·875	7·650

Design load 1·44 kN/m² (dead and imposed load)
Span Table gives effective span from centre to centre of bearings
Plywood skin Douglas fir unsanded sheathing grade
Stringers S2-50 grade processed softwood, stresses and geometrical properties in accordance with CP 112: Part 2: 1971
Assembly Glue with synthetic resin adhesive of resorcinol/phenol formaldehyde type

References

1 SMITH, I. Analysis of plywood stressed skin panels with rigid or semi-rigid connections; TRADA Research report 1/79; The Association 1979

2 SMITH, I. Design of simply supported plywood stressed skin panels with uniformly distributed transverse load; TRADA Research report 2/79; The Association 1979

Information sheet
Timber 16

Timber preservation and fire retardant treatments

This sheet by EZRA LEVIN *describes the various types of timber preservative and their application methods and outlines which to use when. Brief notes on fire retardant treatments follow.*

1 Hazards of decay and insect attack

1.01 Non-durable heartwoods and sapwood generally are subject to decay and insect attack if exposed to conditions raising their moisture content above 20 per cent for prolonged periods. Timbers totally immersed in water, in dock and harbour works, for example, are immune from decay but are at risk from marine borers such as ship worm (see TRADA publication TBL 21).

1.02 In building work above ground the potential hazard can be reduced by good design detail isolating timber members from moisture sources such as rain, ground water and internal vapours, and ensuring adequate ventilation to permit drying out. Regular maintenance is important. Neglect of guttering, faulty roofing or water services, or obstructed ventilators all increase the chances of decay.

1.03 Precautions keeping building timbers at or near the moisture content recommended for a particular end use (eg joists or roof trusses) are important throughout manufacture, transport, site storage and erection. Storage should be under cover, clear of ground and well ventilated. Moisture meter readings taken on building sites show that timber exposed to short periods of rain during erection is wetted on the surface only and quickly dries out once covered over.

2 Where preservative treatment is required

Essential or highly recommended
2.01 Generally for timbers of low durability or durable species with a significant amount of sapwood, exposed to conditions of high decay hazard or insect attack which cannot be prevented by design; or in positions inaccessible to maintenance, especially where a risk of structural failure may be entailed (eg structural members in contact with the soil or ground concrete without damp proof course, or with porous external cladding, or in unventilated cavities subject to ingress of humidity; members in buildings with high humidity processes such as breweries or dye-houses). Statutory requirement for roof timbers in designated areas subject to longhorn beetle attack and for non-durable softwood external cladding; external joinery in general.

To be considered
2.02 Where there is no abnormal risk of attack but maintenance may be difficult or uncertain, preservative treatment may be considered as a measure of insurance against possible future repairs (eg generally in suspended ground floors, roofs and wall framing).

Not required
2.03 In situations where dry conditions are likely to prevail and the risk of attack is negligible (eg generally in internal partitions, inter-storey floors, internal joinery, etc).

2.04 Table II summarises the requirements of BS CP 98 *Preservative treatments for constructional timber*, in regard to materials and methods of preservation related to exposure hazards. This code has been recently revised and re-issued as BS 5268: Part 5[1] and extends the amount of detailed guidance on requirements of treatment for a wide range of building components and other structural uses.

3 The treatment of solid timbers

Effectiveness
3.01 This is influenced by factors such as the preservative product's toxicity; its depth of penetration and the amount retained in the wood; and in the case of unpainted external timbers, its resistance to leaching ie extraction by water action. Penetration and retention depend on the nature of the product, its vehicle or base, the moisture content and natural permeability of the timbers being treated, (see table I) and on the method of application.

Methods of application
3.02 *Diffusion* Freshly cut timbers are soaked in eg boracic acid solution and close stacked for up to six weeks. Complete penetration is possible with fairly permeable timbers, avoiding the need for later treatment of cross cuts, but as the preservatives remain water soluble, the timbers are unsuited to conditions prone to severe leaching (eg ground contact).

3.03 *Pressure impregnation* The timbers are placed in sealed containers and preservative introduced under pressure. Complete penetration is possible with permeable types. With those less so, eg Douglas fir, penetration into the heartwood is less pronounced, but can be improved by puncturing the surface to a depth of 12 to 20 mm at evenly spaced intervals before treatment.

3.04 *Hot and cold tank steeping* Gives good penetration if treatment is sufficiently prolonged. Cold steeping for up to two weeks will give penetration up to 25 mm in permeable species but as little as 2 mm with some resistant timbers. Dipping for 5 to 15 minutes or less will allow shallow penetration in permeable species, giving adequate protection for many uses, but is not recommended where more efficient methods are available.

3.05 *Spraying and brushing* This is only suited to touching up exposed end grain when preserved timber is cut on site. In general, it is important to specify that preservation is carried out after all machining, notching and drilling has

1, 2 *Boron treatment of redwood in Finland*

3 *Hemlock being dipped into Boron diffusion tank in Canada.*

been performed in the shop.

Cost

3.06 Cost varies with the method of application, the preservative, the retention specified, distance to plant and other factors but may be estimated at approximately 6 to 12 per cent of the timber cost using constructional softwood. If cost in use and the likely saving on maintenance are taken into account efficient preservative treatment will be found to be economically worthwhile in most cases where treatment is desirable but not essential (see para 2.02).

4 Treatment of sheet timber materials

Chipboards and fibre building boards

4.01 These are not generally recommended for external use or for internal uses subject to high and persistent humidity. Standard boards are not suitable for effective treatment after manufacture, but fungicides can be included in the manufacturing process itself.

Plywood

4.02 Plywood can be treated by the same preservatives and application methods as those described above for solid timber. BS 3842: 1965[2] *Specification for treatment of plywood with preservatives,* stipulates that only plywoods bonded with WBP type adhesives to BS 1455[3] are suitable for effective preservative treatment and, although listing all application methods including surface coatings by brush or spray, emphasises that only the impregnation methods, eg pressure or diffusion treatment of the veneers will guarantee permanent protection where there is a high hazard of attack by fungi, insects or marine borers. Veneers may be treated before assembly or the plywood itself after manufacture.

5 Fire retardant treatments

The Building Regulations

5.01 Tests carried out by the Fire Research Station (recorded in FRS Note 553)[4] have shown that timbers of a density not less than 400 kg/m² (covering the whole range of structural species with the exception of the lighter Western red cedar) fall into class 3 of flamespread in accordance with BS 476[6]. Part 1.[11] There are however many internal uses for which the Building Regulations require class 1 spread of flame and some internal and external applications requiring class 0 (see Technical study TIMBER 4).

Methods for improving classification

5.02 Processes improving the flame spread classification of timber and wood-based board materials to class 1, and in some cases to class 0 can be broadly divided into three categories:
1 Impregnation with fire retardant solutions of various salts
2 Surface coatings with paints, varnishes, pastes or plasters of special kinds
3 Bonded facings of non-combustible materials such as asbestos felt or thin metal sheets.
Impregnation
5.03 The impregnation method is commonly used in the treatment of solid timber and WBP bonded plywood. With chipboards, fire-retardant salt admixture can be added during manufacture. Chemicals commonly used are ammonium compounds and boric acid, but a number of proprietary treatments add various mixtures of salts for combined resistance to flamespread and to decay.
5.04 Impregnation under pressure is often required for adequate salt retention and here the permeability of the wood species is an important consideration (see table II). Fire retardant salts generally tend to leach out and are unsuitable for external exposure or for internal conditions of high humidity.
5.05 Fire retardant salts greatly reduce the calorific value of timber and thus its contribution to fire, but can reduce strength characteristics and so are best suited for impregnating lightly stressed members such as linings, boardings or sheet materials.
Surface treatments
5.06 Fire retardant silicate paints or varnishes usually consist of soluble glass and pulverised fillers such as chalk, asbestos, ground quartz or sand. They are intumescent, ie swell under the effect of heat, and produce a protective insulating coat. Water resistance is generally poor, preventing their use externally.

Table I *Permeability of timber species to preservative treatment.*

Species	Durability (of heartwood)	Permeability to preservative treatment Heartwood	Sapwood
Softwoods			
Douglas fir	medium	ER	MR
Larch	medium	R	MR
European redwood/Scots pine	low	MR	P
European whitewood/spruce	low	R	MR
Canadian spruce	low	R	MR
Sitka spruce	low	R	MR
Parana pine	low	MR	P
Pitch pine	high	R	P
Western hemlock	low	R	MR
Western red cedar	high	R	MR
Hardwoods			
Abura	low	MR	P
Afara/limba	low	MR	P
African mahogany	medium	ER	MR
African walnut	low	ER	MR
Afrormosia	high	ER	MR
Afzelia	high	ER	MR
Agba	high	R	P
Beech	low	P	P
Guarea	high	ER	P
Greenheart	high	ER	MR
Idigbo	high	ER	MR
Iroko	high	ER	P
Jarrah	high	ER	P
Karri	high	ER	P
Kering/gurjun	medium	R-MR	MR
Oak	high	ER	P
Opepe	high	MR	P
Red meranti	high	R-ER	P-MR
Sapele	medium	R	MR
Utile	high	ER	MR

Terms used in Table I
Durability of heartwood
In conditions conducive to decay such as timber in contact with the ground, durability is classed—low, 5 to 10 years; medium, 10 to 15 years; high, 15 to 25 years. Sapwood of all species is non-durable in these conditions
The degree of permeability
P—permeable: can be completely penetrated under pressure and can be heavily impregnated by the open tank process.
MR—moderately resistant: can be penetrated 6 mm to 9 mm laterally to the grain by 2 to 3 h of pressure treatment.
R—resistant: difficult to impregnate. Prolonged pressure treatment will seldom achieve more than 3 mm to 6 mm lateral penetration.
ER—extremely resistant: will absorb little preservative even under prolonged pressure treatment: no appreciable lateral penetration and small longitudinal penetration.

Table II *Appropriate preservatives and treatments (listed in each case in descending order of preference) for various hazards of attack or decay.*

Suitable treatments for various degrees of decay and insect attack hazards
(Minimum net dry salt retention kg/m³ shown in brackets)

Preservative	Method of treatment	Column 1	Column 2	Column 3	Relevant notes
TO1	P	*	*	*	4
	HC	*	*		4
	S		*		2, 4
	D		*		2, 4
TO2	P				
	HC	*	*		4
	S		*		2, 4
	D		*		2, 4
OS1	P				
	HC				
	S		*		3, 5
	D		*		3, 5, 6
OS2	P				
	HC	*	*		1, 3
	S		*		3, 5
	D		*		3, 5, 6
OS3	P	*	*	*	3
	HC	*	*		1, 3
	S		*		3, 5
	D		*		3, 5, 6
WB1	P	* (8·0)	* (6·4)	* (6·4)	
	HC	*	*		
	S		*		
	D		*		
WB2	P	* (5·3)	* (4·0)	* (4·0)	
	HC	*	*		
	S		*		
	D		*		
WB3	P		* (4·0)	* (4·0)	
	HC		*		
	S		*		
	D		*		
WB4	P				
	DIF		*	* (5·3)	7
	HC				
	S				
	D				

Terms used in Table II
Preservatives
CLASS TO—Tar oil preservatives: 1 Coal tar creosote to BS 144[6] and BS 913; 2 Other coal tar types to BS 3051[5].
CLASS OS—Organic solvent preservatives: 1 Chloronaphthalenes; 2 Metallic naphthenates; 3 Pentachlorophenol its derivates and its zinc and copper salts
CLASS WB—Water-borne preservatives: 1 Copper/chrome to BS 3452[8]; 2 Copper/chrome/arsenate to BS 4072[9]; 3 Fluoride/arsenate/chromate; dinitrophenol to BS 3453[14]; 4 Other single salts such as borax/boric acid†, sodium fluoride and sodium o-phenylphenoxide.
†These salts can also be applied by dipping or spraying the green timber and allowed to soak in by diffusion—an effective method, but the salts are liable to leaching.
Methods of treatment
These are in order of probable effectiveness:
P—pressure impregnation
DIF—diffusion with borates
HC—hot-and-cold open tank treatment
S—steeping   **D**—dipping
Degrees of hazard Columns **1**, **2** and **3** of the table apply as follows:
1 Exterior or interior timbers in ground contact or in contact with foundations below dpc, or above a perforated dpc, or interior timbers exposed to persistent damp atmosphere (eg dye houses, breweries, laundries, tanneries, swimming pools)
2 Exterior timbers generally not in **1**, not painted after treatment, eg exterior structural framing, cladding
3 Where particular protection of timber is required to prevent infestation by house longhorn beetle.
With water-borne preservatives, the recommended minimum retention of salts in kg/m³ dry weight (ie resistance to leaching) is given in brackets. Use of these preservatives may involve kiln drying with a consequent possibility of distortion.
Notes
These appear in the last column and refer as follows:
1 In class OS, solvents with a low boiling point should not be used for hot-and-cold open tank treatment as they constitute a fire hazard.
2 May need to be heated before use.
3 If it is desired to paint the timber, a suitable solvent should be used.
4 In some circumstances the smell or possibility of staining to adjacent materials may be objectionable; moreover the timber cannot easily be painted.
5 European redwood (Scots pine) requires steeping for not less than one hour if intended for weather boarding, and dipping for 1 to 3 minutes if intended for exterior joinery complying to NHB requirements.
6 For protection against house longhorn beetle, class OS preservatives may be used, by dipping for not less than 10 minutes, but the solution must contain not less than 0·5 per cent gamma BHC, dieldrin, or other persistent organochlorine contact insecticide.
7 Exterior timber treated by diffusion with boron compounds must be protected from leaching by the application of paint or varnish.
8 Brushing and spraying methods, generally used for in situ timber can employ any type of preservative.

5.07 Most intumescent silicate paints suffer gradual decomposition and form a powdery efflorescence of soda on the surface. With varnishes it is important to specify products guaranteed to remain clear after prolonged exposure to the atmosphere.

Putties and pastes

5.08 These contain endothermic substances like magnesite or dolomite, often mixed with soluble glass. Non proprietary lime or lime/gypsum plasters can give effective protection to structural members, provided wire mesh (properly secured to the timbers) is used as a ground or base to prevent plaster disintegration at high temperatures.

Bonded non-combustible facings

5.09 Several wood-based sheet materials—plywoods, particle boards and fibreboards—are available with facings of asbestos felts about 0.8 mm thick, lacquered asbestos-glass fibre foil, metal foil, or other proprietary surface coatings tested and certificated to comply with class 1 or class 0 flamespread requirements.

Non-faced boards

5.10 Of the standard wood-based panel products only wood wool slab complying with BS 1105[10] has intrinsic class 0 properties. However, there are now on the market many certificated products, for either impregnation or surface treatments, which will confer class 0 to solid timber or wood-based panels. TRADA wood information sheet no 3 section 2/3[12] lists many such products and supplying or treating companies and many more which attain class 1. Sheet no 7 section 2/3[13] lists a wide range of pretreated boards: plywoods, particle boards and fibre boards, which are available both in class 1 and class 0 of flame spread requirements.

Table III *Summary of the recommendations of* BS 3842: 1965 Specification for treatment of plywood with preservatives.

Preservative	Method of treatment	Preservative types and treatment methods suitable for different conditions of exposure to decay and insect attack (recommended concentration or dry salt retention in kg/m³ in brackets)				Notes
		Column 1	Column 2	Column 3	Column 4	
TO1	P	* (96)		* (96)	* (96)	1
TO2	10D	*		*	*	1, 4
	3SC	*		*	*	1, 4
OS1	10D	*		*		2, 3, 4
	3D				*	2, 3, 4
	3SC	*		*		2, 3, 4
	2SC				*	2, 3, 4
OS2	10D	*		*		2, 3, 4
	3D				*	2, 3
	3SC	*		*		2, 3, 4
	2SC				*	2, 3, 4
OS3	10D	*		*		2, 3, 4
	3D				*	2, 3, 4
	3SC	*		*		2, 3, 4
	2SC				*	2, 4
WB1	P	* (10)	* (16)	* (10)	* (10)	
	DIF	* (10)	* (16)	* (10)	* (10)	
WB2	P	* (6)	* (16)	* (6)	* (6)	
	DIF	* (6)	* (16)	* (6)	* (6)	
WB3	P	* (6)		* (6)	* (6)	
	DIF	* (6)		* (6)	* (6)	
WB5	P	* (8)		* (8)	* (8)	
	DIF	* (8)		* (8)	* (8)	

References

1 BS 5268: Part 5: 1977 Preservative treatments for constructional timber (formerly CP 98)

2 BS 3842: 1965 Specification for treatment of plywood with preservatives

3 BS 1455: 1972 Plywood manufacture from tropical hardwoods

4 Fire Research Station (of the Building Research Establishment) FRS Note 553

5 BS 3051: 1972 Coal tar creosotes for wood preservation (other than creosotes to BS 144)

6 BS 144: 1973 Coal tar creosotes for the preservation of timber

7 BS 913: 1973 Wood preservation by means of pressure creosoting

8 BS 3452: 1962 Copper/chrome wood preservatives and their application

9 BS 4072: 1974 Wood preservation by means of copper/chrome/arsenic compositions

10 BS 1105: 1972 Wood wool slabs up to 102 mm thick

11 BS 746: Part 6: 1975 Fire propagation tests for materials. Part 7: 1971 Surface spread of flame tests for materials

12 TRADA Flame retardant treatments for timber, Wood Information Sheet 3 of section 2/3, the Association 1978

13 TRADA Low flame spread wood products, Wood Information Sheet 7 of section 2/3, The Association 1977

14 BS 5453: 1962 Fluoride/arsenate/chromate/dinitrophenol waterborne preservatives and their application

Terms used in Table III

Class TO tar oil preservatives

TO1—coal tar creosote complying with BS 144[6]

TO2—coal tar oil types of wood preservatives complying with BS 3051[5]

Class OS organic solvent type preservatives

OS1—Chloronaphthalenes (minimum chloride content 46 per cent)

OS2—Copper naphthenate

OS3—Pentachlorophenol and pentachlorophenol/metallic naphthenate mixtures

OS4—Organic derivatives of pentachlorophenol.

Note where non-pressure methods of application are used and the risk of insect attack is high, OS preservatives should be reinforced by one of the following insecticides: 0·5 per cent diedrin, 0·5 per cent gamma BHC or 5 per cent polychloronaphthalene (minimum 46 per cent chloride content). OS preservatives listed in Princes Risborough Laboratory, BRE technical note 24 are also suitable here.

Class WB water-borne type preservatives

WB1—copper chrome complying with BS 3452[8]

WB2—copper/chrome/arsenate complying with BS 4072[9]

WB3—fluoride/arsenate/chromate/dinitrophenol to BS 3453[14]

WB5—borax, boric acid, or a mixture of both

Methods of treatment

P—pressure impregnation of the plywood

10D—dipping of the plywood for minimum 10 minutes

3D—dipping of the plywood for minimum 3 minutes

3SC—three surface coats to the plywood

2SC—two surface coats to the plywood

DIF—diffusion at veneer stage

Moisture content

This should be within the following limits:

Class **TO**—6 to 14 per cent (up to 25 per cent for external uses), both before and after treatment

Class **OS**—6 to 14 per cent (up to 25 per cent for unpainted external use), both before and after treatment

Class **WB**—up to 25 per cent before treatment; 6 to 14 per cent after treatment for internal or painted external use; otherwise up to 25 per cent generally.

Intended use and expected hazards are indicated in columns 1 to 4 of the table:

1 Exterior. Plywood for external roofing or sheathing of buildings, concrete shuttering etc

2 Plywood for use in cooling towers, ie high risk of soft rot attack

3 Interior. High risk. Wall or roof linings in industrial buildings with heavy condensation, eg dye-houses

4 Interior. Protection against decay and insect attack, wall panelling, roof lining, wall or roof linings in agricultural buildings, plywood in contact with fresh plaster, cupboards, fitments, flush doors etc.

Notes

These appear in the last column and refer as follows:

1 Preservatives of class TO normally have a distinctive odour and should not be used near foodstuffs; moreover the plywood cannot easily be painted.

2 Preservatives of class OS also have a marked odour, but with types OS2 and 3 this is mainly from the solvent and largely disappears after a few days. It is more persistent in type OS1.

3 With class OS preservatives, advice should be sought on suitability for painting.

4 Brushing or dipping are acceptable methods of treatment, but are best used when the hazard of attack is less severe, where a shorter life is required, or where treatment can be renewed periodically.

5 The presence of preservative on the surface of the plywood after treatment may interfere with subsequent gluing operations. Users should seek the advice of the adhesive manufacturer.

AJ Handbook
Building structure
CI/SfB (2-)

Section 8
Structural material
Masonry

Section 8

Structural material: Masonry

Scope

Masonry here is taken to mean mortar-bonded stone, block and brick. Technical study 1 notes the various properties of these elements and their resulting effects on detail design and construction and on the form of various building types. Also discussed are the effects on structural form of current building economics and developments, and of the anti-collapse measures required by the 1970 Building Regulations —partly occasioned by the tragedy of Ronan Point.

Technical study 2, broadly outlines the range of the design procedures open to the architect, using the calculation approach of BS CP 111 or the empirical approach of the Building Regulations.

The section will conclude with information sheets containing worked examples and tables.

Authors

The authors of this section are David Adler, BSC, DIC, MICE, also author of Section 2 STRUCTURAL ANALYSIS; and Allan Hodgkinson, MEng, FICE, FIStructE, MConsE, consultant editor of the handbook.

Acknowledgements

Illustrations by courtesy of:
Cement and Concrete Association
Redland Ltd
London Brick Company
British Ceramics Research Association
Brick Development Association

Allan Hodgkinson

David Adler

Technical study
Masonry 1

Material and form

This study by DAVID ADLER *and* ALLAN HODGKINSON *outlines the sources and production of structural stone, brick and block. Notes are given both on the way in which their various properties affect the choice of material and detail design, and on how masonry's general characteristics affect structural form.*

1 Masonry types

Stone

1.01 Once the only form of masonry, stone is now reduced to a comparatively minor role, particularly in its possible engineering applications. It requires considerable work in cutting and dressing and, with the advent of brick and block, has tended to become an applied finish rather than a structural material. Granites, sandstones and limestones are the most common of the structural types remaining (see table I).

Table I Comparative properties of basic structural stones

	Density kg/m³	Ultimate crushing strength MN/m²	Absorption per cent of dry weight
Granite	2500 to 2750	92 to 147	0·09 to 0·55
Sandstone	1950 to 2260	25 to 92	3·50 to 8·00
Limestone	2020 to 2380	10 to 65	4·50 to 11·00

Quarrying

1.02 Stone is obtained from the quarry bed by blasting or splitting with wedges. Types with a granular structure, eg granite, split equally well on any plane. Sandstone and limestone, on the other hand, are built up in layers running parallel to the natural bed or plane on which they were deposited, and as subsequent foldings and upheavals may have moved these layers from their original position, care is needed to ensure that the stone finally used is so positioned that the planes of stratification lie perpendicular to the direction of pressure.

Bricks

1.03 The normal rectangular brick is sized to allow placing with one hand—the length twice the width to permit bonding with bricks lying both parallel and perpendicular to the wall face. Bricks are required to comply with three British Standards, BS 3921 for fired brick earth, clay or shale, BS 187 for sandlime and BS 1180 for concrete. There is a wide range in appearance, strength and durability.

Clay brick manufacture

1.04 Clay bricks are in major use for structural masonry. The dug clay **1** usually contains the essential minerals silica, alumina and kaolinite and perhaps others such as statite feldspar, mica and cordierite. Kaolinite content may amount to 50 per cent. The clay is ground, mixed, screened and passed to a revolving roller mill where water is added to give the plasticity required for pressing or extrusion **2**.

1

2

1 *Dragline excavator quarrying at clay face*
2 *Extruded clay column*

1.05 In the pressed brick process the clay is delivered to a pug mill which presses it into moulds producing a rough clot which is finally shaped by a brick press; a 'frog' or indentation is usually imparted to one face, lightening the brick for transport and handling and providing a mechanical key for the bedding mortar. In the extrusion process the clay passes through a pug mill with a screw extruder, forcing it through a die to produce a column which is then cut to brick size by wires **3**.

1.06 After pressing or wire cutting, the bricks are ready for drying and firing in kilns at a temperature ranging from 950°C to 1220°C. The spontaneous combustion of the organic matter contained in some clays tends to save on fuel but leaves characteristic pores in the brick structure.

3 *Cutting the extruded clay*

Brick sizes

1.07 Before metrication the brick size was established at: *length* 8¾in, *width* 4 3/16 in, *depth* 2⅝in or 2⅞in which lead to a 9in × 3in face module and a width based on multiples of 4½in. The metricated requirement has produced: *length* 215 mm, *width* 102·5 mm, *depth* 65 mm.

The range of density, strength and absorption is shown in table II.

Table II Range of density, strength and absorption figures for the more common brick types

	Density	Ultimate crushing strength M N/m²	Absorption per cent of dry weight
Common and facing bricks	1400 to 1800	3·5 to 70·0	9 to 28
Engineering bricks	1900 to 2500	49·0 to 140·0	0·1 to 7·0

Blocks

1.08 BS 2028: 1364[1] (summarised in table III below) defines a block as a walling unit in which the length, width or height are greater than those of the normal brick. Originally sized to co-ordinate with brickwork, blocks had dimensions 17⅜in × 8⅝in by a variety of thicknesses in one inch modules—the blocks are still available under the pseudo-category of 448 mm × 219 mm. More generally the height should be no greater than either the length or six times the thickness. The proposed full range of sizes from BS 2028 is shown in table IV.

Manufacture

1.09 Technical study CONCRETE 1 lists the materials used in block manufacture. Aerated concrete blocks are produced by sawing from the aerated mass cake, others (eg heavy and lightweight aggregates or clinker blocks) with the aid of moulds or block making machines producing the blocks under pressure.

Table III Summary of main requirements for blocks of BS 2028, 1364: 1968 (published by courtesy of BSI)

Block type	Materials and methods of manufacture	Density of block kg/m³	Minimum average compressive strength N/mm²	Strength—lowest individual block N/mm²	Maximum permitted drying shrinkage (per cent)	Wetting expansion
A	Any combination of materials and methods of manufacture may be used, provided resulting blocks comply with the specification	Not less than 1500	3·5	2·8	0·05	This applies only to block made with clinker aggregates and is to be not more than 0·02 per cent in excess of drying shrinkage value
			7·0	5·6	0·05	
			10·5	8·4	0·06	
			14·0	11·2	0·06	
			21·0	16·8	0·06	
			28·0	22·4	0·06	
			35·0	28·0	0·06	
B	As above	Less than 1500 but more than 625	2·8	2·25	0·07	
			7·0	5·6	0·08	
		Less than 625	2·8	2·25	0·09	
C	As above	Less than 1500 but more than 625	Transverse breaking load is specified (varies with size of block)		0·08	
		Less than 625			0·09	

Note

Type A blocks are for general use in buildings, including walling below damp course level. If solid they must be of dense concrete or one of the denser lightweight aggregates. Type B blocks are for general use in building, including below damp course in internal walls and in the inner skin of external walls. They may be of aerated concrete, hollow or solid cement block, sand and lightweight aggregate, or graded wood particles. As the outer leaf of external walls below damp course, they should be solid, hollow or cellular and must be made with dense aggregates or have an average compressive stress of not less than 7·0 N/mm²

Type C blocks are primarily intended for non-loadbearing walls or very lightly loaded internal partitions.

Table IV *Dimensions of blocks incorporated in latest amendment to* BS *2028, 1364:1968 (published by courtesy of* BSI*)*

Block	Length × height Co-ordinating size (mm)	Work size (mm)	Thickness work size (mm)
Type A	400 × 100	390 × 90	75, 90, 100
	400 × 200	390 × 190	140 and 190
	450 × 225	440 × 215	75, 90, 100 140, 190 and 215
Type B	400 × 100	390 × 90	75, 90, 100
	400 × 200	390 × 190	140 and 190
	450 × 200	440 × 190	
	450 × 225	440 × 215	75, 90, 100
	450 × 300	440 × 290	140, 190
	600 × 200	590 × 190	and 215
	600 × 225	590 × 215	
Type C	400 × 200	390 × 190	
	450 × 200	440 × 190	
	450 × 225	440 × 215	60 and 75
	450 × 300	440 × 290	
	600 × 200	590 × 190	
	600 × 225	590 × 215	

Note
Blocks of work size 448 mm × 219 mm × 51, 64, 76, 102, 152 or 219 mm thick, and 397 mm × 194 mm × 75, 92, 102, 143 and 194 mm thick will be produced as long as they are required.
If blocks of entirely non-standard dimensions or design are required the limits of size or the design shall be agreed. Such blocks shall then be deemed to comply with this standard provided they comply with the other requirements. Types A, B and C are as described under table III.

1.10 *Solid blocks* have end grooves, finger holes or small cavities such that the total voids do not exceed 25 per cent of the volume of the nominal gross block. *Hollow blocks* have cavities passing right through with the cavity volume not exceeding 50 per cent of the volume of the nominal gross block itself. *Cellular blocks* are as above but have one end of the cavity sealed.

1.11 Special blocks can be produced in half lengths with or without cavity closers or, as in the bond beam type, hollowed out to receive the reinforcement bars used to bind several blocks together into a beam or lintel. The hollow is filled with fine concrete once the bars are placed **4, 5**.

2 Mortar

2.01 Mortar connects the masonry elements and binds them into an integrated stable whole. Depending on the strength of the element, mortar will contribute to the strength of the wall and must therefore be chosen with as much care as the element itself. Ideally the mortar should be just strong enough to give the wall its required basic stress value. When exposed it must provide the necessary durability and water resistance (see paras 4.01 to 4.02).

2.02 Modern mortars are composed of cement, lime and sand; masonry, cement and sand; or cement and sand with a plasticiser. Mixes are shown in table v.

4

5

4 *Bond beam blocks in position over window opening being reinforced with 2 bars 12 mm mild steel reinforcement and filled with in-situ concrete*
5 *Blockwork cores containing vertical reinforcement are filled with in-situ concrete*

Table V *Recommended mortar mixes (proportions by volume).*

Direction of change in properties	Mortar designation*	Hydraulic lime/sand	Cement/ lime/sand	Masonry/ cement/sand	Cement/sand with plasticiser
Increasing strength	(i)	—	1:0-¼:3	—	—
but decreasing ability	(ii)	—	1:½:3½ to 4½	1:3 to 3½	1:3½ to 4
to accommodate	(iii)	—	1:1:5 to 6	1:4 to 5	1:5 to 6
movements due to	(iv)	1:2-3	1:2:8 to 9	1:5½ to 6½	1:7 to 8
settlement, shrinkage, etc.	(v)	1:3	1:3:10 to 12	1:6½ to 7	1 to 8

Changing characteristics within any one mortar designation → Increasing resistance to damage by freezing after hardening
← Improvement in bond and consequent resistance to rain penetration

*Types (i) and (ii) should only be used with high strength blocks in walls subject to high loading. Type (iii) should normally be used in external walls subject to severe exposure, eg retaining walls, walls below dpc, parapets and free standing walls, or in walls not subject to severe exposure if there is a danger of early frost action. Type (iv) should be used in walls not subject to severe exposure, eg external walls between dpc and eaves except in exposed situations and internal walls, provided no early frost action is possible. Type (v) may be used in internal walls and partitions provided no early frost action is possible.

2.03 Cement types (considered in detail in Technical study CONCRETE 1, paras 1.02 to 1.11) include:

BS 12   Portland cement (ordinary or rapid hardening)
BS 146   Portland blast furnace cement
BS 915   High alumina cement
BS 4027 Sulphate resisting Portland cement.

As yet there is no British standard for masonry cement and its use should be most carefully controlled throughout any contract by initial and repeated mortar tests and brick cube crushing tests.

2.04 *Limes* may be non-hydraulic (calcium) lime or semi-hydraulic (calcium and magnesium) lime and should conform to the requirements of BS 890: 1966[2]. Lime increases a mortar's workability and because of its higher water retention improves the bonding between mortar and blocks by preventing excessive suction by the blocks. Lime is so very slow-setting that the use of lime sand mortars is not practicable in contemporary building practice, but used with cement it is eminently suitable for almost all masonry.

2.05 *Mortar plasticisers* (air-entraining types). Until a British Standard has been published, manufacturers should submit evidence to show that the material is suitable for the intended purpose. In recent years the use of plasticisers in mortars has become an established practice; they obviate use of lime in the mortar and may effect economies both in capital costs and in labour. Mortar plasticisers are effective either because they entrain extra air in the mortar, or because they increase the proportion of very fine material. Entrapped air has two main effects:

1 It improves plasticity and workability allowing the water content of the mortar to be reduced by up to 50 per cent. This automatically reduces the drying shrinkage and improves the bond with the blocks. Mortar plasticisers entrain within the mix a considerable number of microscopic and disconnected bubbles of air which act as a frictionless aggregate and can be likened to 'ball-bearings' of air. This makes trowelling easier and prevents mortar sticking to the trowel, thus speeding up laying rate.

2 It allows blocks to be laid in frosty weather. When water freezes it expands and in a plasticised mortar this expansion is taken up by the air bubbles instead of forcing outwards and disintegrating the mortar. As the frozen water thaws it returns to its task of hydrating the cement. The frost resistance of the hardened mortar is also increased. By virtue of the disconnected air bubbles it is possible that the resistance of the mortar to rain-penetration is increased by a break in the capillary channels. There may also be an increase in the overall thermal insulation of the wall by a reduction of heat loss through the joints.

2.06 *Aggregate.* Fine aggregate should be free from deleterious substances and comply with the requirements set out for natural sands in BS 1200.

2.07 *Natural sands* should conform to all requirements of BS 1198 to 1200. High proportions of sand in the mortar will reduce its drying shrinkage but may produce a harsh and unworkable mix.

2.08 *Other aggregates* should be sufficiently strong and durable, and free from soluble salts.

2.09 *Water.* See BS 3148, in the case of water supplies of doubtful quality.

2.10 *Pigments* incorporated in mortar should conform to the requirements of BS 1014.

Ready-mixed mortar

2.11 *Delivered wet to site.* Ready-mixed mortars should consist of material conforming to the requirements already indicated and in the proportions specified by the purchaser.

No cement should be added until the mortar is required for use.

2.12 *Bagged ready-mix mortar.* The materials used should conform to the requirements already indicated, and in proportions which should be clearly stated on the label. The material should be packed dry and water added only when required for use.

3 Reinforcement

3.01 Reinforcement may be used to bond, to increase structural stability or to impart tensile values to masonry beams and retaining walls. Bars can run in the cores of blocks, or the cores of brickwork as in the Quetta bond **6**. Relevant British Standards on reinforcement are: BS 405, 785, 4461, 4483[4].

Joint reinforcement

3.02 Joint reinforcement can take the form of flat expanded metal mesh, rolled expanded metal lath or galvanised joint strip of proprietary form. The reinforcement must be adequately protected by a suitable grade of mortar (unless galvanised or non-ferrous) and this requirement may well influence the basic mortar choices mentioned above.

Ties and attachments

3.03 Wall ties are specified in BS 1243: 1964[3]. The type chosen **7** will depend on the expected degree of exposure, the height of the building and the type of loadbearing masonry. Most authorities will now only accept galvanised ties in buildings up to three storeys, stainless steel being the most common answer above this height. The twisted metal tie is suitable where both skins are loaded or are of approximately equal strength and elasticity, but where there is a large difference the butterfly type is preferable as it is capable of flexing to avoid one skin cracking the other.

3.04 A variety of proprietary ties in galvanised strip steel or stainless steel have been marketed for stabilising masonry walls by tying to timber or concrete floors or timber roofs and for tying down timber roofs against wind action.

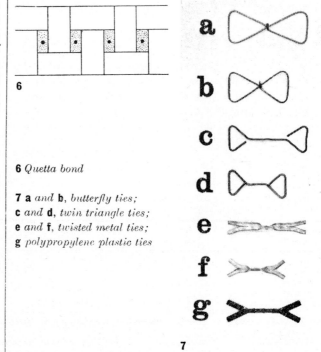

6

6 *Quetta bond*

7 a *and* **b**, *butterfly ties;*
c *and* **d**, *twin triangle ties;*
e *and* **f**, *twisted metal ties;*
g *polypropylene plastic ties*

7

8

4 Durability

Mortar

4.01 The durability of masonry depends on the durability of its components—mortar, stone, brick and block. A mortar able to prevent moisture reaching the bed may be sufficient protection to masonry with adequate face porosity. It may have to withstand frost action during construction, when it is fresh and at its most vulnerable, or during its service life if the masonry lies in an exposed situation. However, only mixes as weak as 1:3:12 are likely to suffer direct frost action after maturing and the stronger the mortar the greater the protection in its early life. Masonry cements and mortars entraining air by means of a plasticiser will have a further frost resistance. Calcium chloride should not be used as the mortar volume is too small to provide significant early heat by hydration. In any case the salt may cause efflorescence and do damage to wall ties. Here Vincol resin is the correct frost inhibitor.

4.02 *Sulphate resisting cements* should be used where sulphate action is expected either from the masonry, or from the soil when the mortar is used below ground level.

Bricks and blocks

4.03 Here durability depends on resistance to frost, moisture penetration and chemical attack. Frost damage occurs when the water in the masonry pores expands on freezing. When the elastic resistance of the masonry cannot accommodate this expansion, fracture occurs and repeated action destroys the material. Obviously the wetter the masonry the greater the danger and the problem will be alleviated by any protection to the face and the bedding joints eg silicone treatment, rendering and use of good mortar. Concrete is an inherently durable material and so are concrete blocks, though a strength of 7 MN/m² is desirable where adverse weather conditions are expected. Engineering bricks give excellent protection where absorption does not exceed 7 per cent but bricks need to be carefully chosen where there is the possibility of saturation occurring (see table II).

4.04 Chemical attack, primarily by sulphates, can be prevented in concrete products by the choice of cement. Certain engineering bricks are noted for their ability to cope with acid concentrations.

8 *Circular chapel, University of Sussex: honeycomb wall in concrete blocks. The blocks are profiled to take coloured glass infill*

9 *Wing to conference centre, Fordham: walls and buttresses of yellow concrete brick*

9

5 Movement

5.01 Movement in the overall structural sense was considered in Technical study SAFETY 1. However, all masonry is subject to movement caused by loading, and changes in temperature and moisture content. Masonry, like other materials, deflects under load due to its part elastic nature. Under sustained load it creeps like concrete. Mortar strength is important here. Creep might continue several times longer with a 1:1:6 mortar than with a 1:¼:3 mortar.

Moisture movement
In brickwork
5.02 After being exposed to the intense heat of the kiln, bricks begin to pick up moisture from the air, until (possibly after a period of years) equilibrium is reached. The moisture input is accompanied by an expansion of 0·1 to 0·2 per cent. Damage has resulted where bricks built as infill to concrete frames have expanded sufficiently to damage the frame itself. Stockpiling will ensure that most of the expansion occurs harmlessly before the bricks have been laid. Contrary to apparent logic, the long term expansion is not accelerated by dipping the bricks in water, nor is it reversed when they dry out. The reversible expansion and contraction of clay products on wetting and drying is in fact quite small.

In blockwork
5.03 Blockwork behaves like concrete and the problem is largely one of contraction rather than expansion. The shrinkage cracking in mortar joints referred to earlier causes more concern in blockwork because of the larger masonry units employed. Control joints should be used at about 6 m to 7 m centres but wall shape is important, as are stiffnesses created by short returns. Block bonding should be avoided.

5.04 There are two categories of moisture movement here:
1 *Irreversible movements* (ie those caused by hydration of the cement and by reaction of carbon dioxide from the atmosphere with the hydrated cement compounds). The chemical combination of cement and water is accompanied by a decrease in the volume of the cement paste unless extra water is available. Shrinkage due to carbonation is caused by the reaction of lime and cement hydrates with carbon dioxide in the atmosphere, probably because water which had previously been bound chemically is released in the reaction and is free to evaporate. Aggregates in the concrete exert a considerable internal restraint on the shrinkage of the cement paste and the higher the aggregate/cement ratio the less will be the resultant shrinkage. Type of aggregate used will also influence the amount of shrinkage that may take place; dense aggregates exert considerably more restraint than lightweight aggregates and reduce shrinkage.
2 *Reversible movements* (ie those caused by moisture changes either in the fine aggregate or in the gel formed by the hydration of cement). The changes in the fine aggregate are not normally serious and may be ignored unless the aggregates contain appreciable amounts of silt, clay or other finely divided materials. Reversible moisture movements will therefore depend on the proportion of cement gel in the blocks, on the porosity of the blocks, and the relative humidity of the atmosphere to which they are exposed. The porosity of the blocks governs the rate at which movement takes place; open-textured types may therefore be affected more rapidly by small daily variations in relative humidity. The tendency of previously dried blocks not to re-expand too readily upon weathering is fortunate, and it follows that blocks dried previously at a moderately low humidity and then protected against water, such as rain, are not likely to undergo any serious expansion when exposed to higher humidities for comparatively short periods.

Thermal movements
5.05 Generally the coefficient of expansion in masonry will be about $5\cdot6 \times 10^{-6}$ per °C. Expansion joint spacing in a south facing wall should be of the order of 12 m while inner skins and internal walls may be continuous up to around 80 m—but any external cavity walls, whether or not south facing, should be jointed at 12 m intervals if directly linked to the internal leaf.

6 Bonding and workmanship

Stone
6.01 Block-in-course is used for heavy work such as harbour walls or embankments. Externally the wall has the appearance of English garden wall bond, internally the wall is bonded to a fairly random arrangement.
6.02 Squared-coursed rubble is used for less massive works. Courses (but not necessarily the stones in them) are of the same height. Again, the wall is randomly bonded internally.
6.03 Ashlar, the highest class of masonry work, is used in string courses, copings, parapets, arches, dock sills and walls. Large, accurately cut stones are laid with joints as small as 4 to 6 mm. Courses are of regular height and usually deeper than 300 mm.

Brickwork
6.04 Single skin structural brickwork is normally laid in stretcher bond **10**; one brick walls in English bond **11**; and thicker walls in variations of English bond.

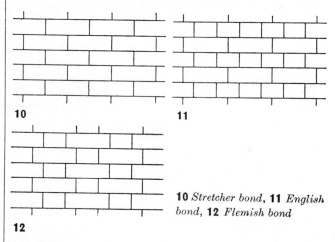

10 *Stretcher bond,* **11** *English bond,* **12** *Flemish bond*

Blockwork
6.05 This is usually single skin and thus laid in stretcher bond.

Workmanship
6.06 Masonry walls have an ultimate safety factor of between 3 and 4 according to laboratory tests. Just how much of that factor exists in the average structure depends on the workmanship quality and the degree of control exerted by the supervisory staff. Individual workmanship has a greater effect on quality in the building of a masonry wall than in any other operation on the building site. Supervision is at its most difficult. There is considerable scope for eg retempering of mortar or failure to fill bedding joints when bonus incentives replace those of pride of performance.
6.07 The British Ceramic Research Association has published a *Model specification for loadbearing clay brickwork*[5] which can be adapted as a general specification for masonry.

7 General approach to form

7.01 Masonry and its component elements are brittle materials, strong in compression and weak in tension. If the masonry is unreinforced it must be employed in forms where no tension can develop, ie restricting its use to members in which the line of action of the resolved forces passes within the mid-third of the member section without crushing the material or causing instability. Reinforced masonry is a composite construction like reinforced concrete but is far less efficient and so tends to be used in the same way as unreinforced masonry.

7.02 Masonry form is therefore restricted to columns, walls and curved surfaces such as arches and though these factors comprised superb structures in the past, curved masonry form has no economical place in current building. But masonry provides vertical enclosure, weathertightness, fire protection and acoustic protection simultaneously, and therefore while of little value in skeleton form, remains the cheapest provision of basic vertical structure for eg residential accommodation and small offices or shops where the structure is unlikely to require modification during its lifetime. If the structure is multi-storey it should follow a standard layout of loadbearing walls at every floor level.

Structural stability

7.03 Here the main concern is for resistance to lateral forces, primarily wind (having assumed that the proportions of the building are such that it will not collapse because of elastic instability). The mechanics of failure have already been discussed in Technical study ANALYSIS 1). It was seen that any compression member, even though well able to resist direct crushing forces, requires a ratio of effective length to radius of gyration such that it cannot buckle. In a rectangular section this ratio can be expressed as effective height compared to least width of member. Obviously a reduction in effective height or an increase in width will increase stability and width increase not only gives greater resistance to deflection but by virtue of the greater section area reduces the critical stress itself.

7.04 'Supported length' is perhaps more appropriate than 'effective height' where side walls acting as buttresses increase the main wall's stability by reducing its effective length. This means that in the consideration of the structure the vertical members are proportioned and disposed to achieve a compromise of quantity and size.

7.05 A word of warning on the design of walls—if the length of a wall is less than four times its width it is classed as a column, assumed to be more vulnerable to collapse and is credited with a greater effective height and with a reduced allowable stress in ratio with its sectional area. This means that the relationship of doors and windows must be carefully considered so as to avoid producing 'columns' which might require a width greater than that of the general run of the wall.

7.06 Failure of masonry structure can occur in several ways. **13** shows failure owing to excessive vertical compression, **14**, **15** owing to gravity and wind forces combined. Wind can also cause direct tensile failure, **16** only possible in the lowest levels of a building, shows failure where one section of masonry has slid on the section below.

Terrace and multi-storey housing

7.07 Two-storey terrace housing **17** on a crosswall system relies on lateral support from front and rear elevation with spine support at selected positions in the length of the terrace. Multi-storey building **18** must be built in cellular

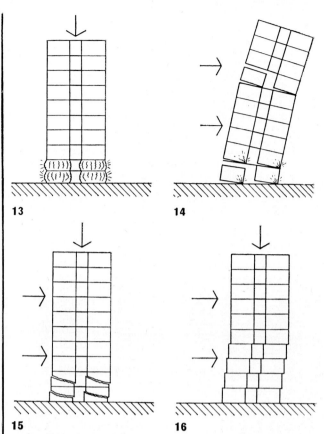

13 *Masonry failure owing to excessive vertical compression;* **14, 15** *gravity and wind forces combined;* **16** *horizontal sliding of sections in lower levels of building*

form, often up to 18 storeys with walls only 230 mm thick. There are a variety of layouts possible between the two extremes and there is nothing to prevent long narrow buildings being carried up to considerable height, provided each cross wall can act as a buttress to carry its share of wind load, and provided there are enough spine walls, staircase and lift complexes to give longitudinal stability.

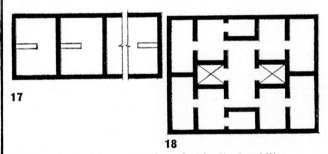

17 *Low rise terrace construction—longitudinal stability provided by front and rear elevations and spine walls (outlined);* **18** *High rise cellular structure—up to 18 storeys with walls 230 mm thick*

1970 Building Regulations

7.08 The design of masonry structures has been complicated by the anti-collapse measures added to the Building Regulations in 1970. The cellular structure discussed in para 7.07 is little affected since an in situ floor slab can easily be spanned two ways to give it two alternative lines of support. Collapse resistance is only required in structures above four storeys. However, four-storey structures are immediately classed as five when carried on columns above a car park area and as car parking layouts rarely accord with the residential layout above, they can be an expensive addition. The most economic residential solution used to be the maisonette block with alternate timber and concrete

19 *High rise structural brick: Baylis Road, Lambeth*

floors between masonry walls. This is now no longer true in buildings higher than four storeys. Industrial buildings are worthy of consideration, particularly the small lettable unit. External walls, stabilised by light bracing in the plane of the roof, and party walls at 340 mm thickness can be employed without the need of buttresses in their length **20**.

Industrialised building

7.09 In the last two decades industrialised buildings have been constructed using large concrete or brick panels as the vertical elements and reinforced concrete panels as flooring members. The vertical elements were designed as large chunks of masonry in accordance with BS CP 111.

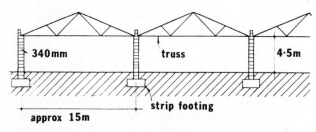

20 *Walls to small industrial buildings may be constructed at 340 mm width without buttresses*

7.10 The Ronan Point disaster occurred at a time of considerable local authority pressure against the social aspects of high building and although much has been done to give greater stability to industrialised structures, the market has evaporated. No doubt other market pressures will arrive to restore the demand.

8 Garden and retaining walls

8.01 Major structures such as dams and high walls are no longer constructed in masonry, but smaller walls in masonry are dealt with by architects and useful advice on such construction will be found in the Brick Development Association Technical note *The design of freestanding brick walls, vol 1, no 5*[6]. Some of the note's suggestions on wall proportions are shown in **21, 22, 23**. (See also Technical study FOUNDATIONS 3.)

8.02 Brickwork has been used successfully below ground in retaining walls to eg basements and manholes, acting either as thick gravity walls or constructed in dimensions small enough to imply that a very flat arch line was formed in the brick section with the resulting thrust retained by cross walls and buttresses. Large manholes were slightly barrel shaped to produce a similar action. The basement structure has been largely replaced by reinforced concrete and the rectangular manhole by spun precast concrete circular sections—which might be regarded as another form of masonry.

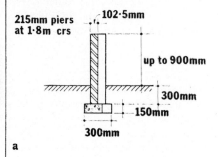

a

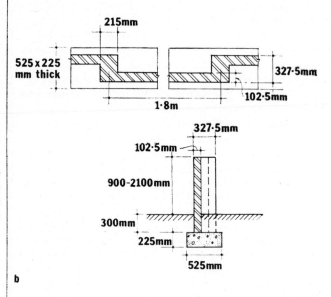

b

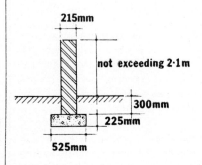

c

21 *Typical garden wall proportions for heights as shown: a with piers; b staggered wall plan and section; c straight wall.*

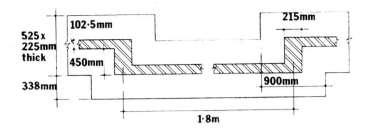

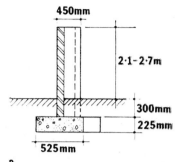

a

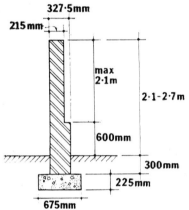

22 *Garden wall*
proportions (cont.):
a *staggered wall plan*
and section; **b** *straight*
wall

b

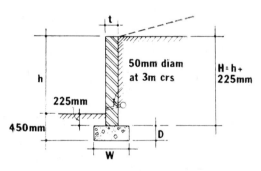

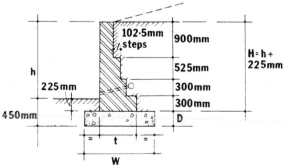

h max	H	t	W	W	D
		(mm)			
900	1125	215	525	525	225
1200	1425	327·5	600	600	225
1500	1725	440	675	900	225
1800	2025	552·5	750	1050	225

No surcharge and slope of retained
earth not greater than 1:10
Safe minimum bearing pressures:
Granular soil 110 KN/m² (1 tonf/ft²)
Cohesive soil 55KN/m² (½ tonf/ft²)
No dpc other than slate
Minimum brick strength 20·5 MN/m²
Mortar mix 1 :¼ :3

23 *Typical retaining wall proportions*

24

9 Development

9.01 Further work is proceeding on the prefabrication of brickwork into both loadbearing and non-loadbearing panels. Automated systems have been devised to cast the panels either horizontally or vertically. Another ceramic development is the production of a storey height single panel—quite an achievement in view of the problems of tolerance, handling and firing.

9.02 The Ronan Point aftermath initially created several problems for the designer of masonry structures but the British Ceramic Research Association and Brick Development Association have risen to the occasion and have carried out experiments primarily aimed at assessing the effects on brick panels of gas explosions but which have yielded much wider information on masonry collapse patterns.

24 *RAF West Drayton; five-storey cross wall*
construction in white concrete blockwork
25 *Dorset water board offices, Poole; loadbearing brick piers*
have reinforced concrete cover. Prefabricated brick slip
panels used as permanent shuttering to in situ edge and
facia beams

25

9.03 The much greater resistance of brick panels to lateral load when under vertical load is leading to methods of eg vertically prestressing masonry.

9.04 High bond mortars based on epoxy resins have been giving large increases in bond and tension compared with ordinary mortar—more than doubling direct load compressive strength. However, as one might expect with epoxies, they are expensive and preferably to be used only where accurately sized masonry units can further benefit from the use of thin joints.

10 Economics

10.01 There are a great many low rise structures where masonry is the cheapest way to build. However, even though 16-storey buildings may be shown to be cheaper in masonry by quantity surveying costing, overall project costing including the advantages of early occupation and thus earlier capital recovery or start of letting, may reveal the slower masonry construction in an unfavourable light.

10.02 There is also the choice of which block or brick to use, further influenced by which of the many types are available—or appear to be available. The stop-go methods which have governed the UK economy for so long have played havoc with all building materials—especially bricks.

10.03 Rival agencies extol the virtues of their own product and the concrete block in particular has received considerable technical advertising attention in recent years. No doubt the lighter weight types, say those below 10 kg per block which can be lifted by one hand, will show a real as well as a 'measured rate' economy over the comparable brick thickness, but it is becoming increasingly difficult to get bricklayers to work with heavy blocks if there is some straightforward 115 mm or 230 mm brickwork not too far away on which a fat bonus can be earned. The entire industry is currently in too much pricing chaos to allow dogmatic comparisons, but if bricklayers' wages continue to rise at 1972/73 rates, and they are only prepared to do simple work, the concrete industry is only too ready to produce alternative construction methods. This country may be in danger of losing yet another craft trade.

References

1 BS 2028 1364: 1968 Precast concrete blocks [Ff]

2 BS 890: 1966 Building lime [Yq1]

3 BS 1243: 1972 Metal ties for cavity wall construction [(21·1) Xt6]

4 BS 405: 1945 Expanded metal (steel) for general purposes [Jh2]

BS 4449: 1969 Hot rolled steel bars for the reinforcement of concrete [Hh2]

BS 4482: 1969 Hard drawn mild steel wire for the reinforcement of concrete [Jh2]

BS 4461: 1969 Cold worked steel bars for the reinforcement of concrete [Hh2]

BS 4483: 1969 Steel fabric for the reinforcement of concrete [Jh2]

5 BCRA Special publication 56. Revised edition 1975. Model specification for loadbearing clay brickwork [(21·1) Fg] BCRA, The Mellorgreen Lane, Hanley, Stoke-on-Trent ST1 4LZ

6 BDA Technical volume 1 no 5 The design of free standing brick walls. 1972 [(21) Fy] o/p 1972. Brick Development Association, 3 Bedford Row, WC1

Acknowledgments

A major source for Technical study Masonry 1 was *Design in blockwork* by Michael Gage and Tom Kirkbride. Architectural Press [Ff] £2·50

Technical study
Masonry 2

Section 8 **Structural material: Masonry**

Calculating structural masonry

In this study DAVID ADLER *outlines two alternative methods of calculating structural masonry—by the analytical method of* BS CP 111 *and by the empirical approach of the Building Regulations. Information sheet* MASONRY·2 *shows examples worked by each method.*

1 Principles in masonry design

1.01 Masonry is mainly used in vertical elements such as walls, columns and piers. A wall is an element whose length is at least four times its width—otherwise it is a 'column'. A 'pier' is basically a column integral with a wall **1**. Reference to walls in the text can be taken to apply to columns and piers also, unless stated otherwise.

Strength

1.02 Strength will depend mainly on the masonry unit used (brick, stone or concrete block) and to a lesser extent on the mortar **2** (see also tables I mortar, II bricks, IIIa and IIIb blocks). The choice between bricks and blocks, and between the various types in each category, is rarely made on structural grounds—there are so many other considerations such as availability, willingness or otherwise of the brick-layers to lay bricks, appearance, thermal and acoustic properties, cost and durability.

1.03 Both masonry units and their mortar are weak in tension, as is the bond between them, and it is therefore normal design practice to assume no tensile strength. The one exception to this is that when considering a wall's resistance to wind pressure, small values of tensile strength are assumed.

Eccentricity

1.04 Bending moments produce both compressive and tensile stresses—the tensile stresses not normally threatening masonry construction since the overall compression on the wall is greater than the negative value of the bending moment, in effect submerging the tensile effect. TS ANALYSIS 2 para 2.09 showed that a moment and a compression on a vertical strut can be resolved into a compressive force at an eccentricity from the centre of area of the horizontal section of the strut. Thus all the horizontal and vertical forces on the wall can be reduced to a single vertical compression at a distance from the centre of the area of the horizontal section of the wall; this distance is called the eccentricity of the load **3cd**. If the eccentricity is such that the resolved force falls outside the wall, the wall will collapse. If the load falls near the edge of the wall section, the area of the wall available to carry the load becomes small, and the stress becomes large, possibly causing the masonry under stress to crush and the wall to fail **3ab**. The main factors affecting eccentricity are:

1.05 *The weight of the wall itself.* This will act through the centre of its section, assuming the wall is of uniform material and has constant thickness. So the heavier the wall in comparison with the other loadings, the less will be the resultant eccentricity.

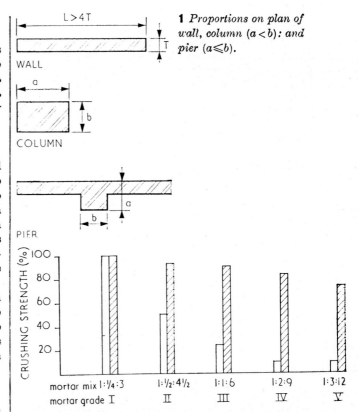

1 *Proportions on plan of wall, column (a < b): and pier (a ⩽ b).*

2 *Relative strengths of various mortar grades (unhatched) compared with brickwork (hatched). Grade mixes shown table* V, TECHNICAL STUDY MASONRY 1.

1.06 *Wind loading on the wall.* This complex problem is covered in para 3.16.

1.07 *Loads applied to the wall.* A panel of masonry wall may have to support another panel in the storey above. If this is thinner and flush with one side of the panel below there will be a resulting eccentricity. The effect of floor and roof loads depends on their position of application which again depends largely on the type of construction. Here BRS Digest (second series) no 61 lays down generally accepted criteria[1]. Reinforced or prestressed concrete floors of moderate span, say less than 30 times wall thickness, bearing on an external wall are assumed to act at the centre of the bearing on the wall **4a**.

1.08 With long, concrete spans more than 30 times the wall thickness, or lightweight floors such as timber joist and boarding, there will be some deflection of the floor and the point of loading will migrate towards the inside of the wall. The recommended allowance to make for this displacement is one-sixth of the bearing width **4b**.

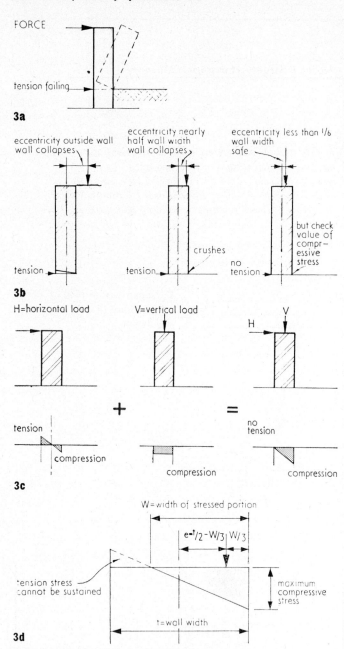

FORCE

tension failing

3a

eccentricity outside wall
wall collapses

eccentricity nearly
half wall width
wall collapses

eccentricity less than 1/6
wall width
safe

crushes

but check
value of
compr-
essive
stress

tension | tension | no tension

3b

H=horizontal load | V=vertical load | H   V

tension | no tension

+   =

compression | compression | compression

3c

W=width of stressed portion

e-t/2-W/3 | W/3

tension stress
cannot be sustained

maximum
compressive
stress

t=wall width

3d

3 *No tensile value is assumed in masonry design* **a, b** *but* **c, d** *all horizontal and vertical forces may be reduced to one vertical compression at a distance (eccentricity) from centre of area of horizontal wall section.*

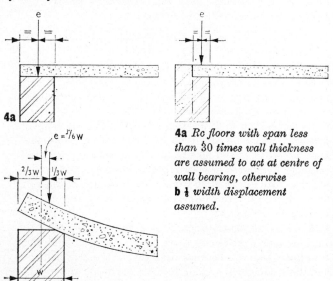

e

e

4a

e = 1/6 w

2/3 w | 1/3 w

w

4b

4a *Rc floors with span less than 30 times wall thickness are assumed to act at centre of wall bearing, otherwise* **b** *⅙ width displacement assumed.*

1.09 Interior walls flanked by roughly equal spans may be assumed to carry loads centrally. However, if one span exceeds the other by more than 50 per cent, the one-sixth displacement rule should be applied **5a**. If a timber floor is supported on joist hangers built into the wall, the BRS Digest recommends that the loading be assumed to act 25 mm out from the wall face **5b**. If the wall is corbelled out to take eg floor or roof, the loads may be assumed to act at the centre of the bearing.

1.10 In the above cases it is assumed that the load is more or less evenly applied along the wall length. Where there is a local concentration of loading—perhaps a steel joist carrying an item of plant—the load is assumed to distribute itself downwards at dispersion angles of 45° **6**.

Eccentricity ratio

1.11 The effects of eccentric loads may be more usefully expressed as a comparison with the width of the wall on which they are acting. Thus *eccentricity ratio* $\dfrac{e}{t}$ is the eccentricity of the load divided by wall thickness. For use of this ratio see para 3.10 this study.

Instability—slenderness ratio

1.12 Quite apart from overturning or crushing, and even if there is no eccentricity, a wall may still collapse owing to instability. The concept of instability of compression members is not really fully understood, although the mechanical effects are well known. A sight deflection of a member seems to increase the bending moment, which again increases the deflection—an ever worsening situation leading to collapse **7**. The three main factors influencing the onset of this process in masonry are load, the height (or sometimes length) of the wall and its thickness. In fact, it is the ratio between the last two and their actual values that is most important:

$$\text{slenderness ratio} = \frac{\text{effective height or effective length}}{\text{effective thickness}}$$

The word *effective* is used because the actual values of size and thickness may be modified in certain circumstances as shown in para 1.13.

5a

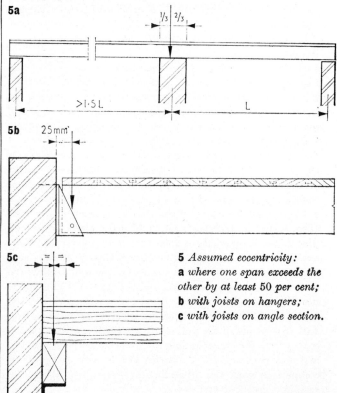

1/3 | 2/3

>1·5 L | L

5b

25 mm

5c

= | =

5 *Assumed eccentricity:* **a** *where one span exceeds the other by at least 50 per cent;* **b** *with joists on hangers;* **c** *with joists on angle section.*

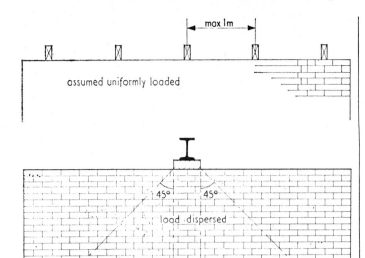

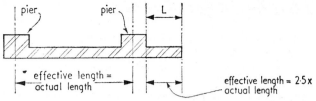

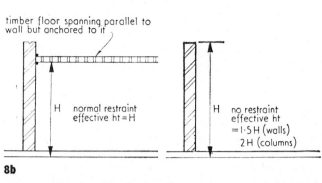

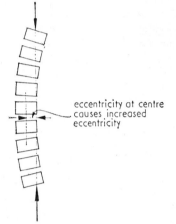

6 *Loading distribution along wall length.*

7 *Collapse owing to instability. Stability depends on slenderness ratio.*

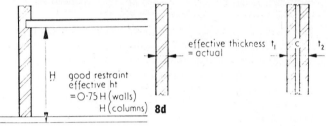

8 *Factors governing* **a** *effective length;* **b,c** *effective height;* **d** *effective thickness* $= \frac{2}{3}(t_1 + t_2)$

1.13 It is necessary to determine the restraints on the wall and how good they are on those physical features apart from the wall itself (other walls, roof, floors) helping to hold it in position. If the main restraints are vertical members at each end, such as reinforced concrete columns or return walls, the wall will tend to span horizontally and slenderness ratio will depend on effective length, ie the distance between the centres of the vertical restraints **8a**. With restraint at one end of the wall only, effective length will be $2\frac{1}{2}$ times the actual length **8a**.

1.14 Walls without end restraint are taken to span vertically and the effective height is used to calculate slenderness ratio. A wall is nearly always assumed to be restrained at its base, by its own weight if by nothing else, and the condition at the top of the wall then determines the value of effective height. This varies from $\frac{3}{4}$ to $1\frac{1}{2}$ times actual height, depending on the degree of restraint **8b,c**. The effective height of columns is taken as 1 to 2 times actual height.

1.15 The effective thickness of ordinary solid walls is the same as the actual thickness; with cavity walls, the value is taken as two-thirds of the aggregate thickness of solid material **8c**.

1.16 Instability is the major problem in loadbearing masonry construction. Reducing slenderness ratio by simply thickening the walls is expensive in material and labour, occupies more space, and may well cause problems at foundation level. Preferably piers should be used at intervals or the plan should be such as to increase stability eg corrugated, chevron or staggered **9**. Arrangement of walls is of prime importance in tall buildings in calculated masonry. It cannot be over-emphasised that economy for such construction starts at the very earliest stage of the design process.

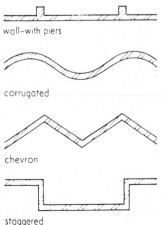

plan form

wall–with piers

corrugated

chevron

staggered

9 *Wall shape and arrangement on plan is of prime importance in tall masonry construction.*

2 Design methods to use

2.01 There are two basic methods commonly used for designing masonry structures: the analytical and the empirical. The analytical method used in this country is based on BS CP 111 *Structural recommendations for load-bearing walls*.[2] The empirical approach is covered in the relevant Building Regulations[3] for the part of the country in which the structure is to be sited.

2.02 It is doubtful whether any local authority will permit different methods to be used in different parts of the same structure; once a method is chosen it will have to be used for all subsequent work. The next question is which method to use when, but there are no hard rules, for each design context has its own peculiarities. As the safety factors in both are relatively high by modern standards, there is unlikely to be any danger of structural failure whichever method is chosen; the choice will more likely depend on economics.

2.03 Most authorities will readily accept applications from architects on the Building Regulations, but will require calculations from a chartered engineer for the analytical method and as the fees so earned are on a reduced scale, it may be hard to find an engineer to do the work.

2.04 Generally speaking, the Building Regulations will give reasonable overall economy in ordinary two-storey housing work, granted that requirements other than structural have to be fulfilled (eg acoustic and thermal insulation). The code of practice is more suited to non-standard, high rise work or to construction using blockwork of other than standard brick widths.

3 Analytical method—BS CP 111

3.01 Design carried out by analytical methods involves use of BS CP 111. If the structure complies with this, it may be assumed safe except for the anti-collapse provisions of D19 of the Building Regulations 1972 (previously called the 'fifth amendment') whose requirements are in para 4.19 this study.

3.02 The current edition of BS CP 111 is Part 2: 1970 incorporating amendment slip no 1 published 11 June 1971. This amendment slip is vital as it constitutes a major revision to the code. Its issue was, however, an interim measure: the whole code was revised as BS 5628 during 1978 to reflect the increasing use of limit state methods in design. Nevertheless, the present code will be an invaluable guide for several years. But it contains many pitfalls for the inexperienced, and a singular lack of explanatory material.

Steps in design

3.03 *Find the total load.* In designing any masonry pier or wall one must first establish the maximum stress occurring in it. To do this the total loading must be found—ie self weight of structure and imposed load from the floors of the building (see BS CP 3: Chapter V[4]). Here is the first difference from the Building Regulations' empirical approach, for there only the broadest classification of building type covers the wide variation of loadings that can occur.

3.04 *Find the maximum eccentricity.* Once the load has been found, it is necessary to find its distribution on the wall. 1) along the length and 2) across its thickness.

1 Theoretically walls can be divided into unit lengths on which the load can be seen to be uniformly distributed. In practice this is usually taken as 1 m, small variations being ignored.

2 Load distribution across the wall thickness, ie eccentricity, is more significant, but can be found as shown in paras 1.07 to 1.10 this study.

3.05 *Hence find maximum actual stress.* Using nomogram 1, p1279, place a straight edge to join wall thickness scale B with loading on scale C and mark the average stress on scale D. Now, join this point with the eccentricity on scale E and extend to read the maximum stress on scale F. Note that the thickness to be used on scale B is the solid thickness of material in the wall (see para 1.17). There are markers on the scale for the common thicknesses of brick solid and cavity walls. Scale A indicates the minimum length of wall for each thickness in the adjoining scale B, at which the full permitted stress may be used. The significance of this will become clear during the second part of the calculation (see para 3.11).

3.06 This involves finding the maximum stress permitted in the wall. Should this value prove to be more than the actual stress just found, the wall is safe; if not it must be redesigned.

3.07 *Find the basic permissible design stress of materials used.* See table II (bricks) and tables IIIa and IIIb (blocks).

3.08 *Find the slenderness ratio.* Basic stress must be modified to allow for the slenderness of the wall. The effective height or length is found as shown in paras 1.12 and 1.13 and also the effective thickness (para 1.14). From these the slenderness ratio may be calculated, as shown in para 1.11. Using nomogram 2, place a straight edge to join the point on scale G (corresponding with the effective thickness of the wall) to the point on scale H corresponding with the wall height on H_1, H_2, H_3 or H_4 (whichever restraint is appropriate). The straight edge will intercept scale M, giving the slenderness ratio value.

3.09 CP 111 lays down maximum allowable values of the slenderness ratio for various types of construction (shown adjacent to scale M):

Table I Required mortar grade strengths (N/mm^2) in accordance with BS 4551. (See grade mixes, table V, TS Masonry 1)

Grade	7-day	28-day
I	11·0	16·0 N/mm²
II	5·5	8·0
III	2·75	4·0
IV and V	1·0	1·5

Table II Working basic stresses for units of standard brick format

Height/Thickness	= or less than 0·75					

Brick class	Strength	Mortar grade I	II	III	IV	V	VI
14	96·5 N/mm² and over	5·85	4·50	3·80	3·10	2·40	1·40
12	82·5	5·19	4·04	3·44	2·79	2·22	1·27
11	76	4·88	3·83	3·28	2·65	2·14	1·21
	75	4·83	3·80	3·25	2·63	2·13	1·21
10·5	72·5	4·72	3·71	3·19	2·58	2·09	1·18
10	69	4·55	3·60	3·10	2·50	2·05	1·15
9	62	4·12	3·27	2·85	2·31	1·91	1·11
8	55	3·69	2·94	2·61	2·13	1·76	1·07
7·5	52	3·50	2·80	2·50	2·05	1·70	1·05
7	48	3·27	2·63	2·35	1·96	1·63	1·00
6	41	2·87	2·33	2·09	1·80	1·51	0·92
5	34·5	2·50	2·05	1·85	1·65	1·40	0·85
4	27·5	2·05	1·70	1·60	1·45	1·15	0·75
3·5	24	1·85	1·58	1·45	1·30	1·05	0·72
3	20·5	1·65	1·45	1·30	1·15	0·95	0·70
2·5	17·5	1·47	1·30	1·20	1·06	0·88	0·66
	17	1·44	1·28	1·18	1·05	0·86	0·65
2	14	1·25	1·15	1·10	1·00	0·80	0·60
1·5	10·5	1·05	0·95	0·95	0·85	0·70	0·55
1	7	0·70	0·70	0·70	0·55	0·49	0·42
	5·5	0·55	0·55	0·55	0·46	0·38	0·34
0·5	3·5	0·35	0·35	0·35	0·35	0·23	0·23
	2·8	0·28	0·28	0·28	0·28	0·21	0·21

Table IIIa *Basic stresses for blocks of face work size* 390 × 190 (*metric modular*)

Thickness	Ratio: height/thickness	Type*	Strength N/mm²	Working basic stresses in mortar grades† N/mm²					
				I	II	III	IV	V	VI
75 mm	2·53	S	3·5	0·70	0·70	0·70	0·70	0·46	0·46
		N	3·5	0·70	0·70	0·70	0·70	—	—
90	2·11	S	3·5	0·70	0·70	0·70	0·70	0·46	0·46
		N	3·5	0·70	0·70	0·70	0·70	—	—
		S	7·0	1·16	1·13	1·12	0·95	0·78	0·68
		S	11·7	1·34	1·24	1·20	1·03	0·86	0·69
		N	14·0	2·50	2·30	2·20	2·00	—	—
100	1·90	H	2·8	0·54	0·54	0·54	0·54	0·40	0·40
		S	3·5	0·67	0·67	0·67	0·67	0·44	0·44
		N	3·5	0·67	0·67	0·67	0·67	—	—
		S	7·0	1·12	1·11	1·09	0·91	0·76	0·65
		N	10·5	2·02	1·82	1·82	1·63	—	—
		N	14·0	2·40	2·21	2·11	1·92	—	—
		N	19·3	3·05	2·65	2·48	2·21	—	—
140	1·36	H	2·8	0·42	0·42	0·42	0·42	0·31	0·31
		H	3·5	0·52	0·52	0·52	0·52	0·34	0·34
		N	3·5	0·52	0·52	0·52	0·52	—	—
		S	7·0	0·91	0·88	0·87	0·74	0·61	0·53
		H	9·0	1·02	0·96	0·94	0·81	0·67	0·55
		S	11·1	1·14	1·05	1·01	0·87	0·73	0·58
		N	10·5	1·56	1·42	1·42	1·27	—	—
		N	19·3	2·37	2·05	1·92	1·72	—	—
190	1·00	H	3·5	0·42	0·42	0·42	0·42	0·28	0·28
		N	3·5	0·42	0·42	0·42	0·42	—	—
		S	7·0	0·76	0·74	0·73	0·61	0·51	0·44
		H	8·2	0·84	0·80	0·79	0·66	0·56	0·46
		S	11·1	1·08	0·98	0·98	0·88	0·72	0·56
		N	10·5	1·26	1·14	1·14	1·02	—	—
		N	19·3	1·91	1·65	1·55	1·39	—	—

Table IIIb *Basic stresses for blocks of face work size* 440 × 215 (*to course with standard brickwork*)

Thickness	Ratio: height/thickness	Type*	Strength N/mm²	Working basic stresses in mortar grades† N/mm²					
				I	II	III	IV	V	VI
75 mm	2·87	N	3·5	0·70	0·70	0·70	0·70	—	—
		S	11·1	1·32	1·23	1·19	1·02	0·85	0·69
		N	14·0	2·50	2·30	2·20	2·00	—	—
90	2·39	N	3·5	0·70	0·70	0·70	0·70	—	—
100	2·15	H	2·8	0·56	0·56	0·56	0·56	0·42	0·42
		S	3·5	0·70	0·70	0·70	0·70	0·46	0·46
		N	3·5	0·70	0·70	0·70	0·70	—	—
		H	10·5	1·29	1·21	1·18	1·01	0·84	0·69
		N	14·0	2·50	2·30	2·20	2·00	—	—
140	1·54	H	2·8	0·46	0·46	0·46	0·46	0·34	0·34
		N	3·5	0·57	0·57	0·57	0·57	—	—
152	1·25	H	9·0	0·98	0·92	0·90	0·77	0·64	0·53
		S	11·1	1·11	1·02	0·98	0·88	0·72	0·56
190	1·13	S	3·5	0·46	0·46	0·46	0·46	0·30	0·30
		N	3·5	0·46	0·46	0·46	0·46	—	—
215	1·00	H	2·8	0·34	0·34	0·34	0·34	0·25	0·25
		N	3·5	0·42	0·42	0·42	0·42	—	—
220	0·98	H	7·5	0·79	0·75	0·74	0·63	0·52	0·44
230	0·93	S	3·5	0·40	0·40	0·40	0·40	0·26	0·26
255	0·84	N	3·5	0·37	0·37	0·37	0·37	—	—
305	0·70	N	3·5	0·35	0·35	0·35	0·35	—	—

*In tables IIIa and IIIb above, block capitals are as follows:
H a hollow block, that is 50 to 75 per cent solid with holes right through, or cellular, with the holes closed at one end.
S a solid block, which is at least 75 per cent solid material.
N a solid block with no voids, although it may be composed of a lightweight aerated material with small voids distributed evenly throughout the volume.
†See Technical study Masonry 1, Table V.

Up to 13 for unreinforced walls of brickwork and blockwork set in cementless mortars grades Vb and VI, except that:
13-20 is permissible for unreinforced walls of brickwork and blockwork using cementless mortars in buildings of not more than two storeys; also for walls less than 90 mm thick in buildings of more than two storeys
20-27 for walls of brickwork and blockwork set in cement mortars other than as above.

3.10 *Hence find basic permissible working stress.* Having found the slenderness ratio (r), and knowing the eccentricity (e) of the loading on the wall, the reduction factor (k_3) can be found from table 4 of the code or from the following formula: $k_3 = 1 - (r - 6)(0·0275 + 0·06 \, e/t)$
However, the reduced stress so obtained may be increased by 25 per cent, provided that the result is not greater than that obtained from the above formula, with e put equal to zero. (This has been allowed for in nomogram 2.) With the

slenderness ratio already found on scale M, find the corresponding point on the graph for the eccentricity ratio e/t on the horizontal scale N. This will lie in a zone with a particular reduction factor. Follow the curves and lines to arrive at the appropriate point on scale P. With a straight edge join this point with the point on scale L corresponding to the basic stress as given in tables II, IIIa or IIIb to obtain the maximum permissible working stress on scale Q.

Special cases

3.11 As mentioned previously, for piers and short lengths of wall, the code requires a further reduction factor to be applied to the basic stress in para 3.07. Using nomogram 2 join the basic stress on scale K with the wall or pier area on scale J to obtain a reduced basic stress on scale L before proceeding further.

3.12 Within one-eighth of the height of the wall from the lateral restraints, the code allows the factor k_3 for slenderness to be ignored. In these zones the maximum working stress permitted will be given directly on scale L.

3.13 Walls of zig-zag form in plan, or stiffened by piers or crosswalls at intervals have reduced slenderness ratios and hence increased permitted stresses under the code. This is beyond the scope of these notes.

3.14 Walls built in random rubble stonework should be designed with permitted stresses three-quarters of those for coursed work in similar material.

3.15 Walls built of more than one material should be assumed to be comprised of the weaker material over the full thickness, or of the stronger material for that material's thickness alone. There may however be added complications owing to differing elasticities, shrinkage and expansion properties.

Wind loads

3.16 BS CP 111 gives very little guidance on the design of walls to resist wind loads. A standard work on this subject is the Clay Products Technical Bureau's technical note *Wind forces on non-loadbearing brickwork panels*[5] by Bradshaw and Entwistle. More recent research, however, suggests that working stresses of 0.14 N/mm² in tensile bending, and 0.28 N/mm² in shear may be permitted.

3.17 The Technical Bureau's note assumes that the walls in question are not carrying any vertical loading other than their own weight, and so will give a conservative result for walls carrying other loads, ie where tensile stresses are reduced if not eliminated—the increase in value of compressive stress owing to wind loading will in most cases be very small.

4 Empirical method—Building Regulations

4.01 The empirical design rules are given in the Building Regulations 1976 for construction in England and Wales, the Building Standards (Scotland) (Consolidation) 1971, the Building Regulations (Northern Ireland) 1977 and the London Building Acts 1972. The 'rules' are largely similar but the London Building Acts cover any type of masonry structure up to a height of 12 m while the Building Regulations only refer to residential buildings up to 3 storeys high. The 'rules' are defined in Schedule 7 of the Building Regulations.

Limit of application

4.02 1 Only to residential buildings with a maximum wall or roof height of 15 m.
2 The width of the building or an outstanding wing of the

building must be at least half the height of the building or wing.
3 The building must have external walls on all but one of its sides or be comprised of sub-divisions so bounded.
4 The area of the building if wholly bounded, or sub-divided so bounded, shall not exceed 70 square metres and 30 square metres if bounded on all but one of its sides.

Materials

4.03 The wall must be constructed of bricks or blocks properly bonded and solidly put together in mortar or of stone flints, bricks or other burnt material laid in mortar otherwise than in horizontal courses.

Design conditions

4.04 1 The wall shall not exceed 12 m in height or length.
2 Each end of the wall is bonded or otherwise securely tied to a buttressing wall, pier or chimney.
3 That the wall in each storey extends to the full height of that storey.
4 That the wall does not support a floor having a span of more than 6 m measured between centres of bearings.
5 Where there is a difference of ground level either side of the wall in its lowest storey, the wall thickness measured as solid or the two thicknesses of a cavity wall shall be at least one quarter of the difference in level.
6 The wall shall not transmit a combined dead and super load exceeding 70 kN/m at its base.
7 The wall shall not transmit or sustain a lateral load other than wind load except in the case of (5) above.

Lateral support

4.05 Excepting certain walls in bay windows (covered in rule 11(2)), all walls must have at each end either a pier, a buttress, a buttressing wall or chimney **10**. A buttressing wall is any wall affording lateral support to the supported wall **11** but it must project from the supported wall a minimum distance of one sixth of its height at any level to the top, or 550 mm, whichever is the greater. In addition a buttressing wall with windows or recesses shall have such openings at least 550 mm from the supported wall unless

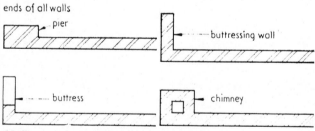

10 *Restraints at wall end.*

the opening is less than 0.6 square metres in area and such windows or recesses are not such as to impair the restraint provided by the buttressing walls.

4.06 A pier or buttress must extend upwards from the base of the wall to the full height of the wall or to the level of support from floor or roof. A pier must project from the wall at least twice the wall thickness and a width of at least 190 mm **12**.

4.07 A chimney used to support a wall must comply with the requirements of a pier in respect of the area of the horizontal section of solid material excluding fire place openings and flues and must have an overall thickness not less than twice the thickness of the supported wall.

Length of wall

4.08 The length of the wall is to be taken between the centres of the piers, buttresses, buttressing walls or chimneys at each end.

Height of wall

4.09 The height of a wall shall be measured from the level of the underside of the floor of a storey (or in the case of a lowest storey from the base of the wall) to the level of the underside of the floor above or roof lateral support.
Any other wall with a gable is measured up to half the gable height and a wall with no gable is to be taken to the highest part, excluding a parapet not more than 1·2 m high.

Thickness

4.10 Table IV gives the minimum required thickness for walls of various total lengths and heights but the minimum thickness shall not in any case be less than one sixteenth part of the storey height. If constructed of stone, flints, clunches of bricks or other burnt or vitrified material the wall shall be one and a third times the thickness quoted in table IV.

Cavity walls

4.11 The cavity wall must be provided, irrespective of its length, with lateral support by every roof which it supports and, if its length exceeds 3 m, with floor lateral support from each floor which it supports.
The leaves at each level must be not less than 90 mm in thickness and tied together by ties complying with BS 1243: 1972 or equivalent, the tie spacings being 900 mm horizontally and 450 mm vertically, staggered. At jambs or openings, in addition, there must be a tie at every 300 mm unless the edge is bonded.
The cavity width shall be not less than 50 mm, and not more than 75 mm except where vertical twist ties are used at 750 mm horizontally and 450 mm vertically when the cavity may be increased to 100 mm. As an alternative to the tie spacings quoted above the area supported by a tie shall be maintained. The sum of the two leaves plus 10 mm shall be not less than the thickness required for a solid wall of the same height and length.

Internal loadbearing walls

4.12 Internal loadbearing walls situated in the lowest storey of a 3 storey building carrying loads from both the upper floors shall be 140 mm thick or equal to other internal walls where the wall thickness plus 5 mm must be greater than half the thickness of an external wall, compartment wall or separating wall of the same height and length.

Annexes

4.13 Rule 14 deals with small buildings and annexes (eg verandas, greenhouses, garages, stores, lavatories—attached to houses) not more than 3 m high and 9 m in width (measured in the direction of the roof span).
Provided the wall is constructed of solid bricks and blocks and subjected to no load other than the wind and roof load and is bonded to piers or buttresses so that the length is reduced to 3 m between supports, its thickness may be a minimum of 90 mm.

Parapets

4.14 The thickness of a parapet wall shall be not less than one quarter of its height.
If the parapet is of solid construction it shall be 190 mm thick or the thickness of the supporting wall under whichever is less. If the parapet is of cavity construction it shall

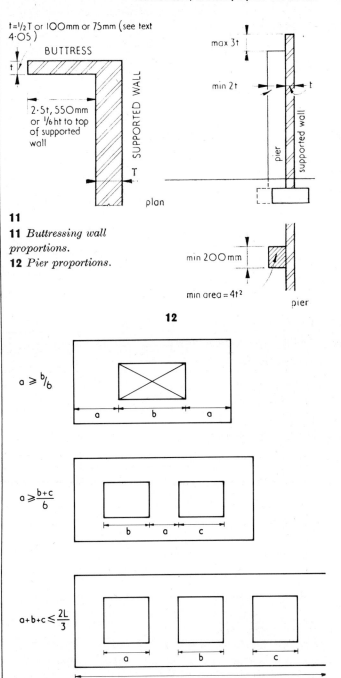

11 *Buttressing wall proportions.*
12 *Pier proportions.*

14 *Proportions for external walls, or walls common to adjoining buildings.*

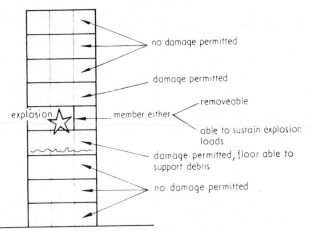

14 *Building Regulations: Requirement D17—anti-collapse provisions for buildings over five storeys including basements.*

Table IV *Minimum wall thickness related to total wall heights and lengths (all in mm) for solid external compartment and separating walls*

Height (m)	Length (m)	Thickness (mm)
< 3·5		190
> 3·5 < 9·0	< 9·0	190
	> 9·0 < 12·0	290 for one storey then 190
> 9·0 < 12·0	< 9·0	290 for one storey then 190
	> 9·0 < 12·0	290 for two storeys then 190

or 1/16th of the storey height or thickness of any other part of the wall which is supported by the part considered, whichever is the greater

Cavity external compartment or separating walls

Inner leaf (mm)	Cavity (mm)		Outer leaf (mm)	
⩾ 90	⩾ 50	⩽ 75	⩾ 90	ties 900 horizontally
⩾ 90	⩾ 50	⩽ 100	⩾ 90	ties 750 horizontally

The sum of thickness of the two leaves plus 10 mm shall not be less than the thickness required for a solid wall

Table V *Minimum wall thickness related to total wall heights and lengths (all in mm) for solid external and party walls*

Height (m)	Length (m)	Thickness (mm)
< 3·5		190
> 3·5 < 9·0	< 9·0	190
	> 9·0	290 for one storey then 190
> 9·0 < 12·0	< 9·0	290 for one storey then 190
	> 9·0	290 for two storeys then 190

Cavity external and party walls

Limiting length and height	Inner leaf (mm)	Cavity (mm)		Outer leaf (mm)
Single storey or top storey of other building Wall height ⩽ 3·0 Wall length ⩽ 7·5	⩾ 75	⩾ 50	⩽ 75	⩾ 90
Two storey building or top two storeys of other building Wall height ⩽ 7·5 Wall length ⩽ 9·0	⩾ 90	⩾ 50	⩽ 75	⩾ 90

be the thickness of the supporting wall under or by conditions previously specified.

Openings and recesses

4.15 The positions and areas of openings and recesses are governed by rule 6 and are illustrated in **13**.

Overhangs

4.16 Parts of a wall must not overhang the wall below if this will impair the stability of any part of the wall.

London Building Acts

4.17 The rules given in the London Building Acts are substantially similar to those summarised above but are not restricted to residential buildings. The wording is however much clearer. Table V illustrates the relationships of height, length and thickness.

Ronan Point—the fifth amendment

4.18 After the progressive collapse failure at Ronan Point, precautionary provisions were added to the Building Regulations. Originally known as the fifth amendment they now have a proper place in the revised regulations as requirement D17. This requirement applies to buildings over four storeys, including basements **14**. There are two alternative criteria: either a structural member must be capable of withstanding a substantial loading over its surface, as from an explosion; or the removal of that member must be shown not to cause further catastrophic damage to the building as a whole. The building is not required to be serviceable after the incident, so dead load, one-third superimposed load and wind load, and the special load of 34 kN/m² need each be multiplied by a safety factor of only 1·05. At this safety factor, the permissible stresses calculated from BS CP 111 may be multiplied by 3·5.

4.19 It is unlikely that many walls could be designed theoretically for a horizontal pressure on one side of 34 kN/m². So the rest of the structure must be designed to

tolerate the removal of any one panel, which is defined as a wall between supports (piers or cross-walls), or from an extremity to a support or to another extremity. The maximum length of wall which needs to be considered removed is 2¼ times the height of the storey (about 7·5 m in the average building).

4.20 Because the damage must be confined within the storey in which the incident occurs and the storeys immediately above and below, special design is not required for the top two storeys of any building. It is however essential to ensure that the fourth slab down (including the roof) is capable of carrying the debris impact load of the material above. Recent experiments by the BCRA[7] found that three storeys down from the top of a building the vertical loading on the brickwork prevented gas explosions having serious effect—hence the revisions to the London Building Constructional bylaws. But there remains the problem of the building's end wall where loading is less and presumably unevenly distributed across wall thickness, and where an alternative path of support is hard to achieve. A supplementary frame or extra wing walls may be needed. Of course the same rules apply to building in other materials, and so the competitiveness of masonry construction generally should not be affected. In fact, the major result of the fifth amendment and Regulation D17 has been to reduce drastically the amount of buildings over four storeys high.

4.21 The subject of progressive collapse in loadbearing masonry should not be left without considering the chance of such a failure occurring in medium rise cross-wall construction in blocks of substantial length. While such a failure has not yet occurred (as far as is known), the designer will be wise to provide at least an occasional bracing wall at right angles to the cross-walls.

B.S. 5628 Part 1. 1978

4.22 BS 5628 is the second limit state code of practice and deals with unreinforced masonry. Plain concrete walls originally covered in CP 111 are now dealt with in CP 110. Fortunately masonry in vertically loaded walls does not lend itself to the same structural performance of the reinforced concrete frame and BS 5628 is a much simpler document to understand and apply than was CP 110. It will therefore probably replace CP 111 by 1982.

The use of limit state enables the degree of risk to be varied by the choice of different partial safety factors. Masonry is obviously built to differing standards of workmanship and the partial factors of safety are varied in the Code to allow for this. However, workmanship in masonry is perhaps the most variable in quality of all constructions and the definitions in the Code of Practice for Walling CP 121: Part 1 leave much to be desired. The requirements for the highest standard are listed at length and are what the normal architect would expect under any circumstances while the ordinary standards are simply described as those not complying with the highest standards. There appears to be a singular lack of logic in the use of extended limit state

calculations and the loose requirement for high quality workmanship. The essential differences between CP 111 and BS 5628 lie in the method of assessment of vertical load carrying capacity calculation and the treatment of lateral load.

Vertical load

4.23 The difference in calculation method is illustrated below by comparing the two codes in the solution of a simple problem in which a wall of block-work 2·5 m high is required to carry a load of 18 kN per m run dead and 32 kN per m superimposed. The block is of 3·5 N/mm² strength with 1 : 1 : 6 mortar.

CP 111

4.24 Assume a block of face work 400 × 200 and 140 mm thickness. If the wall carries a floor producing the load of 50 kN per m then the effective height of the wall is 2500 × 0·75 = 1875 mm and the ratio of height to thickness

Nomogram 1 *Determination of actual stress*

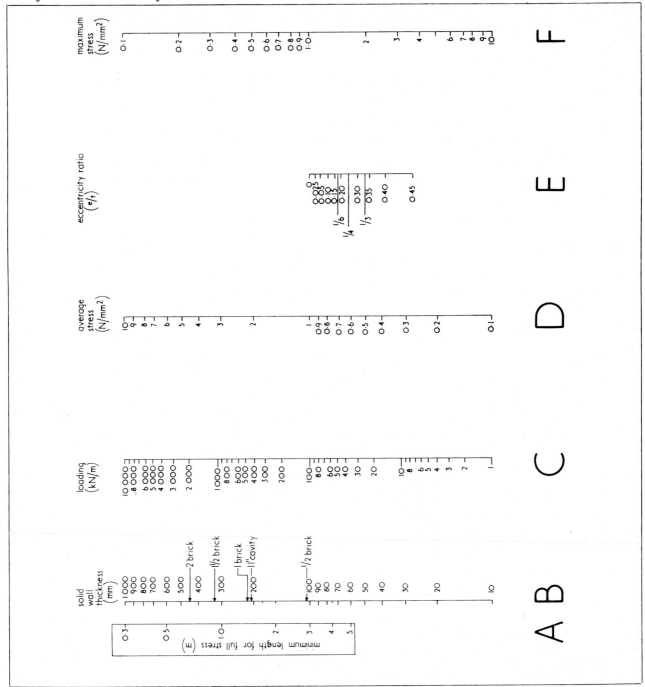

2 *Determination of maximum permissible stress*

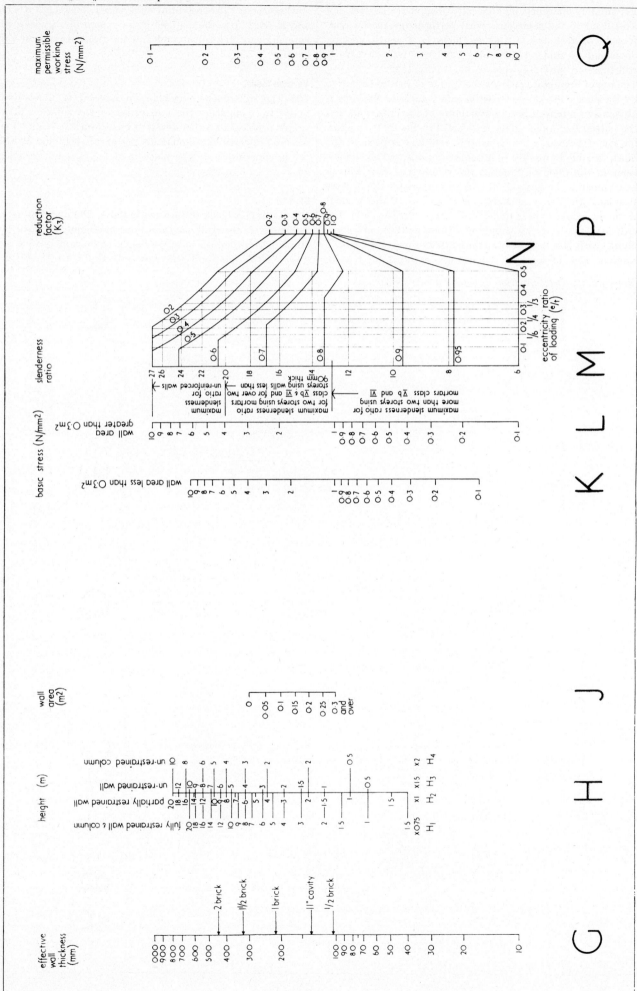

$$= \frac{1875}{140} = 13 \cdot 4 \, .$$

For this ratio the reduction coefficient to be applied to the basic stress of the block is 0·79.

The height to thickness ratio of the block is $\frac{200}{140} = 1 \cdot 43$ and this allows an enhancing factor of 1·54 to be applied to the basic stress. The basic stress from table 3b of CP 111 is 0·35 N/mm². The allowable stress is therefore $0 \cdot 35 \times 0 \cdot 79 \times 1 \cdot 34 = 0 \cdot 37$. The actual stress is $\frac{50 \times 1000}{1000 \times 140} = 0 \cdot 36$ N/mm² so the choice of 140 width block was adequate.

BS 5628

4.25 Again the effective height is $2500 \times 0 \cdot 75 = 1875$ and the slenderness ratio 13·4.

Assume normal standards of construction at site and special control of the manufacture of the blocks. This gives a partial safety factor for material strength γ m = 3·1.

The characteristic compressive strength of the block using 1 : 1 : 6 mortar is obtained from tables 2(b) and (d) of BS 5628:

BS 5628 *Table* 2(b) $h/b = 0 \cdot 6$

Mortar type	Compressive strength of unit (N/mm²)							
	2·8	3·5	5·0	7·0	10·0	15·0	20·0	35 or more
i	1·4	1·7	2·5	3·4	4·4	6·0	7·4	11·4
ii	1·4	1·7	2·5	3·2	4·2	5·3	6·4	9·4
iii	1·4	1·7	2·5	3·2	4·1	5·0	5·8	8·5
iv	1·4	1·7	2·2	2·8	3·5	4·4	5·2	7·3

BS 5628 *Table* 2(d) $h/b = 2 \cdot 0 — 4 \cdot 0$

Mortar type	Compressive strength of unit (N/mm²)							
	2·8	3·5	5·0	7·0	10·0	15·0	20·0	35 or more
i	2·8	3·5	5·0	6·8	8·8	12·0	14·8	22·8
ii	2·8	3·5	5·0	6·4	8·4	10·6	12·8	18·8
iii	2·8	3·5	5·0	6·4	8·2	10·0	11·6	17·0
iv	2·8	3·5	4·4	5·6	7·0	8·8	10·4	14·6

and is pro rata between the tables 2·77 N/mm² = f_k.

The capacity reduction factor f_k is obtained from table 7 of BS 5628 for a slenderness ratio 13·4 and an eccentricity which lies between 0 and 0·05t (where t = wall thickness) and is 0·9. The design vertical load resistance of the wall per 1 m length

$$= \frac{\beta t f_k}{\gamma \, m} = \frac{0 \cdot 9 \times 140 \times 2 \cdot 77 \times 1000}{3 \cdot 1} = 112590 \text{ N } or \text{ } 112 \cdot 6 \text{ kN}$$

The applied design load is determined by reference to the partial safety factor γf.

GK = characteristic dead load; QK = characteristic superimposed load.

For the dead load—design
load $= 1 \cdot 4 \text{ GK } = 1 \cdot 4 \times 18 = 25 \text{ kN/m}$
For the superimposed load
design load $= 1 \cdot 6 \text{ QK } = 1 \cdot 6 \times 32 = 51$

Total $= 76$

It will be seen that the resistance load is higher than the design load and therefore the choice of wall thickness was adequate.

Generally, the factors in BS 5628 have been chosen to give the same order of results as CP 111 for simple vertical loading but higher loads can be carried by using the BS 5628 method where eccentric loads are carried.

Lateral load

4.26 The CP 111 approach to lateral load was noted in Technical Study MASONRY 2 paragraph **3.16** and in information sheet MASONRY 1 paragraph **1.08**.

CP 121 gives certain empirical values for the maximum area of certain walls in table 2.1 related to different zones within the United Kingdom.

CP 121 *Table* 2.1

Wall type	Zone No.	Fixed on three sides	Pinned on one or more of the three supported sides
120 mm solid	1	16·0	12·0
walls or double	2	13·5	10·5
leaf cavity	3	11·0	9·0
	4	9·0	7·5
190 mm thick	1	27·0	22·0
solid walls	2	23·0	18·0
	3	18·0	14·0
	4	13·0	10·0
90 mm thick	1	8·0	6·0
solid walls	2	7·0	5·5
	3	6·0	5·0
	4	5·0	4·0

BS 5628 presents a far more scientific approach based on the results of laboratory work and provides tables of the relationship of flexural strength of a wall panel in the vertical to horizontal and of the relationship of the panel bending moments to the vertical and horizontal. Although BS 5628 defines all the values in the limit state method it is nevertheless possible to use the tables to calculate the solution using the ordinary elastic stress process. The tables are quite extensive and readers are referred to BS 5628 for further study.

References

1 BUILDING RESEARCH STATION Strength of brickwork, blockwork and concrete walls. BRS Digest 61. 1965, HMSO [(21·1) (K)]

2 BS CP 111: Part 2: 1970 Structural recommendations for loadbearing walls, metric units. British Standards Institution [(21·1) (F)] £2.00

3 The Building Standards (Scotland) (Consolidation) Regulations 1971 SI 2052 (S218), HMSO, £1.30; The Building Standards (Scotland) Amendment Regulations 1975 SI 404 (S51), £0.29p; The Building Standards (Scotland) Amendment Regulations 1973 SI 794 (S65), £0.21p [(A3j)]

4 BS CP 3: Chapter V: Part 1: 1967 Dead and imposed loads. British Standards Institution [(K4)] £1.10

5 CLAY PRODUCTS TECHNICAL BUREAU Wind forces on non-loadbearing brickwork panels, Technical note volume 1, no 6. Brick Development Association, 3–5 Bedford Row, London WC1 [(K4f)] o/p

6 ELDER, A. J. Guide to the Building Regulations 1976. London, 6th edition 1978, Architectural Press £8.95 (paperback £5.75)

7 ASTBURY, N. F. et al. Gas explosions in loadbearing brick structures SP 68: 1970 £2.50

8 GLC London Building Acts 1930–1939; London Building (Constructional) Bylaws 1972 [(Ajn)] £1.20 paperback

9 PITT, P. H. and DUFTON, J. Building in Inner London 1976. Architectural Press. £5.50 paperback

Information sheet
Masonry 1

Section 8 **Structural material: Masonry**

Example calculations

Technical study MASONRY 2 *by* DAVID ADLER *compared the analytical masonry design method of* CP 111 *with the empirical design approach of the Building Regulations. This information sheet, by the same author, now illustrates these methods by worked examples. The two articles should therefore be read in conjunction.*

1 Design of external wall by analytical method of CP 111[1]

1.01 Figure **1** shows plan and section of a two-storey domestic building. The walls are to be of cavity blockwork, the trussed roof spanning to plates resting squarely on each wall-head, the intermediate timber floor supported by joist hangers and the ground floor slab separated from the walls around its edges by a 12 mm soft joint. A design for the central wall is also required.

Block thicknesses
1.02 The minimum block thickness allowed in cavity walls is 75 mm (CP 111, clause 308b). but this would be insufficient here and a leaf thickness of, say, 90 mm with a 50 mm cavity will be tried first. Blockwork density varies with composition and strength, but a density of 800 kg/m³ may be assumed at this stage. Later, when actually selecting the make of block to be used, this figure must be checked and the calculation adjusted if necessary.

Wall weight
1.03 The calculations have to be worked in force units. and the blockwork mass density converted accordingly:

mass 1 m³ blockwork = 800 kg

thickness of solid material is 2 × 90 mm = 180 mm

hence 1 m² on elevation has mass 0·18 m × 800 kg

which has a weight of

$$\frac{0 \cdot 18 \times 800 \times 9 \cdot 81 \text{ N}}{1000} = 1 \cdot 41 \text{ kN/m}^2$$

Maximum force on wall and its eccentricity
1.04 The absolute maximum load is not relevant, since the main constraint here in calculating the allowable stress is slenderness ratio and not permissible compressive strength. The factor due to slenderness ratio need not be applied to sections of wall within one-eighth of the wall height above or below lateral supports (CP 111, clause 315f). The load must therefore be found

$$\frac{3 \cdot 5}{8} = 0 \cdot 44 \text{ m up from base (ie half way up foundation)}$$

1.05 Considering first loads applied to the wall centre (the concentric loads)

wall self weight = unit weight × height
= 1·41 kN/m² × (2·9 m + 3·5 m — 0·44 m)
= 8·40 kN/m run

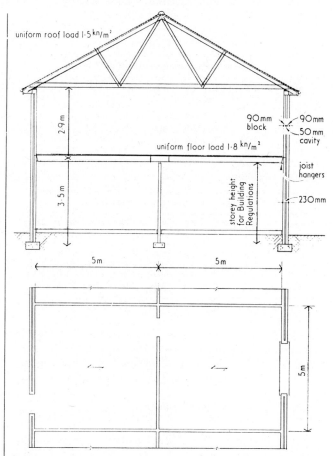

1 *Example plan and section used in typical calculation,* see text

Roof weight
1.06 1 *Materials self weights* (timbers, tiles. ceiling) are found in BS 648[2]. Note that for loads on slope:

$$\text{plan load required} = \frac{\text{roof weight}}{\text{cosine roof pitch}}$$

2 *Superimposed loads*, mainly from snow, are found in CP 3: chapter V: part 1[3].

Assume here roof load on wall to be found at 1·5 kN/m²

load on wall = unit weight × ½ roof span

$$= 1 \cdot 5 \text{ kN/m}^2 \times \frac{10 \text{ m} + (2 \times 0 \cdot 23 \text{ m})}{2}$$

$$= 7 \cdot 85 \text{ kN/m}$$

Hence total concentric load = 8·40 + 7·85 = 16·25 kN/m

Intermediate floor weight

1.07 The intermediate floor load is not concentric. First, the unit loading is found as for the roof. In this case, the sum of dead and superimposed loads will be assumed to be $1 \cdot 8$ kN/m² (including first floor partitions).

Eccentric load = unit weight × $\frac{1}{2}$ floor span

$$= 1 \cdot 8 \times \frac{5}{2}$$

$$= 4 \cdot 50 \text{ kN/m}$$

As shown in Technical study MASONRY 2 (hereafter referred to as TS 2) para 1.09, joist load using hangers is taken to act 25 mm from internal face of wall.

Therefore load eccentricity = 25 mm + $\frac{1}{2}$ wall thickness

$$= 25 + \frac{230}{2}$$

$$= 140 \text{ mm}$$

Wind load

1.08 Wind pressure calculation is based on CP 3: chapter V: part 2[3]. The method is somewhat involved and it is sufficient here to take an average figure of 500 N/m². (Depending on the part of the country and the building's exposure, exceptional cases apart, figures could vary between 250 and 1500 N/m².) Applying this figure to the lower larger wall panel gives a bending moment in the middle of the panel. Though this is not strictly the point under consideration, the resulting approximation is on the safe side.

1.09 TS 2, para 3.16 noted that research work had established values for wind load moments on unloaded panels. In this example, the panel is loaded and the tensions do not apply, but the compression stresses should be added to those resulting from the vertical loads.

1.10 The value of the bending moment factor depends on the ratio of panel edges and on whether the edges of the panel are continuous, ie 'continuous' with adjacent panels. All four edges must be held in position to comply with the theory. In this case, there is a panel above and on either side but the lower edge rests on the strip foundation and this edge is thus not continuous.

1.11 The ratio of the panel sides is $\frac{5}{3 \cdot 5} = 1 \cdot 43$

Three sides of the panel are continuous (the bottom one not so), so referring to the graph W in nomogram 3, top line, a panel ratio of 1·4 indicates a bending moment factor of 16.

Bending moment = $\dfrac{\text{wind force} \times \text{least panel dimension}}{\text{bending moment factor}}$

$$= \frac{(500 \times 3 \cdot 5) \text{ N/m run} \times 3 \cdot 5}{16}$$

$$= 383 \text{ Nm/m run}$$

$$= 383 \text{ kNmm/m run}$$

Find total resultant eccentricity

1.12 The resultant eccentricity of all the loads on the wall has now to be found. This is done by taking moments (see Technical study ANALYSIS 1, para 2.06 et seq) about the central axis of the wall.

Adding the vertical loads:

total concentric load (para 1.06) = 16·25
eccentric intermediate floor load = 4·50
total resultant = 20·75 kN/m

moment of resultant = moment of concentric load
+ moment of eccentric load
+ moment of wind load

but moment of concentric load is zero since it acts as point around which moments are being taken. Hence

resultant × its eccentricity

$$= \frac{(\text{eccentric load} \times \text{its eccentricity}) + \text{wind moment}}{\text{total resultant}}$$

$$= \frac{(4 \cdot 50 \times 140) + 383}{20 \cdot 75}$$

$$= 48 \cdot 8 \text{ mm}$$

the eccentricity ratio = $\dfrac{\text{eccentricity}}{\text{overall wall thickness}}$

$$\frac{48 \cdot 8}{230} \text{ mm}$$

$$= 0 \cdot 21$$

Hence find actual stress

1.13 To find the actual stress in the wall at this level, turn to TS 2, nomogram 1. Scale B requires the solid wall thickness which is $2 \times 90 = 180$ mm. Join this point with the point on scale C corresponding to the total load of 20·75 kN/m, found above, and extend to cut scale D at an average stress value of 0·116 N/mm². Join this point with the point on scale E corresponding to the eccentricity ratio of 0·21, and extend to read the maximum stress on scale F of 0·27 N/mm².

The block and mortar required

1.14 To find the type of block and mortar that will carry this stress, turn to TS 2, nomogram 2. Here, scale G refers to effective wall thickness. For a cavity wall, this is 2/3 of the actual thickness of solid material (TS 2, para 1.14), ie here is $2/3 \times 180 = 120$ mm. Join this point on scale G with the point on scale H_2 corresponding to the effective height of the wall. (TS 2, para 1.13). In this case, the wall is fully restrained, so the actual height, 3·5 m, is read on scale H_1 and transferred across to scale H_2 to a value of 2·6. The line joining 120 mm on scale G with this point is extended to cut scale M at a slenderness ratio of 21·9. It is seen from the comments against this scale that this value is above the limit for two-storey buildings using mortars weaker than grade Vb and VI (ie, cementless mortars).

1.15 The eccentricity ratio of 0·21 is projected upwards from the horizontal scale N to meet the projection of 21·9 horizontally from scale M at a point which can be seen to be about midway between the curves representing 0·4 and 0·5. Following these round, scale P is cut at a reduction factor value of 0·44.

1.16 The maximum permitted working stress that is required is 0·27 N/mm². Join this point on scale Q with the point on scale P 0·44 found above and extend back to cut scale L at a value of 0·61 N/mm². This is the basic stress that is required. Referring to TS 2 tables IIIa and IIIb, it can be seen that a block of 3·5 N/mm² strength (using two 90 mm skins) in a mortar equal to or better than grade IV, ie stronger than grade Vb above, will suffice.

2 Design of internal wall by CP 111

2.01 First try a single 90 mm skin in the type of block above. On nomogram 2, join 90 mm effective thickness on scale G with the point on scale H corresponding to 3·5 on scale H_1—the wall being fully restrained. The extension of this line falls outside the slenderness ratio scale M, showing that the wall is too thin for the height.

2.02 Repeating the process for a wall 100 mm thick, the slenderness ratio reading is just on scale M at 26·3. Following round the curves, the reduction factor on scale P is 0·43. Using a 3·5 N/m² block with grade IV mortar it will be seen from TS 2 table IIIa that the basic stress is 0·67 N/mm² for a block of face size 390 mm × 190 mm. Table IIIb shows a slightly higher value for the 440 mm × 215 mm face size —this is owing to the effect on the shape factor. Using the lower value, join the point on scale L with the point just found on scale P to read a maximum permissible stress of 0·29 N/mm² on scale Q.

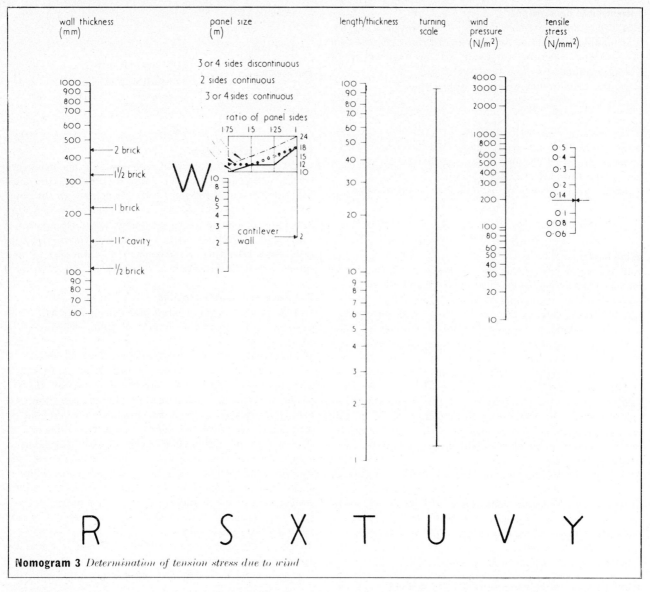

Nomogram 3 *Determination of tension stress due to wind*

2.03 The unit weight of this 100 mm block wall is

$$\frac{0 \cdot 100 \text{ m} \times 800 \times 9 \cdot 81}{1000} \text{ (as para 1.03)} = 0 \cdot 79 \text{ kN/m}^2$$

All the load on this wall is concentric, and at a level $\frac{1}{8}$ of the height up from the base,

floor load   = unit load × span
               = 1·8 kN (from para 1.07)
               × 5   = 9·0 kN/m
self weight   = unit weight × height
               = 0·79 × (3·5 — 0·44)   = 2·42 kN/m
               total load   = 11·42 kN/m

2.04 On nomogram 1, join the point 100 mm wall thickness on scale B with 11·42 kN/m on scale C and extend to cut scale D at an average stress of 0·11 N/mm². As the load is all concentric, the maximum stress is the average stress, and as this is less than the permissible stress of 0·29 N/mm² found above, the design is satisfactory. In fact, if a block of strength 2·8 N/mm² is used with a basic stress from TS 2 table IIIA of 0·54 N/mm², the permissible stress from nomogram 2 is found to be 0·23 N/mm², which is still more than the actual stress.

Cavity wall with brick external leaf
2.05 If it is desired to construct the external walls with a brick external leaf and block internal leaf, there are two approaches to the problem. As the brick is likely to be considerably stronger than the blocks (most bricks have crushing strengths in excess of 20 N/mm²), it might be

advantageous to assume that the external leaf carries all the loads. The effect of the inner leaf can still be taken into account when calculating the slenderness ratio (TS 2, para 1.14). However, this method is not suited to the example here, where the intermediate floor is carried by the internal leaf only. The alternative is to assume that the whole wall is composed of the weaker material. (Not stated explicitly, but implied in CP 111, clause 314b).

3 Design of external and internal wall by the Building Regulations

3.01 The same example may now be used to illustrate the empirical design approach of the Building Regulations 1972, schedule 7[5] as described in TS 2, paras 4.01 to 4.17.
3.02 Total height of wall 3·5 + 2·9 = 6·4 m
Length of wall between buttresses = 5 m
type of building- rule 7. (TS 2, para 4.07).
3.03 The term wall thickness in the building regulations is taken as the solid material thickness, ie. 100 + 100 = 200 mm. From TS 2, table V, walls to rule 7 must have

$$\text{thickness} = \frac{\text{storey height}}{16}$$

Storey height will be less than the 3·5 m shown **1**, since the regulations measure the dimension from the top of the foundation to the underside of the joist above (TS 2, para 4.06). Assuming a value then of 3·2 m

wall thickness required $\dfrac{3 \cdot 2 \times 1000}{16} = 200$ mm as above.

3.04 From TS 2, table V, the internal wall must be such that

$$\frac{\text{storey height}}{\text{wall thickness}} = 32$$

$$\therefore \text{thickness} = \frac{3 \cdot 2 \times 1000}{32}$$

$$= 100 \text{ mm as above.}$$

4 Non-loadbearing panels

4.01 It is often necessary to check the stability of a non-loadbearing block infill panel, eg, in a reinforced concrete frame in a tall building. The example panel is as shown **2** and the wind pressure (from CP 3: Chapter V: Part 2[4]) is 100 N/m².

4.02 The panel size is the length of the shorter side of the panel, ie, 2 m. The ratio of the panel sides is $\frac{3}{2} = 1 \cdot 5$.

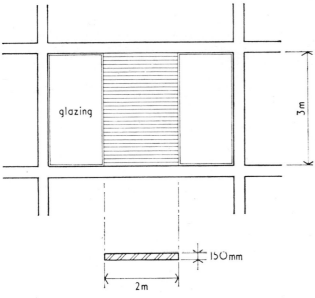

2 *Example non-loadbearing masonry panel, see calculation in text*

4.03 On nomogram 3, join the 150 mm point on wall thickness scale R with the 2 m point on panel size scale S. Extend to cut scale T at an $\frac{L}{t}$ value of $13 \cdot 3$. Join this point to 1000 N/m² value on wind pressure scale V. Note where this cuts the 'turning' scale U. Find the point on the horizontal scale W corresponding with the ratio of panel sides $1 \cdot 5$, and extend downwards to meet the line representing two sides continuous (see para 1.10, this study) at a resulting value on scale X of 12. From this point on scale X, draw a line to the point on scale U and extend to cut the tensile stress scale Y at a value of $0 \cdot 1$ N/mm². As this is less than the $0 \cdot 14$ N/mm² permitted (TS 2, para $3 \cdot 16$) the design is satisfactory.

4.04 Repeating the exercise for a half-brick panel of the same elevation it is found that the result is now $0 \cdot 16$ N/mm². This exceeds the allowable stress and is not permissible.

5 Garden wall design

5.01 It is required to find the permissible height for a 225 mm thick brick garden wall, if the expected wind pressure is 600 N/m².

Analytical method of CP 111
5.02 The wall has no load other than its own weight, so the method for non-loadbearing walls is applicable (see para 4).

Using nomogram 3 and starting on the right-hand scale Y, join the allowable stress $0 \cdot 14$ N/mm² with the point on scale X corresponding to a factor of 2, ie for a cantilever wall as marked. Note the point where this line cuts scale U. From the 600 N/m² wind pressure point on scale V, draw a line through the point on scale U to cut scale T at an $\frac{L}{t}$ value of $8 \cdot 8$. Join this point with the point on wall-thickness scale R corresponding to a one brick wall to obtain a maximum height of $1 \cdot 98$ m.

Shear stress check
5.03 The total horizontal force on one metre length $= 600 \times 2 = 12 \cdot 00$ N. The shear stress on the lowest bed joint will therefore be

$$\frac{\text{total force}}{\text{area}} = \frac{1200}{225 \text{ mm} \times 1000 \text{ mm}} = 0 \cdot 005 \text{ N/mm}^2$$

This is less than the $0 \cdot 28$ N/mm² allowed and hence the wall is satisfactory.

References

1 BS CP 111: Part 2: 1970 Structural recommendations for loadbearing walls, metric units [(21·1) (K)] £2.00

2 BS 648: 1964 Schedule of weights of building materials [Yy (F4)] £2.00

3 BS CP 3: Chapter V: Part 1: 1967 Dead and imposed loads. [(Ajr)] £1.10

4 BS CP 3: Chapter V: Part 2: 1970 Wind loads [(K4)] £1.30

5 The Building Regulations 1976, SI 1676, HMSO [(A3j)] £3.30; The Building (First Amendment) Regulations 1978, SI 723, HMSO £0.60p

6 BRICK DEVELOPMENT ASSOCIATION Technical note vol 1, no 6. Wind forces on non-loadbearing brickwork panels. The Association, 3–5 Bedford Row, London, WC1R 4BU [(21·3) Fy (K4f)] o/p

7 Low Rise Domestic Construction in Brick—1—Structural Design. Brick Development Association, 1979, £2.00

8 External walls: Design for Wind Loads. Brick Development Association, 1978

9 Brickwork retaining walls. Brick Development Association 1977

Technical study
Composite structures 1

Composite structures: definition and types

Many composite structural types tend to be categorised under the heading of one or other of their constituent materials, or have become so well known as to be regarded as materials themselves. Reinforced concrete and GRP, both discussed elsewhere in the handbook, are examples of this. Consequently, DAVID ADLER confines the scope here to explaining the ways in which composite action can occur and indicating some of the more common composite types.

1 What is composite?

1.01 All structures are to a greater or lesser extent composite. Everything attached to the structural elements of a building, including the loads carried, make alterations to the characteristic behaviour of the structure. For example, the cladding on a building frame usually makes a substantial contribution to the stiffening.

1.02 One way to economise in building is to find ways of utilising these hidden reserves of strength; another is to avoid using material wastefully. Often both objectives can be met by combining different structural elements and/or materials so as to use the strength of one to compensate for the weakness of the other. In such hybrids—composite structures, or composite structural elements—the essential factor is that the resulting strength is greater than the sum of the strengths of the components. If this is not so, the structure is not composite, only combined **1**.

1.03 It is vital that the materials comprising the composite should be fully compatible under all the conditions likely to be met during the life of the structure of which they form a part. A piece of timber supplemented by an aluminium strip suddenly subjected to heat from the sun would most likely distort or break its bond with the metal.

2 Types of composite structure

Reinforced concrete
2.01 Many composites are so well known that their nature is either not realised, or is forgotten. The commonest of these is reinforced concrete. Of the two materials used here, the thin steel bars are strong in tension and though strong in compression are liable to buckle. Concrete, strong in compression and with little or no tensile resistance, is by far the cheaper of the two materials and so for compression purposes it is cheaper to use more of the concrete rather than less of the expensive steel. However, where steel is used in compression the concrete surround helps to prevent buckling **3**.

2.02 Concrete can be reinforced with other materials including bamboo, fibre composites (see Technical study STRUCTURAL INNOVATION 1) and glass. At the other end of the scale are the straw reinforced earth walls common to old cottages, or in less developed countries the use of vegetable material to reinforced mud.

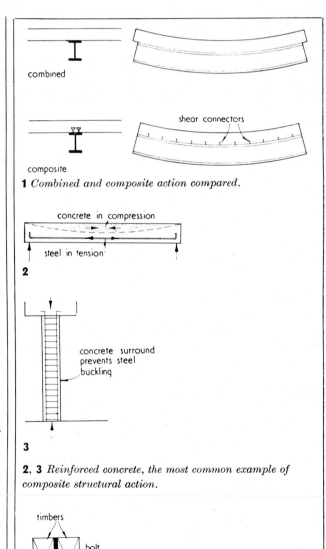

1 *Combined and composite action compared.*

2

2, 3 *Reinforced concrete, the most common example of composite structural action.*

4 *Flitch plate beam, timber and steel composite.*

5

Flitch plates

2.03 Before the use of cast iron or steel in structures became common, timber was strengthened with wrought iron plates. These 'flitch plate' beams were used either with the steel carrying the load, and the timber restraining it from buckling, or with the load shared between the timber and the steel **4**. Although flitch plates are rare in modern construction, they have their uses in strengthening or removing the sag from old timbers in existing buildings. Steel ties can be tensioned to take up load already on the member, or the member jacked and the steel plates added **6**.

Slabs and panels

2.04 Composite action is employed in the channel-reinforced woodwool slabs used for roofing. The channel increases bending resistance while being restrained by the woodwool from twisting **7**.

Blockboard

2.05 Here plywood acts as a compression and tension skin to the cheap bulk timber between. The loads are transferred through the essential medium of the glue.

2.06 Egg-box infill panels are becoming popular in many cladding situations. Two skins of plasterboard (itself a composite material) are separated by cellular cardboard. Facings of suitable materials can provide finished cladding units without further treatment **8**. Similarly, hardboard can be made to span by combining it with strip timber, provided that sufficient nails or glue are used to connect the materials firmly.

GRP

2.07 Analogous to reinforced concrete is grp glass reinforced plastic. Strands of glass fibre supply the tensile resistance to the more brittle resin matrix. The composite material can itself act compositely, eg as cladding panels adding stability to a structural frame **5**.

5 Grange farm, Harrow. Here GRP, itself composite, also acts compositely with timber. The panels, providing racking resistance. each contain timber stiffeners, so that once assembled the timber provides a continuous frame.

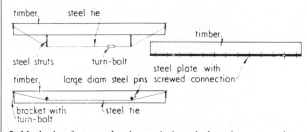

6 Methods of strengthening existing timbers by composite action of steel.

7 Channel reinforced wood wool slab.

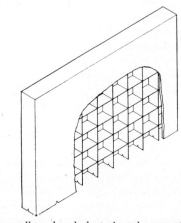

8 Cellular cardboard and plasterboard composite infill panel.

9

2.08 Such composite materials can be used without requiring any deep understanding of their nature. There are, however, two forms of composite construction frequently used nowadays where an appreciation of the mode of action is important—the combination of reinforced concrete with structural steel, and the combination of masonry with structural steel or reinforced concrete.

Reinforced concrete—structural steel

2.09 A common form of construction is reinforced concrete floor slabs supported on a structural steel frame. When the whole building is of reinforced concrete, it is usual to design the beams as I-beams, with the slabs acting as compression flanges spanning between the columns **10a**. There is no reason why the same method should not be applied using structural steel beams **10b**. In most cases a mechanical bond must be provided between the beam and the slab to carry the horizontal shear and this is the mechanism by which the action becomes composite. Shear connectors for this purpose commonly take the form of special studs welded onto the top of the beam and cast into the concrete of the slab **9**, **10b**, **11**. Alternative methods are shown **12**.

2.10 The design of such simply supported composite beams is covered in BS CP 117[1]. The method involves finding the depth of the neutral axis for the combined beam and slab. The neutral axis will usually be within the slab depth and at the collapse load all the concrete above this level will be at the ultimate compressive stress allowed (4/9 of the concrete cube strength) while the steel beam will all be at the yield stress for the particular steel grade employed. As the total compression must be equal to the total tension, and as either multiplied by the lever-arm between them must equal the applied ultimate moment, it is a simple matter to check that a particular arrangement is safe. The code also gives tables allowing the strength of the shear connectors to be ascertained.

Masonry—reinforced concrete/steel

2.11 The traditional design method for a lintel supporting a masonry wall was to consider the lintel as carrying a triangular portion of the wall while the rest of it arched over. This is shown in **13** where the brickwork outside the triangle

9 *Raith bridge, river Clyde. Welded studs enable steel sections to act compositely with concrete deck.*

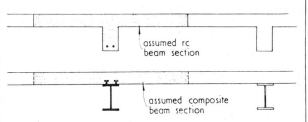

10 *Reinforced concrete floor* **a** *rc down-stand* **b** *with structural steel beams acting compositely.*

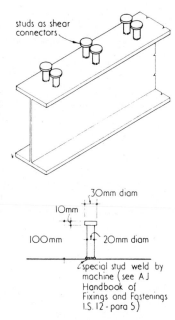

11 *Studs act as shear connectors between beam and rc placed above.*

exerts a compressive force on the flanking columns. Work at the Building Research Station over the last 20 years has confirmed that an arching action does occur, involving composite action between the beam acting in tension, while the wall or part of it takes the compression. Because most wall panel heights are of the same order of magnitude as the span or length of the wall, the classic bending theory cannot apply (see Technical study ANALYSIS 1 para 3.40). An empirical design method has been given in BRS research paper 13[2] and current paper 26/69[3].

2.12 In contrast to the steel-concrete construction, there is normally no need here for any special bond between the beam and the panel, nor for particular precautions during masonry construction. The composite action grows as quickly as the weight of the masonry increases.

2.13 However, the effect of window and door openings in the masonry does have to be considered. There is substantial composite action with large openings near to the supports, whereas small openings and openings near mid-span have little appreciable effect.

2.14 The design rules for a span of L **14** are as follows:

1 the minimum height of the wall must be $0 \cdot 6 \times L$

2 the beam depth must be between $\dfrac{L}{15}$ and $\dfrac{L}{20}$

3 the beam reinforcement should be designed for a bending moment of

$\dfrac{WL}{100}$ for a plain panel or where openings occur only at midspan

and $\dfrac{WL}{50}$ for a wall with openings near the supports

4 if a steel section is to be used instead of a reinforced concrete beam, it should be encased in concrete, and designed to carry a moment of $\dfrac{WL}{50}$

5 If a load, eg from a floor, is applied at the level of the beam, it is necessary to tie the beam to the panel above with some reinforcement to carry tension, such as strips of expanded metal in the vertical joints, or rods in the cavities of hollow blocks.

2.15 New forms of composite action are being utilised all the time. A particular field at the moment is the use of relatively light cladding materials to stabilise a framework, particularly against racking forces. The roof of the new Covent Garden Market Building at Nine Elms has been designed in this way, in conjunction with research work at Salford University.

References

1 BS CP 117: 1965. Composite construction in structural steel and concrete: Part 1 Simply supported beams in building [(2–) Gy (K)] 90p

2 NATIONAL BUILDING STUDIES Research paper 13. Studies in composite construction Part 1 The composite action of brick panel walls supported on reinforced concrete beams. 1969, HMSO [(23)] O/P

3 BUILDING RESEARCH STATION Current paper 15/69 1959. Studies in composite construction. An elastic analysis of wall beam structures. J. R. Colbourne [(2–)] 15p

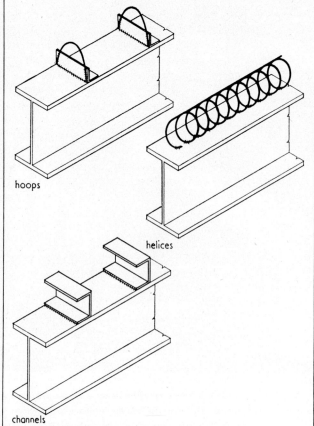

hoops

helices

channels

12 *Types of shear connector.*

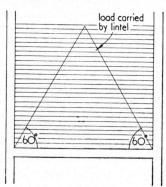

13 *Composite action between masonry panel and concrete frame. Area of brickwork outside triangle assumed to arch between flanking columns.*

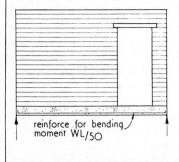

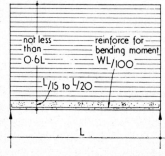

14 *Design rules for masonry panels on rc beams:* **a** *plain panel;* **b** *panel with opening.*

AJ Handbook
Building structure
Cl/SfB (2-)

Section 10
Structural
innovation

BP North Sea drilling rig Sea Quest, located east of Aberdeen.

Section 10

Structural innovation

Scope

This section of the handbook is concerned with 'progress' — progress in building techniques over the last decade or so and also the areas of progress that can be realistically forecast. Of course, new ideas and innovations can only happen within a framework of environmental and practical limitations and the first study outlines some of these. What are the factors determining where and how we build? What improvements can be expected in the strength of materials? How high the structures? How far the spans? The second study illustrates the theme, touching on points of architectural excitement and novelty.

Authors

The authors are G. M. E. COOKE and WILEM FRISCHMANN of Pell Frischmann & Partners; SIMON DOWLING of Simms Group Research & Development Ltd; and MALCOLM QUANTRILL, head of the department of architecture at the Polytechnic of North London.

Gordon Cooke

Simon Dowling

Wilem Frischmann

Malcolm Quantrill

Technical study
Structural innovation 1

Section 10 **Structural innovation**

The environmental
and practical constraints

This study relates the current possibilities for innovation in building with the relevant environmental and social contexts and with the practical constraints imposed by materials strength and structural form. The major part of the article is by G. M. E. COOKE *of Pell Frischmann & Partners, and the sections on structural and material constraints by* SIMON DOWLING, *a design engineer with Simms Group Research and Development.*

1 Environmental constraints to innovation

1.01 The constraints placed upon structural innovation are not solely confined to limitations in the physical properties of materials; although this is perhaps the major constraint at present it is fortunately being rapidly eroded by technological advance. Social, environmental and ecological constraints also exist: they gain increasing importance as population densities increase, more natural resources are consumed, and increasing environmental pollution becomes incompatible with the need to preserve satisfactory living conditions.

Transportation
1.02 Many of our current environmental problems emanate from the sudden increased mobility of man and materials that occurred during the 19th century. The siting of sprawling railway termini in central urban areas in the 1850s not only eliminated thousands of homes but also provided physical and social barriers between adjoining residential areas which were downgraded by the effects of noise and smell of trains. As a result, innovatory approaches were necessary to solve related structural problems such as design against airborne and structure-borne sound transmission, and design of structures to raft over the railway area providing homes and recreation spaces while retaining the transport system below.
1.03 The problem of the conflicting environmental and economic needs of society is repeated again today, as illustrated by grave public concern over the siting of airports near residential areas. Such environmental constraints again call for solution by structural innovation, eg offshore construction.

Shortage of natural resources
1.04 Ecological constraints begin to be felt through the growing shortage of natural resources, notably natural aggregates, but innovatory forces are already at work to develop alternative materials, perhaps utilising synthetics.
1.05 The demand for aggregates has accelerated rapidly in recent years and is currently running at about 200 million tonnes p.a. It is becoming increasingly difficult to supply the country's requirements from natural sources and there is a danger that shortage may inflate their price and the construction cost seriously impair the nation's social and economic position. This is not helped by the present public concern over the increasing acreage of land required for mineral working and the environmental disturbance and loss of amenity which follows.
1.06 The Secretary of State, aware of the implications of regional if not national shortages of natural aggregates, has set up urgent studies into demand, resources and alternative materials and methods with the principal objective of ensuring that high quality natural aggregates are not wastefully employed. It is expected that waste materials such as colliery shale, china clay waste, various slags, clinker and ashes could be used to manufacture artificial aggregates. The limiting factor, however, is transport cost, which is high in relation to the low material cost at source. The transport of 'crude' minerals, mainly aggregates, accounted for 26 per cent of total road freight in 1968, and there being no wish to increase this usage it is likely that long-haul transport will be by rail or water. Also it makes good sense to site production plants near to the indigenous raw material sources rather than transport the materials to production plants far away, often in other countries. This trend is not only likely to lead to reduced costs in finished products leaving the works but also, importantly, will make basic building products and the accompanying technology available in areas previously devoid of them. Hence innovation may accelerate in areas such as the under-developed countries, and new structural forms may emerge as a result.

Regulation authorities
1.07 Building Regulations, which properly serve the interests of public health and safety, often act as a brake on structural innovation. It is often said that it is easier to innovate within the framework of fire regulations abroad than it is in the UK **1**. This may be true, but not because of a lack of ideas in the UK but rather because it is more difficult to prove to the regulation authorities that the idea is safe in practice under all likely conditions. Where unprotected structural steel members are used in buildings of very low fire load, tests can show that a complete burn out would not lead to critical temperatures in the steel.[1] In some European countries such data have been used especially to obtain waivers from the regulation authorities but in the UK it has often been inadmissible as the worst possible yet

1 *Michelson building, Newport Beach, California. Entire structural system of water filled, unprotected steel.*

realistic combination of parameters has not been examined in the test programme. Also, some overseas authorities are empowered to enter buildings to examine conditions and have hazards such as abnormally high fire loads removed, enabling the building use to be monitored and guarding against changes inappropriate to the fire protection system.

1.08 Structural innovation in the UK is governed by three main sets of fire statutes covering Inner London, elsewhere in England and Wales and Scotland. Before 1963 some 1700 separate sets of local authority building by-laws made the national acceptance of a new design concept or building product virtually impossible. Now with the greater level of expertise centred in a few government departments, and with back-up of the Joint Fire Research Organisation, it is easier to obtain waivers on a national basis. For example this has already been achieved in the case of open-sided multi-storey steel framed car-parks[2].

1.09 But there remain a number of anomalies operating against structural innovation. The first is the requirement that steelwork placed outside the building facade should have the same measure of fire resistance as that inside. This is clearly wrong as the regulation fire ratings apply to steelwork at the seat of the fire whereas external steelwork protected against flame impingement can remain below the critical temperature of 550°C. Admittedly an application for a waiver can be made which may be successful, but the unknown outcome and delay involved is often untenable to the building designer who will find it difficult to convince the client of the sense of departing from the norm when possible delays in completion may result.

1.10 Another anomaly which is difficult to remove because of strong emotive forces at work is the increase in the required fire ratings with increased height of buildings. It can be argued that if the construction of a low building is adequate to resist a complete burn-out, then the same construction is adequate for tall buildings. The counter argument, based on emotion, not scientific fact, asserts that an extra factor of safety should be provided. Whichever view is taken, it is an unfortunate fact that the provision of extra fire resistance for structural materials such as steel adds to the cost and thereby acts as a brake on structural innovation[3].

Building life—economics

1.11 Another aspect of design which is an important constraint on, and is reflected in, current structures is the often unquestioned need for longevity. Clearly, the type of material chosen to last say 60 years—the minimum life of a domestic dwelling to warrant a building society mortgage —will be very different from the material required for temporary (perhaps disposable) low cost structures. Hand in hand with longevity is the need to cater for unforeseeable changes in use and this is increasingly provided for by designing the structures to give maximum internal flexibility for planning and services. The problem is to forecast the long term benefit of such buildings and to set this against the higher first cost of the structure.

1.12 This poses the important question as to how a project is funded. The present policy of major purchasing powers (for example that of local authorities) tends toward maximising the short term benefit from its cash outlay. The classic example is the usage of architectural stainless steel which is seldom specified, although it can often be clearly proven by discounted cash flow exercises that high initial costs more than offset the virtual absence of maintenance costs over the following decades. A change in purchasing policy would alter this anomaly and hence building form.

Emphasis on research

1.13 Again, and compared with the advances resulting from the design of eg spacecraft, building construction has made comparatively minor demands on new material research. The use of weathering steels and carbon fibre reinforcement, for example, are all interesting developments but they appear to have tactical rather than strategic implications in structure as they do not modify the basic approach to engineering design. Plastics, on the other hand, make an impact in unconventional structural forms and are an integral part of developments in inflatables and stressed skin structures.

1.14 It is not surprising that nature should provide the best examples of fully developed materials and structural forms, so we must continue to seek other primitive and natural antecedents for future solutions to our more sophisticated environmental problems. The question of scale is of course relevant in the adaptation of simple, small scale examples to complex large scale developments. An elegant solution easily loses its elegance in magnification. This is illustrated in Frei Otto's olympic stadia at Munich where the demands of the support system tend to obscure the elegance of the original tent model.

1.15 There are also the important practical constraints: the properties of materials and the limitations to structural scale.

2 Structural constraints to innovation

2.01 All structural systems are governed by three main factors: shape, erection technique and material properties. Ingenuity in design can usually solve the problems arising from shape and erection and it is therefore the material, with its inherent properties of strength and weight, which will impose the limitations.

Scale effect

2.02 If the size of a structure is increased, say by multiplying all its linear dimensions by a factor of m, then its volume and hence weight will increase by m³, while the cross-sectional area of the base supporting it increases by m² only. Thus either the proportions must be altered to increase the cross-sectional area (which is how the problem is normally overcome in nature—compare ant's legs with elephant's legs) or the strength must be increased.

2.03 Increase of strength alone, while maintaining the original proportions and modulus of elasticity of the material, may not be acceptable for three reasons:

1 The increased stress will be accompanied by increased deflections which though quite safe may affect secondary systems and finishes and indeed in buildings be alarming to any occupants

2 The 'strength' of structures prone to failure through elastic instability, ie columns, frames, panels, shells and suchlike liable to buckling, is in no way increased by an increase in the strength or permissible stress of the material

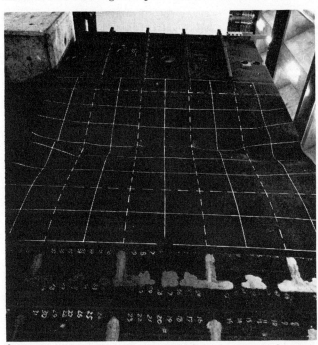

2

3

2 *Elastic instability: failure by buckling of steel plate compression flange (dotted line shows location of stiffeners behind). Similar failure led to the West Gate Bridge collapse.*
3 *Tacoma Narrows bridge collapse 1940. Coincidence of natural and forcing frequencies.*

used, but only by an increase in the modulus of elasticity **2**

3 If the size, weight and strength are increased, but not the modulus, the stiffness-to-density ratio of the structure will be reduced and the natural resonant frequencies lowered. This can cause trouble should these coincide with any naturally occurring forcing frequencies, perhaps from earth tremors, or from wind gusting as in the case of the Tacoma Narrows bridge collapse **3**.

3 Material constraints—the theoretical limits

3.01 The concept of a material's breaking length is helpful in appreciating maximum attainable spans. The breaking length of a vertically suspended bar of constant cross section is defined as the length at which the bar fractures at its point of suspension under the effect of its own weight. This length, which is independent of the size and shape of cross section, can be expressed in terms of the ultimate tensile strength and the density (or specific gravity) of the material; in other words, its strength-to-density ratio. Table I shows breaking lengths for various materials of uniform cross section. If, however, the cross section of a freely suspended vertical bar is adapted to suit the varying

Table I Breaking lengths for various materials

Material	Ultimate strength (kg/mm²)		Density (kg/m³)	Breaking length (kilometres)	
Lead	1·4		11 514	0·122	
Ordinary mild steel BS 4360, grade 43	40		7880	5	
Structural aluminium HE30-TF	30		2727	11	
High strength structural steel BS 4360, grade 55	70		7880	9	
Pine wood	10		505	20	
High tensile steel wire	220		7880	28	
Silk	—		—	44	
Glass fibre/polyester resin composite (70 per cent fibre)	84		1920	44	
Carbon fibre/polyester resin composite (40 per cent fibre)	75		1555	48	
Carbon fibre	302		1760	173	
S-glass fibre	460		2525	184	
	Tension*	Compression		Tension	Compression†
Ordinary structural concrete	0·35	3·5	2220	0·16	1·6
Ultra high strength concrete	2·53	38	2830	0·9	13·5

*Assuming that tensile strengths are 1/10 and 1/15 of compressive strengths for ordinary and ultra high strength unreinforced concrete respectively.
†Theoretical maximum length of unreinforced concrete column assuming no buckling occurs.

stresses due to self weight which increase up to point of suspension, then the breaking length can theoretically become infinite. This has important practical implications for long-span suspension bridges of the future.

Actual and theoretical behaviour

3.02 Structural materials fall into the two general categories brittle, eg ceramics, and ductile, eg mild steel. The former can be stretched elastically to a strain of only 0·01 per cent followed by immediate failure with no plastic deformation, the latter may be stretched to around 1 per cent strain, followed by considerable plastic deformation before failure **4**.

3.03 The reasons for this have only been understood for the last 20 years. Brittle materials have surface imperfections—cracks and steps which cause local increases in stress. The stress concentration at the tip of a crack increases as the radius of curvature of the tip of the crack reduces, and thus a crack so small as to be almost invisible under a microscope may induce a local stress say 100 times greater than the average stress. This quite simply means that if the average stress is greater than $\frac{1}{100}$ the theoretical maximum

4 *Crack propagation from machined notch in ductile steel subject to bending.*
5 *Fatigue crack growing from machined notch in brittle material.*
6 *Classic instacne of crack propagation. The tanker Schenectady after failure.*

stress, the crack will propagate and the material fail **5, 6**.
3.04 Here emerges the fundamental difference between brittle and ductile materials. The former have no way of dealing with the situation. The atomic bonds at the crack tip will break one after another and the crack propagate. However, in ductile material such as mild steel, the material at the tip can yield plastically and by thus increasing the tip radius of curvature, reduce the stress concentration. In metallic structures, particularly mild steel, this process occurs round corners, bolt holes or any point of high stress. The material simply yields locally and the stress becomes more evenly distributed. Provided a reasonable factor of safety is applied it is neither critical for parts to fit exactly nor for the designer to know the precise location of peak stresses. This ability of ductile metals to absorb all manner of maltreatment without suffering anything worse than often imperceptible local distortion makes them safe and very popular with engineers. They are understandably said to possess 'toughness', yet paradoxically it is their weakness which permits the necessary plastic flow.

7 *Comparison of cast copper cylinders before and after compression, showing deformation of ductile crystal grains.*

4 Material types

Fibre composites

4.01 Undoubtedly the greatest scope for improved stiffness-to-weight ratios lies in the field of composites, where whiskers or fibres are embedded in a matrix. These supply reinforcing stiffness and have a 'stitching' effect which retards crack propagation in the matrix material. Whiskers, usually around 0·002 mm in diameter and up to 20 mm in length, have a near perfect crystalline structure, again reducing the tendency for crack propagation with resultingly high strengths—up to 2100 kg/mm² with graphite and silicon carbide. However, their short length and the very high cost of forming a composite, exclude them from all but special (for example, high temperature) applications at the present. Fibres, on the other hand, can be produced in any length but owing to their polycrystalline structure, and hence relatively uneven surface, have strengths limited to about 450 kg/mm².[4]

4.02 Until a few years ago, almost the only composite was glass reinforced plastic. Today, glass fibre is only one of a range of reinforcing materials including asbestos, boron, carbon, silicon carbide and silicon nitride. Matrix or binding materials incorporating these are equally varied and include epoxy and polyester resins, cobalt, aluminium, nickel, nylons and polypropylene. Commercially, the organic resins, eg cold curing polyester, are the most common.

4.03 The properties of fibre-reinforced materials which can be varied by the engineer are tensile strength, Young's modulus and density. Table II lists these for the more important materials. It is particularly interesting to note

Table II Properties of fibre materials

Fibre	Density (kg/m³)	Tensile strength kg/mm²	Young's modulus kg/mm² × 10³	Specific strength kg/mm²	Specific modulus
E glass	2575	655	7·45	138	2·92
S glass	2525	460	8·9	183	3·56
Boron	2385	285	39	119	16·5
Carbon	1760	300	23·5	176	13·5
Tungsten	19 595	410	41·8	21	2·16
Molybdenum	10 300	230	36·8	22	3·6
Steel	7820	425	20·5	54	2·65

the wide variations in specific strengths and moduli, the factors most affecting performance.

Carbon fibre

4.04 The carbon fibres used over 50 years ago by Edison for electric light bulb filaments were fragile and had to be handled with great care. Today, they are produced with tensile strengths as high as 300 kg/mm² and with their very low densities possess strength-to-density ratios roughly 1¼ times that of the strongest steel yet made in the laboratory. However, these figures relate to the fibre alone and it cannot be used in this form; in practice it is carried in a matrix, epoxy or polyester resin **8**. At the Royal Aircraft Establishment, Farnborough[5], tensile strengths of 75 kg/mm² have been achieved using 40 per cent fibre in conventional polyester resin. With a specific gravity of 1·5, this composite represents a strength-to-weight ratio 2¼ times that of high tensile steel wire. The breaking lengths shown in table I clearly illustrate the superiority of carbon fibre/resin composites in relation to other materials.

8 *Fractured surface of composite using carbon fibres in silica matrix. Magnification approximately 1:250.*

Though current prices make its use prohibitive to the construction industry, the excellent properties should lead to extended demand and consequent lowering of production costs, especially in the aircraft industry and later for building applications.

4.05 Perhaps the most publicised use of carbon fibre composite was in the original fan blades of the compressor stage of the Rolls-Royce RB211 gas turbine[7]. Each blade, which is 1200 mm long and 300 mm wide, was made of laminated pre-impregnated fibre moulded under heat and pressure. The stress arises from centrifugal load and bending, but owing to the high specific strength, the weight and hence the centrifugal force were less than with a conventional metal blade. However the composite was prone to a form of brittle failure induced by bird strikes, for example, and the blades are now of titanium alloy.

4.06 The importance of achieving high strength-to-density structures will have practical implications in long span suspension bridges of the future. In detail this is illustrated by the Anglo-French Concorde. While its speed is twice that of a Boeing 747, its passenger capacity is only a fifth. The small payload, which represents 5 to 7 per cent of the total weight of this very high cost aircraft, means that even small weight savings are of enormous economic importance —a saving of 150lb allows another fare-paying passenger to be carried, increasing receipts by 1/ per cent.

4.07 Weight savings can be, and are being, achieved in a number of high performance aircraft by replacing or reinforcing materials with carbon fibre composites. The excellent properties of unidirectional strip reinforcement is used by BAC in fuselage floor beams with consequent weight savings of 3 to 4lb for every 1lb of fibre used. Weight savings are also important in airborne equipment. The Plessey telescopic cabin, housing a complete radar installation, is carried by helicopter **9**. A total weight reduction of 20 per cent on 1000lb was achieved by carbon fibre reinforcement of the bulkheads.

9 *Plessey air transportable radar cabin. Vital weight reduction achieved by use of carbon fibre composite.*

4.08 It is, of course, a mistaken belief that carbon fibre composites will always be black and will always contain carbon fibre alone. The most exciting and commercially variable designs adopt the upgrading concept—this is based on the recognition that it makes good technical and economic sense to upgrade the performance of present construction materials, such as metal, wood, asbestos and glass reinforced plastics, by judicious fibre reinforcement. It is claimed that the carbon-glass body of the Ford GT40 racing car was a major factor in the success scored in the 1968 and 1969 endurance trials at Le Mans, whereas the previous all-glass reinforced body had been heavy, noisy and suffered from fatigue **10**.

4.09 Setting aside the present high cost of producing the fibres and moulding the resin composites, there are many structural forms in which the composite can score over alternative materials. Of paramount importance is the need to design the structural shape specifically to exploit the advantages of the material, while minimising the effect of any detrimental properties, including creep of resin matrix.

10 *Le Mans winning Ford GT40. Success attributed to strong, lightweight carbon-glass body.*

It is necessary to guard against emulating the familiar structural forms associated with, say, steel or concrete.

4.10 It is an unfortunate fact that the emphasis on development so far has been confined to the high cost areas, namely the high strength and temperature requirements predominantly found in the aerospace industry. A review of available literature shows that little serious thought has been given to structural applications in building and civil engineering. I-shaped beam sections could be produced from fibre composites using the extrusion process in which the continuous fibres are pre-wetted with resin and formed to the appropriate shape before entering a heated die for final curing. Alternatively, the poor creep behaviour of the resin and resulting changes in section shape can be overcome by employing a thin walled steel member, possibly cold roll formed, onto which is adhered fibre in areas where high tensile forces are expected. With tubular structures the fibre could be adhered unidirectionally on the outside. In such applications the steel would require to be designed for any higher compressive loads resulting from the stress redistribution. The creep-free properties of steel sections would prevent long-term stress redistribution, avoiding increased deflection with the passing of time.

4.11 Radio masts are another application where a tapered carbon fibre composite tube might replace the conventional lattice steelwork structure. The dominant problems with steel masts at the moment are high aerodynamic drag and icing in bad weather; the carbon fibre mast could be formed by building up over an expanding former using pre-impregnated rovings (fibre strands) to produce a light, low drag structure. The most notable property of carbon fibre—its high specific tensile strength—makes it suitable for ropes and suspension cables.

Glass fibre composites

4.12 Glass fibre reinforced plastic also has a high strength-to-density ratio, derived from the high tensile strength (355 kg/mm²) of E-glass fibre. Unfortunately the modulus of elasticity is much lower than with steel and carbon, so glass resin composites are only one-seventh as stiff as steel. However, excessive deflection can be minimised by careful choice of structural shape[7].

4.13 Notable applications of glass fibre resin composites are the Concorde nose cone **11** which is the hottest part of the air frame, the British Rail 'Inter-City' high speed passenger coaches, lifeboats on the QE2, and aerial shrouds for the upper parts of 300 m high television masts.

4.14 Developments in glass fibre reinforced composites have not been limited to resin matrices: concrete and hemi-hydrate plaster matrices have been examined in depth by the Building Research Establishment. Initial work was based on ordinary E-glass and hemi-hydrate plaster. Structural components such as hollow floor slabs and double leaf internal partitions were successfully developed but have not yet been commercially exploited. Now, under the joint sponsorship of BRE and Pilkington Bros, it is

11 *Anglo-French Concorde. Nose-cone of high temperature resisting glass fibre and other composites used for structural weight reduction.*

hoped to develop an alkali resisting glass for use with Portland cement to give a high strength, crack free, glass reinforced concrete where previously only steel reinforcement was possible.

4.15 BRE work has mainly been based on randomly dispersed glass fibre, contents around 5 per cent weight of concrete, which provides a workable mix which can be sprayed, rendered or spun. At these low glass contents, the glass is contributing little stiffness—only 10 per cent improvement in Young's modulus. Short lengths of pipe have been produced using a rotating mould into which the glass fibre cement slurry is sprayed. Spray techniques have also been used to render high suction block walls, providing a layer which is highly impermeable to water. The rendering offers an extremely high resistance to mechanical impact and moreover the fire resistance properties have suggested use of the composite as permanent formwork for concrete providing a fire resisting soffit **14**.

4.16 New design procedures will have to be developed for these composites, depending upon the function of the glass. Short randomly distributed fibres, in addition to crack-stopping, enhance spalling resistance in fire, while bundles of parallel strands offer tensile reinforcement in place of steel reinforcement for longer span structures.

4.17 As with any other material, fibre composites should only be used if they can do a job possible in no other material; better than another material at the same total cost; or as well as any other material but at a lower total cost.

12

13

12 *Australia Square, Sydney: the world's tallest lightweight concrete building.* **13** *Stand, Doncaster racecourse. The large canopy was only feasible through developments in prestressed lightweight concrete.*

14 *Glass fibre composite as permanent formwork provides additional impact and fire resistance to finished structure.*

Concrete

4.18 Fortunately, the strengths of two most important construction materials, steel and concrete, continue to increase as a result of intensive research and development and, in the case of concrete, improved processing on site. From the low strength 'wet' 'pea-soup' concretes of the 1920s have developed concretes with compressive strengths of 40 kg/mm² and at extra cost high strength (100 kg/mm²) Portland cement concretes. But these practical achievements fall far short of the 350 kg/mm² strengths obtained in the laboratory[8].

4.19 As concrete has a low tensile strength, it is inappropriate to consider its tensile breaking length. However, on the theoretical assumption that no buckling occurs, the maximum height of an unreinforced uniform column which crushes at its base can be calculated, eg 2·7 km for a high strength concrete which corresponds to a 3·6 km span.

4.20 The exceptionally bad winter of 1962/63 emphasised the need for improved concrete production and curing methods on site, including heating water and aggregate, protecting freshly laid concrete by well insulated shuttering and overhead cover and the use of rapid hardening additives, plasticisers and methods of air entrainment.

4.21 Reinforced concrete strength can increase further with the use of high tensile reinforcement, allowing stresses 50 per cent higher than with mild steel at approximately 15 per cent extra cost. Prestressing developments minimise the effect of concrete's low tensile strength, maximising the strength-to-density ratio and increasing possible spans.

4.22 Marginal advantages result from density reductions using lightweight aggregates, eg expanded clay and pulverised fuel ash, or from employing foaming agents producing cellular concretes **12**. Research continues in this field. Of the three types of lightweight concrete—'no fines', 'aerated' (using hydrogen or carbon dioxide bubbles), or 'lightweight' aggregate the last offers most to the structural innovator interested in long spans. Recent work by the Cement and Concrete Association on concrete, finding strengths using different aggregates, now provides information on an area of design previously distinguished by trial and error. The very large canopy of the stand at Doncaster Racecourse **13** was feasible only through developments in prestressed lightweight concrete.

Steel

4.23 Mild steels for use in building structure were limited to tensile strengths of below 30 kg/mm² several years ago.

Now, high strength structural steels in common use are more than twice as strong (eg 70 kg/mm²) and commercial steel for wire production can be seven times as strong (220 kg/mm²). Despite these major advances, there is still scope for further development of commercial steels as only about 10 per cent of the theoretical maximum strength (2000 kg/mm²) is being used at the moment. While laboratory samples with tensile strengths of 1000 kg/mm² have been achieved, there are difficulties in obtaining commercial quantities with the required near-perfect crystal structure and physical properties such as ductility, weldability and resistance to brittle fracture.

References

1 COOKE, G. M. E. and BUTCHER, E. G. Structural steel and fire. Proceedings of conference on steel in architecture. 1969, British Constructional Steelwork Association, Hancock House, 87 Vincent Square, London, SW1P 2PJ [(2–) Yh2 (R)] o/p

2 MINISTRY OF HOUSING AND LOCAL GOVERNMENT The Building Regulations. Multi-storey car parks. MOHLG Circular 17/68. 1968, HMSO [223 (A3j)] o/p

3 CONSTRUCTIONAL STEEL RESEARCH AND DEVELOPMENT ORGANISATION Internal report on fire regulations. G. M. E. COOKE. CONSTRADO, London SW1 [(R1) (Ajk)] o/p

4 WINTERS, R. F. Newer engineering materials. 1969, Macmillan [Yy] o/p

5 WATT, W. and PHILLIPS, L. N. Carbon fibres for engineering applications. *Proceedings of the Institution of Mechanical Engineers*, 1970, November. IME, 1 Birdcage Walk, London SW1 [Yy]

6 LANGLEY, M. Carbon fibre composites. *The Chartered Mechanical Engineer*. 1970, February. IMF, 1 Birdcage Walk, London SW1 [Yy] o/p

7 MAKOWSKI, Z. S. Plastic structures: a series of articles on trends and developments in architectural engineering. *Systems, building and design*. 1968, October, November and December [(2–) Yn6] o/p

8 HARRIS, A. J. High strength concrete: manufacture and properties. *The Structural Engineer*. 1969, November [Yq]

9 STEELE, B. R. Glass fibre reinforced cement. Current paper CP 17/72. 1972, September. Building Research Establishment, Department of the Environment, Bucknalls Lane, Garston, Watford, Herts, WD2 7JR [Ym1] Free

Technical study
Innovation 2

Section 10 **Structural Innovation**

The architecture
of excitement

*Following the previous study's outline of the material and
environmental limits to innovation, the theme is now
illustrated with examples from the international field of civil
engineering and building. The authors are* GORDON COOKE,
MALCOLM QUANTRILL *and* WILEM FRISCHMANN.

1 Bridges

1.01 Developments in bridge design show, perhaps more
than in any other sphere of structural design, how exciting
structures have been determined by developments in
materials and methods.[1] The experience gained from construction failures has also provided valuable information in
a field of design which is constantly breaking new ground.[2]
The Tay railway bridge disaster of 1879 showed that
wind effects could be severely underestimated and as a
result design wind pressures were increased. The Tacoma
Narrows failure in 1940 stressed the importance of model
testing in wind tunnels—now a commonplace practice.
Brittle fracture caused by the poor welding of high strength
steel led to the collapse of Kings Bridge, Melbourne, while
the recent collapses of steel plate box girder bridges show
the need for care in design and in erection to guard against
the buckling of highly stressed steel plates.

1.02 Bridge construction also graphically illustrates the
relationship between scale and structural system. With
spans up to 300 m girders of many forms compete, spans
from 300 to 550 m require lattice cantilever girders and
steel arches, and those exceeding 600 m require suspension
systems. The 1400 m maximum span to be achieved with
the Humber steel suspension bridge is nowhere close to the
limit. In the absence of any externally applied load, it is
theoretically possible to span a distance of 35 km using a
constant section steel wire of strength of 200 kg/mm².
According to Schleyer it is possible to span over 100 km
using the same steel, but with varying section. To achieve
the minimum force in a catenary cable requires a sag equal
to one-third of the span so that ultra-long span bridges will
be accompanied by high cable-support towers. Lighter
bridge deck construction and higher strength catenary
cables now permit spans well in excess of 1400 m. Similar
factors apply to tall buildings and towers. For example, the
high strength-to-weight ratio of masts stabilised with ropes
has allowed heights in excess of 500 m **1a, b**.

Stressed concrete ribbon spans

1.03 The 135 m catenary roof spans of the Jumbo Hangar
at Frankfurt **2** are one of the world's largest concrete span
systems but they by no means exploit the full technical
potential of this form of construction[3] (in building the
constraint is more likely to be one of economics). Professor
Finsterwalder, the designer, had earlier proposed the
stressed ribbon concept for a bridge with main spans up to
460 m for the proposed Lake Geneva crossing, but so far
only a foot bridge in Freiburg, West Germany, with a
maximum span of 42 m has been built using this principle **3**.

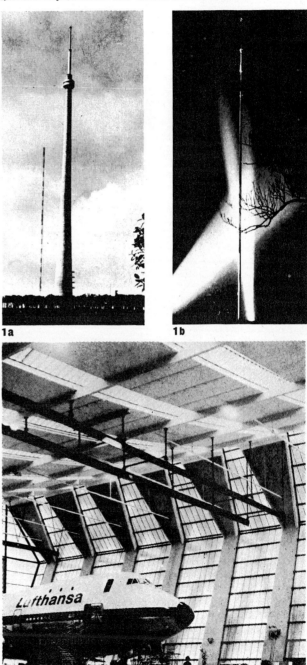

1a   **1b**

2

1 *Left, the Independent Broadcasting Authority's 360 m
self-supporting concrete tower, Emley Moor, Yorkshire—the
highest concrete structure in the UK. Right, the Authority's
420 m cylindrical steel tower, Belmont, Lincolnshire.*
2 *Boeing 747 hangar, Frankfurt: 1970. Prof. Dr Ing.
Finsterwalder. Prestressed concrete 135 m catenary roof
spans.*

3

4

Cable stayed concrete bridges

1.04 The Wadi Kuf bridge in Libya **4** designed by Morandi is the longest main span cable-stayed concrete bridge so far built and its 282 m makes it the third longest concrete span in the world. Morandi's bridges are instantly recognisable, incorporating A-framed portals and inclined pier legs supporting, in this case, a single box girder bridge deck of varying depth with single pairs of forestay and backstay cables. The moving-form cantilever method of construction was particularly suitable in view of the deep gorge. The Maracaibo bridge **7** is perhaps the best known of Morandi's cable stayed bridges. It stretches roughly 9 km across water some 30 to 45 m deep to the bottom of the silt and includes five main spans of 235 m. Factors influencing the choice of system included reduced maintenance costs in relation to a steel construction. Finsterwalder's prestressed concrete ribbon represents the simplest, if not the most familiar, form of modern suspension bridge—catenary cables, suspension rods, stiffening girders and a deck of conventional construction are combined into a single structural member.

Cable-stayed steel bridges

1.05 Cable-stayed steel bridges were extensively developed in post-war Germany. The Wye bridge at Chepstow is a fine example of simple, lightweight trapezoidal box construction supported by isolated towers with single cables. It has a main span of 230 m, but spans of 600 m are feasible with this form of construction.

Box decks

1.06 Conventional suspension bridge design has seen major changes in recent years due to the development of the

5

lightweight trapezoidal box deck, sophisticated model testing techniques and the help of the computer for analysing complex stress systems. The Severn bridge **5** is a good example of modern design. Completed in 1966, it spans 972 m with a remarkably low weight per metre run of 4000 kg compared with 50 000 kg for New York's Verrazano Narrows 1280 m span—the world's longest suspension bridge **10**. The dramatic reduction in the Severn bridge dead weight stems from the use of high strength steel plate, both in the towers and, more importantly, the trapezoidal box deck. Model testing was extensive. Towers were checked for aerodynamic stability in various stages of completion and in their free standing state. The river floating characteristics of the roadway box sections in erection and the effect of wind on individual sections when being lifted were also determined[1] **8**.

3 *Footbridge in prestressed concrete 'ribbon', Freiburg, West Germany. Dyckerhoff & Widman.*
4 *Cable stayed concrete bridge by Riccardo Morandi, Wadi Kuf bridge, Libya. The 282 m main span is the longest of this type.*
5 *Severn Bridge, completed 1966. Freeman Fox and Partners. Modern box deck construction, spans 972 m weighing only 4000 kg per metre run—one-twelfth the weight of the Verrazano Narrows Bridge, New York (see 10).*

6

7

8

9

10

6 *The Humber Bridge currently under construction.*
Freeman Fox and Partners. It will span 1396 m using
the steel boxdeck principle.
7 *Maracaibo Bridge, Venezuela. 500 identical precast,*
prestressed, concrete approach girders used four abreast for
the 50 m spans of the 9 km crossing, with five main spans

of 235 m.
8 *Severn Bridge. Box section being lifted to close side span.*
9 *Scheme for English channel suspension bridge—1973.*
Pell Frischmann & Partners.
10 *Verrazano Narrows bridge, New York, opened 1964 at*
1280 m the world's longest completed span.

11 *Competition winning proposal for Messina Straits crossing, 1970. Positive buoyancy submerged bridge anchored to the sea bed by steel cables. Alan Grant & Partners.*

1.07 The potential for increased spans is by no means fully exploited. The Humber bridge currently under construction in the UK is based on the Severn bridge method of construction but with a main span of 1396 m it is roughly 40 per cent longer **6**. In 1950, Steinman produced a design for a bridge over the Straits of Messina, with a main span of 1524 m capable of carrying a road and double track railway. Frischmann has proposed an English Channel suspension bridge 36 km in length **9**. Starting between Folkestone and Dover, it would span to the proposed Common Market industrial city and port founded on the existing sand banks in mid-channel, and continue to the French coast between Boulogne and Calais. The proposed bridge would have main spans of 2400 m to allow unimpeded passage of shipping in the deep water lanes.[5] An alternative solution, particularly suited to earthquake zones, is a positive buoyancy submerged bridge anchored to the seabed by steel cables **11**.

1.08 It has been seen that the strength-to-weight ratio is the main determining factor for structural spans, and that for each material there is an optimum shape: arch for concrete, suspension for steel. Limiting spans for concrete and steel have not been reached, nor are they likely to be while strength-to-weight performances improve. It is interesting that as far back as 1930 Freyssinet proposed a concrete arch of 1000 m—and this is three times the span of the world's largest concrete bridge yet built. With steel bridges, which in recent years have seen a remarkable increase in structural economy brought about by the box deck development, one can expect significant increases in spans. Spans of $2\frac{1}{2}$ km are being talked about, which would open up a new market for steel suspension bridges where, for example, large unrestricted widths of navigable water must be maintained. Over the next decade the motorway system which has emerged more or less simultaneously in most European countries may be extended over the larger natural barriers, over the sea, connecting islands and continents.

2 Tall buildings

2.01 There has been a continuous trend in recent years towards building tall structures, but planning permission difficulties in the UK have limited these to around 160 m. Heights of 60 m to 120 m are common, with an exception in the 200 m high headquarters of the National Westminster Bank now under construction in the City of London[6] **12**.

2.02 Construction of buildings up to 45 m in the early post-war period merely reflected extensions of traditional construction methods. However, the increasing cost of land in central urban areas and the need to preserve more open areas for circulation and gardens at ground level have encouraged new methods of design and construction. Areas of development include:

1 The use of ready-mixed concrete on restricted city sites and the quality control, mixing and placing of large quantities of concrete

2 The design and development of efficient formwork, eg table or slip forms and many other types of rapidly assembled shuttering

3 Industrialised methods of precasting techniques on and off site applied to structure and cladding, reducing site work at high levels; and also the use of dry finishes to speed construction

4 The design and erection of structure and cladding, avoiding the use of traditional independent scaffolding, and the development of hanging scaffolding, electrically powered suspended platforms, safety nets, etc

5 The use of cold worked and hot rolled high tensile steel reinforcement with indentations for improving bond, and of high strength friction grip bolts

6 The use of high strength structural steelwork and the availability of a wider choice of deep and heavy universal sections, and also the use of composite steel and concrete construction

7 The use of lightweight fire encasements

8 Structural analysis by model studies and electronic computers **14**

9 The use of large diameter piles.

2.03 Methods of construction have tended to follow rather than initiate new structural concepts. The logic of keeping activities such as casting concrete floors at site level led to Adler's invention of the Jackblock lifting technique first used in a 17-storey block of flats at Barras Heath, Coventry, in 1964. The roof is cast in a jig at ground level and after curing is lifted by hydraulic jacks, situated underneath the core walls, to first floor level. The uppermost floor is then constructed in the same jig, the structure raised again by a storey height to make room for the next floor to be cast, and so on until the building reaches its full height.

Stability

2.04 Tall buildings require special consideration to ensure lateral stability against horizontal wind forces. The five main criteria are:

1 Critical stresses in the structure

2 Lateral sway which could reach a point of discomfort to the occupants or affect the structural fatigue life

3 Wind excited oscillation causing resonance if the natural frequency of vibration of the building coincides with that of the excitation force

4 Local collapse causing general instability

5 Vibration in the complete structure including windows, and cladding.

12

13

14

Resilient mountings

2.05 An innovation allied to the requirement for stability is that of resiliently mounted buildings, where steel reinforced rubber springs eg at foundation level, permit buildings to be constructed over vibration sources such as railways. An example is the seven-storey Holiday Inn at Swiss Cottage, London, which is supported by 240 low-creep rate springs under the action of column loads of up to 200 tonnes **13**. The lounge is only 15 m above the Euston express railway track[7].

2.06 Resiliently mounted structures must be designed and constructed carefully if they are to comply with requirements of vibration and acoustic insulation, wind loads and sway frequencies induced by wind gusts during a gale, and the codes of practice for structural stability, material soundness and fire resistance. The spring stiffness must be tailored to suit the individual column loads and building mass: some buildings in the UK and abroad have actually experienced greater vibrations because of badly designed resilient mounts.

2.07 The tall buildings constructed in the early post-war period were essentially an extension of rigid structural frames similar to those of American skyscrapers such as the 245 m-high Woolworths Building and also the 380 m-high Empire State Building completed in 1932. These buildings were based on the steel cage skeleton system, and their stability depends on the rigid or monolithic connection between column and beam. In practice the connection involves site welding, friction grip bolts or, for reinforced concrete frames, very complicated design arrangements of reinforcement—often hard to achieve in construction.

2.08 With this clear emphasis on high-stiffness structures, the aim should be to design a structural system, with the necessary perforations, which derives its stiffness from inherent structural forms rather than excessive use of material. Two structural categories can be identified, those where the wind is resisted by shear walls often in the form of lift and service cores **15-17** and those based on a wind resisting facade **18, 19** seen in its purest form in the thin walled cooling tower. In the core type, other structural elements can be attached to the core by pin joints because they carry mainly gravity loading and can be prefabricated. Where the structural material is concrete, precasting can save construction time and site labour (especially of skilled craftsmen). It can also be advantageous if the required quality of finish cannot be achieved with in situ work, if joints between elements are carefully designed to meet erection and other requirements, and if the design is of a repetitive nature, allowing quantity production of standardised parts.

12 *Model of proposed 200 m high headquarters to National Westminster Bank. Engineers Pell, Frischmann & Partners.*
13 *Holiday Inn, Swiss Cottage, London. J. H. A. Crockett & Associates. Steel reinforced rubber springs at foundation level cope with railway vibration.*
14 *Perspex model of National Westminster Bank ready for test with electrical strain gauges.*
15 *Headquarters US Steel, Pittsburg, Harrison & Abramovitz & Abbe. First design for water-filled structural system uses Cor-Ten weathering steel. Stability from triangular-plan braced core and horizontal structural zones including 'top hat' bracing which reduces lateral deflection.*

15

16

17

Lifts in tall buildings

2.09 As the height of tall buildings increases, the need for an alternative form of vertical transportation becomes greater. Conventional lift systems installed in very tall buildings require floor areas too large to be acceptable—vertical transport in tall buildings (today 60 storeys, tomorrow 200) uses up to 20 per cent of the total building volume, the percentage initially increasing with the building height. This is because conventional rope and pulley lift systems cannot operate flexibly: only one lift car can be used in each shaft and this is aggravated by the peak hour traffic which tends to be at ground level, causing serious congestion. One system proposed would use the linear induction motor principle to give variable direction transport—vertical, horizontal and inclined—using computer-controlled cars which can switch from one shaft to another. Such a development would remove one of the present real obstacles to the tall building complex.

Cleaning and maintenance

2.10 The cleaning and maintenance of tall building facades also requires careful consideration. Access equipment for maintenance must facilitate inspection of the external fabric, replacement of glass, removal or replacement of plant and maintenance of sealants. Automatic devices are necessary for the more frequent job of cleaning. For the 200 m-high National Westminster Bank Headquarters, the window cleaning equipment will be fully automatic to give monthly cleaning of the glass and three monthly cleaning of the mullions and frames. The equipment, operated by one attendant at roof level, can clean several hundred square feet per minute and costs are low compared with labour intensive systems.

2.11 Problems associated with environmental services and the effects which tall buildings have on surrounding areas—wind turbulence, loss of sunlight and daylight—are difficult to resolve. Recognising that very tall buildings can use techniques adapted from other modern structures, it can be shown that buildings up to 3 km high are technically feasible before weight, wind sway, and thermal movement become limiting factors.[8] Such buildings would provide accommodation for complete communities, up to half a million people, with employment, education and all other indoor needs. Social rather than structural implications may be the limiting factors.

18

3 Lightweight tension structures

3.01 The largest spans of all are achieved with what Frei Otto calls 'minimal structures of the tension kind'[9]—a minimal structure being one which achieves a certain structural performance with minimum use of material and labour (see AJ 9.5.73 p1141). Hence lightweight, tent-like reinforced membranes are hung on a boundary system of masts and cables, covering large areas at low cost **20**. Soap bubble and rubber membranes at model scale suspended in eg a warped frame, have led to a better understanding of minimal surface structures. Membrane stresses at all points in a soap bubble are uniform and by observing the precise shape of bubbles the designers can reproduce full size tension structures where, importantly, stress concentrations in the fabric are at a minimum. In addition to membrane structures and suspended roofs, tension structures include pneumatic buildings where the skin is the tension element and the enclosed air the compression element, prestressed cable net structures acting like a membrane with holes and reinforced areas, and bridges. Grid domes are really minimum surface tension structures inverted, and so act in compression with the restraining edge cables replaced by beams or arches.

16 *World Trade Center, New York, architect, M. Yamaski; engineers, Worthington, Skilling, Helle & Jackson. Twin 412 m towers use a rigid frame of close spaced columns linked to central core by prefabricated lattice steel floor units.*
17 *Marina Towers, Chicago, Bertrand Goldberg Associates. Internal concrete service cores providing lateral stability were constructed ahead of in situ floors by central climbing crane.*
18 *John Hancock Center, Chicago, Skidmore Owings & Merrill. Lateral stability mainly from diagonally braced facade and $1\frac{1}{2}°$ inward slope with rising height.*

19

20

19 *Sears Roebuck Tower, Chicago. To be completed* 1974.
483 m high, it will be the world's highest office complex.
The 'bundle-tube' concept comprises nine 25 m sq units
rising to different heights. No core, instead facade rigidly
connected to composite steel/concrete floor decks with
additional stability from horizontal zones given over to wind
bracing (see dark bands on illustration).
20 *Cable-net structure, Munich, architect Gunter Behnisch*
& Partners, engineer Frei Otto.
21 *Air supported transparent pvc dome, reinforced by*
cable nets, designed by Arthur Quarmby for 20th Century Fox

Tension structure dams

3.02 The use of cable networks with membrane infills for
retaining fluids and solids has also been considered. Dams
would utilise a system of horizontal cables tied into the
sides of the valley with the membrane on the upstream side.
The main holdback force could be reduced by laterally
inclined cables tied to the upstream river bed. Similar
techniques are possible for retaining walls or snow fences in
mountainous areas. Land reclamation on a large scale may
be feasible using distended membranes.

Pneumatic structures

3.03 These are lightweight tension structures with minimal
surface supported by pressurisation of the air contained[10] **21**.
The first proposals for an inflated building structure was by
F. W. Lanchester of London who submitted a patent for
'An improved construction of tent for field hospitals,
depots and like purposes' in 1917. In fact, Lanchester's field
hospital was not achieved until it came into use in the
Vietnam war, half a century after the filing of his patent.
In the interim there had been military interest in inflatable
structures at first during the 1939-45 war with the development of the barrage balloon and subsequently in the production of air-inflated bridges and fascines. In the construction industry an inflated membrane had been used as

21

the 'shuttering' for a Buckminster Fuller dome and for
sheltering an electrical transformer under construction.
There are also numerous examples of storage buildings,
sports facilities and exhibition buildings, even offices and
computer buildings, where the control of the internal
environment is of particular concern, culminating in the
sophisticated Fuji Group pavilion at Osaka in 1970 and the
Cathedral at San José in Costa Rica. Lanchester's provision
for the large span dome was not realised until the third and
final design for the US pavilion at the Osaka Expo.

3.04 The shape and size of spherical pneumatic structures is
governed by the simple equation, membrane stress equals
half the excess internal pressure times the radius of curvature. Hence, with the same pressurisation, low domes have
higher membrane stresses than high ones of the same span
and smaller radius. To cover large areas and yet limit
membrane stresses, it is often desirable to introduce low
points, membrane ribs and cable networks. Large roofs with
reduced heights can be achieved using low points where
tension ground anchorages introduce small radii of curvature
in the membrane and hence lower membrane stresses, while
also enabling rainwater to be discharged through tubes into
an internal drainage system. The structures, similar to
vaults and shells, have the advantage of being translucent.
Also, instead of solid internal supports designed to support
heavy buckling loads, only thin internal ties are required.

22 *Sydney Opera House. Ove Arup & Partners.*
Computer-aided design for concrete shell construction.

Using membrane ribs, the sharp straight roof valleys allow the flattest of pneumatic shapes.

3.05 For pneumatic structures with radii of curvature greater than 50 m, it is usually necessary to reinforce the fabric materials (themselves with strengths up to 25 tonne/m) with cables, or to use double layer fabric laminates with strengths up to 35 tonne/m. Cable networks divide the dome surface into many small, highly curved membranes, with resultingly lower stresses. However, with very large structures the membrane is relegated to the sole function of air containment, with the tensile loading collected by a fine net and transmitted to a course network of cables. Between the economic minimum span of 10 to 20 m and the theoretical maximum come the problems of snow and wind loading, but discounting these and allowing a woven fabric membrane with a tearing stress of 30 kg/m² and self-weight loading only of 1 kg/m², the maximum theoretical diameter is 3000 m in a minimal pressure inflated hemisphere, with 796 m for a three-quarter sphere in similar conditions. Actual maximum spans are in fact much smaller as wind and snow loads are greater than the internal pressure, and permissible stresses must be kept well below the tearing stress.

Air walls

3.06 Also deriving from the air-displacement school is the development of air walls, a method of eliminating the external membrane of a structure by introducing a curtain of hot (or cold) air which serves to control the internal environment. For effective implementation, this system also requires the air locks advocated in Lanchester's 1917 patent. The work of Etkin and Goering in this field was presented to the Royal Society's conference on architectural aerodynamics in London, 1970.

4 Shells

4.01 These include all three-dimensional monolithic structures, formed or moulded and regardless of size, where the wall thickness is small in relation to span and the form is normally achieved by double curvature. This heading covers developments as diverse as The Sydney Opera House **22**, Arthur Quarmby's trackside huts for British Rail, Christopher Riley's work on plastic shells at Liverpool University, Hildebrand's and Schultze-Fielitz's Tainer Homes for Biafran children at Santonne, and Paolo Soleri's earth-formed vaults at Arcosanti, Arizona. Again there is the problem of lack of continued interest and investment, already identified in the field of pneumatics, which has also struck again recently in the British hovercraft industry. This obstacle to development has meant only sporadic outbursts of activity such as the Tainer example and the GLC involvement with moulded plastic shells for external cladding.

4.02 Such projects in the UK will only prosper given Government backing and the full support of long term testing, eg by the Building Research Establishment. In the present climate, and bearing in mind the economic failure of the Cumberland Hovercraft factory due to non-payment of a promised Government development grant, one is compelled to sit back and watch the poaching of British invention. The American development for Fibreshell, a new method of producing building units evolved from the techniques developed in the marine and aerospace industries, seems another promising breakthrough.[11] It is clear that this work parallels that being done elsewhere and developments, however successful in any one country or industry, are bound to profit from shared information and resources.

5 Glass reinforced plastic

5.01 It is unfortunate that while the strength-to-density ratio of glass reinforced plastic is high, its modulus of elasticity is roughly one-seventh that of steel, preventing its use where deflection rather than strength is the limiting design criterion—undoubtedly one of the reasons why GRP spans have been limited and in the UK do not exceed 18 m in conventional buildings. The largest all-plastic structures are GRP 'radomes' spanning up to 42 m, and research on pyramidal roof units suggests that roof spans up to 36 m are feasible if slight double curvature is acceptable. Creep, high cost, difficulties in quality control and poor fire resistance all limit the use of structural GRP. Rigidity can be achieved using eg folded plate construction, but to obtain sufficient penetration resistance requires certain minimum skin thicknesses which are relatively expensive in all but large structures. Nevertheless, the Italian architect Renzo Piano has produced some exciting roof structures, notably a prestressed GRP factory roof in Genoa. This consists of an opaque GRP membrane supported at points on vertical spreaders stabilised by prestressed steel cables.

6 Space structures

6.01 Steel and aluminium alloy complete as tubular space structures for the large span, double-layer grids or braced domes of exhibition halls and sports stadia.[12] The 200 m clear span steel dome of the Harris County Sports Stadium in Houston, US, provides seating for 66 000. The tubular space structure of the Heathrow Jumbo Jet Hangar,[13] has a span of 170 m, ie 1¼ times that of the Frankfurt prestressed concrete ribbon hangar, and yet supports over 500 tonnes of equipment contained within the roof **24**.

6.02 Space dome diameters could theoretically reach 1½ km, allowing for imposed weather loadings—those proposed by Buckminster Fuller contain complete towns with housing for, say, 20 000 inhabitants but keeping industrial noise and air pollution outside the envelope **25**. With internal pressurisation, the size of enclosure is perhaps unlimited. An air-conditioned environment free of heavy rain and strong winds would require lower strength and therefore lower cost structures, smaller drainage systems and cheaper weatherproofing, and would entirely dispense with individual domestic heating systems[14] **25**. In environmental terms, the concept might be more appropriate to countries with extreme climatic conditions—places where the temperature is rarely below 40°C or above freezing, or where humidity is 100 per cent. Owing to the complexity of space structure analysis, the structures of today would have been insoluble a few years ago. The introduction of the electronic computer gave the required impetus. Examples which demonstrate the range of application include a project for a cliff hotel in Japan by Mo-Rin, Okada & Shima; Le Ricolais' hangar in Central Africa; and Cedric Marsh's aluminium structure for the Anhembi Park exhibition hall at San Paolo, Brazil. The Le Ricolais building is of special interest because he employs a two-curvature network in a convex reaction tension grid. The San Paolo building is outstanding for its size—it covers 17 acres. The structural calculations were worked out by the Waterloo University Computer Unit at Ontario. 48 000 separate components are involved, with 10 different sizes of aluminium pipe in the frame, varying from 6 mm to 150 mm, the whole being assembled on the ground and lifted in one operation.

7 Suspended buildings

7.01 The concept of suspended structures has developed over the past 15 years, particularly as applied to office building. The central service core is of reinforced concrete,

23

24

25

23 *CEGB Power Station, Pembroke: completion of four GRP flues contained in 213 m high concrete chimney.*
24 *Boeing 747 hangar, London Heathrow, engineer Prof. Z. S. Makowski. The 170 m × 84 m tubular section space frame roof has 46 m clear span and supports over 500 tonnes of equipment.*
25 *Space domes could theoretically span up to 1½ km.*

usually·constructed using sliding formwork, and is designed to carry all vertical and lateral loads with the floors suspended from the top of the building using steel tension members around the perimeter. The concept has much to commend it. Only the structural core needs to be designed to carry all the major loads including those arising from earthquakes; it also allows the main accommodation to be terminated above ground level, leaving useful pedestrian circulation space below. Some designers have expressed this suspension concept in dramatic visual terms by featuring the rooftop cantilever structure, as in the 12-storey office block for Phillips Nederland NV in Eindhoven and the new office building for BP Belgium **26**. These buildings use structural steelwork and prestressed concrete cantilevers respectively.

7.02 The concentration of structural support at strategic points on the site allows full exploitation of plan flexibility in the suspended decks of the buildings at the upper levels. But the main advantage of this form of structure is in the combination of tower and bridge. The tower relates to particular restricted sites defined by vehicular traffic routes, with the possibility of developing air rights over existing traffic routes and industrial sites in urban centres, bringing the full potential of bridge suspension structure into play.

7.03 Werwecka's 'boulevard building' is in simple terms the application of the traditional boulevard concept on a multi-level basis, using the air space above existing railway and vehicular traffic routes, with both suspended and cantilevered structures involved. So far, with the possible exception of Place Bonaventure, Montreal, little has been done on a large scale to develop the potential of air space except over railway terminals and concourses. Yet in attempting to solve central urban traffic and living problems on a truly comprehensive basis it would be absurd to neglect the potential of air space use and the attached benefit of transport/pedestrian segregation.

8 Demountable and portable structures

8.01 The Victorians excelled in the field of portable buildings, from the elaborate cast-iron palaces exported to African potentates to Paxton's Crystal Palace which remains the outstanding example of sophisticated prefabrication in exhibition buildings. A visit to Otto's structures at the 1972 Olympiad perhaps confirmed the elegance of Paxton's solution, spans notwithstanding.

8.02 Tubular scaffolding is one of the most flexible of the demountable systems available in building construction. Whether as scaffolding, formwork or centering, this type of re-usable component is an essential factor in building, and it requires only a change of emphasis to allow the re-usable component to become an integral part of the building form rather than merely a step in the building process. This principle is central to Buckminster Fuller's geodesic structures and has applications in the construction of other space frames. It is as the building frame that this form of demountable system has the widest application, ie the frame as opposed to the enclosing membrane. The roof and walls still have to be added and the problems of weatherproofing make the demountability of those elements more difficult to·achieve.

8.03 The French architect Jean Prouvé and his collaborators have been particularly active in this area of rapid dry erection and easy demountability. Their emphasis is on the use of commercially available standardised components and the development of the industrialised process in building. The history of the industrialised building process over the past 50 years has paralleled that of pneumatic structures in that the designers and promoters have not found support from consumers at either governmental or general public level except in times of national emergency or in the promotion of prestige.

8.04 Prouvé's slogan that 'the population explosion is such that pseudo-traditional construction, even when planned, cannot produce enough', contains an element of truth. The difficulty in realising the potential of industrial production has always centred on the variety of conditions met on site—ie topographic, climatic, etc—a problem not en-

26

27

26 *BP Office building, Belgium 1964, engineers Bureau Bakker & Dicke. In suspended buildings the compression on the core removes any chance of tensile forces developing on the core's windward side.*

27 *Church at Eller, Dusseldorf. Eckhard Schulze-Fielitz. Design, intended as cheap, flexible system suited to similar churches and halls, incorporates tubular space frame clad in polyester and translucent plastic panels.*

28

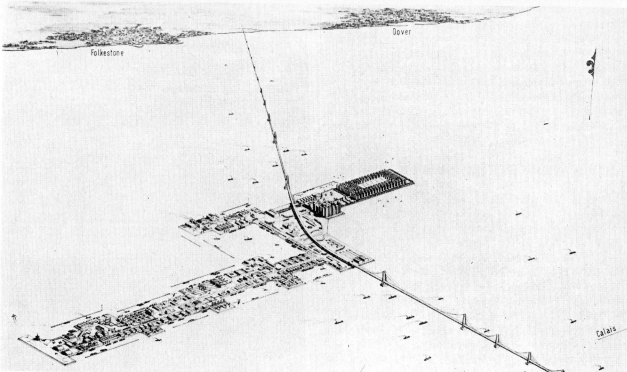

29

30

28 *The world's largest man-made island, 'Port Island',*
Kobe, Japan. To be completed in 1975 its 4·36 km² will
integrate port and urban facilities, creating a trading city.
29 *The bridge (illustrated* **10**) *could access a 52 km²*
artificial island on the existing chalk banks in the English
Channel. Pell, Frischmann & Partners.
30 *Globtik Tokyo, completed 1973—at 477 000 DWT,*
currently the world's largest ship. Tankers of this size and
larger will inevitably require the novel docking facilities
above.

countered in other standardised forms of industrial assembly. Use of standardised components to clothe special frames is evidence of the potential in this area of design. Outstanding examples of this use are found in Prouvé's vacation house at Beauvallon and the Maison Madame Jaoul at Mingueim, and his design for standardised service station structures for the Total oil company. In the last of these, a core building with a circular plan accommodates attendants, ancillary units and, sometimes, living quarters above. These have a steel infrastructure clad entirely with polyester panels with built-in windows and doors similar to those of buses and trains. A second structure comprises a 23 200 m² exhibition hall with restaurant for 1000 guests. There the main structure is of built-up plate girders on tubular columns carrying a lattice space frame. The vertical enclosure is achieved by steel mullions with clip-on glass and vitreous enamel panels bonded with neoprene gaskets.

8.05 Eckhard Schultze-Fielitz had similar aims in his design for a church at Eller, near Dusseldorf **27**. The brief was restricted to identifying an industrialised building components system suitable for constructing churches and other small-scale halls. The system had to be strictly priced, quick to erect and dismantle, and provide the highest degree of flexibility of form, elevations and choice of materials. The Mero tube space frame system was used, with variable tube wall thicknesses to ensure maximum economy. Cladding was by laminated plastic sheet.

9 Offshore construction

Artificial islands
9.01 With the accepted land shortage in heavily populated countries and the need to divorce certain industrial and commercial operations from residential areas, serious thought is being given to building 'artificial' islands in areas where it is shallow enough to allow the economical construction of the sea defence wall and infill, yet deep enough for deep water access tanker **28**, **29**. Many environmental and pollution problems could be solved in this way.

9.02 Concentrating on island sites in the southern area of the North Sea, Schreuder and Stigter have made detailed studies of the physical and nautical design conditions, the marine and civil engineering constructional aspects, the problems relating to sea defences and sea currents and the proposed locations and cost implications. Multi-purpose islands equipped for centralised waste processing, industrial and related activities, oil terminals and offshore airfields could be constructed at reasonable cost: £40 to £50/m² is claimed for small islands of 50 hectares, dropping to £12/m² for a 1000-hectare site. The islands would particularly benefit the Netherlands which has problems of waste disposal and has no wish to add to its 1000 land-based dumping sites.

Channel bridge
9.03 The channel bridge, industrial city and deepwater port project **29** is an interesting application of the artificial island concept to the existing chalk banks in the middle of the English Channel. The 52 sq km island would provide:
1 64 km of deepwater shoreline where vessels of all draughts could dock, particularly large oil tankers, ore carriers and container vessels
2 A positive navigational aid to shipping
3 Ideal conditions for enormous safe underground storage facilities for oil and petrol
4 Ultra-modern power generating plant feeding on the oil and gas stored nearby
5 A location for industries which are increasingly the source

of objection on the mainland owing to pollution factors, including smoke, smell, dirt, noise, and traffic. Large steel plants, refineries, chemical plants, power plants, and airport facilities could be safely integrated.

Offshore airport
9.04 Another interesting proposal will (if it comes to fruition) provide an airport in the Bay of Osaka about 10 km offshore in a water depth of 20 m. It is planned to relieve the excess burden of air traffic on the present Osaka International Airport and could be used by B747 and L1011 aircraft weighing up to 350 tonnes with minimum noise pollution of the mainland. The proposals for the 4800 m long by 1800 m wide airport cover three methods of bed-based construction, namely steel pipe pile system, dyke and fill, and pile and fill. There is also a system where a floating airport is bounded by a separate floating breakwater to lessen the influence of the tidal current and waves. The body structure of the airport is composed of cellular steel boxes each 150 m long, 50 m wide and 10 m deep and the whole assembly is moored to the bed with steel chains and sinkers. The advantages of a floating airport are:
1 The selection of the construction site is unaffected by water depth
2 Ground base improvement is unnecessary
3 Water pollution by the traditional earth filling process is avoided
4 Earthquake hazard is much reduced
5 Any future ground sinking has no effect
6 The space below the surface deck can be used for storage tanks, cargo handling facilities and car parking, leaving the surface deck clear for essentials—runways and aircraft manoeuvring areas.

9.05 Problems to be resolved are methods of assembly, anchorage, maintenance over the 20- to 30-year life and water breaking facilities. The provision of reliable access is of fundamental importance to a transportation system. Alternatives include: air—aeroplanes, helicopters, STOL and VTOL; sea—ordinary ships, hydrofoils, hovercraft, fixed bridge or floating bridge; seabed—underwater tunnel.

Offshore drilling rigs
9.06 Although the basic problems of drilling in the ocean are the same as on land, it is only in the last decade that technological advance has provided the means of working in such hazardous areas as the North Sea. Methods are constantly strained to the limit as demonstrated in the collapse of the jack-up rig Sea Gem.

9.07 The problems of operating above the surface of a turbulent sea, with the uncertain underwater factors between the control and the object of the operation, appear to be insuperable. A fairly obvious answer is to install the production plant at the point where the project emerges from the seabed and the first underwater well-heads were installed as early as 1959. Of the 100 or so now in operation, some are placed as deep as 200 m, installed by divers and submersibles, or by remote control with guidelines. The North American Rockwell Corporation and Mobil Oil are working on a seabed production 'satellite' which will be the collection point for a number of well-heads. Ultimately, the system will incorporate a giant floating barge, acting as surface store and control centre for a string of seabed satellites.

Offshore storage tanks
9.08 Whether the recovery is achieved on the seabed or at surface level, the problem of bulk storage is important. The

31

32

high. Operating on the water displacement principle, it is bottomless and always full of either water or oil or a combination of the two. It is filled by putting the oil above the water on which it floats. The additional weight of the oil on the water creates an imbalance of pressure which forces the water out of the tank through openings in the wall.

10 Computers

10.01 Computers have an enormous potential range of applications covering design (and particularly construction) aspects of architecture and civil engineering. They are normally used by large organisations able to justify the high cost. It is of course a commonplace to say that computers only provide a modified form of the input data, but this is worth repeating as this has important repercussions in practice. So the computer cannot create a design but it can produce a limited range of alternative designs in certain circumstances. The quality of the results depends on the type of programme and, importantly, the expertise of the programmer who may have to make value judgements.
10.02 With building systems, for example, computer aided design can provide the following advantages:
1 Many more designs can be assessed
2 Logic for relationships of structure and service zones within available space is improved

Ekofisk structure off Stavanger, Norway **31**, consists of an island reservoir in the form of a vertical cylindrical concrete tank with a total height of 90 m, of which 20 m projects above the water. The diameter is 95 m, and the tank's outer wall (breakwater) is perforated to reduce wave shocks. The prestressed 6 m deep hollow box floor was cast in an excavated dry dock and the first 8 m of the breakwater walls (holes temporarily closed) was added before the whole was floated into the deep water of the fjord. Once located in position, the cellular floor was progressively ballasted to the seabed.
9.09 The undersea oil storage tank at Dubai in the Persian Gulf **32**, completed in 1969, has a capacity of some 77 280 000 litres and is located 97 km off shore in 60 m of water. It weighs over 15 000 tonnes, is 82 m in diameter and 61 m

33 *Saturn V assembly building, Cape Kennedy, Florida. This, the world's most capacious 'room', 175 m high and 147 m wide, is constructed of massive steel portals. Air-conditioning prevents an indoor climate of clouds and rainfall.*

3 Checks on performance standards can be done rapidly, eg daylighting, heat requirements, floor space
4 Design process is quicker
5 Communication between members of design team is improved after initial learning period
6 There is greater accuracy and consistency of data, eg schedules and bills.

10.03 Computers have made the biggest impact, however, in structural design. Design of even a modest structure with say 100 members, could involve the solution of 600 arbitrary constants—and solution of the 600 simultaneous equations so formed would be quite impossible by normal hand calculation methods. The electronic computer takes a few seconds. Today, problems involving the solution of 10 000 simultaneous equations are not unusual.

10.04 Perhaps the greatest benefit from the computer is the reassurance given to the engineer that assumptions made in his elementary analysis were reasonable. This is particularly true in the field of shell roofs and three-dimensional space structures.

10.05 With highly complex structures, tall buildings for example, it may be extremely difficult to express the structure in mathematical terms, thus preventing the computer being used to its fullest advantage. In such cases, model analysis is often used so that strains may be observed and stresses deduced with the model loaded under various conditions.

11 Transfer of technology from space research programmes

11.01 Many claims have been made about the technological spin-off from the American space research programme. While it is clear to all that man has triumphed on technological frontiers, only a little of the knowledge has been applied outside the space field. NASA applications to date have been concerned with new bio-engineering instruments, compact new electronic parts, new alloys, new adhesives, new lubricants. They have devised ingenious methods of shaping and joining metals and built pumps to handle torrents of volatile fuels. They have broken new ground in materials technology by fashioning a heat shield to withstand the intense heat of re-entry into the earth's atmosphere. Fuel cells have been developed from a laboratory curiosity to a serious contender for driving future exhaust-free automobiles and lighting homes.

11.02 But serious application of space research results to building and engineering has been limited. True, there have been isolated instances such as the adaption of the NASA self-balancing beam for use in dam construction and the use of developments in the design of thin-walled cylinders for prefabricated metal buildings, but these perhaps illustrate how very specific and limited the spin-off benefits of research can be. The US Office of Technology Utilisation has been set up to identify innovations and prepare documentation on them.

11.03 The irony is that so much investment, so many resources, were devoted to solving problems well outside the scope of man's environmental predicament on earth. Yet the space projects served the useful purpose of showing how single-minded human endeavour could succeed and that is always an optimistic note. Were the same resources applied to tackling environmental crises, radical changes might be effected—but always depending on man's commitment and actual involvement. For this reason, environmental reform and management must be seen as a prestigious and challenging activity. For two decades we have concentrated on space heroism. It is now necessary to redress the balance by structuring the emergence of environmental heroes.

References

1 OVERMANN, M. Roads, bridges and tunnels. London, 1968, Aldus Books [18] £2.25
BECKETT, DERRICK. Bridges, 1970, Paul Hamlyn, 42 The Centre, Feltham, Middx. [182] o/p

2 HAMMOND, R. Engineering structural failures. 1958, Odhams [(2–) (S)] o/p

3 BRAUN, W. M. Jumbo hangar for Frankfurt. *The Consulting Engineer*, 1970, October [247] p50

4 'The Severn Bridge', Associated Bridge Builders. Prudential House, 28-40 Blossem Street, York. 1966 [182]

5 Suspension bridge and man-made island scheme for channel crossing. *International Construction*, 1972, September [82] p30

6 London's National Westminster Tower, *New Civil Engineer*, 1972, May [186] p30

7 SOWRY, J. A. The railroad runs under the middle of the house. *Rubber Developments*, vol 25, no 4, 1972, National Rubber Producers Research Association [111]

8 FRISCHMANN, W. W. Tall buildings. *Science Journal*. Associated Iliffe Press Ltd, 1965, October [9 (F4h)] p62

9 ROLAND, C. Frei Otto structures. London, 1972, Longman [(2–)] £7.50

10 DENT, R. Principles of pneumatic architecture. London, Architectural Press, 1971 [9 (2–5)] o/p
PRICE, C. and NEWBY, F. Air structures: a survey. London, 1971, HMSO [(2–5)] £4·25

11 Fibreshell transfer technology. *Architectural Design*, 1970, October [(2–5)] p406

12 MAKOWSKI, Z. S. Survey of recent three-dimensional structures. *Architectural Design*, 1960, January [(2--) (F4j)] p10

13 *Tubular Structures* vol 15, 1970, March. British Steel Corporation Tubes Division [(2–) (Ih2)]

14 Space Structures. *Building with Steel*. no 11, 1972, August, British Steel Corporation [(2–5) (Yh2)]

Acknowledgements

Cement and Concrete Association
Plessey Avionics and Communications
British Steel Corporation
R.P. Structures Ltd
National Physical Laboratory
The Institute of Building
The Natural Rubber Producers' Research Association
British Aircraft Corporation Ltd
The British Petroleum Co Ltd
Globtik Tankers Ltd
Pilkington Brothers Ltd
United Press International
Shell International Petroleum Company Ltd
US Bureau of Shipping
Ford Motor Co Ltd
Imperial College London
Elkalite Ltd
Bissell August Associates
Alan Grant & Partners

Freeman Fox & Partners
US Steel Corporation
Guinness Superlatives Ltd
Aerofilms Ltd
20th Century-Fox
Independent Broadcasting Authority
The Welding Institute
Royal Aircraft Establishment, Farnborough
Chicago Bridge and Iron Company

34 '. . . *so much investment, so many resources, devoted to problems well outside man's environmental predicament on earth. Yet the space projects served the useful purpose of showing how single-minded human endeavour could succeed.*

AJ Handbook
Building structure
CI/SfB (2-)

Design guide

Design guide

Building structure

Form
This section consists of:
Design guide
Appendix—Building Acts and Regulations
     —Standards and codes of practice
     —Specialist advisors and professional bodies
     —Index to handbook

Author
The author of the design guide is HERBERT WILSON CEng,
FICE, MCOnSE

Scope
Structural design will usually begin only after the broad
decisions on building orientation and function have been
taken and the architect may find the AJ Building type
guide relevant to his own particular project helpful before
embarking on the design sequence proper

How to use the guide
The guide broadly outlines the range of decision-taking in
structural design, considering factors of, eg, site and building
function, environment, servicing, and loading, in order to
arrive at structural form. In practice, of course, these factors
will overlap and design will seldom be the linear process
shown below. But we hope that architects will be able to
follow the broad decision sequence and adapt its details to
suit the particular needs of their own scheme. Some may
stick to the suggested procedures quite closely while others
may follow their own well-tried methods while using the
guide as a useful checklist.

Appendixes
Following the Building Acts and Regulations is a brief list
of the more important standards and codes of practice as
they relate to the different aspects of structural design.
The list is not exhaustive and more detailed references
appear as appropriate throughout the handbook.

Contents of guide:
Data required
1 Data from client
2 Statutory requirements
3 Data from site and locality
4 Functional requirements
Design
5 Analyse functional requirements
6 Analyse structural requirements
7 Analyse structural elements
8 Consider structural form
Contract: drawings, documents, supervision
9 Working drawings
10 Tender documents
11 Contract stage

Appendix:
1 Legislation
2 Some useful standards and codes of practice
3 Specialist advice

Data required

1 Data from client

1.01 Building type and purpose

The purpose of the building and the activities it is to house will impose specific structural requirements and limitations. Identify these requirements.

1.02 Structural form

Classify the structural form required—simple enclosure of space, cellular enclosure, monumental or symbolic, single purpose (studio, exhibition hall, tower, silo, warehouse) see Technical study STRUCTURES 2 paras 3.01-5.04 and Technical study ANALYSIS 2.

1.03 Building life

Temporary or limited life buildings may be expendable, salvageable or demountable for re-use. Phased construction involving a family of separate units built at different times could impose a form of standardisation and a development plan with special regard for services (engineering). Change of use, if unavoidable, and flexibility of loading patterns is an expensive structural commodity. Fully discuss with client and seek a reasonable balance. Consider the effects of flexibility or change of use on floor loadings, spans, column spacings and clear heights, also future services and demountable partitions and non-loadbearing walls which could be removed etc.

1.04 Imposed floor loadings

The client's present or future use of the building may create greater imposed loads than the statutory loadings—which it should be noted are minimum values. Establish these loads and any dynamic loads as soon as possible, with the client's agreement.

1.05 Services

The provision, storage and distribution of the engineering services for environmental and possibly for manufacturing purposes will have corresponding structural effects such as plant rooms, tanks, stores, circulating ducts, double floors etc. See AJ Handbook Building Services and Circulation. AJ.1.10.69 to 16.9.70 and Technical study STRUCTURES 2 paras 8.01 and 8.02. Discuss with the client the problems of attendance, maintenance and access. Are emergency supplies required?

1.06 Cost

The cost limits may be expressed in different ways at this stage, eg total capital expenditure, cost per functional unit, cost per square metre etc.
Establish a cost plan and with this data examine the relationship between cost and time of construction and the implications of off-site work, specialised construction, industrialised systems and components, negotiated contracts, pre-ordering etc.

1.07 Client's preferences

Some clients will have a wide experience of building and will have decided views and preferences but check with all clients to find their opinions on association with specific contractors or specialist subcontractors and suppliers. Check if client has any preferences in structural materials—timber, concrete, steel, masonry—and discuss their relative merits.

2 Statutory requirements

2.01 Planning

Check the structural implications of the following before outline planning permission is sought:
- Public utilities and wayleaves—above and below ground.
- Access to site and future improvements.
- Building line and improvement line if any.
- Height restrictions—planning, or direct—as imposed by local airports or beamed telecommunications.
- Check local restrictions on types of fuel, chimney heights, smoke emission etc.
- Check local restrictions concerning building materials.

2.02 Building legislation and structural safety

Make a thorough check of the Building Regulations on all matters concerning structure. A clear understanding of the regulations as they affect the project will at this stage be very rewarding.
Consult, as necessary, Building Regulations, the London Building Act, codes of practice (see references at end of this guide) with specific attention to:
- Progressive collapse—Regulation D19 (see Technical study MASONRY 2, para 4.22 to 4.25 and Technical study SAFETY 1, para 2.01 to 2.05).
- Fire protection, compartmentation and fire resistance. Spread of fire and means of escape. Relate degree of fire protection required to any preferences for structural materials. (See Technical study SAFETY 3 and Information sheets SAFETY 1-4).
- Consider alternative materials.
- Check statutory requirements for thermal and sound insulation.

3 Data from site and locality

3.01 Structural feasibility

In theory, almost any building can be constructed on almost any site, but fairly obvious considerations may render such decisions impracticable.

Check the following considerations where applicable to the site: (see Technical study FOUNDATIONS 1 para 6.01 to 6.24)

Sound rock
- Building size and arrangement not critical.
- Extensive excavation is uneconomic.

Soft rock
- Not critical for medium rise buildings, but high rise buildings may require thick rafts.
- Very soft rock may have to be removed and backfilled. Basements likely to be uneconomic.

Sands and gravels
- High rise buildings may require piling.
- Medium rise buildings require a good load dispersion. A ground water table is likely to be in evidence which will make excavation (basements) expensive.

Glacial drifts
Not critical for most buildings, but piling and excavation may be difficult if large boulders are present.

Clays (stiff)
- Large buildings will require piling.
- Basements help to reduce foundation pressures and may be economic.
- Differential settlement will be a problem and loading should be balanced as far as possible and if necessary movement joints introduced into the structure.

Soft silt and clay
- Difficult for all buildings.
Small buildings can be carried on lift rafts and large buildings can be piled provided deposits are not greater than 25 m over firm strata.
- For deeper deposits, piling becomes very expensive. Basements help by providing buoyancy.

Peat and marsh, made up ground and areas subject to subsidence
- These are critical conditions and no general observations can be made other than that foundation will be expensive and the design of the foundation should be carried out with the help of a specialist.

3.02 Subsoil survey

This will be essential for all but the simplest projects (see Technical study FOUNDATIONS 1 para 7.01 to 7.21). It is best undertaken at a stage when the structural objectives are better known. Before survey is carried out check by general observation of the site for evidence of:
- peaty, marshy or waterlogged ground
- made-up ground, felling or tipping
- surface irregularities, subsidence or slip
- condition of adjacent property, evidence of settlement problems
- previous earthworks, pits, mine workings, culverts, pipes, tunnels, well borings
- underground services
- chemical pollution, particularly sulphates (which may be present naturally) but also nitrates and chlorides. (See Technical study FOUNDATIONS 1 paras 7.22 and 8.01 to 8.11).

Sources of further information may include:
- geological survey
- local authority records
- local mineral workings
- recent well borings
- adjacent open excavations and cuttings

3.03 Site contours and features

Prepare contoured plans of site including the following information beyond site limits:
- limits of slopes within 20 m
- positions of trees within 20 m
- water courses within ½ mile
- ponds and lakes within ½ mile
- quarries within ¼ mile
- adjacent buildings within 250 m
- roads and railways within 250 m
- wells within ¼ mile

3.04 Climate

Obtain information about general regional condition but make further local investigations for exposed site particularly in remote areas.

Check prevalence of extreme wind speeds, especially when associated with low temperatures.

Check possibilities of flooding and relate this to method of construction —ie piles instead of open excavation. (See Technical study FOUNDATIONS 2 para 2.02 to 2.10).

3.05 Sources of nuisance and hazards

from the site

Ascertain the local sensitivity level towards noise or vibrations emanating from the site during demolition, construction and subsequent use of the building and relate this to constructional methods and degree of enclosure.

to the site

Assess the possibility of transmission of noise or vibrations from areas surrounding the site and if the nuisance is severe initiate special investigation. (In areas where earthquakes are a known risk, specialist advice is essential). Check for any source of pollution, such as fumes or dirt, saline pollution in coastal or estuary locations, contamination by wastes from adjacent sites, any of which may influence choice of structural materials, or the degree of protection of such materials.

3.05 Resources

Check local building methods and materials. Note any local difficulties in the supply of building materials. Check the availability of labour, skilled or unskilled. In certain instances these matters can influence choice of materials and the proportion of the work to be fabricated away from the site.

Most of the following items fall within the routine planning procedures of the architect but they all have structural implications and are therefore essential to this checklist.

4 Functional requirements

4.01 Movement of people and materials

Note from the required circulation patterns any element which could be exploited structurally, for example
horizontally: tunnels, corridors, bridges, walkways, runways, conveyors and cranes
vertically: lifts, hoists, flues, chimneys, ducts
horizontally and vertically: staircases, ramps, escalators, chutes, bunkers.

4.02 Environmental requirements

Some are statutory and some arise from activities within the building, or are specific requirements of the client. (See Technical study STRUCTURES 2 paras 8.01 and 8.02), but note the following:

Thermal insulation

Check the required standards of the building envelope and relate to structural and cladding materials.

Mechanical heating and cooling

Assess the structural implications of plant, distribution and transference.

Ventilation, natural and mechanical

Establish demands for natural ventilation on the building envelope, and where mechanical ventilation is required assess structural implications of plant etc.

Structural isolation

Check if structural isolation is necessary in special areas to prevent the transmission to or from the area of sound or vibration. (See Technical study STRUCTURAL INNOVATION 2 paras 2.05 and 2.06).

4.03 Services

For general guidance on the relationship of structure and services (see AJ Handbook Building Services and Circulation) and make specific checks of the following:

Water

Assess structural demands of supply distribution and storage. Note any special requirement, eg cooling, high pressure.

Refuse disposal

Examine the provisions for refuse disposal, including trade waste.

Plumbing

Assess structural influences of drain and pipe runs, eg space for falls, bends access etc between floors and in vertical ducts. Check details of storage tanks, delay tanks and pumping units, particularly in connection with basements and foundations.

Electrical

Note any special provision for heavy equipment. Check need for structural isolation or insulation because of noise and vibration from transformers and alternators.

Other services

Note any provision for special services such as steam, gas, compressed air, refrigeration and services for industrial processes.

4.04 Structural loadings:

Statutory and actual

Record all statutory or, where in excess, actual loadings relative to use of space in the building. (See Technical study STRUCTURES 2 paras 6.01 to 6.13 and references at end of guide).

Wind

Record statutory loadings amended as necessary to suit local conditions.

Snow   Record statutory loadings (see references) and amend as necessary to suit local conditions, particularly in northern and exposed areas. Note that concave roof shapes and efficient thermal insulation can result in accumulations of snow far in excess of statutory loadings.

Special static loading   Assess any static loadings (permanent or temporary) not covered by general requirements, eg storage or plant, during construction and use.

Dynamic loading   Assess dynamic loadings (horizontal and vertical arising from cranes, lifts and hoisting machinery) unbalanced reciprocating machines and testing machines, vehicular traffic and fork lift trucks etc.

Complex loadings   Specialist advice is required for conditions of complex loadings. Define areas where heavy static load is combined with dynamic effects, because under these conditions the structure may be required to be heavy (to minimise problems associated with acceleration from dynamic loads) and very stiff (to minimise displacement). Problems of flutter and resonance may arise with very tall buildings (in which structure and envelope must be compatible) and in long span structures (see Technical study STRUCTURAL INNOVATION 1 para 2.03) and structures subject to very high wind speeds. The effects of earthquakes are complex, and similar loadings can arise from large forging hammers and drop stamps.

Design

5 Analyse functional requirements

5.01 Planning

In most cases, planning largely determines the structural form, but the complementary process is the accommodation of structural elements within that form.

Examine the spatial demands of the various structural elements on the building shape, considering in detail the basic structural necessities in providing:

● working platforms
● enclosure
● a load-transference system down to foundation
● a foundation system to transfer loads to the earth. (See Section 4 FOUNDATIONS AND RETAINING STRUCTURES and Technical study STRUCTURES 2 para 7.01 to 7.03).

5.02 Services etc

Identify the spatial and structural limitations imposed by circulation, services and amenities (see AJ Handbook of Building services and circulation).

5.03 Appearance

Consider the relationship between envelope and structure within the following alternatives:

● structure and enclosure combined in that the structure is also the envelope
● structure external to building envelope
● structure visible but integral with building envelope
● structure concealed
(See AJ Handbook Building enclosure)

6 Analyse structural requirements

6.01 Structural subdivision

Consider the sources of building movements and determine the degree of structural subdivision required. Identify those areas requiring complete isolation.

6.02 Limiting dimensions

Determine any limiting or critical dimensions for structural elements and seek amendment of these dimensions if necessary to structural efficiency. Check that the structural deformations are within statutory limitations and compatible with limiting dimensions and other building elements.

6.03 Site limitations and erection

Do the limitations of the site, or problems of access and erection, impose any structural restrictions?

6.04 Cost and contract

Assess the advantages in terms of time and cost of
● a comprehensive contract
● a separate foundation contract
● a separate structure contract
Check the allowance for structure in the overall cost plan against any preliminary ideas on structure and compare the costs of the various structural materials.

7 Analyse structural elements

7.01 Form and materials

At this stage, the limitations imposed on the structure should be fully understood and decisions made on form and choice of materials.

Consider each structural element, floor, beam, wall, column etc, in terms of the requirements and limitations imposed on it and study the interrelationship of all the elements to discover any predominating influences on choice of structural form or materials within the following guidelines.

8 Consider structural form

8.01 Foundations, basements and retaining structures

The foundation is an integral part of the structure and, except for simple structures on very good ground, the design of foundations should be carried out with due regard for the degree of influence the foundations may have on the form of structure, and vice versa. (See Information sheet FOUNDATIONS 1).

In certain conditions, the degree of influence may be considerable, for example in structures where loads have to be concentrated on deep foundation elements or where varying foundation loadings create consequential problems of differential settlements. With basic site data and preliminary ideas on building form, consider possible foundation structure:

Shallow
● Individual or pad footings, strip footings and rafts. (See Technical study FOUNDATIONS 2 paras 4.01 to 5.16 and Information sheets FOUNDATIONS 4 and 5).

Deep
● basements and retaining structures, buoyant basements
● piles, bored (replacement) or displacement (driven). (See Technical study FOUNDATIONS 2 paras 7.01 to 7.11).

Composite
● strip footings and short bored piles
● basement rafts
● piles, rafts

8.02 Site investigation

With the design objectives of the foundation determined, the preliminary site appraisal should be confirmed, on all but the most simple jobs, with a fully specified and directed site investigation. Specialist assistance will be necessary in drafting scope of investigations.

8.03 Superstructure

Although for convenience in this checklist superstructure is listed after foundation, it should be clear that these matters are in parallel and decisions are mutually definable. Consider structural form most suited to requirements.

8.04 Solid structures (and elements of structure)

Loadbearing walls with slab or slab and beam floors, see Technical study STRUCTURE 2. Concrete, see Technical study CONCRETE 2. Timber, see Technical study TIMBER 3. Masonry, see Technical study MASONRY 2.

8.05 Skeletal structures

Basically, skeletal structures are an assembly of beams, struts and ties (or tendons) in:
● frames having rigid structural elements
● frames having rigid elements combined with flexible tendons or cables. See Technical study STRUCTURES 2.

Tendons and cables are the most efficient of structural elements as they have none of the stability problems associated with rigidity and are usually constructed from high grade steel, but there is no restriction on choice of material for the other elements of skeletal structures.
● Concrete, see Technical study CONCRETE 2 paras 2.01 to 2.19.
● Steel, see Technical study STEEL 2 paras 1.01 to 3.16.
● Timber, see Technical study TIMBER 3 paras 3.01 to 6.14.

8.06 Surface structures

Space is contained by the structure as a three-dimensional entity; the structure has the double function of providing enclosure and structural support. See Technical study STRUCTURES 2. Although concrete, as a cast-in-place material, is particularly suitable for these shapes, other materials in certain applications are equally acceptable.
● Concrete, see Technical study CONCRETE 2 paras 9.11 to 9.22.
● Timber, see Technical study TIMBER 3 paras 7.01 to 7.10.
● Masonry, see Section 8 Structural material MASONRY.

8.07 Composites

If the selected structural form is a composite of a number of types of construction, check the structural logic of the points of continuity from one system to another and confirm that the load flow pattern is efficient. The structural deformations of each system should be compatible because the loaded structure will assume a definite shape — notwithstanding the theoretical concepts of stress and strain of individual elements of structure. See Technical study STRUCTURES 2 and Section 9 COMPOSITE STRUCTURES.

8.08 Selection of form and material

Evaluate possible alternatives in terms of:
- functional requirements: check against client's brief
- structural requirements: check against loading pattern, stress grades, sizes, fire resistance, maintenance
- statutory requirements: check that these are complied with
- site limitations: check constructional problems
- specialist consultants: check that requirements are met
- cost: check that proper balance is maintained between structure and other parts of the cost plan.

8.09 Detailed design

With decisions on form and materials made, the detailed design of the building elements can be started.

Note that, although structural safety is of prime importance, most of the faults in building structures arise from lack of consideration at the detailing stage of movements in buildings.

Contract: drawings, documents, supervision

Working drawings, tender documents, specifications, bills of quantities, are component parts of an exercise in precise communication. Ambiguities, sloppiness of direction and incomplete information are the roots of future troubles and claims.

9 Working drawings
9.01 Check lists:

Loadings

Check all loadings, particularly plant and storage areas (see references).

Dimensions

Check critical dimensions with architect and engineering services consultants.

Fire resistance

Check fire resistance requirements (See Technical study SAFETY 3).

Deflection and movement

Relate structural deflections to non-structural building elements, partitions and cladding, and similarly check effects of movement joints or provisions for differential settlement.

Local authority

Liaise with the local authority and determine its requirements for submission of structural information.

Progressive collapse

Make a special study of the progressive collapse requirements.

Foundations

Collate all data from trial holes, site investigation etc, and review foundation proposals and design.

10 Tender documents
Local authority

Complete basic agreements with local authority.

10.01 Site data

Draft, for inclusion in the bill of quantities, all the data concerning the site necessary to the tenderers' understanding of the work.

10.02 Special features

Draft, for inclusion in the bill of quantities, details of any special features of the structure on erection.

10.03 Work below ground

Explain any special requirements in connection with the work below ground, such as pumping or dewatering, old foundations and services, adjacent structures, underpinning and back filling. Check the provisions for watertightness of work below ground, joints, impervious membranes, water bars, etc. Check that the logic of the proposed protective measures is complete and not open-ended, and see that the bills of quantities contain all relevant information.

10.04 Specialist work

Check all specialist items of structure, eg precast concrete, prestressed concrete, steelwork etc, and see that the documentation of such subcontracts is in accordance with the terms of the main contract. Check that the relationship between specialist subcontractor and main contractor is fully described in the bills with particular reference to the scope of the work, assistance or service to be provided by main contractor and acceptance of work at start and finish from one to the other.

10.05 Specification

Examine the structural content of the bills and check all specification clauses and descriptions.

10.06 Services

Secure, if possible during the tender stage, all information on the final structural demands of the services (including lifts) and check against working drawings. If substantial amendments are to be made, try to advise the tenderers before submission date.

11 Contract stage
11.01 Checklist:

Local authority

Ensure that all local authority approvals are secured in good time.

Clerk of works

Advise the client on the appointment of a clerk of works and/or a resident engineer in good time before the start of work on site. (For any substantial structure recommend the appointment of a qualified and experienced resident engineer.)

Quality control

Agree, before the start of work, with the resident engineer and the contractor, the measures to be taken by the contractor for 'quality control' and establish a formal system for programming, inspections, approvals and records.

Setting out

Initiate the contractor's survey of the original site conditions, adjacent buildings, services etc, and provide him with the basic information and reference points for his setting out.

Foundations

Check the conditions at foundation formation levels and co-relate with site investigation data, making any design changes necessary to suit actual conditions as disclosed.

Water table

Check the level of the standing water table, if any, as soon as disclosed and correlate with the site investigation data.

Protection

Check that the contractor is providing sufficient protection to exposed excavations against disturbance, flooding or drying out etc.

Piles

If piles are being used, check that adequate supervision is being exercised by the specialist contractor or consultant. See that accurate records of dimensions of piles and penetrations of piles are kept. If cast-in-place piles are being employed, check that all necessary precautions are being taken to ensure that the piles are true to cross-sectional dimensions and that 'necking' or 'voiding' does not occur.

Levels

Establish a permanent levelling reference point, and check at intervals during construction the varying degrees of settlement of parts of the structure, and relate these measurements to the anticipated behaviour of the structure.

Tests

Check that all routine tests of accuracy and quality are being followed.

Construction loading

Check that the contractor's methods of construction do not impose loads on the structure greater than those for which the structure was designed, with due regard for the age of the elements under consideration.

Appendix

1 Legislation
England and Wales
The Building Regulations 1976 SI 1676 HMSO £3.30; The Building (First Amendment) Regulations 1978. SI 723 HMSO £0.60p

London
GLC London Building Acts 1930–1939 and London Building (Constructional) By-laws 1972, GLC [(Ajn)] £1.20 (paperback)

Scotland
The Building Standards (Scotland) (Consolidation) Regulations 1971 SI 2052 (S218) HMSO £1.30; The Building Standards (Scotland) Amendment Regulations 1975 SI 404 (S51) £0.20p; The Building Standards (Scotland) Amendment Regulations 1973 SI 794 (S65) £0.21p [(Adj)]

2 Some useful standards and codes of practice
BS CP 3: Chapter v: Part 1: 1967 Dead and imposed loads [(K4)] 75p

BS CP 3: Chapter v: Part 2: 1970 Wind loads [(K4)] £1·30

BS 648: 1964 Schedule of weights of building materials [Yy (F4)] £1·30

BS 476: Part 3: 1958 Fire tests on building materials and structures. External fire exposure roof tests [Yy (R4) (Aq)] 60p

Foundations
BS CP 101: 1972 Foundations and substructures for non-industrial buildings of not more than four storeys [(1–)] £1.35

BS CP 102: 1963 Protection of buildings against water from the ground [(12)] £2.00

BS CP 2001: 1957 Site investigations [(11) (A3s)] £3.00

BS CP 2004: 1972 Foundations [(Ajr)] £6.90

Concrete
BS CP 110 The structural use of concrete, Part 1: 1972. Design, materials and workmanship £6.90

BS CP 114: Part 2: 1969 Structural uses of reinforced concrete in buildings [(2–) Eq4 (K)] metric units £4.40

Steel
BS 449: Part 2: 1969 The use of structural steel in building [(2–)] Yh2 (K)] metric units £3.00

Timber
BS CP 112: Part 2: 1971 The structural use of timber [(2–) Yi (K)] metric units £3.00

Masonry
BS CP 111: Part 1: 1964, Part 2: 1970 Structural recommendations for loadbearing walls [(21.1) (K)] £2.00 each

Composite structures
BS CP 117: Part 1: 1965 Composite construction in structural steel and concrete. Simply supported beams in building [(2–) Gy (K)] £1.35

More detailed references appear as appropriate within the handbook.

3 Specialist advice
Agrément Board, PO Box 195, Bucknalls Lane, Garston, Herts

Association of Consulting Engineers, 87 Vincent Square, London SW1P 2PH

BISRA—Corporate Development Laboratories of the British Steel Corporation, Hoyle Street, Sheffield 3, S. Yorkshire

Brick Development Association, Woodside House, Winkfield, Windsor, Berks SL4 2DP

British Ceramic Research Association, Queens Road, Penkhull, Stoke-on-Trent, Staffs ST4 7LQ

British Construction Steelwork Association Ltd, Hancock House, 87 Vincent Square, London SW1P 2PJ

British Engineering Brick Association, Grove House, Sutton New Road, Birmingham 23

British Plastics Federation, 5 Belgrave Square, London SW1

British Precast Concrete Federation, 60 Charles Street, Leicester LE1 1FB

British Ready-Mixed Concrete Association, Shepperton House, Green Lane, Shepperton, Middlesex

British Reinforcement Manufacturers Association, 15 Tooks Court, London EC4A 1LA

British Wood Working Federation, 82 New Cavendish Street, London W1M 8AD

Building Research Establishment, The Princes Risborough Laboratory, Aylesbury, Bucks

Cement and Concrete Association, 52 Grosvenor Gardens, London SW1W 0AQ

Clay Products Technical Bureau of Great Britain Ltd, see Brick Development Association

Concrete Society, Terminal House, Grosvenor Gardens, London SW1W 0AJ

CONSTRADO (Constructional Steelwork Research and Development Organisation), NLA Tower, 12 Addiscombe Road, Croydon, Surrey CR9 3JH

Construction Industry Research and Information Association, 6 Storey's Gate, London SW1P 3AU

Federation of Concrete Specialists, 60 Charles Street, Leicester LE1 1FB

Federation of Piling Specialists, Dickens House, 15 Tooks Court, London EC4A 1LA

Institution of Civil Engineers, Great George Street, London SW1P 3AA

Institution of Structural Engineers, 11 Upper Belgrave Street, London SW1X 8BH

Iron and Steel Institute (see Metal Society)

Metal Society, 1 Carlton House Terrace, London SW1Y 5DB

National Physical Laboratory, Queens Road, Teddington, Middlesex TW11 0LW

Plastics and Rubber Institute, 11 Hobart Place, London SW1

Reinforcement Manufacturers' Association (see British Reinforcement Manufacturers Association)

Timber Research and Development Association, Stocking Lane, Hughenden Valley, High Wycombe, Bucks HP14 4ND

Index

Index to Handbook of Building Structure

The handbook comprises 10 sections, design guide and index. Each section contains technical studies (TS) usually followed by information sheets (INF). The design guide is not referred to in the index as it contains no information additional to that contained in the rest of the handbook—it is mainly a procedural checklist.

The document referred to in each of the following entries is identified in the following manner: type of document/reference keyword/number of document. For example: TS COMPOSITE 1 = technical study COMPOSITE STRUCTURES 1; or INF TIMBER 6 = information sheet TIMBER 6. To find the document referred to, scan the top strips of the pages of the relevant handbook section; the title of each document, with numbers of paragraphs appearing on that page, is printed at the top outer corner.

Section title		Reference keywords, page	Abbreviated keyword used in index
Section 1	Building Structure: General	STRUCTURE 7	STRUCTURE
Section 2	Structural analysis	ANALYSIS 27	ANALYSIS
Section 3	Structural safety	SAFETY 71	SAFETY
Section 4	Foundations and retaining walls	FOUNDATIONS 99	FOUNDATIONS
Section 5	Structural material: Reinforced concrete	CONCRETE 157	CONCRETE
Section 6	Structural material: Steelwork	STEEL 183	STEEL
Section 7	Structural material: Timber	TIMBER 211	TIMBER
Section 8	Structural material: Masonry	MASONRY 317	MASONRY
Section 9	Composite structures	COMPOSITE STRUCTURES 345	COMPOSITE
Section 10	Structural innovation	STRUCTURAL INNOVATION 349	INNOVATION

H. E. Warne 1025 27.3.74 3ET84914A